PROGRESS IN
BORON CHEMISTRY

VOLUME 3

PROGRESS IN
BORON CHEMISTRY

VOLUME 3

Edited by
R. J. BROTHERTON
Research Supervisor
U.S. Borax and Chemical Corporation
Los Angeles, California
and
H. STEINBERG
Vice President, Technical Department
U.S. Borax and Chemical Corporation
Los Angeles, California

PERGAMON PRESS

OXFORD · LONDON · EDINBURGH · NEW YORK
TORONTO · SYDNEY · PARIS · BRAUNSCHWEIG

Pergamon Press Ltd., Headington Hill Hall, Oxford
4 & 5 Fitzroy Square, London W.1
Pergamon Press (Scotland) Ltd., 2 & 3 Teviot Place, Edinburgh 1
Pergamon Press Inc., Maxwell House, Fairview Park, Elmsford, New York 10523
Pergamon of Canada Ltd., 207 Queen's Quay West, Toronto 1
Pergamon Press (Aust.) Pty. Ltd., 19a Boundary Street, Rushcutters Bay, N.S.W.
2011, Australia
Pergamon Press S.A.R.L., 24 rue des Écoles, Paris 5e
Vieweg & Sohn GmbH, Burgplatz 1, Braunschweig

First edition 1970

Library of Congress Catalog Card No. 64-13501

Printed in Hungary

08 013080 1

CONTENTS

CONTRIBUTORS TO VOLUME 3

A. FINCH and
P. J. GARDNER

Royal Holloway College, Englefield Green, Surrey, England

D. S. MATTESON

Department of Chemistry, Washington State University, Pullman, Washington

B. M. MIKHAILOV

N. D. Zelinsky Institute for Organic Geochemistry, Moscow, U.S.S.R.

H. NÖTH

Institut für Anorganische Chemie der Universität, Marburg, Germany

W. G. WOODS and
R. J. BROTHERTON

U.S. Borax Research Corporation, Anaheim, California

PREFACE

THE large number of publications and books in recent years dealing with the chemistry and applications of boron and its compounds has consolidated the position of boron chemistry as a scientific area of broad interest. The need for authoritative up-to-date reviews is thus more apparent than ever.

Volume 3 of this series offers a critical treatment of five areas of boron chemistry which merit review at this time. Each chapter covers an area of research in which substantial progress and significant discoveries have been made during the past few years.

1

OXIDATIONS OF ORGANIC SUBSTRATES
IN THE PRESENCE OF BORON
COMPOUNDS

by W. G. Woods and R. J. Brotherton

U,S. Borax Research Corporation, Anaheim, California

CONTENTS

I. INTRODUCTION

The use of boric acid to favor alcohol formation in the air oxidation of hydrocarbons has been known for many years and this process has been improved and utilized commercially in recent years in the Soviet Union, France, Germany, Japan, the United Kingdom and the United States.

1

This dramatic new use of boric acid and related boron–oxygen derivatives has resulted in numerous publications, largely in the form of patents. To illustrate the volume of literature in this burgeoning field, a preliminary version of this review in 1964 contained 74 references while the present one has nearly 300. Because of commercial interest in the preparation of secondary alcohols and nylon intermediates using oxidations in the presence of boron compounds, most of the research was performed in industrial laboratories.

The earliest reports of boric acid as a coreactant (the word "catalyst" is not strictly correct) in hydrocarbon oxidation are contained in patents issued to I. G. Farbenindustrie A.-G.[40, 154] and to A. Riebecksche Montanwerke A.-G.[108, 109] with the earliest application date being 1928.[108] The effect of boric oxide on the oxidation of coal[145] was noted in a 1941 German paper. After the early work in Germany, Japanese workers reported[131–3, 178–85] extensive studies on shale oil oxidations in the presence of boric acid during World War II. Further research in Italy in 1949[223–5] was followed by Russian work which is continuing to appear in the literature and, in more recent years, by a series of West European and United States reports mainly in the form of patents.

The extent of the Russian work is indicated by the appearance of many publications since 1956. The strong possibility exists that the oxidation of hydrocarbons in the presence of boric acid is being used in the Soviet Union on an industrial scale. A pilot plant was built and operated at Shebekino[29] and a description of a production plant has appeared.[27] The intensity of interest in the United States and Europe is clearly shown by the voluminous patent literature, with one company alone claiming over 1000 applications.[8]

There are now a number of industrial plants in the United States and Europe using boric acid oxidation processes. Several companies are licensing cyclohexane oxidation processes in the presence of boric acid as a route to nylon 6 and 66 intermediates.[6, 10] A similar process for the production of secondary alcohols from linear hydrocarbon fractions is operating in the United States[8] with another pilot plant under construction.[9] The secondary alcohols are used in reactions with ethylene oxide to give non-ionic, biodegradable detergents. Other commercial uses which are on stream or contemplated are the oxidation of cyclododecane to intermediates for the production of nylon 12[142] and the production of phenol from cyclohexane via cyclohexanol–cyclohexanone.

This review covers the literature dealing with the oxidation of organic substrates mostly in the liquid phase with oxygen-containing gases or other oxidizing agents in the presence of various boron compounds. Comprehensive tables of all reported experimental data are included in the appropriate sections of the chapter. An attempt has been made to convert units, conditions, etc., to a uniform basis for convenient comparison. The related areas of catalytic gas phase oxidations and the utility of boron derivatives as stabilizers in polymers and other substrates also are included. The journal literature and available patents have been covered through the end of 1967.

II. MECHANISMS OF OXIDATION REACTIONS

A. *General*

Saturated hydrocarbon autoxidation in the presence of boron compounds is a complex process sensitive to changes in reaction variables. The mechanism, therefore, has many facets. Fortunately, a large body of experimental information is available on saturated hydrocarbon autoxidation in the absence of boron compounds which can serve as a basis for the present discussion. The simplest approach to the influence of added boron compounds is to assess their effect on each of the known stages of autoxidation. This method has been used wherever feasible, but the many unique aspects of boron chemistry introduce entirely new features into the situation. A further restriction is levied by the lack of important experimental data obtained from a mechanistic point of view. Only a handful of publications whose prime objective has been mechanism elucidation have appeared; much of the available data is found in patents. As a consequence of these considerations, this mechanism evaluation is partially systematic and partially fragmentary. An attempt has been made to point out specific areas where more experimental information is needed for further clarification.

The data used in this section have been obtained almost entirely from studies which employed linear saturated paraffins or cycloparaffins as substrates. As a result, the mechanistic conclusions apply only to these classes of substances.

Since mechanistic information is lacking for olefin and alkylaromatic oxidations, these areas are not covered in this section (see Sections III, D and III, E below).

B. *Experimental Observations*

1. *Overall Kinetics*

As in any discussion of mechanism, the kinetic order and the effect of reaction conditions on overall rate are of major importance. Hydrocarbon oxidation in the presence of boron compounds suffers from a deficiency of data in this area. The acquisition of reliable kinetic data is hampered by heterogeneity when boric acid is used.

Bashkirov[17] reported that the rate of conversion of alkanes to alcohols is independent of the speed of the oxidation. His study utilized an n-alkane fraction composed of C_{15} to C_{17} paraffins. A more detailed rate study used n-hexadecane as the substrate and showed that the rate of alcohol formation, for a given flowrate of gas, was independent of the degree of conversion.[192] The amount of oxygen consumed, the refractive index of the system, and the amount of n-hexadecane consumed all varied linearly with time up to 45 per cent conversion. It was concluded, therefore, that the oxidation was of zero order. Zero order kinetics also was found using synthetic petroleum ("synthol", b.p. 275–320°C) as the substrate. At 165–170°C, the rate of oxidation of synthol to alcohol was 0.19–0.25 mole per cent/minute.[192] These rates were measured at a single temperature (165–170°C) and oxygen pressure (3–4 per cent oxygen in nitrogen).

Another study with the same substrate yielded data over a range of temperatures (140–175°C) which were used to evaluate an "effective activation energy" of 24 ± 0.7 kcal/mole.[186] This value can be compared with 29 kcal/mole found for the uncatalyzed air oxidation of n-hexadecane in the 120–160°C range.[279]

The presence of boron compounds apparently does not change the essential free radical-chain reaction character of alkane autoxidation.[186] Induction periods are observed with and without boron present[192] and the reaction is autocatalytic under both conditions.[186]

It has been postulated that esterification of secondary alcohol by the boric acid is the slowest step in the reaction sequence and therefore determines the overall rate.[2]

2. *Oxygen Pressure Dependence*

Rates of autoxidation of a variety of compounds in the absence of boron compounds are found to be independent of the total oxygen pressure, provided that this pressure is greater than 50–100 mm.[267] Hydrocarbon oxidations in the presence of boron compounds generally have been carried out using nitrogen containing 3–20 per cent of oxygen gas. Oxidations performed under a total of one atmosphere pressure might be expected to show oxygen pressure dependence at the lower end of this range.

A study was made of the relationship between oxygen content of the oxidizing gas and the selectivity towards cyclohexanol in the oxidation of cyclohexane.[2] The results showed that the selectivity decreased as the oxygen content was increased. A similar decrease in alcohol yield was encountered as the total pressure of the oxidation system was raised.[2] No results are available relating the overall rate to oxygen partial pressure; such data would be of interest.

3. *Inhibition by Boron Compounds*

It is important to ascertain whether boric acid and other boron compounds are catalysts or inhibitors when present during hydrocarbon autoxidation. Although the literature contains some conflicting reports, the weight of evidence points to inhibition by boron compounds.[186]

Bashkirov[17] recognized that the role of boric acid is not limited to its ability to esterify the alcohol products and protect them against further oxidation; boron compounds are deeply involved in the mechanism of the oxidation.

The inhibitory effect of boric acid and its esters on normal paraffinic hydrocarbon oxidation was stated by Chertkov[62] and demonstrated by Freidin.[90] When boric acid, in small increments, was introduced into kerosene undergoing air oxidation at 170°C, termination of the oxidation occurred when the 1.25 per cent level of boric acid was reached. Tri-n-butyl borate gave complete inhibiton at the 0.2 per cent level under similar conditions. This inhibition was overcome by the addition of 0.02 per cent of manganese naphthenate.[90] Boric acid inhibited the air oxidation of a petroleum fraction containing traces of aromatics.[70] At 173°C, boric acid almost fully inhibited the air oxidation of butylbenzene.[71] A careful study of cyclododecane oxidation revealed that boric acid retarded the rate.[47] Boric oxide, when added during the autoxidation of n-octylbenzene, practically stopped the reaction.[76] Essentially no peroxide developed during an autoxidation of n-octylbenzene in the presence of isobutyl metaborate.[76] Triisobutyl borate, on the other hand, did allow some build up of peroxide in the same system.

The inhibitory influence of boron–oxygen compounds has been observed in some closely related free radical processes. For example, boron and boric oxide inhibited the air oxidation of graphite at a level of 0.1 mole per cent.[200] This reaction, performed at 700–800°C, also was inhibited by phosphorus, but a large number of other elements catalyzed it. This unique behavior of boron and phosphorus was related to their ionization potential and electron affinity.[200] Boric acid, boric oxide, trigonal ortho- and metaboric acid esters, and boron phosphate all inhibited the peroxide-catalyzed cure of polyester resins.[278] This inhibition, resulting in soft castings, was observed with methyl ethyl ketone peroxide curing at room temperature as well as with benzoyl peroxide curing at 100°C.[278]

In contrast to the above reports, Veselov et al.[266] claimed that boric acid increased the rate of oxidation of "synthol", a synthetic paraffin fraction. It should be pointed out, however, that a lower partial pressure of oxygen was utilized here than in air oxidations. One worker reports[224] that boric acid had no effect on the rate of oxidation of paraffin hydrocarbon.

One report[266] gives the influence of the boric acid level on autoxidation rate. At amounts near the minimum required to esterify a theoretical alcohol yield, boric acid accelerated the oxidation rate and did not improve selectivity. Excess boric acid was required to reduce the rate and improve the selectivity towards alcohol product.

The latter reports reflect the fact that air oxidations are difficult to reproduce; this also is shown by the variability in yields and product distribution found under apparently identical conditions (see Table 1). Traces of peroxides, metal ions, aromatic impurities, or inhibitors in the substrate can markedly alter induction times, overall rate, and the proportion of products.

The use of tri-n-propyl borate at 180°C with air as the oxidant gave a more rapid formation of alcohol than a similar normal paraffin oxidation using boric acid with 3–4 per cent oxygen in the gas stream.[74]

It was recognized by Chertkov[62] that two key rates are involved here. First is the conversion of hydrocarbon to hydroperoxide and its decomposition products and, secondly, the esterification of the alcohols by boric acid. The selectivity of the process will be a maximum when these rates are equal. Since the esterification is considerably slower[2] than the autoxidation rate, the oxygen content of the gas stream is kept low to balance these two rates.

Selectivity for alcohols was generally thought to be due to esterification of ROH by boric (or metaboric) acid. The resulting ortho- or metaboric acid ester was assumed to be less susceptible to further oxidation than the free alcohol. This has been born out in several cases. Trimethyl borate (I) has been found to be quite resistant to oxidation, even under 200 psi

$$B(OCH_3)_3$$
(I)

of oxygen pressure at 200°C in the presence of mercuric acetate.[92] Cyclododecyl metaborate (II) and cyclododecyl orthoborate (III) both were observed to absorb very little oxygen at 160–180°C after several

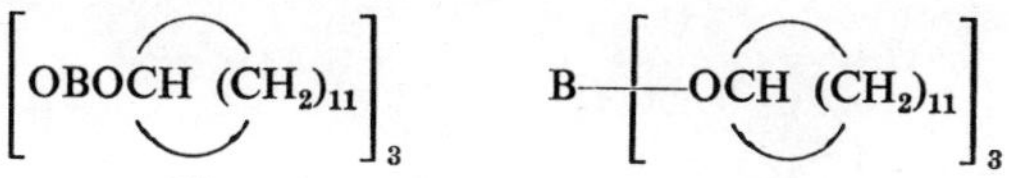

$$\left[OBOCH\ (CH_2)_{11} \right]_3 \qquad B \left[OCH\ (CH_2)_{11} \right]_3$$

(II)   (III)

hours of having air passed through them.[47

Since further oxidation of boric esters would involve hydrogen atom abstraction at the α-carbon, the radical generated from the ester presumably would be less stable than that derived from the alcohol; i.e. species (IV) would be less stable than (V) (equations (1) and (2)). It has now been shown[176] that t-butoxy radicals at 127°C abstract the α-H of *sec*-butil

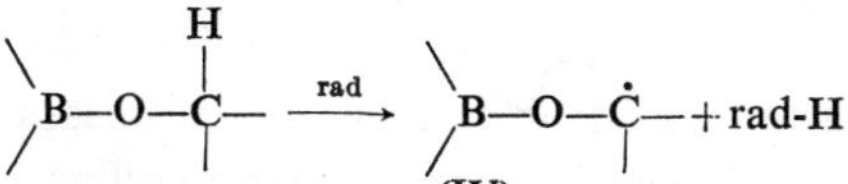

$$\underset{(IV)}{B-O-\overset{\overset{\displaystyle H}{|}}{\underset{|}{C}}-} \xrightarrow{rad} B-O-\overset{\cdot}{\underset{|}{C}}- + rad\text{-}H \qquad (1)$$

$$\underset{(V)}{HO\overset{\overset{\displaystyle H}{|}}{C}} \xrightarrow{rad} HO\overset{\cdot}{C} + rad\text{-}H \qquad (2)$$

alcohol with a rate of more than twice that found for *sec*-butyl borate ($k_1/k_2 = 2.3$, equations (3) and (4)).

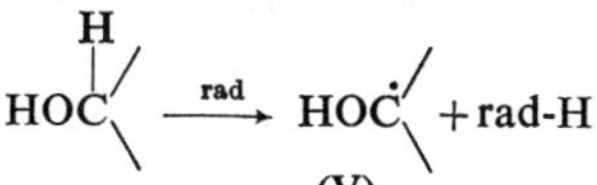

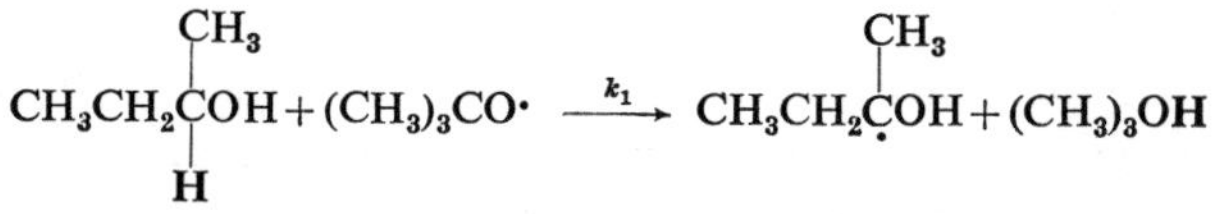

$$CH_3CH_2\overset{\overset{\displaystyle CH_3}{|}}{\underset{\underset{\displaystyle H}{|}}{C}}OH + (CH_3)_3CO\cdot \xrightarrow{k_1} CH_3CH_2\overset{\overset{\displaystyle CH_3}{|}}{\underset{\cdot}{C}}OH + (CH_3)_3OH \qquad (3)$$

$$CH_3CH_2\overset{\overset{\displaystyle CH_3}{|}}{\underset{\underset{\displaystyle H}{|}}{C}}OB\left(\overset{\overset{\displaystyle CH_3}{|}}{OCHCH_2CH_3}\right)_2 + (CH_3)_3CO\cdot$$

$$\xrightarrow{k_2} CH_3CH_2\overset{\overset{\displaystyle CH_3}{|}}{\underset{\cdot}{C}}OB\left(\overset{\overset{\displaystyle CH_3}{|}}{OCHCH_2CH_3}\right)_2 + (CH_3)_3COH \qquad (4)$$

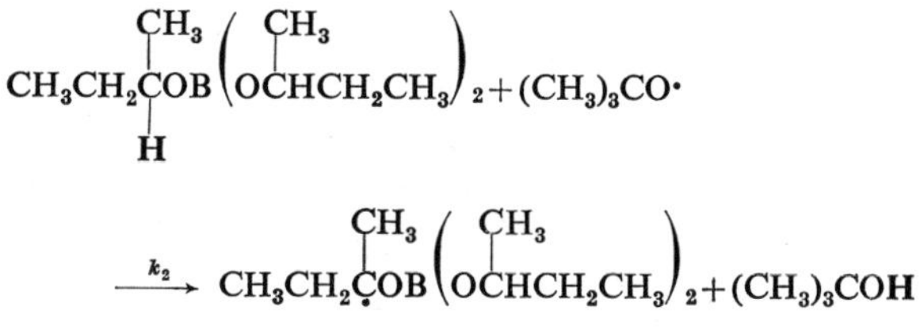

Further work[176] using t-butoxy radicals generated from t-butyl hypochlorite has provided the following relative reactivities:

Substrate	Rel. rate

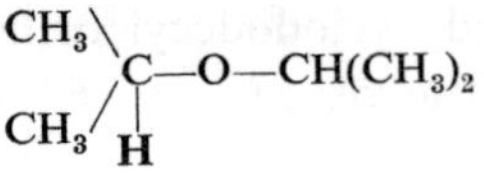

$$\underset{CH_3}{\overset{CH_3}{\diagdown}} \!\!\! \underset{H}{\overset{}{C}} \!\!\! -O-CH(CH_3)_2 \qquad 10.6$$

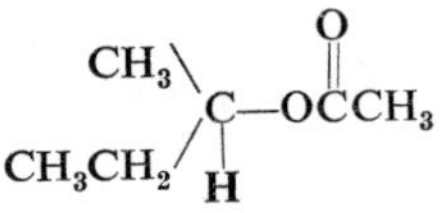

$$\underset{CH_3CH_2}{\overset{CH_3}{\diagdown}} \!\!\! \underset{H}{\overset{}{C}} \!\!\! -O\overset{O}{\overset{\|}{C}}CH_3 \qquad 3.0$$

$$\underset{CH_3CH_2}{\overset{CH_3}{\diagdown}} \!\!\! \underset{H}{\overset{}{C}} OB \left(O\underset{\overset{\displaystyle |}{CH_3}}{\overset{}{C}}HCH_2CH_3 \right)_2 \qquad 1.5$$

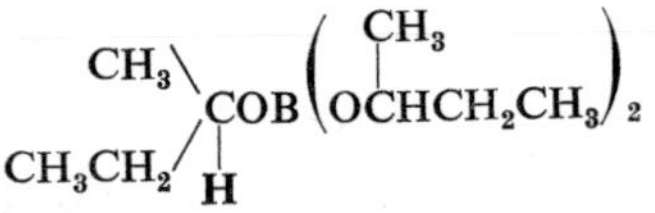

Again, the α-H of the boric acid ester exhibits low reactivity towards radical abstraction. The relatively higher reactivity of the acetic acid ester is significant, in that carboxylic acids are not as effective as boric acid in directing autoxidation (see Section II, B, 7 below).

Many boron compounds have utility as antioxidants in oils, polymers, and other substrates (see Section VI below). 10-Hydroxy-10,9-boroxaro-phenanthrene (VI) is an excellent example of this class of boron derivatives; it is a very effective antioxidant for white oil at 150°C.[46] It is significant, however, that VI did not inhibit the AIBN[†]-catalyzed autoxidation

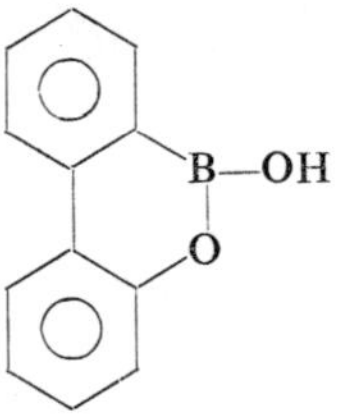

(VI)

of cumene at 60°C. This fact, coupled with the observed superiority of ester (VII) over (VI) as a white oil antioxidant at 150°C (rel. inhib. time of 25.7), was taken as evidence that the hydroxyl group of (VI) was

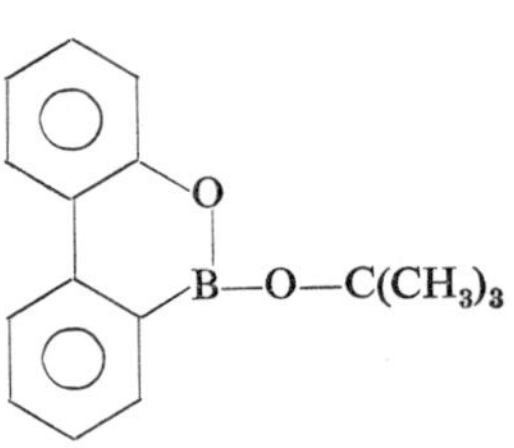

(VII)

† Azobisisobutyronitrile, a free radical initiator.

not necessary for its inhibitory effect. The cumene result also indicated that the $\diagdown$B—OH group did not serve as a hydrogen atom source; i.e. the hydrogen was not abstracted by peroxy radicals (equation (5)). A similar

$$\diagdown B\text{—OH} + RO_2^\bullet \;\;\diagup\!\!\!\!\times\!\!\!\!\diagdown\;\; \diagdown B\text{—O}^\bullet + RO_2H \tag{5}$$

conclusion was reached regarding the interaction of (VIII) and t-butoxy radicals at 127°C.[176] As measured by the t-butyl alcohol/acetone ratio, reaction (6) did not occur to any significant extent. Similarly, the same product ratio from decomposition of di-t-butyl peroxide was unaffected

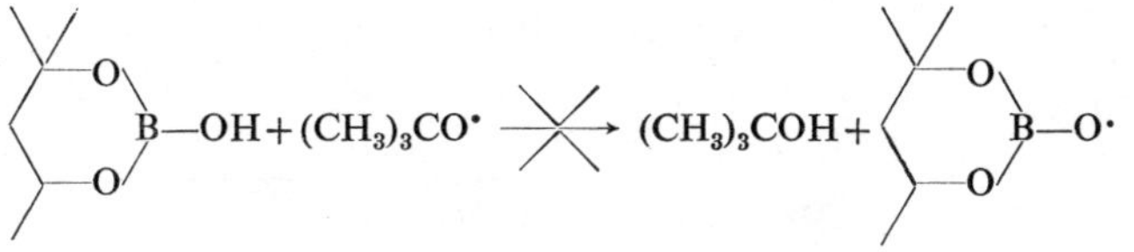

$$\text{B—OH} + (CH_3)_3CO^\bullet \;\;\diagup\!\!\!\!\times\!\!\!\!\diagdown\;\; (CH_3)_3COH + \text{B—O}^\bullet \tag{6}$$

(VIII)

by the presence of (VI).[46]

These results suggest that orthoboric acid, metaboric acid, and alkoxy-(hydroxy) boranes probably do not function as hydrogen atom sources for alkoxy or peroxy radicals during hydrocarbon oxidation. It would be of interest, however, to verify these conclusions kinetically using deuterated boric acid.

4. The Effect of Transition Metal Ions

A number of oxidation catalysts have been tried as a means of reducing induction times and increasing the overall rate of hydrocarbon oxidations in the presence of boric acid. Although manganese resinate,[135] potassium permanganate[132, 183] and cobalt and iron naphthenates can accelerate the rate, selectivity is reduced. This loss of selectivity for alcohol also was found when both boric acid and cobalt naphthenate were used to catalyze the breakdown of cyclododecyl hydroperoxide.[255] The yield of cyclododecanol was less than when boric acid was used by itself, but it was still higher than without any catalyst present. The effect of cobaltous ion in the absence of boron has been studied in some detail.[42]

5. The Effects of Structure and Temperature

Considerable data are at hand to confirm that the relative reactivity of various positions in hydrocarbons is in the order: tertiary > secondary > primary. (See Section III, B, 2, (b) below). This is the order expected for a free radical mechanism. Quantitative relative reactivities have not been determined and would be of interest for comparison with values obtained in the absence of added boron compounds.

Hydrocarbon oxidation to favor alcohol formation usually is carried out at 165–175°C. Autoxidation to produce carboxylic acids, on the other hand, usually is performed at 105–115°C. It has been suggested[62] that hydroperoxide intermediates are at a very low steady state concentration in the alcohol system, due both to the elevated temperature and to the low partial pressure of oxygen in the oxidizing gas stream. The low temperature in the acid system would allow formation of vicinal bis-hydroperoxides which then would give 2 moles of carboxylic acid (equation (7)). This is one of the reasons that low oxygen partial pressures in

$$\text{RCH}_2\text{CH}_2\text{R}' \xrightarrow{O_2} \underset{\underset{\text{OOH}}{|}}{\text{RCHCH}_2\text{R}'} \xrightarrow{O_2} \underset{\underset{\text{OOH} \quad \text{OOH}}{| \quad |}}{\text{RCH} \text{—} \text{CHR}'} \xrightarrow{\frac{1}{2}O_2} 2\,\text{RCOOH} + \text{H}_2\text{O}$$

$$(7)$$

the oxidizing gas enhance the selectivity towards alcohol product.

In the range of 140–240°C, n-paraffin air oxidations in the presence of boric acid esters gave increased alcohol yields as the temperature was increased.[73] Increased selectivity for alcohol also was found in cyclohexane oxidation as the temperature and residence time were increased.[2] In practice, however, the maximum temperature is determined by the melting point of metaboric acid and the thermal instability of the alkyl metaborate product.

6. *Boron Species Present during Autoxidation*

There is little doubt that boric acid exists essentially as metaboric acid under the conditions of hydrocarbon autoxidation (140–170°C) (equation (8)). Orthoboric acid (m.p. 170.9°C) is converted to metaboric

$$3\,\text{B(OH)}_3 \xrightarrow{\Delta} (\text{HOBO})_3 + 3\,\text{H}_2\text{O} \qquad (8)$$

acid at temperatures above 100°C. Metaboric acid exists in three polymorphic forms: (I), m.p. 236°C; (II), m.p. 200.9°C; and (III), m.p. 176.0°C.[177] A eutectic between orthoboric and metaboric acid (III) has m.p. 158.8°C. These melting points restrict the useful temperature range for oxidations in the presence of boric acid; too high a temperature can result in melting and agglomeration of the metaboric acid with consequent clogging of equipment.

Even more highly dehydrated forms of boric acid such as "pyroboric acid" ($H_2B_4O_7$) have been suggested as being present.[21] If boric oxide is used originally, water formed in the oxidation can convert it to the metaboric form (equation (9)). Esterification of alcohol product would be expected, consequently, to produce metaboric acid esters (equation (10)),

$$3 B_2O_3 + 3 H_2O \xrightarrow{\Delta} 2 (HOBO)_3 \qquad (9)$$

provided sufficient boric acid were present initially; production of alcohol in excess of one mole per boron could result in orthoborate formation (see Section III, B, 2 (c) below). A study of cyclododecane oxidation tended to bear out this idea. In that study[47] the absolute cyclo-

$$2 (HOBO)_3 + 6 ROH \longrightarrow 2 (ROBO)_3 + 3 H_2O \qquad (10)$$

dodecanol yields were put in ratio to the boric acid present as a function of conversion. At a boric acid content of 6 per cent and a conversion of 33.3 per cent, a maximum of 1.18 moles of alcohol were produced per mole of boric acid present. This ratio strongly indicated the formation of largely cyclododecyl metaborate in the oxidate. Specifically, the result can be rationalized by a mixture of 91 mole per cent meta and 9 mole per cent orthoborate. If all the alcohol were produced in the form of its metaborate, the theoretical alcohol : boric acid ratio would be 1.0. It is significant that no free alcohol as such was observed in the oxidate; hydrolysis was required to release it from the metaborate. Further verification of this point is provided by the isolation of alkyl metaborate from the oxidation of n-akanes at 160–200°C in the presence of boric acid or boric oxide.[239] Other workers have isolated only orthoborates from this type of system, probably because of the thermal instability of the alkyl metaborates (see Section III, B, 2, (c) below).

7. The Effect of Acids other than Boric Acid on Autoxidation

If the selectivity afforded by the presence of boric acid in hydrocarbon autoxidation were due solely to its acidity and ability to esterify alcohols, it is possible that other acids might function in a similar capacity. This possibility has been examined by several workers and was recognized in an early German patent[69] which claimed the use of antimony, arsenic and phosphorus pentoxides, benzoic anhydride and other weak organic and inorganic acids and anhydrides as beneficial additives in hydrocarbon air oxidations.

Although silicic and antimonic acids enhanced the yield of alcohols during paraffin hydrocarbon autoxidation, they were less efficacious than boric oxide, metaboric acid, or pyroboric acid.[17] Arsenic acid, arsenic trioxide, and a combination of arsenic trioxide and borax all decreased the yield of cyclododecanol when they were present during the decomposition of cyclododecyl hydroperoxide at 140°C.[47] Replacement of boric acid with arsenious acid in the oxidation of paraffins also gave negative results.[180] Paraffin wax oxidation provided lower hydroxyl yields in the presence of arsenic trioxide than in the presence of boric acid or boric oxide.[108]

Phosphoric acid caused the formation of a dark solid when it was present during a paraffin air oxidation at 170–175°C.[252] A similar effect was observed when ethyl orthophosphate was used. Dehydration of any alcohol formed apparently took place to give olefins which rapidly polymerized. Similar results were found when arsenesulfonic acids were employed.[252]

Acetic acid can be introduced into the air stream during oxidations,[40] but larger amounts of carboxylic acids are found in the oxidate.[223] For example, acetic acid was introduced continuously into a paraffin oxidation in order to maintain its level at 2–3 per cent. Analysis of the oxidate gave a hydroxyl number of 35–40, an ester number of 20–25, and an acid number of 5–6; the unsaponified residue gave an OH number of 40–45. Boric acid, at a level of 3 per cent, gave an oxidate with hydroxyl number 58, ester number 26, and acid number 6; the unsaponified residue had OH number 62. Results comparable to those found with acetic acid were reported[252] using propionic acid or a mixture of acetic, propionic and butyric acids. Fatty acids give poor results in paraffin oxidation.[180]

Although the lower aliphatic carboxylic acids reduced the overall oxidation rate and did not give as high a selectivity for alcohols as did boric acid, their usage was felt to offer possibilities for commercialization.[252]

Strong carboxylic acids such as chloroacetic and oxalic also were tested.[252] Chloroacetic acid at the 5 per cent level gave a higher ester and carbonyl content and a lower hydroxyl number than a comparable oxidation using boric acid. Although oxalic acid gave a hydroxyl number in the unsaponified product near that found with boric acid, the carbonyl content was much higher.[17] In the oxidation of cyclododecane, oxalic or benzoic acid had little effect on product distribution as compared with that found with no additive present.[47] Phthalic anhydride actually reduced the yield of cyclododecanol and increased the amount of ketone produced.[47]

Boric acid appears to have just the right degree of Lewis acidity to direct alkane autoxidations towards alcohol product. Other Lewis acids are not effective. Aluminium ethoxide, for example, gave immediate inhibition when it was added to cyclododecane undergoing oxidation.[47] Similarly, aluminium and chromic hydroxides were found to be impractical as additives.[184]

8. *Boric Acid and Derivatives as Catalysts for Hydroperoxide Decomposition*

Esterification of hydroxyl groups by boric acid has been considered to be the major reason for increased selectivity towards alcohols in hydrocarbon oxidation in the presence of boric acid. Considerable evidence is at hand, however, that orthoboric acid, metaboric acid and their esters catalyze the decomposition of hydroperoxides. In particular, the decomposition is directed towards alcohol formation, thereby possibly accounting for the enhanced selectivity in hydrocarbon oxidations in the presence of boric acid.[90] In fact, the inhibition of autoxidation rate by boric oxide has been attributed to acceleration of alkyl hydroperoxide decomposition and a decreased rate of radical formation.[186]

A number of workers have noted that, during hydrocarbon autoxidations in the presence of boric acid, the hydroperoxide level in the system is considerably lower than a similar reaction without boric acid. In the case of cyclododecane, for example, the maximum peroxide content achieved during autoxidation at 145–160°C is halved by the presence of boric acid.[47] A similar decrease in hydroperoxide level was observed when boron compounds were added in the autoxidation of n-octylbenzene.[76]

A number of mechanisms have been proposed for hydroperoxide decomposition in the presence of trigonal boron compounds. One scheme[252] involves an initial association between hydroperoxide and boric acid in which the boric acid functions as a Lowry–Brönsted acid (equation (11)). Homolytic decomposition of this complex is envisaged (equation

$$\textbf{ROOH} + \textbf{B(OH)}_3 \rightleftharpoons \underset{\underset{\textbf{H}}{|}}{\textbf{RO}\cdots\textbf{O}\cdots\textbf{H}\cdots\textbf{OB(OH)}_2} \qquad (11)$$

(12)) to give rise to alkoxy radicals. The alkoxy radicals subsequently

$$\underset{\underset{\textbf{H}}{|}}{\textbf{RO}\cdots\textbf{O}\cdots\textbf{H}\cdots\textbf{OB(OH)}_2} \longrightarrow \textbf{RO}\bullet + \textbf{H}_2\textbf{O} + \bullet\textbf{OB(OH)}_2 \qquad (12)$$

would abstract hydrogen from the hydrocarbon to produce alcohol. The fate of the dihydroxyboryloxy radical was not specified.[252] This association was postulated as becoming important after the oxidation had produced enough alcohol to carry boric acid into solution.

Dimitrov[73] and others have found that boric acid esters are about as effective as boric acid itself in directing n-paraffin oxidations towards alcohol. The absence of B—O—H groups in such systems casts doubt on the interaction shown in equations (11) and (12) above. These workers found that a series of esters (n-propyl to n-cetyl) gave selectivity for alcohol, even at concentrations equivalent to only 0.2 per cent boric acid.[73] They concluded that these boric acid esters chiefly function to catalyze the decomposition of hydroperoxides to alcohol. A mechanism not involving hydrogen bonding was proposed (equation (13)). An ionic

$$ROOH + B(OR')_3 \longrightarrow \left\{ \begin{matrix} RO\cdots O : B(OR')_3 \\ | \\ H \end{matrix} \right\}$$

$$\downarrow$$

$$RO\cdot + HO\cdot + B(OR')_3 \tag{13}$$

decomposition pathway also was suggested, but rejected as unlikely.[73]

The question arises as to whether reaction (13) proceeds to give a peroxyborane species with liberation of alcohol. This possibility of peroxyborane formation has been suggested by several workers.[251] Alcohol product could then be formed from this peroxyborane intermediate.[251] Alternatively, the peroxyborane would lead directly to the boric acid ester, without free alcohol ever being formed. This proposal received some support by the absence of any free alcohol in the oxidate from cyclododecane in the presence of boric acid.[47] In general, however, hydroperoxides and orthoborates do not give rise to trialkylperoxyboranes[241] (equation (14)). Equilibria such as (14) are known to lie largely on the left,[67] although it is possible to synthesize trialkylperoxy boranes by this method under special conditions,

$$3\ ROOH + B(OR')_3 \ \overset{\longrightarrow}{\longleftarrow}\ B(OOR)_3 + 3\ R'OH \tag{14}$$

Aside from the low peroxide levels during autoxidations with boron compounds, evidence for boron-catalyzed decomposition is based on direct observations using isolated hydroperoxides. For example, the boric acid-catalyzed rate of decomposition of cyclohexyl hydroperoxide was measured and found to be twice that of the uncatalyzed rate.[2] Cyclohexyl borate gave a six-fold increase in the same decomposition rate.

A more definitive study has been made by Broich and Grasemann using pure cyclododecyl hydroperoxide.[47] Firstly, the rate of decomposition was markedly accelerated (at 140 or 160°C) by the presence of boric acid; this acceleration was significant at 160°C and quite marked at 140°C. Secondly, the cyclododecanol:cyclododecanone product ratio was increased three- to five-fold by the presence of boric acid. At 160°C, the ol:one ratio went from 0.84 to 2.6 by the addition of boric acid. The decompositions were carried out in cyclododecane as the solvent. In the presence of boric acid, the cyclododecanol-cyclododecanone yield exceeded 100 per cent based on starting cyclododecyl hydroperoxide. This latter observation led the authors to propose that "free active oxygen" was released during the boric acid catalyzed hydroperoxide decomposition. This released oxygen was formulated as atomic oxygen which subsequently could attack cyclododecane to provide more alcohol and ketone. The possibility of singlet oxygen formation has not been examined. Equations (15)–(17) show a modified form of their proposed decomposition mechanism. No intermediate steps in transformation (17) were suggested.

$$\left(\begin{array}{c} HO \\ | \\ -B-O- \end{array}\right) + ROOH \;\rightleftharpoons\; \left(\begin{array}{c} OH \\ | \ominus \\ -B-O- \\ \oplus| \\ O-O-R \\ | \\ H \end{array}\right) \tag{15}$$

$$\begin{array}{cc} HO \\ |\ominus \\ -B-O- \\ \oplus| \\ O-O-R \\ | \\ H \end{array} \;\rightleftharpoons\; \begin{array}{c} \overset{\oplus}{HOH} \\ |\ominus \\ -B-O- \\ | \\ O-O-R \end{array} \tag{16}$$

$$\begin{array}{c} H\diagdown\overset{\oplus}{\underset{|\ominus}{O}}\diagup H \\ -B-O- \\ | \\ OOR \end{array} \;\longrightarrow\; H_2O + \begin{array}{c} -B-O- \\ | \\ OR \end{array} + \overset{..}{\underset{..}{O}} \colon \tag{17}$$

The question of the intermediacy of peroxyboranes (equation (17)) is of importance. It has been inferred[251] that alkyl peroxyboranes are the chief intermediates in these alkane autoxidations and that they subsequently give rise to the boric acid ester. In the case of cyclohexane, this concept was supported[43] and it was proposed that cyclohexyl hydroperoxide or the cyclohexylperoxy radical could give rise to cyclohexyl perborate. The idea has been put forward[186] that the direct interaction of alkylperoxy radicals with the metaboric acid (perhaps leading to peroxyboranes) accounts for the inhibitory effect of the boric acid on hydrocarbon oxidation. As was discussed above, it is quite unlikely that the direct reaction of hydroperoxides and boric acid (or metaboric acid) can give rise to peroxyboranes.[241]

Several other acids such as dodecanedicarboxylic acid, arsenious oxide, trichloroacetic acid, phthalic anhydride and phosphorous pentoxide did not direct the decomposition of cyclododecylhydroperoxide towards alcohol at 140°C.[47] Apparently the Lewis acidity of trigonal boron uniquely allows boric acid and its esters to alter the normal course of this decomposition.

The fact that the cyclododecanol:cyclo odecanone ratios found in the boric acid catalyzed decomposition of cyclododecylhydroperoxide are essentially the same as the ol:one ratios observed in the autoxidation of cyclododecane in the presence of boric acid[47] can be taken as compelling evidence that boric acid functions largely to direct the course of hydroperoxide breakdown in such autoxidations.

Boric acid, boric oxide, tetraacetyl biborate (IX), "monoisobutyl

$$(CH_3COO)_2B—O—B(OOCCH_3)_2$$

(IX)

borate" (presumably isobutyl metaborate), and small amounts of triisobutyl borate strongly catalyzed the decomposition of cumyl hydroperoxide (X) at 136°C in cumene solvent.[76]

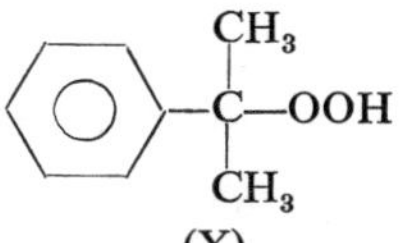

(X)

Considerable tar accompanied the increased decomposition rate of (X). Large amounts (50–75 weight per cent) of triisobutyl borate had little effect on the rate. A mixture of tri-n-propyl borate (2.1 grams) and (X) (5.2 grams) was stable for 3 hours at room temperature and then for 2 hours at 70°C, at which point an exothermic reaction occurred.[67] Phenol was recovered as a major product from the latter experiment. Tri-n-propyl borate at a level of about 30 weight per cent did not markedly catalyze the decomposition of t-butyl hydroperoxide at 60°C or of pinane hydroperoxide at 80°C.[67]

The products from the boron compound-catalyzed decomposition of (X) (phenol and acetone) correspond to those formed by strong acid catalysis. These results, then, support the ionic mechanism for hydroperoxide decomposition catalyzed by metaboric acid and its esters. Furthermore, this type of catalysis by boric oxide leading to increased phenol content accounts for the inhibition of alkane oxidations when alkyl aromatic impurities are present.[114] For example, diethylbenzene inhibited the autoxidation of n-decane only when boric acid was present.

A number of reports have been published concerning the decomposition of hydroperoxides[79] and hydrogen peroxide[160] by boron trifluoride. Peroxytrifluoroacetic acid–boron trifluoride etherate is an effective oxidant in organic synthesis.[52, 102-6, 268] A mechanistic discussion of these reactions is beyond the scope of this review.

C. *Discussion*

Overall, autoxidation can be broken down into three essential stages: initiation, propagation and termination. Each of these stages are examined below with regard to how boron compounds might influence the course of the autoxidation at each phase.

1. *Initiation*

Any substance which can generate a free radical fragment can serve as an initiator of autoxidation. In the case of hydrocarbon oxidation in the absence of added initiator, peroxides are generally responsible for the initiation. These peroxides may be present in the hydrocarbon substrate at the outset, or they can be produced by a direct bimolecular reaction between the hydrocarbon and oxygen.

Hydroperoxide decomposition to give initiating radicals is believed to be bimolecular (equation (18)). At the temperature employed in hydrocarbon oxidations in the presence of boron compounds (160–180°C), spon-

$$2\,ROOH \longrightarrow H_2O + RO\cdot + ROO\cdot \tag{18}$$

taneous hydroperoxide formation is likely and a supply of initiating radicals is assured. Continuous introduction of an external initiator during a hydrocarbon oxidation in the presence of boric acid has not been studied.

2. *Propagation*

The usual propagation steps written for autoxidation are shown in equations (19) and (20). These two reactions constitute the chain reaction;

$$R\cdot + O_2 \longrightarrow RO_2^\cdot \tag{19}$$

$$RO_2^\cdot + RH \longrightarrow ROOH + R\cdot \tag{20}$$

the length of the chain is measured by the average number of cycles that R·
goes through before it or the peroxy radical are destroyed by some termination process. It is likely that the chain length in hydrocarbon autoxidation in the presence of boron compounds is short.

The rate of reaction (19) is some 10^4 to 10^8 times that of reaction (20).
This means that the overall rate-determining propagation step is the abstraction of a hydrogen atom by the peroxy radical to give hydroperoxide.
It is known that peroxy radicals are the primary products in the autoxidation of alkanes at moderate temperatures.[214] Since peroxy radicals
are poor hydrogen abstractors, reaction (20) effectively limits the rate at
which alkanes can be oxidized.

It remains to rationalize the retarding effect of boric acid on the overall
rate of alkane oxidation. The role of various boron compounds as inhibitors of free radical processes is discussed in Section II, B, 3 above.
Orthoboric acid, metaboric acid, and alkoxy(hydroxy)boranes do not
function as hydrogen atom sources that could interfere with reaction (20)
by competing with the substrate for the peroxy radicals (see Section

II, B, 3). Since $\diagdown$B—OH groups also are unreactive towards alkoxy

radicals, any competition of hydroxyboron compounds for the peroxy
radical in reaction (20) is ruled out. Another route to inhibition of reaction (20) might involve interaction of the peroxy radical with the electron-
deficient boron atom. Other evidence exists suggesting that trigonal boron
compounds can complex free radicals. For example, it has been proposed[12]
that a growing vinyl polymer radical can complex with a trialkylborane
to give a non-inhibitable species, although later work casts doubt on this
conclusion.[164] It was proposed that dialkylperoxyboranes (R_2BOOR)
form a much weaker complex. Peroxy radical displacements on trialkyl-
boroxines also have been suggested,[65] as have alkyl radical displacements on trialkylboranes.[98]

In the case of hydrocarbon oxidation in the presence of boric acid,
a formally quadrimolecular reaction has been suggested (equation (21)),
resulting in the production of hydrogen peroxide and oxygen.[47] This

$$2\,RO_2^{\cdot}+2\,HOBO \longrightarrow 2\,ROBO+H_2O_2+O_2 \qquad (21)$$

chain disruption by metaboric acid would lead to termination as well as to
formation of alkyl metaborate. Although no details of the discrete mechanistic steps implied by reaction (21) were given,[47] initial attack of $RO_2^{\cdot}$
on the boron atom was suggested. Speculation on this reaction is presented in the section on termination (II, C, 2) below.

It has been noted that water can have a deleterious effect on oxidations in the presence of boron compounds. Careful control of the water concentration is important in order to achieve optimum results.[88] Water is known to retard the oxidation of alkaneboronic acids and to retard the oxygen-induced polymerization of vinylboron compounds. Coordination of the trigonal boron atom with a water molecule would reduce its ability to form complexes with oxygen or other radicals (equation (22)) or to function effectively as a Lewis acid catalyst for hydroperoxide decomposition. This may be the major reason for the necessity of removing water

$$
H_2O + \begin{array}{c} X \\ \diagdown \\ B-X \\ \diagup \\ X \end{array} \rightleftharpoons \begin{array}{c} X \\ \diagdown \quad - \quad + \\ X-B-OH_2 \\ \diagup \\ X \end{array} \tag{22}
$$

during alkane and cycloalkane oxidations in the presence of boron compounds.

3. *Termination*

Several termination reactions are known for autoxidations in the absence of boron additives (equations (23), (24) and (25)). The relative importance of reaction (23) increases and that of (25) decreases as the oxygen pressure

$$2\,RO_2^{\cdot} \longrightarrow \text{non-radical products} \tag{23}$$

$$RO_2^{\cdot} + R^{\cdot} \longrightarrow \text{non-radical products} \tag{24}$$

$$2\,R^{\cdot} \longrightarrow \text{non-radical products} \tag{25}$$

is increased. Let us examine the influence of metaboric acid and its esters on each of these termination reactions.

A commonly written pathway for reaction (23) involves cyclic transition state (XI) leading to equimolar amounts of alcohol and ketone (equation (26)). This termination mechanism is clearly unimportant in the presence

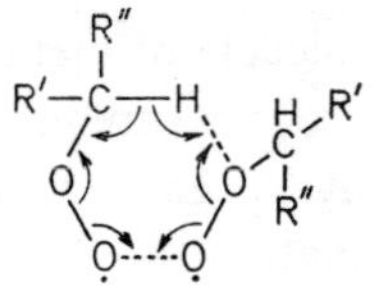

(XI)

$$
2\,R'-\underset{\underset{R''}{|}}{\overset{\overset{H}{|}}{C}}OO^{\cdot} \longrightarrow R'-\underset{\underset{R''}{|}}{C}HOH + R'-\underset{\underset{R''}{|}}{C}{=}O + O_2 \tag{26}
$$

of boron compounds since the alcohol: ketone product ratios far exceed unity.

Another route for the bimolecular destruction of peroxy radicals provides two alkoxy radicals and a molecule of oxygen (equation (27)). The alkoxy radicals can simply abstract a hydrogen from the hydrocarbon to

$$2\,RO_2^{\cdot} \longrightarrow 2\,RO^{\cdot} + O_2 \tag{27}$$

give alcohol (equation (28)). In the presence of metaboric acid, a sequence

$$RO^{\cdot} + RH \longrightarrow ROH + R^{\cdot} \tag{28}$$

involving two peroxy radicals and two HOBO units has been suggested[47] (see Section II. C, 2, equation (21)). The mechanism of this termination step might involve a tetroxide (XII. equation (29)), which would interact

$$2\,RO_2^{\cdot} \longrightarrow ROOOOR \tag{29}$$
$$(XII)$$

with metaboric acid to give rise to oxygen, hydrogen peroxide and the metaboric acid ester (equation (30)). A novel cyclic transition state (XIII) can

$$ROOOOR + \tfrac{2}{3}(HOBO)_3 \xrightarrow{\;k_t\;} 2\,ROBO + H_2O_2 + O_2 \tag{30}$$

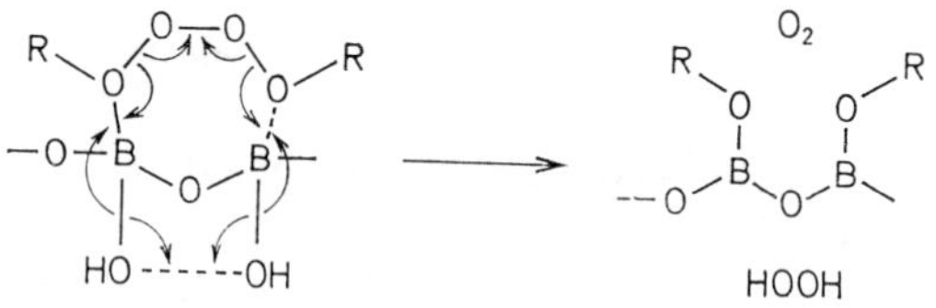

(XIII)

be visualized for this transformation. The kinetics of this process should be termolecular since one molecule of metaboric acid interacts with two peroxy fragments. Careful rate measurements would be needed to evaluate k_t. The detection of hydrogen peroxide in such an autoxidation system would provide confirmation for this mechanism.

Returning now to a consideration of termination reaction (24), there is no evidence to support or deny its involvement in the presence of boron compounds. If a dialkyl peroxide were formed, it could homolyze to give two secondary alkoxy radicals at the temperatures involved, ultimately providing alcohol and some ketone (equations (28), (31) and (32)). Isolation and identification of a dialkyl peroxide from an autoxidation system

$$R^{\cdot} + RO_2^{\cdot} \longrightarrow ROOR \longrightarrow 2\,RO^{\cdot} \tag{31}$$

$$2\,\underset{R''}{\overset{R'}{\diagup}}CH{-}O^{\cdot} \longrightarrow \underset{R''}{\overset{R'}{\diagup}}C{=}O + \underset{R''}{\overset{R'}{\diagup}}CH{-}OH \tag{32}$$

in the presence of a boron compound would help to clarify this possibility. Direct attack of an alkoxy radical on boron also must be considered.[143]

Finally, coupling reaction (25) can be disposed of as unimportant. This was shown by the very small amount of dicyclohexyl found in a careful analysis of the oxidate from cyclohexane in the presence of boric acid.[2]

D. *Summary*

Saturated aliphatic and alicyclic hydrocarbon oxidation in the presence of boric acid and its esters is a free radical chain reaction. As such, it exhibits an induction period, autocatalysis, and an activation energy comparable to that found for autoxidation without boron. The overall kinetics are zero order, but no individual initiation, propagation, or termination rate constants are known. Boron compounds inhibit the overall rate, perhaps by catalyzing termination or by destroying peroxy radicals to disrupt propagation. At the reaction temperatures commonly used, the boron is almost entirely in the form of metaboric acid and its esters.

The process gives a very high selectivity for alcohol, which is released by hydrolysis of the metaboric acid ester product (equation (33)). The selectivity is attributed to the oxidative stability of the metaborate and to the

$$(OBOR)_3 + 6\,H_2O \longrightarrow 3\,ROH + 3\,B(OH)_3 \tag{33}$$

selectivity during decomposition of hydroperoxides by the metaboric acid. Hydroperoxide decomposition probably occurs by an ionic mechanism and produces the same alcohol : ketone product ratio as is found in the hydrocarbon oxidate.

Additives other than boric acid are not successful because they catalyze undesirable side reactions or because their esters are not oxidatively stable.

III. OXIDATIONS WITH MOLECULAR OXYGEN

A. *Process Studies*

1. *Introduction*

The standard procedure for hydrocarbon oxidation consists of passing air or a dilute oxygen in nitrogen stream through a suspension of boric acid in the heated substrate. Numerous patents claim only general improvements in this standard process for the liquid phase oxidation of aliphatic and alicyclic hydrocarbons. Most of these patents are concerned either with pretreatment of the borate (usually boric acid) used or its recovery from the system, and this section will concentrate primarily on these subjects. Some additional general modifications and process-related studies are described in Sections III, B and C.

2. *Boric Acid Preparation*

In most air oxidations of organic compounds which have been reported, boric acid, H_3BO_3, is the preferred boron compound. However, these reactions are normally carried out at about 160°C or above, and boric acid is never present as such during the reaction. Dehydration of boric acid occurs stepwise according to the following sequence. The existence of pyroboric acid as a stable compound has never been demonstrated experimentally, but it is frequently referred to in the literature on hydrocarbon oxidation.

$$H_3BO_3 \xrightarrow[-H_2O]{(1)} \frac{1}{n}(HOBO)_n \xrightarrow[-\frac{1}{4}H_2O]{(2)} \frac{1}{4}[H_2B_4O_7] \xrightarrow[-\frac{1}{4}H_2O]{(3)} \frac{1}{2}B_2O_3$$

boric acid†   metaboric acid‡   pyroboric acid   boric oxide

The temperatures required for steps (1), (2) and (3) depend to a large extent on conditions such as pressure, physical state, nature of the atmosphere, etc., but step (1) can generally be assumed under normal conditions to occur at 100–150°C and steps (2) and (3) above 150°C. It is very difficult to remove the last traces of water to obtain pure boric oxide.

†H_3BO_3 is referred to as boric acid throughout this chapter, although the name orthoboric acid is often used for this compound.

‡Metaboric acid is often written as the cyclic trimer (XIIIa) or but is probably more correctly described as a polymer, $(HOBO)_n$.

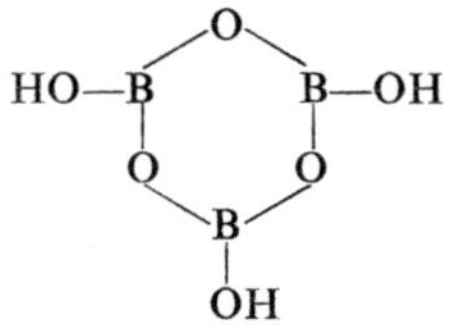

It is obvious that boric acid cannot exist in oxidations carried out above 160°C, and that the exact nature of the boron species depends upon the specific reaction conditions regardless of the state of hydration of the boric acid added to the reaction mixture. In the tables in this chapter, boric acid and metaboric acid are noted as described in the original papers or patents, but it is understood that the nature of the boron species may change during the course of the reaction.

The condition of the initial boric acid added and the manner in which it is added to the reaction mixture and dehydrated are important in oxidation reactions as shown in a series of Russian papers. The results of these oxidations of a liquid paraffin, b.p. 275–320°C, are given in Table 1.[252, 260, 263] The general conclusions from this work were that optimum conversions to alcohols are obtained with finely dispersed, partially dehydrated boric acid. The increased dispersion of the finely divided material appears to result in improved contact between the hydrocarbon and the boron compound. The effect of particle size is more evident when smaller quantities of boric acid are used (2.4 per cent). This is because the use of excess boric acid (4.0 per cent) compensates for the effect of decreased particle size. The particle size factor was also noted with boric oxide and boric acid previously dehydrated at 170°C. It was shown that the yields of alcohols produced increase using from zero up to 2.0–2.2 per cent boric acid and essentially level off after that. Maximum yields were obtained using a dehydrated form of boric acid containing 20.4 per cent boron. One of the major reasons for the increased yields observed with this partially dehydrated material and also with boric oxide is that the agglomeration of boric acid which always occurs during dehydration is avoided if most of the water is removed previously. This agglomeration leads to less complete reaction and lower yields of alcohols. The partially dehydrated boric acid is more effective than boric oxide which is a fused product with a relatively low surface area.

The following formula was developed for determining the quantity of boric acid necessary to combine with the alcohols formed:[260]

$$a = 0.0011 \, k\alpha\beta bc$$

a = amount of boric acid required (weight per cent of hydrocarbon),
k = coefficient of excess boric acid,
α = factor to account for boric acid losses,
$\beta = \frac{1}{3}, \frac{2}{3}$ or 1 for calculation of tri-, di- and monoalkyl borates,
b = oxidate yield, weight per cent,
c = hydroxyl number of oxidate, mg.KOH/g.

For example, if boric acid loss is neglected ($\alpha = 1$) then for 100 per cent oxidate yield with a hydroxyl number of 70, the theoretical quantity of boric acid ($k = 1$) is 2.56 per cent assuming the alcohols are converted to trialkyl borates ($\beta = \frac{1}{3}$). It was found that approximately 5 per cent of the boric acid added is lost in these reactions. In this case the α factor used in the equation above should be 1.05 and the amount of boric acid needed is 2.1 per cent.

In laboratory studies and subsequently in pilot plant work, the dehydration rate of boric acid was shown to be important in determining particle size.[187, 263] To obtain a finely divided product, a mixture of 25 per cent fresh boric acid, or regenerated boric acid from the oxidation process, and 75 per cent of the paraffin is heated gradually (1.5–2.5°C per minute) at 200–300 mm pressure to 165°C.[263] Heating at 165°C for 10–20 minutes results in crystals 20–50 microns in size. Heating up to 170°C leads to agglomeration of boric acid and poor conversion to alcohols.

In another study of boric acid dehydration for use in a continuous commercial process, it was reported that similar heating at 140–145°C produced a finely divided product (<70 microns) of optimum particle size and quality for continuous operation.[194]

In an Esso patent a means of continuously introducing maximum amounts of boric acid to the oxidation reactor is described.[172] This is accomplished by continually passing a portion of the reactor product mixture through a bed of solid boric acid and recycling this to the oxidation reactor. Boric acid is more soluble in the oxygenated hydrocarbon mixture than in the pure hydrocarbons and more boric acid is dissolved into the reactor by this means.

The importance of prior dehydration of boric acid in increasing oxidation rates and the proportion of alcohols formed has been noted by other investigators[280] who also found that alcohol formation was favored by raising the reaction temperature and reducing reaction times. Moderate increases in alcohol yields were noted when the amount of boric acid was increased. Other weak acids such as acetic and propionic acids also increased alcohol yields, but they were not as effective as boric acid derivatives.[252] (See Section II, B, 7).

Process modifications involving the handling and dehydration of boric acid have been described in a number of patents. In most cases these process modifications are part of continuous oxidation processes. Crystallization of large particles of boric acid from the reaction mixture has been described followed by dehydration to finely divided metaboric acid in cyclohexane.[187] The ability to use this large particle size boric acid as a feed to the dehydration step reportedly eliminates filtration and handling problems. Stamicarbon N.V. has patented a cyclohexane oxidation process in which a finely divided boric acid is prepared by mixing an aqueous solution of boric acid with cyclohexane and then removing water by evaporation.[238] The resulting finely divided material is reported to give improved yields and selectivity. A related modification is the continuous stepwise dehydration of boric acid at elevated temperatures and pressures in the presence of the original hydrocarbon.[218] In some cases dehydration to metaboric acid is incomplete. For example, dehydration of boric acid to a 50–50 mixture of boric acid and metaboric acid at 140–165°C has been patented.[219] Dehydration in two steps has been claimed in another patent: free water is removed first at about 160°C followed by water bound as boric acid at approximately the same temperature.[156] A slurry containing finely divided partially dehydrated boric acid is obtained by azeotropically removing water from a dispersion of boric acid in the hydrocarbon to be oxidized.[175]

Russian workers have discussed the dehydration and ultimate recycle of boric acid on a commercial scale.[264] It was found early that excess boric acid leads to plugging in preheaters, and this was corrected by using higher reaction temperatures and removing water continuously to drive the reaction to completion. In the dehydration of boric acid (after hydrolysis) mixed with hydrocarbons and water it has been noted that boron compounds are carried over and deposit on heat exchangers, reboilers, etc.[215] This can be eliminated by adding a monohydric alcohol to the mixture to be dehydrated.

It has been observed that problems arise during crystallization of boric acid from the aqueous extract from hydrocarbon oxidation. In a normal apparatus boric acid tends to build up on the metal surface of the crystallization zone.[94] When the crystallization zone is lined with glass or ceramic materials, boric acid crystallizes in the solution rather than on the walls. Smooth surfaces such as Teflon, polished steel and glass fiber reinforced polyester resin were not effective in this regard.

3. Boric Acid Recovery

A number of process patents have described methods for the hydrolysis of borate esters formed in hydrocarbon oxidations and for the recovery of the resultant boric acid. The normal method for alcohol isolation was originally the addition of sufficient water to hydrolyze the borate esters present followed by centrifugation of the organic–aqueous boric acid slurry. This process can be improved significantly by adding sufficient water to saturate the organic mixture and form separate water–organic layers.[190] Hydrolysis is usually accomplished with hot water,[181, 183] but sodium bicarbonate solution,[139] alcoholic potassium hydroxide,[223] and, in some cases, aqueous methanol have been used. In the latter case boron is removed as the methyl borate–methanol azeotrope.

Formation of borate esters from the alcohols occurs in these reactions, and more volatile by-products and unreacted starting material can be removed by distillation.[269] After hydrolysis, boric acid is recovered and recycled; recoveries as high as 97 per cent are claimed.[29] In one process modification boric acid is crystallized and separated and the aqueous mother liquor recycled and reused for hydrolysis of the oxidation product mixture.[277]

When boric acid is recovered from the oxidation mixture, it is often contaminated with organic compounds such as water-soluble carboxylic acids. These organic contaminates can reportedly be removed by extraction of the aqueous boric acid by organic solvents, distillation with steam, or adsorption on activated carbon.[95] In a related claim the oil–water phases are separated after hydrolysis, and the water layer is extracted with cyclohexanol to remove adipic acid and other water–soluble carboxylic acids.[107]

It has been claimed that recycle boric acid is not as effective as fresh boric acid in hydrocarbon oxidations.[189] Washing the recycle boric acid from cyclohexane oxidation with water, cyclohexanol, cyclohexanone or cyclohexane apparently improves its efficiency by removing unknown impurities.

In one suggested approach to boric acid recovery, low molecular weight alcohols are added to the total oxidation mixture.[48] The resulting mixture of the low-boiling alcohols and borate esters (e.g. methanol–methyl borate azeotrope) is distilled and hydrolyzed to boric acid with the stoichiometric amount of water.

It has been suggested that esters formed in the oxidation should be hydrolyzed at 275°F.[96] The resulting water layer is then added directly to hot fresh hydrocarbon for the boric acid dehydration step.

A laboratory apparatus for the high temperature hydrolysis and subsequent separation of boric acid has been reported.[265] Other related modifications in cyclohexane oxidation have also been described,[188, 220-1] including hydrolysis at 95–98°C followed by two water-wash steps.[285] A subsequent two-step cooling process of the aqueous layers has been used to recover solid boric acid.[284]

A series of French patents issued to Institute Français du Petrole des Carburants et Lubrifiants is concerned with process improvements in the oxidations of linear and cyclic paraffins.[116] These patents, which were issued after completion of this manuscript, include isolation of boric acid, purification of boric acid and other related modifications.

In a Continental Oil Company process a paraffin mixture such as C_{11}–C_{15} hydrocarbons is oxidized under standard conditions except that sufficient boric acid or boric oxide is used to convert the alcohols formed to metaborates.[239] The patent claims a separation method in which the unreacted hydrocarbon–metaborate mixture is atomized at reduced pressure and the hydrocarbon separated by stripping with a low-boiling C_1–C_4 alkane.

4. Heat Conservation

Several process modifications have been suggested whereby the heat produced by the oxidation reaction can be used to advantage elsewhere in the system.

In a general description of boric acid recovery, the use of hot vapors from the reaction mixture to heat fresh feed and to dehydrate boric acid is claimed.[38] Heat transfer and reactor deposits are reported to be eliminated by vaporizing a portion of the hydrocarbon (in this case cyclohexane) directly into the reaction zone.[155] In another process modification the water–hydrocarbon vapors from cyclohexane oxidation are condensed by adding them directly to cool, unreacted cyclohexane.[88] A minor modification has been patented in which recycle hydrocarbons from the final separator are heated by contact with hot hydrocarbon vapor from the reaction.[89] In one patent, improved yields of the desired oxidation products (particularly in cyclohexane oxidation) are obtained when gaseous hydrocarbons and water are recovered from the oxidation reaction mixture, cooled and separated.[155] The hydrocarbon fraction is then heated in a separate vessel and the hot vapors used to heat the reactor slurry of boric acid in cold hydrocarbon.

Imperial Chemical Industries has patented a cyclohexane oxidation process improvement in which the oxygen-containing gas is dissolved in the hydrocarbon in an absorber unit.[111] This separate unit is held at lower temperatures and pressures than the reaction vessel.

5. Control of Water Level

A critical factor in these oxidations is the water concentration (see Section II, C, 2 above). Attention has been given to methods for continuous water removal during the reaction. In the case of cyclohexane the partial pressure of water is usually controlled by continuous removal of a mixture of water and hydrocarbons.[270] This mixture is separated and the hydrocarbon recycled. The desired water vapor pressure is defined as less than P where P is represented by the equation

$$\log_{10} P = 0.0175T - 1.85 \qquad\qquad (34)$$

$$T = \text{reaction temperature, °C.}$$

This definitive equation limiting the water vapor partial pressure was subsequently modified as follows:[274]

The partial pressure p should be greater than P, but less than P'.

$$\log_{10} P = 0.0175T - 1.85 \qquad\qquad (35)$$

$$\log_{10} P' = 0.0107T + 0.068 \qquad\qquad (36)$$

The preferred description for the maximum value of P' is given by

$$\log_{10} P' = 0.0112T - 0.259.$$

The presence of small quantities of water is claimed to destroy the beneficial effect of added boron compounds in hydrocarbon oxidations.[90]

6. *Continuous versus Batch Processes*

A mathematical analysis of the boric acid modified oxidation of cyclohexane has been carried out.[237] The results indicate that a batch or plug flow reactor gives better selectivity at a given conversion than does a continuous stirred reactor. Low conversions give optimum yields; the best result from a batch reaction is 85 per cent selectivity at 15 per cent conversion, with selectivity defined as the mole fraction of cyclohexanol plus cyclohexanone in the oxidate.

B. *Oxidation of Aliphatics*

1. *General*

A large number of aliphatic compounds ranging from low molecular weight compounds such as hexane to macromolecules such as polyethylene have been oxidized in the presence of boric acid. Of most commercial interest is the oxidation of linear paraffins containing 10–16 carbon atoms. Liquid phase oxidation of these materials in the presence of boric acid results in selective oxidation to secondary alcohols which are isolated as their borate esters. Hydrolysis of these esters yields secondary alcohols which are added to ethylene oxide to give biodegradable non-ionic detergents. Alternatively the borate esters can be pyrolyzed to linear olefins which are used to alkylate benzene in the preparation of biodegradable anionic detergents. In this section oxidations of linear paraffins are summarized in Tables 1 through 4. Table 1 includes oxidations in the presence of boric acid, metaboric acid, pyroboric acid and boric oxide. Table 2 includes the few reported reactions in which organic borates were used. Table 3 describes reactions in which various types of additives have been utilized to improve the normal boric acid oxidations. In Table 4 the results of some interesting experiments on the effects of ultrasound are summarized.

2. Oxidations in the Presence of Boric Acids and Boric Oxide

(a) *General conditions.* Reaction conditions including temperature, borate concentration, oxygen concentration and reactor design can be varied to optimize yields of the desired products (usually alcohols). For example, Bashkirov reports[17] the following optimum conditions for the oxidation of paraffins:

(i) A concentration of 5 per cent boric acid or boric oxide.
(ii) An oxidizing gas stream which contains 3–4 per cent oxygen in nitrogen.
(iii) An oxidation temperature of 165–170°C.
(iv) A gas flow rate of 500 liters per kilogram-hour.
(v) The height of the hydrocarbon layer should be minimized.
(vi) The size of the gas bubbles must be adjusted to maximize the yield of alcohols.

Bashkirov and his co-workers[17] employed a gas velocity of 0.02 meters per second in their reactor. They found maximum alcohol yields (but not necessarily maximum conversions) with 3.2 per cent oxygen in the nitrogen stream, a temperature of 165°C, a hydrocarbon concentration of 74 mole per cent, a specific consumption (flow rate) of 500 liters per kilogram-hour, a height of hydrocarbon layer of 25 millimeters and a gas distributor hole size of one millimeter. Variables must be adjusted for each specific system, however, because all variables are interrelated. For example, Scipioni[223] found that an air flow rate of 60 liters per hour using 400 grams of paraffin at 165°C in a 70-cm tube of 40 mm diameter with a fine gas dispersion gave best results.

TABLE 1. OXIDATIONS OF LINEAR PARAFFINS IN THE PRESENCE OF BORIC ACIDS AND BORIC OXIDE

Substrate	Boron compound	Wt. (%)	Temp. (°C)	Time (hr)	Gas, % O$_2$ in N$_2$	Flow rate (l./kg-hr)	% Conv.	Products	Ref.
n-Hexane[a]	H$_3$BO$_3$	10	165	4	4	210	~7	80% hexanols, 15% hexanones	220
n-Hexane[b]	H$_3$BO$_3$	10	165	4	4		7	86% (hexanols + hexanones)	270
1-Chlorodecane	H$_3$BO$_3$	12	140–150	5	Air			Hydroxy 1-chlorodecanes	138
n-Dodecane	Na$_4$B$_2$O$_7$	10	165		Air	300	30	44% alcohols, mostly secondary	122
Tridecane	H$_3$BO$_3$	5	165–170	4	3–3.5	1000	54	71% alcohols, 18.1% ketones	30, 32
2-Methyldodecane	H$_3$BO$_3$	5	165–170	4	3–3.5	800	53	77.8 mole % alcohols (71% secondary, 29% tertiary)	36
Tetradecane	H$_3$BO$_3$	5	165–170	4	3–3.5	1000	51	69.5 mole % alcohols	30, 32
2,3-Dimethyldodecane	H$_3$BO$_3$	5	165–170	4	3.5	800–1000		73 mole % alcohols (6.3% primary, 56.8% secondary, 36.9% tertiary)	34
2,6,10-Trimethylundecane	H$_3$BO$_3$	5	165–170	4	3.5	800–1000	52	78.2% alcohols (no primary, 40.8% secondary, 59.2% tertiary)	34
Pentadecane	H$_3$BO$_3$	5	165–170	4	3–3.5 Air	1000		67.0 mole % alcohols	30, 32
n-Pentadecane	H$_3$BO$_3$		165–170					49% alcohols, 31% acids, 10% ketones	128
n-Pentadecane	H$_3$BO$_3$		165–170		3.5–5.0			67% alcohols, 12% acids, 21% ketones	128
Hexadecane	H$_3$BO$_3$	5	165–170	4	3–3.5	1000	62	79.9 mole % alcohols	30, 32
Hexadecane	H$_3$BO$_3$	5	165–170	4	3.5	1000	62–67	67–70 mole % alcohols	127

Substrate	Boric compound		Temp. °C	Time				Results	Ref.
8-Methylpentadecane	H_3BO_3	5	165–170	4	3–3.5	800	55	71.4 mole % alcohols (72% secondary, 28% tertiary)	36
Tricosane ($C_{23}H_{48}$)	H_3BO_3	—	200	5	Air	—	61	~55 mole % alcohols	179
C_{11}–C_{14} Paraffins	$(HOBO)_3$	1.6	160		Air	120	18	72% alcohols	244
Paraffins (C_{15}–C_{16} mean)	H_3BO_3	5	165	4–5	Air		30	80% alcohols, mostly secondary	93
Synthol, C_{15}–C_{17} Paraffins	H_3BO_3	5	165	4	Air	500	—	OH no. 91.3; Acid no. 36.5; Carbonyl no. 62.0; Ester no. 42.0	17
Synthol[c], C_{15}–C_{17} Paraffins	H_3BO_3	5	165	4	Air	500	—	OH no. 92.5; Acid no. 15.0; Carbonyl no. 35.0; Ester no. 10.2	17
Synthol, C_{15}–C_{17} Paraffins	H_3BO_3	5	165	4	3	1000	—	OH no. 96.2; Acid no. 5.0; Carbonyl no. 15.0; Ester no. 4.2	17
Paraffin petroleum fraction, av. mol. wt. 235	H_3BO_3	12	180	6	Air	21	—	Boric acid becomes sticky, poor results	146–8
Paraffin petroleum fraction, av. mol. wt. 235	75% B_2O_3 + 25% H_3BO_3	11.8	180	6	Air	21		21 mole % alcohols	146–8
Paraffin petroleum fraction, av. mol. wt. 235	B_2O_3	12	196	5	Air	21	—		146–8, 196
Paraffin, b.p. 275–320°C	H_3BO_3	—	170–175	2	—	700	—	OH no. 55–62; Acid no. 5–6; Carbonyl no. 20; Ester no. 20–30	252
Paraffin, b.p. 275–320°	$(HOBO)_3$	—	170–175	2	—	700	—	OH no. 71–75; Acid no. 10.5–11; Carbonyl no. 22; Ester no. 26–32	252
Paraffin, b.p. 275–320°C	$H_2B_4O_7$	—	170–174	2	—	700	—	OH no. 81–89; Acid no. 10–16; Carbonyl no. 23–31; Ester no. 28–34	252

TABLE 1. *(cont.)*

Substrate	Boron compound	Wt. (%)	Temp. (°C)	Time (hr)	Gas, % O_2 in N_2	Flow rate (l./kg-hr)	% Conv.	Products	Ref.
Synthetic paraffin, b.p. 270–320°C	H_3BO_3	5	165	—	3–4	—	—	70 mole % alcohols	150
Synthetic paraffin, b.p. 270–320°C	H_3BO_3	15–20	185	—	Air	—	65–70	70–75 mole % alcohols	150
Synthol, b.p. 280–320°C	H_3BO_3	5	165	4	3.2	500	—	79.5 mole % alcohols	17
Synthol, b.p. 280–320°C	H_3BO_3	5	165	4	5.0	500	—	73.2 mole % alcohols	17
Synthol, b.p. 280–320°C	H_3BO_3	5	165	4	9.9	500	—	64.6 mole % alcohols	17
Synthol, b.p. 280–320°C	H_3BO_3	5	165	4	15.0	500	—	42.0 mole % alcohols	17
Synthol, b.p. 280–320°C	H_3BO_3	5	165	1.5	3.5	1000	—	65.0 mole % alcohols	17
Synthol, b.p. 280–320°C	H_3BO_3	5	165	3.0	3.5	1000	—	73.1 mole % alcohols	17
Synthol, b.p. 280–320°C	H_3BO_3	5	165	4.5	3.5	1000	~50	63.8 mole % alcohols	17
Synthol, b.p. 280–320°C	H_3BO_3	5	165	6.0	3.5	1000	—	62.4 mole % alcohols	17
Synthol, b.p. 280–320°C	H_3BO_3	5	165	4	3.5	300	—	77.5 mole % alcohols	17
Synthol, b.p. 280–320°C	H_3BO_3	5	165	4	3.5	500	—	79.0 mole % alcohols	17
Synthol, b.p. 280–320°C	H_3BO_3	5	165	4	3.5	1000	—	71.2 mole % alcohols	17
Synthol, b.p. 280–320°C	H_3BO_3	5	165	4	3.5	1500	—	67.0 mole % alcohols	17

Substrate	Catalyst		Temp. (°C)					Product	Ref.
Synthol, b.p. 280–320°C	H₃BO₃	5	165	4	3.5	1000	—	75.2 mole % alcohols	17
Synthol, b.p. 280–320°C	H₃BO₃	5	165	4	3.5	1000	—	69.1 mole % alcohols	17
Mineral oil, b.p. 300–350°C	H₃BO₃	—	165–170	4–5	Air	500	—	Alcohols	72
Normal paraffins from urea adducts	B₂O₃	—	200	—	Air	150	—		280
Ethanol-extracted shale oil paraffin, m.p. 45.5°C	H₃BO₃	5	200	3	Air	360	—		133
Shale oil paraffin, m.p. 46.3°C	H₃BO₃	5	200	3	Air	600	—	96.5 mole % alcohols	131
Shale oil paraffin, m.p. 46.3°C	H₃BO₃	10	200	7	Air	600	—		133
Paraffin, m.p. 52°C, mol. wt. 350	H₃BO₅	4	165–180	7–8	Air	150	—	45% (max yield) alcohol	223
Hard paraffin wax	B₂O₃	4.8	175	3	Air	1000	44	Alcohols	154
Polyethylene	H₃BO₃	3	160–170	4	10	720	~80		136

[a] This reaction was carried out at 12 atm pressure.
[b] This reaction was carried out at 170 psig pressure.
[c] This reaction was carried out at 450 psig pressure.

In a paper presented at the Fifth World Petroleum Congress in 1959, Bashkirov and Kamzolkin discussed in detail the optimum conditions for preparation of higher molecular weight alcohols by liquid-phase oxidations using air or oxygen.[28] As noted above, best results were obtained with nitrogen–oxygen mixtures containing 3–4 per cent oxygen at 165–170°C in the presence of boric acid. The alcohols formed have the same basic carbon chain structures as the original hydrocarbons. Secondary alcohols predominate in the product mixtures, and all possible secondary alcohols are formed.

In a basic Russian patent,[26] a general procedure for the air oxidation of hydrocarbons to alcohols is outlined. Suggested conditions are 5 per cent boric acid, air at 500 liters per kilogram-hour, and temperatures of 120–200°C for 3–5 hours. Removal of olefins and aromatics before oxidation is recommended. A continuation of this patent claims the use of 3–5 per cent oxygen in nitrogen at 1000 liters per kilogram-hour.[20] This patent also claims recycling boric acid in a suspension of the hydrocarbon to be oxidized. Another extension of the basic patent describes a process modification in which the oxidation is carried out at 300–400 mm pressure.[61] A continuation of the earlier Russian patent discloses a general separation method for the products of hydrocarbon oxidation.[25] The total product is saponified by alcoholic base. The resulting basic solution is extracted with benzene, and alcohols are obtained by fractionation of this benzene extract.

The oxidation of normal paraffins to alcohols has been studied in some detail by Germain and Cognion.[93] An attempt was made to systematically study process variables in order to determine optimum conditions for this reaction. The paraffin fraction used contained C_6 to C_{18} molecules with a mean molar mass of C_{15}–C_{16}. In a study of temperature using 5 per cent boric acid for 4 hours at 400 liters per kilogram-hour of air, the optimum temperature was found to be 160–170°C. The reaction became uncontrollable above 175°C. It was also shown that 4 hours is about the optimum duration under these conditions. Alcohol yield fell off rapidly at longer reaction times. The rate of air flow was investigated at 75–245 liters per kilogram-hour. A definite maximum conversion to alcohols was noted at approximately 150 liters per kilogram-hour. The boric acid quantity was studied from 0 to 10 per cent with optimum conversion to alcohols observed at about 5 per cent. The authors concluded that the optimum conditions for the oxidation of this type of hydrocarbon mixture to alcohols are a temperature of 160–170°C, 5 per cent boric acid, 4–5 hours reaction time, and 150 liters per kilogram-hour of air. Under these conditions conversions of 25–30 per cent are attained to give a product containing 80 per cent alcohols in which secondary alcohols predominate. The alcohols formed essentially retain the linear hydrocarbon structure of the starting paraffins. These results are comparable to those reported by Bashkirov et al., above.

The air oxidations of mineral oil fractions (b.p. 200–350°C) have been studied by Dimitrov and Stankova with and without added boric acid.[72] Optimum conditions for alcohol preparation appear to be 165–170°C at 500 liters per kilogram-hour air and 4–5 hours reaction time.

The optimum conditions for the oxidation of shale-oil paraffins are reported to be 200°C with a flow rate of 600 liters per kilogram-hour.[131, 183] One shale oil is claimed to have given 96.5 per cent yields of a product with a hydroxyl number greater than 100, of which 40 to 60 per cent had a hydroxyl number of about 170 to 180.[131]

(b) *Alcohol structure.* The relative reactivity of the primary, secondary and tertiary positions of hydrocarbons in oxidation reactions has been mentioned (see Section II, B, 5 above). More comprehensive coverage of the pertinent data is given here.

Linear paraffins have been oxidized in the liquid phase with oxygen–nitrogen mixtures at 165°C to give a mixture containing 72.4–88.6 per cent secondary and 11.4–27.6 per cent primary alcohols which was converted to detergents.[144]

In these paraffin oxidations the major components of the product mixture are secondary alcohols. In the early stages of the reactions, the oxidation products are principally alcohols which decrease in amount as ketones, acids and condensation products are formed.[225] In one Russian paper it was shown that the alcohols produced by the oxidation of paraffins contain about 15 per cent glycols and small amounts of hydroxyketones.[126] During studies on the oxidation of synthol (b.p. 250–270°C) in the presence of boric acid, it was noted that increased conversions and increased amounts of monofunctional products led to increased bifunctional products, glycols, etc.[125] Prolonged reaction time usually increases the proportion of acids formed,[280] although increasing the oxidation time with shale oil paraffins is claimed not to affect the product composition.[133]

Liquid phase oxidation of monomethyl paraffins in the presence of boric acid gives a mixture of secondary and tertiary alcohols. The latter comprise about 25–30 mole per cent of the product.[36] In all cases, the amount of primary alcohol produced is very small. Extension of this study to dimethyl- and trimethyl-substituted paraffins was described in a subsequent paper.[34] The results showed clearly that under comparable conditions the amount of tertiary alcohols formed increased with the number of tertiary carbon atoms or methyl groups in the original hydrocarbon. For example, the amounts of tertiary alcohols produced were approximately 27, 37 and 60 mole per cent for monomethyl-, dimethyl- and trimethyl-substituted compounds of approximately the same chain length. No evidence for cleavage of the hydrocarbon chains was found in these oxidations.

It has been shown that the alcohols obtained in these oxidations retain the carbon skeleton present in the hydrocarbon substrate.[31, 32, 36] The secondary alcohols give a mixture of acids on oxidative cleavage indicating that all possible position isomers are present in comparable amounts.[31, 32] Similarly, the ketones formed in the oxidation of linear paraffins have the same carbon chains as those of the original paraffins. All of the possible position isomers are present in the product mixtures.[127] In one case the alcohols obtained were dehydrated to the corresponding olefins and the olefins hydrogenated back to the original hydrocarbons thus demonstrating retention of the original carbon skeleton[30] (equation (37)).

$$CH_3(CH_2)_{m+n+1}CH_3 \xrightarrow[H_3BO_3]{O_2} \xrightarrow{H_2O} CH_3(CH_2)_m\overset{\overset{\displaystyle OH}{|}}{C}H(CH_2)_nCH_3$$

$$\downarrow -H_2O$$

$$CH_3(CH_2)_{m-1}CH{=}CH(CH_2)_nCH_3$$

$$\xleftarrow{H_2/cat}$$

(37)

A claim has been made that the use of air at 185°C in the presence of 15–20 per cent boric acid gave better yields and conversions[150] than the optimum conditions described by Bashkirov. In this case a product containing 60 per cent primary alcohols and 40 per cent secondary alcohols was reported, in complete disagreement with all other work which indicates that very little primary alcohol is produced. The same investigators also claim that reduced flow rates of air rather than nitrogen-diluted air gives good yields of alcohols.[280] Optimum conditions were 150 liters per kilogramhour of air at 200°C using boric acid.

(c) *Pyrolysis of borates to give olefins.* The intermediate alkyl meta and orthoborates produced from saturated alkane oxidation can be pyrolyzed to provide alkenes (equation (38)). Such a sequence constitutes the con-

$$CH_3(CH_2)_{m+n+1}CH_3 \xrightarrow[H_3BO_3]{O_2} \left[CH_3(CH_2)_m\overset{\overset{\displaystyle (CH_2)_nCH_3}{|}}{C}HOBO \right]_3$$

$$\downarrow$$

$$CH_3(CH_2)_{m-1}CH{=}CH(CH_2)_nCH_3 + CH_3(CH_2)_mCH{=}CH(CH_2)_{n-1}CH_3$$

(38)

version of a linear paraffin to a mixture of straight chain olefins.

Veselov and co-workers[261] found that use of excess boric acid leads to less thermally stable boric acid ester residues. This causes problems during the distillation of unreacted hydrocarbons due to decomposition of the borates to give olefins. These authors conclude that only enough boric acid should be used to convert all of the alcohols present to orthoborates; this avoids the formation of less stable metaborates. The percentages of boric acid used can be reduced from the normal 4 per cent to approximately 2.5 per cent in the pilot plant air oxidation of gasoline fractions.[262] Alcohol yields were not significantly affected. Better results are obtained at this lower boron level if the boric acid is finely dispersed.

Boric acid esters obtained by the oxidation of C_{11} to C_{15} linear paraffins can be pyrolyzed at 360°C to produce C_{11} to C_{15} olefins.[99] The pyrolysis temperature can be reduced to 250°C by adding excess boric acid.[1] As noted above, the metaboric acid ester produced in the latter case would be less thermally stable than the orthoborate. Similarly, a paraffin oxidized at 220°C in the presence of boric acid gives an unspecified olefin (iodine value above 50) when subsequently treated with an "acid earth".[181]

(d) *Effect of aromatic impurities.* The presence of aromatic impurities in paraffins inhibits oxidation. This may account for the improved results with shale oils which had been extracted with ethanol prior to oxidation.[133] Similar removal of aromatics was required in the oxidation of paraffins separated by means of urea adducts.[280] The inhibitory effect of aromatics is also discussed in Section II, B, 8 above and Section III, D below. In studies at 165–170°C and 300 liters per kilogram-hour of air the oxidation of a crude oil (b.p. 270–320°C) to alcohols in the presence of boric acid was shown to be strongly inhibited by aromatics in the oil.[70] The authors suggest that this inhibition is due to esters formed by reactions of boric acid and aromatic alcohols produced during the initial stages of the oxidation.

Aromatics have been shown to strongly inhibit the oxidation of n-dodecane and n-decane with molecular oxygen in the presence of boric oxide.[114] With n-decane only 0.45 per cent aromatics almost completely inhibited oxidation. It was concluded that phenols produced by oxidation of the aromatics are the true inhibitors not the aromatics in the feed.

Oxidations of aliphatic hydrocarbons in the C_{10}–C_{30} range by molecular oxygen have been improved by distilling the unreacted hydrocarbon from the borate esters formed.[244] Partially dehydrated boric acid is used at 160°C. In the examples given, a C_{11}–C_{14} hydrocarbon fraction containing 1.5 per cent aromatics was oxidized. Best results were obtained when the distilled unreacted hydrocarbon-ketone fraction was recycled back through the oxidation process.

(e) *Miscellaneous.* In a series of 1955–7 patents to Kendall Refining, air oxidation of petroleum fractions in the presence of boric acid mixtures is described.[145-7, 196] The products were not identified but presumably alcohols were produced and the produce mixture contained the borates derived from the alcohols.

α-Olefin polymers with molecular weights of 500 to 300,000 are hydroxylated conveniently without degradation by air oxidation in the presence of boric acid.[136]

In a secondary reference, esterification of C_7–C_9 primary and secondary alcohols with phthalic anhydride has been investigated.[130] The alcohols were prepared originally by hydrocarbon oxidation in the presence of boric acid.

Air oxidation of linear hydrocarbons is usually carried out in the presence of boric acids, but it is claimed in the patent literature that sodium borates are also effective.[122] However, when 10 per cent borax is used in a typical oxidation reaction, the yield of secondary alcohols produced (44 per cent) is relatively poor compared to reactions using boric acid.

n-Hexane has been oxidized with an oxygen–nitrogen mixture in the presence of boric acid;[220, 270] an improvement in the hydrolysis procedure was claimed.

Tricosane is converted (55 per cent) to boric acid esters by air oxidation at 200°C for 5 hours.[179] Oxidation of a synthetic paraffin cut under the optimum conditions gave a 40 per cent conversion to a produce found to contain 70 per cent alcohol.[150] The liquid phase air oxidation of n-pentadecane by molecular oxygen in the presence of boric acid at 165–170°C yielded a mixture of oxygenated products containing 49 per cent alcohols, 31 per cent acids and 20 per cent carbonyl compounds.[128] With a nitrogen–oxygen mixture containing 3.5–5 per cent oxygen the yields were 67 per cent alcohols and 12 per cent acids.

Tri-, tetra-, penta- and hexadecane all exhibit similar oxidation rates and give analogous product mixtures when oxidized under the same conditions.[32] The air oxidation of C_{10}–C_{30} aliphatic hydrocarbons in the presence of partially dehydrated boric acid has been discussed in a Halcon patent.[243] Increasing the length of the oxidation reaction does not increase the yield of alcohols, but results in more acid formation.[223]

Shiman *et al.* studied air oxidations of high boiling fractions from the unsaponifiable products from a prior gas phase paraffin oxidation.[231] A 10 per cent excess of boric acid was used in the air oxidation at 110–120°C for 2–3 hours. After hydrolysis of the borate esters formed, the alcohols were separated (12.7 per cent yield) and then sulfated to give detergents.

In the only reported example of air oxidation of a halogenated aliphatic, chlorodecane has been oxidized to hydroxy–chlorodecanes in the presence of boric acid in a stream of air.[138]

3. Oxidations in the Presence of Organic Borates

Although most reported liquid phase oxidations of aliphatics have been carried out in the presence of boric acid or its partially dehydrated forms such as metaboric acid, several cases of process improvements using organic borates have been reported (see Table 2).

The substitution of borate esters for boric acid in air oxidations of hydrocarbons has been reported to minimize problems which can occur during isolation of the desired products and also to require smaller amounts of the boron compound.[193] Monoalkyl borates (presumably metaborates) are preferred.

Replacement of boric acid by tri-n-propyl borate in the liquid phase air oxidation of a paraffinic hydrocarbon (synthol, b.p. 270–320°C) has been investigated by Dimitrov et al.[74] Using much lower flow rates than in typical boric acid reactions, the oxidation rates were increased two- to three-fold. Alcohol yields were optimum at 165–185°C, 300 liters per kilogram-hour of air, and 5–10 weight per cent tri-n-propyl borate. Lower temperatures and higher flow rates gave increased amounts of carbonyl compounds.

Air oxidation of a linear hydrocarbon fraction in the presence of tri-n-butyl borate resulted in a 65 per cent yield of alcohols after hydrogenation of the ketones in the initial product mixture.[235] Tri-n-propyl borate has been used in similar oxidations of C_5–C_{12} hydrocarbons to provide good yields of secondary alcohols.[234]

In one example a relatively low conversion (3.1 per cent) of n-dodecane was obtained when tri-n-propyl borate was used in the presence of cobalt and manganese acetates.[281] No comparable reaction in the absence of the metal salts was reported.

The liquid phase oxidation of n-dodecane in the presence of butoxy boroxines (metaborates) and orthoborates has been claimed.[229] Only small differences were observed in comparative oxidations with n-butoxy-boroxine and isobutoxyboroxine. Addition of a trace amount of cobalt did not improve yields or conversions. At comparable boron concentrations, oxidations with isobutoxyboroxine yielded products with significantly higher alcohol/ketone ratios than did triisobutoxyborane. This suggests the use of boroxines or metaborates if products containing a higher proportion of alcohols are desired.

TABLE 2. OXIDATIONS OF LINEAR PARAFFINS IN THE PRESENCE OF ORGANIC BORATES

Substrate	Boron compound	Wt. (%)	Other	Temp. (°C)	Time (hr)	Gas, % O₂ in N₂	Flow rate (l./kg-hr)	% Conv.	Products	Ref.
n-Decane	$B(OCH_2CH_2CH_3)_3$	54		165	5.5	4	105	19	40.2% n-decanols plus 5.5% decanediols	234
n-Dodecane	$B(OCH_2CH_2CH_3)_3$	54	Co and Mn acetates	125	4	4	300	3.1	68.8% dodecanols	281
n-Dodecane	$(CH_3CH_2CH_2CH_2OBO)_3$	10		182	4	10		16.1	91% dodecanols+6% dodecanones	229
n-Dodecane	$\left(\genfrac{}{}{0pt}{}{CH_3}{CH_3}CHCH_2OBO\right)_3$	10		182	4	10		14.2	84% dodecanols+4% dodecanones	229
n-Dodecane	$(CH_3CH_2CH_2CH_2OBO)_3$	15.2	0.006 wt. % cobalt as cobalt octanoate	180		10.5		16	80% dodecanols+9.5% dodecanones	229
n-Dodecane	$B\left(OCH_2CH\genfrac{}{}{0pt}{}{CH_3}{CH_3}\right)_3$	21		165				11	3.7/1 dodecanols/dodecanones wt.ratio	229
n-Dodecane	$\left(\genfrac{}{}{0pt}{}{CH_3}{CH_3}CHCH_2OBO\right)_3$	9		165				13	12.0/1 dodecanols/dodecanones wt. ratio	229
Straight chain hydrocarbons, b.p. 200–300°C	$B(OCH_2CH_2CH_2CH_3)_3$	3		140–150	—	Air	300	66	28 mole % alcohols	235
Synthol (b.p. 270–320°C)	$B(OCH_2CH_2CH_3)_3$	5–10		165–185			300		Optimum for alcohol formation	74

4. *Oxidations in the Presence of Boric Acids and Boric Oxide Plus Other Additives*

In most of the numerous publications and patents on the oxidation of aliphatic hydrocarbons, the use of boric acids or boric oxide is recommended. In several publications, beneficial effects are implied or claimed for a variety of additives (Table 3).

The effect of temperature on the oxidation of paraffinic hydrocarbons in the presence of manganese naphthenate catalysts has been investigated by Tsyskovskii and Shcheglova.[250] After an initial oxidation stage, boric acid (5 per cent) was added. In a series of experiments with n-decane, n-undecane and n-tridecane it was shown that the proportion of primary alcohols produced decreased with increasing temperatures from 140° to 170°C. The total alcohol yields were relatively constant with temperature, although the reaction times used were significantly lower at higher temperatures. The average molecular weights of the primary alcohols obtained decreased with increasing temperature. No evidence was given for improvements in the oxidation as a result of the added metal salts.

A process has been patented for the production of carboxylic acids from C_6–C_{40} hydrocarbons.[227] This is a two-step process involving initial oxidation to an alcohol–ketone mixture in the presence of boric acid and manganese naphthenate followed by nitric acid oxidation to carboxylic acids.

The oxidation of paraffin waxes has been reported in the presence of boric acid plus manganese resinates.[135] Increasing the amount of boric acid from 0.2 to 12.5 per cent resulted in lower yields of carboxylic acids.

Manganese naphthenate has been added to an oxidation of a petroleum fraction in the presence of boric oxide[146-8] and sodium oxalate added to a paraffin wax oxidation with boric acid.[108]

An Imperial Chemical Industries' patent describes the oxidation of various linear paraffins in the presence of boric acid and hydrocarbon-soluble transition metal carboxylates.[281] As shown in Table 3 the alcohol yields in these reactions were good (70–85 per cent). It is claimed that the transition metal derivatives lead to shorter induction periods and reaction times.

Somewhat contradictory evidence has been presented in the oxidation of n-undecane.[114] Oxidation in the presence of boric acid led to a product with an alcohol/ketone ratio of 24/1. With added cobalt stearate this ratio dropped to 0.7/1.

Potassium permanganate has been added to several oxidation reactions involving borates. In an early (1944) Japanese patent, air oxidation of a

TABLE 3. THE EFFECTS OF ADDITIVES ON THE OXIDATIONS OF LINEAR PARAFFINS IN THE PRESENCE OF BORIC ACIDS AND BORIC OXIDE

Substrate	Boron compound	Wt. (%)	Other	Temp. (°C)	Time (hr)	Gas, % O_2 in N_2	Flow rate (L./kg-hr)	% Conv.	Products	Ref.
n-Decane	H_3BO_3	5	Mn naphthenate	140	8.5			—	—	250
n-Decane	H_3BO_3	5	Mn naphthenate	150	5			—	32.4 mole % primary alcohols	250
n-Decane	H_3BO_3	5	Mn naphthenate	160	3.5			—	13.9 mole % primary alcohols	250
n-Undecane	H_3BO_3	5	Mn naphthenate	140	7			55.3	46.2 mole % primary alcohols	250
n-Undecane	H_3BO_3	5	Mn naphthenate	150	5			—	35.1 mole % primary alcohols	250
n-Undecane	H_3BO_3	5	Mn naphthenate	160	3			53.5	13.9 mole % primary alcohols	250
n-Undecane	H_3BO_3	5	Mn naphthenate	170	2.5			54.8	14.5 mole % primary alcohols	250
n-Tridecane	H_3BO_3	5	Mn naphthenate	140	7			55.6	40.2 mole % primary alcohols	250
n-Tridecane	H_3BO_3	5	Mn naphthenate	150	5			48.5	22.9 mole % primary alcohols	250
n-Tridecane	H_3BO_3	5	Mn naphthenate	160	3			53.0	20.5 mole % primary alcohols	250
n-Octane[a]	B_2O_3	5	Co naphthenate	180	2	1	1000	10	68.7% octanols	281
Dodecane	B_2O_3	10	Co and Mg acetates	165	5	4	300	16	70.8% dodecanols	281
n-Dodecane	B_2O_3	2.4	Ni stearate	165	4	Air	82.5	19.2	79.2% dodecanols	281
n-Dodecane	B_2O_3	2.4	Ru stearate	165	4	Air	82.5	6.3	70.9% dodecanols	281
Hexadecane	H_3BO_3	9.9	1.3% Mn naphthenate	160±3	5.25	Air	400	48	59 mole % alcohols	227

Substrate	Reagent		Catalyst	Temp. (°C)	Time	Gas			Products	Ref.
n-Eicosane	B_2O_3	2.4	Ni stearate	165	5	Air	82.5	31.8	84.5% eicosanols	281
Petroleum fraction[b], mol. wt. 150	B_2O_3	11.5	0.09% Mn naphthenate	155	7	Air	31	—	12 mole % alcohols	146–148
Refined paraffin waxes	H_3BO_3	0.2–12.5	0.2% Mn resinate	—	—	—	—	—		135
Paraffin wax, m.p. 54°C	H_3BO_3	5.5	2.8% Na oxalate	180–200	2	Air	—	—	42% crude alcohols	108
Hentriacontane ($C_{31}H_{64}$)	H_3BO_3	5	0.1% $KMnO_4$	200	—	Air	—	—	40 mole % alcohols	185
C_{10}–C_{14} Linear paraffins	H_3BO_3	5	0.2% $KMnO_4$	200	—	Air	—	22	H_3BO_3 added to product followed by pyrolysis to 56% yield of olefins	1
C_{11}–C_{15} Paraffins, 97% linear	H_3BO_3	5	0.1% $KMnO_4$	200	3	Air	560	21.5	Products pyrolyzed to olefins	99
C_{11}–C_{15} Paraffins	H_3BO_3	5	0.1% $KMnO_4$	200	3	Air	—	21.5		99
Paraffin, m.p. 46.3°C	H_3BO_3	10	0.1% $KMnO_4$	200	3	Air	10	—	Alcohols	169
Shale oil paraffin, m.p. 47.6°C	H_3BO_3	5	0.1% $KMnO_4$	200	3	Air	—	42	58% (alcohols and ketones)	182
Shale oil paraffin, m.p. 49°C	H_3BO_3	5	0.1% $KMnO_4$	200	5	Air	600	—	OH no. 130	183
C_{11}–C_{14} Paraffins	$(HOBO)_3$	1.6	0.2% t-BuOOH	160	—	Air	120	18	63% alcohols	244
C_{11}–C_{14} Paraffins	$(HOBO)_3$	2.2	0.7% cyclohexane 0.15% t-BuOOBu-t	160	2.7		110	22	70% alcohols	244

a Reaction was carried out at 400 psi pressure.
b Reaction was carried out at 1.7 atm pressure.

paraffin (m.p. 46°C) was disclosed in the presence of 10 per cent boric acid and 1 per cent potassium permanganate.[169] Heating for 3 hours at 200°C and 10 liters per kilogram-hour of air resulted in formation of alcohols.

Oxidations of shale oil fractions,[182–3] C_{10}–C_{15} paraffins,[1, 99] and a C_{31} paraffin to alcohols have all been carried out with added potassium permanganate.

Organic peroxy compounds have been added to oxidations of C_{11}–C_{14} paraffins in the presence of metaboric acid.[244]

In a 1931 patent an air oxidation of hard paraffin wax in the presence of boric oxide was described.[118] The oxidation was carried out at 160–175°C with air saturated with acetic acid. After 3 hours, 40–44 per cent yields of unidentified alcohols were obtained. In the absence of acetic acid only about a 10 per cent alcohol yield was reported.

In another early patent (1934) oxidations of paraffins and waxes to alcohols in the presence of boric acid were reported.[110] In one case, a mixture of 6 per cent boric acid and 3 per cent sodium oxalate was used and in another 4 per cent boric acid plus 2 per cent sodium chloride. Air was passed through at 180–200°C to give unreported yields of alcohols.

5. *The Effects of Ultrasound on Oxidations in the Presence of Boric Acid*

The effects of ultrasound on the air oxidation of a synthetic hydrocarbon, b.p. 275–320°C, in the presence of 5 per cent boric acid has been investigated (see Table 4) over the 300–1000-kc range.[53] Maximum yields of alcohols were obtained in the 300–600-kc range with increases over the yields in the absence of ultrasound being approximately 15–20 per cent. Oxidation rates also increased significantly. The beneficial effect may be due to disintegration of the boric acid particles and consequent exposure of a fresh, reactive surface.

C. *Oxidation of Alicyclic Compounds*

1. *Introduction*

It is probable that oxidations of alicyclic compounds, and cyclohexane in particular, have received more attention than any other area of research on the oxidation of hydrocarbons in the presence of boron compounds. This is true because the oxidation products of cyclohexane and of cyclododecane can be used as intermediates in the manufacture of nylon. A mixture of cyclohexanol and cyclohexanone is obtained by cyclohexane oxidation with boric acid present. This mixture can be oxidized directly

TABLE 4. THE EFFECTS OF ULTRASOUND ON OXIDATIONS OF A SYNTHETIC
HYDROCARBON, b.p. 275–320°C, IN THE PRESENCE OF BORIC ACID[a]

Ultrasound	Time (hr)	OH no.	Acid no.	Carbonyl no.	Ester no.
None	1	40.2	6.4	18	12
None	2	58.5	10.5	25.8	22.4
300 kc ultrasound, 30 ma to generator	1	45–49	5.9–8.2	—	13–29
300 kc, 30 ma to generator	1.5	52–74	6.2–10	—	19–31
300 kc, 30 ma to generator	2	59–81	7.4–15	21–32	22–44
300 kc, 12 ma to generator	1	48–58	4.8–5.6	—	22–44
300 kc, 12 ma to generator	1.5	57–68	6.1–7.8	—	27–29
300 kc, 12 ma to generator	2	68–74	10–10.4	29–35	35–40
600 kc	1	40–42	3.8–5.5	18–22	17–20
600 kc	2	63–65	7–11.4	22–29	21–28
750 kc	1	41–48	3.4–6	21–25	18–27
750 kc	2	58–60	6–8	30–32	28–39
1000 kc	1	31–39	3.8–5.3	14–20	10–15
1000 kc	2	40–45	6–7.4	18–26	13–25

[a] Conditions used in all experiments were 4 per cent boric acid and air circulating at 700 l./kg-hr and 170°C.[14]

to adipic acid, which then is combined with hexamethylenediamine to give nylon 66 (equation (39)). Alternatively, the cyclohexanol is dehydrogenated to cyclohexanone which is converted to the oxime. Beckmann

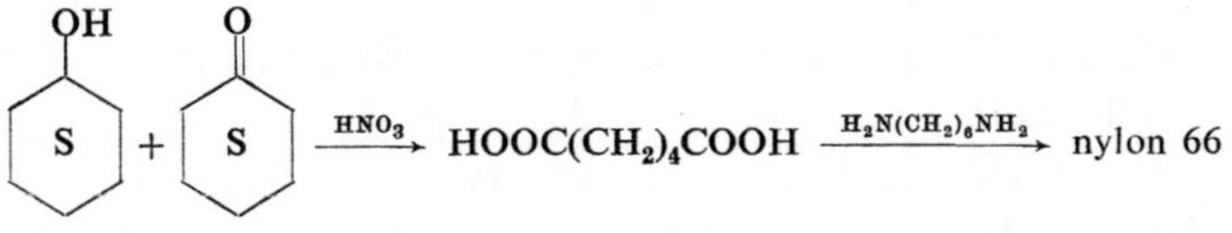

$$\text{OH} \quad \text{O} \qquad \xrightarrow{\text{HNO}_3} \text{HOOC(CH}_2)_4\text{COOH} \xrightarrow{\text{H}_2\text{N(CH}_2)_6\text{NH}_2} \text{nylon 66} \qquad (39)$$

rearrangement provides ε-caprolactam which is utilized to prepare nylon 6 (equation (40)). Similarly, cyclododecane leads to nylon 12. Many of

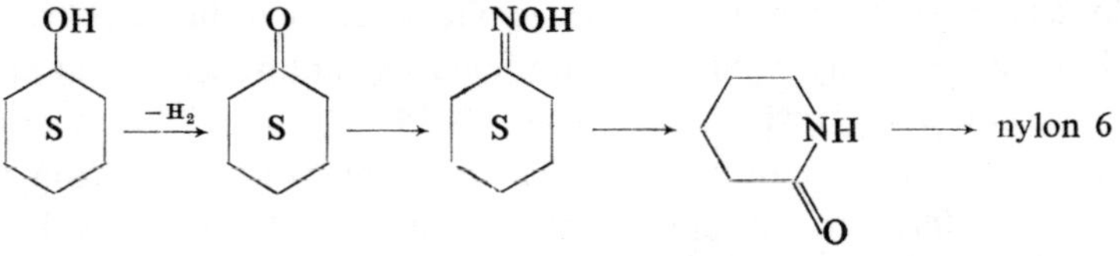

$$\text{OH} \xrightarrow{-\text{H}_2} \text{O} \longrightarrow \text{NOH} \longrightarrow \text{NH} \longrightarrow \text{nylon 6} \qquad (40)$$

the papers discussed in Section III, A on processes also involve cyclohexane as the substrate. In most of these papers the process modifications involved handling or reprocessing of the boron compound. In this section on oxidations of alicyclics, some of the patent literature discussed is also concerned with process modifications, but the discussion appeared to fit more appropriately here. These results are summarized in Tables 5 and 6.

A number of patents have been issued in which an oxidation of cyclohexane to a cyclohexanol–cyclohexanone mixture in the presence of boric acid is included as an initial step. The major patent claims, however, describe subsequent conversions of the oxidate and do not pertain to the hydrocarbon oxidation itself. Halcon International Inc. holds patents of this type on the conversion of cyclohexane to phenol[273], cyclohexyl-amine,[15, 16] ε-caprolactam,[15, 100] adipic acid[39, 100, 272, 276] and cyclo-hexanone.[275] Imperial Chemical Industries[112–13, 173] and du Pont[166] have also patented processes leading to adipic acid.

2. *Cyclohexane Oxidations* (Table 5)

A number of cyclohexane oxidation patents have been issued to Halcon International Inc. The primary claims in these patents all involve boric acid or metaboric acid. In a basic patent, the liquid phase oxidation of C_5–C_7 alicyclic compounds has been claimed.[271] This appears to be a general patent for this type of reaction and covers a variety of reaction conditions. A similar basic patent covering the general conditions for cyclohexane oxidations was filed by Imperial Chemical Industries[78] and later amended.[174] Many of the patents issued subsequently claim process variations of the general procedures described in these basic disclosures. In a Halcon International Inc. patent a composition has been claimed from the oxidation of cyclohexane with oxygen in the presence of metaboric acid.[276] The product composition covered a range of cyclo-hexanol, cyclohexanone and other oxidation products. The existence of this and the myriad other patents noted are illustrative of the complex patent claims in this area, and it is almost impossible to unravel them completely and clearly.

In another patent, Halcon has claimed that selectivity in the oxidation of cyclohexane to cyclohexanol plus cyclohexanone can be improved by small, but economically significant, amounts by adding small quantities of aromatic compounds which boil in the 80–145°C range.[86] In reactions run under standard conditions cyclohexane conversion rose from 91.5 per cent with 0.001 per cent added benzene to 94.6 per cent with 0.5 per cent benzene. With cyclohexane containing 1 per cent or 2 per cent benzene, oxidation is severely inhibited, so it is clear that added aromatics are advantageous only within rather narrow limits. Inhibition of oxidation by aromatics is also discussed in Sections III, B and III, D. It was shown that the inhibitory effects of aromatics above 1 per cent can be offset to some degree by adding very small amounts of initiators such as t-butyl peroxide and cyclohexanone.

A process in which cyclohexane is oxidized in separate stages in the presence of metaboric acid has been patented by Halcon International Inc.[216] A Japanese patent describes a general process for oxidation of cyclohexane to cyclohexanol and cyclohexanone.[165]

Imperial Chemical Industries has claimed that conversions in oxidations of cyclohexane to cyclohexanol–cyclohexanone mixtures by oxidation in the presence of borates can be increased significantly by addition of small amounts of alkaline earth compounds (preferably magnesium).[91] This modification of borate oxidation was disclosed in the patent, but no specific example was given. It was also found that oxidation rates of various hydrocarbons including alicyclics and cyclohexane in the presence of boron compounds are increased by the addition of 50–100 ppm of transition metal carboxylates.[281] Cobalt, manganese, rhodium and copper naphthenates were used.

Imperial Chemical Industries has patented cyclohexane oxidations in which the ratio of boron compound to cyclohexane has been lowered to significantly below the necessary stoichiometric ratio for the oxidation reactions observed.[121] Yields are not lowered by using these smaller amounts. Boric acid was used in most examples but cyclohexylborate was also noted.

Imperial Chemical Industries discloses the use of borate esters of the product aliphatic alcohols in the oxidation of aliphatic compounds.[112, 120] The examples given are with cyclohexane using ortho- and metaborates in relatively small amounts (0.2–6.4 weight per cent of cyclohexane). A small quantity of cobalt naphthenate was also present in these runs. The examples suggest that orthoborates lead to less selective oxidation to cyclohexanol. There do not seem to be significant differences in conversion and selectivity between cyclohexyl borates and metaboric acid. The major advantage of this disclosure is stated to be the use of a borate ester which is much more soluble in the reaction mixture than boric acid or metaboric acid.

Stamicarbon N.V. has patented the use of metaborates and $\frac{4}{3}$ boric acid–alcohol borates in cyclohexane oxidations.[240] Improved solubility in the reaction mixture was claimed.

TABLE 5. OXIDATIONS OF CYCLOHEXANE

Boron compound	Wt. (%)	Other	Temp. (°C)	Pressure	Time (hr)	Gas, % O$_2$ in N$_2$	Flow rate (l./kg-hr)	% Conv.	Products	Ref.
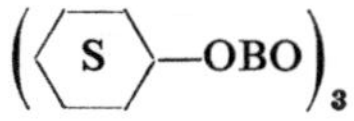 $\left(\text{S}-\text{OBO}\right)_3$	0.2	2.5 ppm cobalt naphthenate	160	400 psig	1	5	800	10.1	75.3% (1.25 cyclohexanol/cyclohexanone)	120
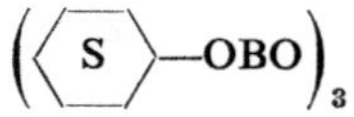 $\left(\text{S}-\text{OBO}\right)_3$	0.6	2.5 ppm cobalt naphthenate	160	400 psig	1	5	800	9.8	82.3% (1.42 cyclohexanol/cyclohexanone)	120
$(\text{HOBO})_3$	0.5	2.5 ppm cobalt naphthenate	160	400 psig	1	5	800	10.0	80.8% (1.45 cyclohexanol/cyclohexanone)	120
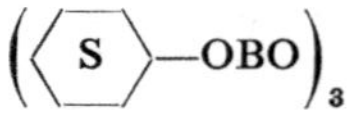 $\left(\text{S}-\text{OBO}\right)_3$	0.6	2.5 ppm cobalt naphthenate	170	400 psig	1	5	800	9.3	83.7% (1.61 cyclohexanol/cyclohexanone)	120
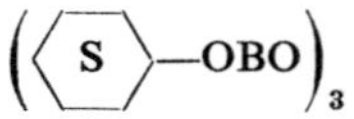 $\left(\text{S}-\text{OBO}\right)_3$	2.5	2.5 ppm cobalt naphthenate	160	400 psig	1	5	800	7.2	74.1% (1.55 cyclohexanol/cyclohexanone)	120
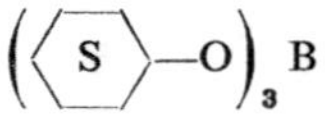 $\left(\text{S}-\text{O}\right)_3\text{B}$	2.5	2.5 ppm cobalt naphthenate	160	400 psig	1	5	800	6.5	76.8% (1.12 cyclohexanol/cyclohexanone)	120
None		2.5 ppm cobalt naphthenate	160	400 psig	1	5	800	9.2	70% (1.37 cyclohexanol/cyclohexanone)	120

H₃BO₃	16.6	Initial dehydration to (HOBO)₃ under N₂	165–167	120 psig	5	4	89	~13	82% cyclohexanol, 3% cyclohexanone	221, 270, 272, 273, 275
H₃BO₃	11	Initial dehydration to (HOBO)₃ under N₂	165–167	120 psig	5	4	89	~13	76% cyclohexanol, 4% cyclohexanone	272, 273, 275
H₃BO₃	5.5	Initial dehydration to (HOBO)₃ under N₂	165–167	120 psig	5	4	89	~13	60% cyclohexanol, 15% cyclohexanone	272, 273, 275
None			165–167	120 psig	5	4	89	—	30% cyclohexanol, 30% xyclohexanone	273
(HOBO)₃	12		165	120 psig	4	Air	89	8		15, 188
H₃BO₃	9.1		200		4	4	140	6	76% (cyclohexanol+cyclohexanone)	274
H₃BO₃	14.2		165	9.4 atm	6	4	85	11	82 mole % cyclohexanol+3% cyclohexanone	274
	20.4		190	14 atm	0.9	Air	162	12	68 mole % cyclohexanol+4% cyclohexanone	274
	4.8		180	225 psi	3	1.0	1010	6.4	75.5% cyclohexanol+13.8% cyclohexanone	78

TABLE 5. (cont.)

Boron compound	Wt. (%)	Other	Temp. (°C)	Pressure	Time (hr)	Gas, % O_2 in N_2	Flow rate (l./kg-hr)	% Conv.	Products	Ref.
B_2O_3	2.4		180	300 psi	3	1.0	1010	7.9	68.3% cyclohexanol+ 14.2% cyclohexanone	78
H_3BO_3	10		150	400 psi	2.5	1.0	820	10	48% cyclohexanol+ 18.3% cyclohexanone	78
(HOBO)₃	16.6		165–167	9.4 atm	5	4	89		Products not isolated	39
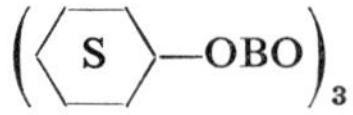 (⟨S⟩—OBO)₃	6.4	50 ppm Co as naphthenate	155	9.4 atm	4	5	320	8	Products not isolated but oxidized to 83% yield of adipic acid	112
None	—	25 ppm Co as naphthenate	160	200 psig	1.0	5	800	9.2	70% (1.37 cyclohexanol/cyclohexanone)	121
H_3BO_3	0.2	25 ppm Co as naphthenate	160	400 psig	1.0	5	800	9.8	78% (1.4 cyclohexanol/cyclohexanone)	121
H_3BO_3	0.5	25 ppm Co as naphthenate	160	400 psig	1.0	5	800	7.6	78.2% (1.45 cyclohexanol/cyclohexanone)	121
(HOBO)₃	0.2	25 ppm Co as naphthenate	160	400 psig	1.0	5	800	10.2	78.2% (1.26 cyclohexanol/cyclohexanone)	121

(HOBO)₃	0.5	25 ppm Co as naphthenate	160	400 psig	1.0	5	800	10.0	80.8% (1.45 cyclohexanol/cyclohexanone)	121
(S—OBO)₃	0.2	25 ppm Co as naphthenate	160	400 psig	1.0	5	800	10.1	75.3% (1.25 cyclohexanol/cyclohexanone)	121
(S—OBO)₃	0.6	25 ppm Co as naphthenate	160	400 psig	1.0	5	800	9.8	82.3% (1.42 cyclohexanol/cyclohexanone)	121
(HOBO)₃	5	50 ppm cobalt naphthenate	160	120 psi	1.6	air		10.3	77.9% cyclohexanol	173
H₃BO₃	0.1	50 ppm cobalt naphthenate	160	120 psi	1.6	air		10.7	73.6% cyclohexanol	173
H₃BO₃	0.4	50 ppm cobalt naphthenate			1.4	air		10.4	74.6% cyclohexanol	173
B(OCH₃)₃	20	Trace of cobalt naphthenate	150	400 psi	2	2.5	800	9	51% cyclohexanol, 7% cyclohexanone	281
B(OCH₃)₃	20	Trace of manganese naphthenate	150	400 psig	2	2.5	800	15	54% cyclohexanol, 3.4% cyclohexanone	281
B(OCH₃)₃	20	Trace vanadium naphthenate	150	400 psi	2	2.5	800	11	52.9% cyclohexanol, 11.3% cyclohexanone	281
B(OCH₃)₃	20	Trace vanadium naphthenate	130–138	400 psi	3.25	2.5	800	6	66.7% cyclohexanol, 7.6% cyclohexanone	281

TABLE 5. (cont.)

Boron compound	Wt. (%)	Other	Temp. (°C)	Pressure	Time (hr)	Gas, % O_2 in N_2	Flow rate (l./kg-hr)	% Conv.	Products	Ref.
$B(OCH_3)_3$	20	Trace rhodium stearate	150	400 psi	2.25	2.5	800	13	59.5% cyclohexanol, 9.8% cyclohexanone	281
$B(OCH_3)_3$	20	Trace copper palmitate	150	400 psi	2.25	2.5	800	10	66.9% cyclohexanol, 6.2% cyclohexanone	281
$B(OCH_3)_3$	20	Trace vanadium naphthenate	150	400 psi	1.5	5	800	8	63.5% cyclohexanol, 7.8% cyclohexanone	281
$B(OCH_3)_3$	20	Trace vanadium naphthenate	150	400 psi	2.5	2.5	800	13	51.1% cyclohexanol, 13.8% cyclohexanone	281
$B(OCH_3)_3$	20	Trace of cobalt, vanadium naphthenate mixture	140	400 psi	1.5	2.5	800	7	57.7% cyclohexanol, 20% cyclohexanone	281
$B(OCH_3)_3$	20	Trace cobalt, vanadium naphthenate mixture	130	400 psi	2.75	2.5	800	6	62% cyclohexanol, 15% cyclohexanone	281
$B(OCH_3)_3$	20	Trace cobalt, vanadium naphthenate mixture	130	400 psi	0.5	2.5	800	6	61.4% cyclohexanol, 18.5% cyclohexanone	281

Compound		Additive	Temp.	Pressure					Products	Ref.
(HOBO)₃	5	Trace cobalt naphthenate	150	120 psi	1	air		11.6	32.6% cyclohexanol, 28.9% cyclohexanone	281
H₃BO₃	10		200		4	4		6	75.7% (cyclohexanol + cyclohexanone)	270
B₂O₃	25.6	100 ppm cobalt naphthenate	190	2000 psig	0.92	4		12	67.9% cyclohexanol, 3.9% cyclohexanone	270
H₃BO₃	11.4	0.18 wt. % cyclohexanone	163–170	120 psig		air	53	7.6	80% (17.3 cyclohexanol/cyclohexanone)	270, 271
H₃BO₃	5.7	0.18 wt. % cyclohexanone	163–164	125 psig		air	53	7.8	76% (4.7 cyclohexanol/cyclohexanone)	270, 271
B₂O₃	3	0.18 wt. % cyclohexanone	168–174	150 psig		air	53	6.7	80% (3.9 cyclohexanol/cyclohexanone)	270, 271
H₃BO₃	16.6	0.18 wt. % cyclohexanone	165–167	120 psig		4	53	14.4	90% (3.1 cyclohexanol/cyclohexanone)	270, 271
(HOBO)₃	5		165	120 psig		air		8		15
(HOBO)₃	8	0.001 % benzene	165	125 psig		8			84.6% cyclohexanol, 6.9% cyclohexanone	86
(HOBO)₃	8	0.1 % benzene	165	125 psig		8			92.1% cyclohexanol, 2.7% cyclohexanone	86
(HOBO)₃	8	0.5 % benzene	165	125 psig		8			88.9% cyclohexanol, 5.7% cyclohexanone	86

TABLE 5. *(cont.)*

Boron compound	Wt. (%)	Other	Temp. (°C)	Pressure	Time (hr)	Gas, % O₂ in N₂	Flow rate (l./kg-hr)	% Conv.	Products	Ref.
(HOBO)₃	4–8	1–5% benzene	165	125–175 psig		8–10			Reaction strongly inhibited	86
(HOBO)₃	3.3	3% benzene, 0.004% di-t-butyl peroxide, 0.012% cyclohexanone	165	175 psig				8.9	85.5% (cyclohexanol + cyclohexanone)	86
(HOBO)₃	4	2% benzene, 0.004% di-t-butyl peroxide, 0.012% cyclohexanone	165	175 psig				11.6	85.3% (cyclohexanol + cyclohexanone)	86
(HOBO)₃	8	Recycled	165	125 psig		8				217
Pinacol biborate	33.3	0.25% cyclo-hexanone	165	9.5 atm		5	297	6.3	3/1 cyclohexanol/cyclohexanone	166
Pinacol biborate	33.3	0.25% cyclo-hexanone	165	6.8 atm		10	149	6.9	5.5/1 cyclohexanol/cyclohexanone	166

Catalyst		Additive							Products	Ref.
(HOBO)₃	10	0.25% cyclohexanone	165	7.8 atm		air	149	9.0	9/1 cyclohexanol/cyclohexanone, colored product, carbonaceous deposits	166
Pinacol biborate	21	0.28% cyclohexanone	185	1.7 atm		air	165		3.6/1 cyclohexanol/cyclohexanone	166
(HOBO)₃	15	0.28% cyclohexanone	185	1.7 atm		air	165		5.9/1 cyclohexanol/cyclohexanone, colored product, carbonaceous deposits	166
None	—	0.28% cyclohexanone	185	1.7 atm		air	165		0.21/1 cyclohexanol/cyclohexanone	166
Pinacol biborate	20.7	0.28% cyclohexanone	165	7.1–7.6 atm		5	212		3.8/1 cyclohexanol/cyclohexanone	166
H₃BO₃	5	50 ppm cobalt naphthenate	160	9.4 atm	1.6	air	150	12.2	77.9% (cyclohexanol+cyclohexanone)	113
(S—OBO)₃	12.3		164–165	9 atm	2	5	200	10	66% cyclohexanol+14.5% cyclohexanone	240
(S—OBO)₃ +1/2 B₂O₃	12.3		164–165	9 atm	2	5	200	9	73% cyclohexanol+14% cyclohexanone	240
Trace cobalt naphthenate — H₃BO₃	1.5	Trace cobalt naphthenate	160	120 psi	1	air		8.6	45% cyclohexanol, 29% cyclohexanone	281
B₂O₃	5		170	400 psi	2	1.0	1000	3.25	76.7% cyclohexanol	281

TABLE 5a. OXIDATION OF ALKYL CYCLOHEXANES

Substrate	Boron compound	Wt. (%)	Other	Temp. (°C)	Pressure	Time (hr)	Gas, % O_2 in N_2	Flow rate (l./kg-hr)	% Conv.	Products	Ref.
Methylcyclohexane	$(HOBO)_3$	9.1		160	4.3 atm		8	180	10		274
Methylcyclohexane	H_3BO_3	16.6		175–180	140 psig		air	53	8.3	83% (methylcyclohexanols + methylcyclohexanones)	271
Methylcyclohexane	H_3BO_3	10		160	62 psig	5	4	205	7	62.1% methylcyclohexanols	273
Methylcyclohexane	H_3BO_3	10		170	70 psig	2	4	205	6.7	50.5% methylcyclohexanols	273
Methylcyclohexane	$(HOBO)_3$	8.6		160	4.3 atm	2.3	8		~7	68.8 methylcyclohexanols	221
Methylcyclohexane	$(HOBO)_3$	10		160	62 psig		8		10	26% 1-methylcyclohexanol, 17% 2-methylcyclohexanol, 28% 3- and 4-methylcyclohexanols, 6% methylcyclohexanones	270

Methylcyclo-hexane	(HOBO)₃	7.1	10 ppm Co naphthe-nate, 0.03% acetalde-hyde	160	62 psig	4	4	147	4.3	43.0% methyl-cyclohexanols	273
n-Amylcyclo-hexane	H₃BO₃	5		165		4	3.5	1000		OH no. 87.0	33
n-Heptylcyclo-hexane	H₃BO₃	5		165		4	3.5	1000		OH no. 62.0	33

du Pont has patented the use of pinacol boric anhydride (XIV) in the air oxidation of cyclohexane.[166] This borate is recommended because boric acid,

$$
\begin{array}{ccc}
\underset{|}{CH_3} & & \underset{|}{CH_3} \\
CH_3-\underset{|}{C}-O & \diagdown & O-\underset{|}{C}-CH_3 \\
& BOB & \\
CH_3-\underset{|}{C}-O & \diagup & O-\underset{|}{C}-CH_3 \\
\underset{}{CH_3} & & \underset{}{CH_3}
\end{array}
$$

(XIV)

metaboric acid, etc., are solids which give slurries in the reaction zone and lead to handling problems. The acids also promote undesirable decomposition reactions, whereas pinacol borate is stable to carbonization and oxidation.

In other patents a process modification was claimed in which the initial step is water removal and formation of a cyclohexyl borate ester.[3-5] A mixture of metaboric acid and cyclohexane is added to the ester continuously in the presence of oxygen. The reaction is operated on a continuous basis with constant addition, take-off and recycle of hydrolyzed boric acid.

Institut Français du Petrole des Carburants et Lubrifiants has several patents which are directly related to cyclohexane oxidation. In one case the recirculating or recycle gases are passed into the condensed cyclohexane feed at 5–25°C higher than the reaction temperature.[115] This continuous preheating step results in a more easily controlled reaction, higher yields and expenditure of less overall energy than the use of the usual operating conditions. For example, oxidation of cyclohexane without recycling the gases gave a 66 per cent conversion to a product mixture containing 75 per cent of 5/1 cyclohexanol/cyclohexanone mixture. Using recycle gases for heating the feed resulted in 62–69 per cent conversion to a product containing 90.5 per cent of a 9/1 cyclohexanol/cyclohexanone product. Recycling the gases reduced the energy input from 25,730 kcal/mole to 14,900 kcal/mole.

The problem of handling a saturated water solution of boric acid obtained in cyclohexane oxidation is discussed in a Halcon International Inc. patent.[217] This aqueous phase can either be recycled to the hydrolysis step or it may be dehydrated and the resultant residual boron compounds used in the oxidation reaction. With cyclohexane it was shown that both dehydration and recycling to the hydrolysis step led to decreased yields due to buildup of impurities on the solid boric acid produced. The patent claims that this buildup can be eliminated by either continually bleeding off some of the saturated aqueous phase or by extracting the aqueous phase with solvent (usually cyclohexanol or cyclohexanone in cyclohexane oxidations). This was illustrated by the following yield values taken after four successive batch runs at 165°C, 125 psig and 8 per cent oxygen in nitrogen.

	Successive yields, per cent (cyclohexanol + cyclohexanone)			
	1	2	3	4
1. Normal operation	89	87	83	79
2. With bleeding of aqueous phase	89	87	85	85
3. Aqueous phase extracted with 1/1 cyclohexanol/cyclohexanone	89	87	87	87

It is obvious that yields normally drop off on reuse of the boric acid but appear to stabilize at high values when the modifications disclosed in the patent are used.

Similar oxidations of alkylcyclohexanes are summarized in Table 5a;[33, 221, 270-1, 273-4] the results parallel those obtained with cyclohexane itself.

3. Other Cycloaliphatics

Because of the use of its oxidation products as intermediates in the production of nylon 12, the oxidation of cyclododecane in the presence of boron compounds has been investigated in a number of laboratories (see Table 6). The few reported oxidations of other cycloaliphatics are also included in this table.

It has been reported that industrial air oxidation of cyclododecane (presumably by Chemische Werke Hüls A.-G.) is carried out at 150–160°C with approximately 6 per cent boric acid.[142] Reaction is carried to 20–25 per cent conversion and a yield of about 82 per cent of a 6–8/1 cyclodecanol/cyclododecanone mixture.

In a series of publications Bashkirov et al. have described an investigation of the liquid phase air oxidation of cyclododecane in the presence of boric acid. With 5 per cent boric acid present, cyclododecane conversion was 30–35 per cent to a product containing 80 per cent cyclododecanol.[18, 19, 21] The product ratios obtained by oxidation of cyclododecane with and without boric acid are noted below.

TABLE 6. OXIDATION OF CYCLOALIPHATICS OTHER THAN CYCLOHEXANE

Substrate	Boron compound	Wt. (%)	Temp. (°C)	Time (hr)	Gas	Flow rate (l./kg-hr)	% Conv.	Products	Ref.
Cyclooctane[a]	H_3BO_3	3.9	100	3.5	Air	—	20	35 mole % cyclooctanol+35 mole % cyclooctanone	59
Cyclododecane	None	—	160	3.5	Air	354	21	30 mole % cyclododecanol	139
Cyclododecane	H_3BO_3	6.0	160	5.25	Air	354	21	83 mole % cyclododecanol	139
Cyclododecane	H_3BO_3	5.7	160	5.25	Air	250–270 cc/min	27.2	81.0 mole % cyclododecanol	139, 161
Cyclododecane	H_3BO_3	5.7	160	6.75	Air	250–270 cc/min	21.4	83.0 mole % cyclododecanol	139, 161
Cyclododecane	H_3BO_3	7.5	165	4	4% O_2 in N_2		38	90 mole % cyclododecanol	282
Cyclododecane	H_3BO_3[b]		90–100	4–5	Air	33	33	79.8% yield (5/1 cyclododecanol-cyclododecanone)	58
Cyclododecane	H_3BO_3	5					30–35	~80% cyclododecanol, 8% cyclododecanone	18
Cyclododecane	H_3BO_3	5	165	4	Air	600	33	86% cyclododecanol	22
Cyclododecane	B_2O_3[c]	8	170	1.5	Air	106	38	92% cyclododecanol	198

Substrate	Additive	Amount	Temp	Time	Gas	Flow	Conversion	Product	Ref.
Cyclododecane	B_2O_3[c]	8	170	3	Air	106	37	76% cyclodo-decanol	198
Cyclododecane	B_2O_3	10	160	3	Air	400	42	85.7% cyclodo-decanol	196
Cyclododecane	$B[OC(CH_3)_3]_3$	18.4	155–160	2.0	Air	250–270 cc/min	20.6	77.5 mole % cyclododecanol	139, 161
Cyclododecane	$B[OC(CH_3)_3]_3$	18.4	155–160	2.1	Air	250–270 cc/min	22.2	77.5 mole % cyclododecanol	139, 161
Cyclododecane	$B[OC(CH_3)_3]_3$	18.6	160	3.5	Air	324	46.4	62 mole % cyclododecanol	139, 161
Cyclododecane	$B(OCH_2CH_2CH_2CH_3)_3$	18.6	160	3.5	Air	324	37.0	42 mole % cyclododecanol	139, 161
Cyclododecane	$(CH_3CO)_2BOB(OCCH_3)_2$	8.75	170–190	1	O_2			8/1 to 15/1 cyclodo-decanol/cyclodo-decanone ratios	80
Cyclohexadecane	B_2O_3[c]	8.3	170	1.5	Air	83	40	92% cyclohexa-decanol	198
Cyclohexadecane	B_2O_3[d]	8.3	170	1.5	Air	83	35	86% cyclohexa-decanol	198
Cyclohexadecane	B_2O_3	10	160	3	Air	400	40	90.5% cyclohexa-decanol	196

[a] 0.2 per cent cobalt naphthenate.
[b] Added as saturated solution.
[c] Cylindrical pellets.
[d] Powdered.

	Cyclododecanol	Cyclodode-canone	Acids and polyfunctional oxygenated product
With boric acid	8.0	0.8	1.2
Without boric acid	1.8	2.7	2.4

It is obvious that the use of boric acid restricts further oxidation and favors the production of cyclododecanol. Several of the papers in this series[123, 129, 282] are more concerned with the subsequent treatment of the cyclododecane oxidation products and are of minor interest in this chapter. A Russian patent based on this work claims an 86 per cent yield of cyclododecanol based on cyclododecane reacted.[22]

In a patented oxidation reaction under similar conditions, a product containing approximately 80 per cent of a 5/1 mixture of cyclododecanol/cyclododecanone was isolated.[58] Conversion of cyclododecane was approximately 33 per cent. The patent claims noted specifically that the process was improved by continuously adding a saturated solution of boric acid to the reaction vessel.

In another patent Halcon International Inc. claims the oxidation of cyclododecane in the presence of boric acid, but experimental details were not included.[272]

An Esso patent is specifically oriented toward the oxidation of cyclododecane obtained from the hydrogenation of cyclododecatriene.[55, 96] Boric acid is used as the catalyst.

Modification of boric oxide or boric acid (not included in any examples) by forming compact molded particles reportedly leads to improved cyclododecanol yields.[196] The compact particles have high surface areas. In a subsequent patent the use of a cylindrical reactor and good stirring were also claimed to improve yields, although the data does not appear to show significant improvements.[198]

Compounds other than boric oxide and boric acid can be used in the oxidation of cyclododecane. For example, Esso has patented the use of tri-n-butyl and tri-t-butyl borates as well as boric acid.[139, 161] Liquid phase air oxidations of cyclododecane at 160°C using ~ 14 mole per cent of tri-t-butyl borate resulted in a 46.4 per cent conversion to a product containing 62 per cent cyclododecanol compared to a 37 per cent conversion to 42 per cent cyclododecanol with tri-n-butyl borate. Use of boric acid gave conversions and alcohol yields similar to tri-t-butyl borate, but the boric acid runs required 5–7 hours compared to 2 hours with tri-t-butyl borate.

A du Pont patent describes the use of "boron acetate" or tetraacetoxy biborate,

$$\left(\underset{CH_3CO}{\overset{O}{\parallel}}\right)_2 BOB \left(\underset{OCCH_3}{\overset{O}{\parallel}}\right)_2,$$

in the oxidation of cyclododecane.[80] Products containing alcohol to ketone ratios of 8/1 to 15/1 are claimed.

The general oxidation of C_8–C_{12} alicyclics has been claimed by Chemische Werke Hüls.[59] The patent also describes a continuous reactor for cyclododecane oxidation.

D. *Oxidation of Aromatic Compounds*

Oxidation of aromatic compounds, and alkylbenzenes in particular, have been described by several workers. This section includes oxidations of such aromatic substrates in the presence of boron compounds. The results of these oxidations by molecular oxygen are summarized in Table 7. In general, oxidation of hydrocarbons in the presence of borates appears to be inhibited or at least retarded by aromatic compounds, although the presence of trace amounts of aromatics was claimed to improve the conversions in some cases.[86]

In one report diisopropylbenzene is oxidized in the presence of boric acid to give unspecified hydroxylated products.[139] In a more detailed paper Bashkirov *et al.* noted that aromatics contents of greater than 0.5 per cent inhibit oxidation significantly.[27] The same authors have subsequently published several papers on the direct oxidation of aromatics.[23, 24, 77]

Oxidations of several alkylbenzenes were investigated under a range of conditions at temperatures from 165–250°C and a boric acid concentration of 5 per cent.[24] In general it was observed that boric acid inhibits oxidation and leads to the formation of dark resinous products. Variations in processing conditions had no effect on the boric acid retardation, and the use of initiators such as metal salts and peroxides was ineffective. Oxidation is facilitated by increased alkyl chain length in the alkylbenzene. The effects of boric acid were demonstrated by adding various quantities to an octylbenzene oxidation which was already in progress. The oxidate darkened immediately with obvious resinification. Conversion was approxi-

TABLE 7. OXIDATIONS OF AROMATIC COMPOUNDS BY MOLECULAR OXYGEN

Substrate	Boron compound	Wt. (%)	Temp. (°C)	Time (hr)	Gas	Flow rate (l./kg-hr)	% Conv.	Products				Ref.
								Acid no	Ester no	Carbonyl no.	Hydroxyl no.	
Mixed xylenes[a]	$[(CH_3)_3CO]_3B$	19.8	137–139	2	Air	148	10	2 methyl benzyl alcohols/1 aldehyde				139, 161
Pseudocumene[a]	$[(CH_3)_3CO]_3B$	19.8	137–139	2	Air	148		Mostly alcohols				139
n-Butylbenzene	None		173	1	Air	500		4.9	9.7	22.7	10.4	71
n-Butylbenzene	None		173	3	Air	500		20.8	29.1	39.3	—	71
n-Butylbenzene	None		173	5	Air	500		42.9	55.4	57.4	—	71
n-Butylbenzene	$(CH_3CH_2CH_2O)_3B$	15.4	173	1.5	Air	500		5.3	13.3	4.4	28.0	71
n-Butylbenzene	$(CH_3CH_2CH_2O)_3B$	15.4	173	3	Air	500		6.1	11.7	6.8	23.2 (alcohol was α-hydroxy-butylbenzene)	71
n-Butylbenzene	H_3BO_3	5	173	1	Air	500		Completely inhibited				71
Diisopropylbenzene	H_3BO_3	4.7	160–175	4.25	Air	208		Hydroxylation in side chain				139
n-Octylbenzene	H_3BO_3	5	165–250	5 max.	O_2, air, or mixture	300–800	7	Some primary alcohol				24
Octylbenzene	B_2O_3	2.5	170	4	Oxygen	800		2.8	0	3.0	10.5	77

Substrate	Additive				Gas								
Octylbenzene	$(CH_3CO)_4B_2O$	5	170	4	Oxygen	800		1.6	1.7	3.0	13.1		77
Octylbenzene	Isobutyl monoborate[b]	8	170	4	Oxygen	800		Completely inhibited					77
Octylbenzene	$[(CH_3)_2CHCH_2O]_3B$	20	170	4	Oxygen	800		2.6	16.5	7.2	50.7		77
Octylbenzene	$[(CH_3)_2CHCH_2O]_3B$	50	170	4	Oxygen	800		2.8	18.0	9.1	64.3		77
Octylbenzene	None		170	4	Oxygen	800	37	165	35	94	0		77
n-Alkylbenzenes	H_3BO_3	5	165–250	5 max.	O_2, air, or mixture	300–800	—	> 95% secondary alcohols					24

[a] 0.01 per cent cobalt naphthenate.
[b] Probably the metaborate $[(CH_3)_2CHCH_2OBO]_n$.

mately 7 per cent to a product mixture containing only 7 per cent alcohols. The alcohol fraction was shown to be composed of 97.3 per cent secondary and 2.7 per cent primary phenyloctyl alcohols. In an attempt to stabilize the presumed intermediate alcohols which form in the octylbenzene oxidations, the reactions were carried out in the presence of an esterifying agent, acetic anhydride. With this modification conversion reached 61 per cent (compared to 7.2 per cent with boric acid). The products were esters of essentially the same secondary (94.4 per cent) and primary (5.6 per cent) phenyloctyl alcohols obtained with boric acid.

In a subsequent paper the effects of a variable aromatics content (0–4 per cent) in two paraffin distillate fractions were examined under standard oxidation conditions (3.5 per cent oxygen in nitrogen, 165–170°C, 5 per cent boric acid, 4 hours oxidation time).[23] The hydroxyl number of the final product dropped from 90 in the absence of aromatics to 20–30 with 3–4 per cent aromatics. Extensive darkening and resinification occurs when the aromatics content is above 1 per cent. Increasing the reaction time resulted in a further hydroxyl number reduction and resinification.

In a series of runs with synthol containing 0.15–1.0 per cent of pure aromatic compounds the following increasing order of inhibitory effectiveness was noted: tetralin, diphenyl, ditolylmethane, decylbenzene, naphthalene and butylnaphthalene. At a concentration of only 0.15 per cent butylnaphthalene, the hydroxyl number of the product reached 40 after 4 hours compared to 93 with aromatics-free synthol. In the absence of boric acid, the presence of aromatics in large amounts did not significantly inhibit oxidations at 165–170°C.

Bashkirov *et al.* also studied the effect of aromatics on reactions of hydrocarbons with molecular oxygen in the presence of boric oxide, monoisobutyl borate†, triisobutyl borate and tetraacetyl diborate.[77] Oxidation of octylbenzene was studied at 170°C over a 4-hour period in oxygen at 800 liters per kilogram-hour. The results (see Table 7) showed that oxidation is strongly inhibited by monoisobutyl borate and tetraacetyl diborate as well as boric oxide. Oxidation proceeds satisfactorily in the presence of triisobutyl borate even at concentrations as high as 20 and 50 per cent. In fact it was shown that alcohol and ester contents in the product increase slightly with borate concentration. The alcohols were separated and shown to be a mixture of phenyloctyl alcohols. With 50 per cent triisobutyl borate, 37 per cent conversion is obtained with

---

†Probably isobutyl metaborate, $[(CH_3)_2CHCH_2OBO]_n$.

an alcohol yield of 64–68 per cent, so alkylbenzenes can be successfully converted to phenylalkyl alcohols under these conditions.

Oxidation of mixed xylene isomers in the presence of tri-t-butyl borate resulted in 10 per cent conversion to a mixture containing 67 per cent methylbenzyl alcohols and 33 per cent aldehydes.[139, 161] No phenols or acids were obtained, but significant amounts of toluic acid were formed in the absence of the borate. Comparable oxidations of p-xylene and pseudocumene in the presence of tri-t-butyl borate also yielded mostly alcohols.[139] Oxidations of methylbenzyl alcohol in the presence of methyl borate and of diisopropylbenzene in the presence of boric acid yielded, respectively, a xylene diol and unidentified alcohols. These oxidations (except for diisopropylbenzene) were carried out in the presence of 0.01 per cent cobalt naphthenate.

The air oxidation of n-butylbenzene was found to be completely inhibited by the addition of 5 per cent boric acid at 173°C.[71] Oxidation in the absence of borates leads to a complex mixture of acids, esters and carbonyl compounds with initial formation of alcohols which disappeared with time. Use of an amount of tri-n-propyl borate equivalent to 5 per cent boric acid resulted in controlled oxidation with the formation of significant amounts of alcohols of which α-hydroxybutyl benzene is apparently the major constituent. It was postulated that formation of mixed aromatic propyl borates stabilized the alcohols formed to further oxidation.

In an isolated patent claim, the oxidation of styrene to styrene oxide as the major product was reported at 150°C and 250 psig of oxygen.[92] Styrene conversion was 51 per cent using trimethyl borate as a solvent.

A number of related oxidations (usually in the gas phase) of aromatic compounds in the presence of boron-containing catalysts are discussed in Section V.

E. *Oxidation of Olefins by Molecular Oxygen*

Relatively few examples of liquid phase olefin oxidations by molecular oxygen or air have been reported. In many cases a wide range of products is obtained thereby reducing the practicability of these reactions. These olefin oxidations are summarized in Table 8.

1. *Linear Olefins*

The oxidation of 1-dodecene has been studied in detail in the presence of boric acid under conditions which are comparable to those normally

TABLE 8. OXIDATION OF OLEFINS BY MOLECULAR OXYGEN

Substrate	Boron compound	Wt. (%)	Other	Temp. (°C)	Pressure	Time (hr)	Gas, % O_2 in N_2	Flow rate (l./kg-hr)	% Conv.	Products	Ref.
$CH_2=CH_2$	H_3BO_3									Formaldehyde	117
$CH_2=CH_2$	$B(OCH_3)_3$	467	2.5% CH_3CHO	220	300 psig	0.167	Oxygen	Used closed vessel	20	28% ethylene oxide	92
$CH_3CH=CH_2$	$B(OCH_3)_3$	210	0.2% $Hg(OAc)_2$, CH_3CHO	180–190	200 psig	0.33	Oxygen	Used closed vessel	—	CH_2—$CHCH_3$ major product— CH_3CHO, CH_3OH, CH_3OCCH_3, H_2O,	92
$CH_3CH=CH_2$	$B(OCH_3)_3$	357	1.4% CH_3CHO	200	200 psig	0.083	Oxygen	Used closed vessel	30	$HCOOH$, CH_3CCH_3, CH_3COOH, CH_2—$CHCH_3$ major product	92
$CH_3CH=CH_2$	$B(OCH_3)_3$	760		200	200 psig	0.083	Oxygen	300 g/kg	39	>45% propylene oxide	92
2-Methyl-2-pentene	$B(OCH_3)_3$	250	1.25% CH_3CHO	150	300 psig	0.167	Oxygen	Used closed vessel	42	52% 2-methyl-2,3-epoxybutane	92
2-Methyl-2-pentene	H_3BO_3	—		170–180	8–10 atm		Oxygen	—	—	Unsaturated alcohols	139

Cyclohexene	H_3BO_3	25.4		80–105		160	Oxygen	100–200	~ 75	33%  cyclohexen-OH, 67% HO-cyclohexen-O-cyclohexen-OH 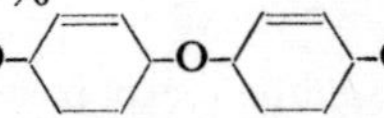	171
α-Pinene	H_3BO_3	10		154		58	Air	110–120	1–2	Mixture of pinene alcohols	171
Cyclooctene	H_3BO_3	5.3		155–165		3.5	Air	125–250	—	Cyclooctenol (50%)	139, 140
Cyclooctene	None	0		155–165		3.5	Air	125–250	—	Epoxide (55%); alcohol and ketone (15%); other (35%)	139, 140
2-Ethyl-hexene-1	H_3BO_3	6.7		120		120	Air	74–80	8	Mixture of unsaturated alcohols	171
1-Dodecene	H_3BO_3	12.8		185		6.5	Air	335	<10	Mixture of dodecenyl alcohols	171
1-Dodecene	H_3BO_3	5–7		165–175		0.5	4	800	17.3	79.6% alcohols; 20.4% ketones	35
1-Dodecene	H_3BO_3	5–7		165–175		1.0	4	800	20.6	78.2% alcohols; 21.8% ketones	35
1-Dodecene	H_3BO_3	5–7		165–175		1.5	4	800	23.8	77.1% alcohols; 19.8% ketones; 3.1% acids	35
1-Dodecene	H_3BO_3	5–7		165–175		2.0	4	800	26.5	77.2% alcohols; 16.4% ketones; 6.4% acids	35
1-Dodecene	H_3BO_3	5–7		165–175		3.0	4	800	30.7	73.7% alcohols; 13.2% ketones; 13.1% acids	35

TABLE 8. (*cont.*)

Substrate	Boron compound	Wt. (%)	Other	Temp. (°C)	Pressure	Time (hr)	Gas, % O_2 in N_2	Flow rate (l./kg-hr)	% Conv.	Products	Ref.
1-Dodecene	H_3BO_3	5		165		3.0	4	800	~16	19% epoxide plus a mixture of alcohols, ketones, esters, etc.	124
2-Dodecene Cyclododecene	H_3BO_3 H_3BO_3	5–7 5.3		165–175 155–165		3.0 3.5	4 Air	800 125–250	30.7 36	Mostly n-octyl alcohol Cyclododecenol (50%) cyclododecene oxide (40%)	35 139, 140
Cyclododecene	B_2O_3	5.3	0.01% cobalt naphthenate	155–165		3.5	1.6	125–250	—	Cyclododecenol	139, 140
n-Tridecene-6	H_3BO_3	5		165		3.0	4	800	~22	26% epoxide plus a mixture of alcohols, ketones, esters, etc.	124
n-Heptadecene-8	H_3BO_3	5		165		3.0	4	800	~26	34% epoxide plus a mixture of alcohols, ketones, esters, etc.	124
Tripropylene	H_3BO_3	11	0.31–0.36% cobalt naphthenate	60–82	13 atm	2	Air	285	—	Mixed glycol borates containing 0.7–1.1% boron	151
Tripropylene	H_3BO_3	11	0.31–0.36% cobalt naphthenate	106–132	13 atm	1	Air	285	—	Mixed glycol borates	151

Substrate	Oxidant		Catalyst	Temp.	Pressure	Time	Gas			Product	Ref.
Tripropylene	H_3BO_3	11	0.31–0.36% cobalt naphthenate	161–165	13 atm	2	Air	285	—	Mixed glycol borates	151
Tetrapropylene	H_3BO_3	5.4	0.29% cobalt naphthenate	130–140	20 atm	3.0	Air	285	—	Mixed glycol borates containing 0.54% boron	151
Tetrapropylene	H_3BO_3	10	0.34% cobalt naphthenate	130–140	13 atm	2.0	Air	285	—	Mixed glycol borates containing 0.56% boron	151
Tetrapropylene	H_3BO_3	7.8	0.28% iron naphthenate	70–80	13 atm	3.0	Air	285	—	Mixed glycol borates containing 0.45% boron	151
Tetrapropylene	H_3BO_3	7.8	0.28% iron naphthenate	135–145	13 atm	2.0	Air	285	—	Mixed glycol borates	151
Tetrapropylene	H_3BO_3	7.8	0.25% lead naphthenate	70–80	13 atm	3.0	Air	285	—	Mixed glycol borates containing 0.53% boron	151
Tetrapropylene	H_3BO_3	7.8	0.25% lead naphthenate	135–145	13 atm	2.0	Air	285	—	Mixed glycol borates	151
Tetrapropylene	H_3BO_3	7.8	0.7% cumene hydroperoxide	95–105	13 atm	8.0	Air	285	—	Mixed glycol borates containing 0.60% boron	151
Representative refinery olefins	H_3BO_3	11	0.2% cobalt naphthenate	85–95		46	Oxygen	6.7	—	Mixed glycol borates containing 0.7–0.8% boron	151

TABLE 8. (cont.)

Substrate	Boron compound	Wt. (%)	Other	Temp. (°C)	Pressure	Time (hr)	Gas, % O₂ in N₂	Flow rate (l./kg-hr)	% Conv.	Products	Ref.
Representative refinery olefins	H_3BO_3	11	0.2% cobalt naphthenate	90–105		8–28	Air	133–200	—	Mixed glycol borates containing 0.5–1.2% boron	151
Isoprene-isobutylene copolymer, 3 mole % unsatn., as 10 wt. % soln. in bromobenzene	H_3BO_3	7.0	1% $CoBr_2$	150		3.0	Air	—	—	Hydroxylated rubber containing 0.9–1.0% oxygen	141

used for paraffin oxidations.[35] The production of acids, carbonyl compounds and alcohols was followed as a function of time. The major product fraction was composed of mixed alcohols which, at increasing conversions of 17.3 to 30.7 per cent, dropped from 79.6 per cent of the total oxidate to 73.7 per cent. At 27 per cent conversion the alcohol fraction (75 per cent of total) was found to contain n-nonyl alcohol, isomeric secondary dodecenols and 1,2-dodecanediol in a molar ration of 2.2 : 1.0 : 0.6. Approximately 60 per cent of the secondary dodecenol fraction was 3-dodecenol. Similar oxidation of 2-dodecene yielded n-octyl alcohol as the major product. These results illustrate that olefin oxidation in the presence of boric acid leads to a variety of products. The three most important general reactions are shown in equations (41) to (43) using a terminal olefin as an example.

$$-CH_2CH_2CH{=}CH_2 \longrightarrow -CH_2OH \qquad (41)$$

$$-CH_2CH_2CH{=}CH_2 \longrightarrow -CH_2\overset{\text{OH}}{\underset{|}{C}}HCH{=}CH_2 \qquad (42)$$

$$-CH_2CH_2CH{=}CH_2 \longrightarrow -CH_2CH_2\overset{\text{OH}}{\underset{|}{C}}H{-}\overset{\text{OH}}{\underset{|}{C}}H_2 \qquad (43)$$

Another patent disclosure describes the oxidation of 1-dodecene in the presence of boric acid to a mixture of alcohols which were not identified.[171] Similar oxidation of 2-ethylhexene-1 appeared to give a mixture of isomeric olefinic alcohols.

In a Russian paper the concentrations of olefins, epoxides, peroxides, alcohols, carbonyl compounds, esters and acids were followed as a function of time in the liquid phase oxidation of 1-dodecene, n-tridecene-6 and n-heptadecene-8, both in the absence and presence of boric acid.[124] The rate of olefin consumption is markedly reduced in the presence of boric acid, presumably because of a decrease in hydroperoxide concentration. The major reaction path was postulated to be as shown in equation (44) after initial formation of the hydroperoxide. This mechanism effectively

$$R_1CH{=}CHCH_2R_2 + R_1CH{=}\overset{\text{OOH}}{\underset{|}{C}}HCHR_2$$
$$\downarrow \qquad (44)$$
$$R_1\overset{\diagdown\diagup}{\underset{O}{CH{-}CH}}CH_2R_2 + R_1CH{=}\overset{}{\underset{\overset{|}{OH}}{C}}HCHR_2$$

limits the yield of epoxide to 50 per cent unless the reaction is carried out in the presence of a compound more easily oxidized than the olefin.

The higher stability of internal epoxides resulted in a greater yield of epoxide from heptadecene-8 relative to the other olefins.

Oxidation of 2-methyl-2-pentene in the presence of boric acid yielded unsaturated alcohols which were not identified.[139]

A Shell patent[230] claims a general process for the oxidation of aliphatic, alicyclic and polymeric olefins to epoxides. The olefin is dissolved in a suitable saturated hydrocarbon solvent and oxidized in an oxygen–nitrogen stream in the presence of alkoxy boroxines. A 74 per cent conversion of 1-dodecene to 1,2-epoxydodecane was obtained, and a number of other examples were reported including oxidation of propylene and other low molecular weight olefins under pressure.

Another patent disclosure includes a large number of examples of oxidations of linear olefins to the corresponding epoxides.[92] In all cases, borate esters are used as solvents and are not claimed to alter the course of the reaction. The advantages claimed for the added borates are to provide a general liquid phase process and to allow oxidation in the absence of catalysts, although an aldehydic initiator is usually added. This by inference suggests that the reaction is modified to some extent by the borates. In addition to typical linear aliphatic monoolefins, both styrene and butadiene were claimed to be oxidized to the corresponding epoxides, although experimental details of these reactions were not given. Also some unusual borates were claimed, such as dimethyl ethyl borate, which has never been isolated.

In a very early patent (1923), conversion of ethylene to formaldehyde was reported at 375°C in oxygen in the presence of boric acid.[117] This is a vapor phase reaction and is somewhat unrelated to the liquid phase oxidation processes which are the main subject of this discussion. Several vapor-phase catalytic oxidations of olefins are discussed in Section V.

A series of olefinic mixtures from petroleum refinery streams have been oxidized in the presence of boric acid to give mixtures of glycol borates[151] (equation (45)).

$$3\ RCH{=}CHR' \xrightarrow[\text{H}_3\text{BO}_3]{\text{O}_2} \begin{array}{c} R \\ \diagdown \\ CH{-}O \\ | \qquad \diagdown \\ \qquad\qquad B{-}O{-}CH{-}CH{-}O{-}B \\ | \qquad \diagup \qquad\quad |\quad\ | \\ CH{-}O \qquad\quad R\ \ R' \\ \diagup \\ R' \end{array} \begin{array}{c} R \\ \diagup \\ O{-}CH \\ \diagdown \quad | \\ \qquad O{-}CH \\ \diagdown \\ R' \end{array} \tag{45}$$

These reactions were all carried out in the presence of metal naphthenate catalysts.

Oxidation of an isoprene and isobutylene copolymer containing some unsaturation resulted in a hydroxylated rubber.[141] The rubber was not degraded and could be cross-linked with isocyanates. A comparable oxidation in the absence of boric acid resulted in significant polymer degradation and insertion of only a few hydroxyl groups in the chain.

An interesting oxidation reaction of acetylenic glycols is illustrated in equation (46).[49] The reported yield of the lactide was 86 per cent. The

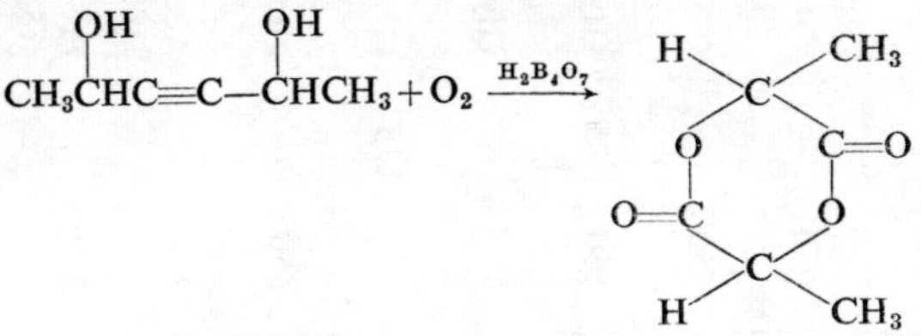

$$CH_3\overset{\displaystyle OH}{\underset{\displaystyle |}{CH}}C{\equiv}C-\overset{\displaystyle OH}{\underset{\displaystyle |}{CH}}CH_3 + O_2 \xrightarrow{\text{H}_2\text{B}_4\text{O}_7}$$

(46)

oxidation was carried out at 110°C in the presence of oxygen after previously converting the glycol to a borate by reaction with "pyroboric acid".

2. Cyclic Olefins

Oxidation of cyclooctene in the presence of boric acid yields mainly cyclooctenols of unspecified composition.[139, 140] In the absence of boric acid, significant amounts of polymers, epoxides and ketones are obtained. Oxidation of cyclododecene under similar conditions resulted in a product comprised of 50 per cent 3-hydroxycyclododecene and 40 per cent cyclododecene oxide (equation (47)). 3-Hydroxycyclododecene was also

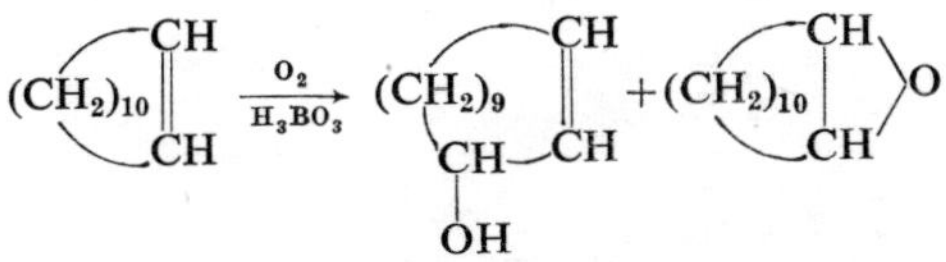

$$(CH_2)_{10}\overset{\displaystyle -CH}{\underset{\displaystyle -CH}{\big\|}} \xrightarrow[\text{H}_2\text{BO}_3]{\text{O}_2} (CH_2)_9 \overset{\displaystyle -CH}{\underset{\displaystyle \underset{\displaystyle OH}{\overset{\displaystyle |}{CH}}-CH}{\big\|}} + (CH_2)_{10}\overset{\displaystyle -CH}{\underset{\displaystyle -CH}{\big\langle}}O \quad (47)$$

the major product when boric acid was replaced by methylborate or boric acid with added cobalt naphthenate.

Oxidation of cyclohexene in the presence of boric acid over a long period of time (160 hours) at 80–105°C resulted in approximately 75 per cent cyclohexene conversion.[171] The product mixture contained 33 per cent 3-hydroxycyclohexene (2-cyclohexenol) and 67 per cent of the hydroxy-cyclohexenyl ether (XV). Oxidation of d,l-pinene yielded a mixture of isomeric alcohols of d,l-pinene.

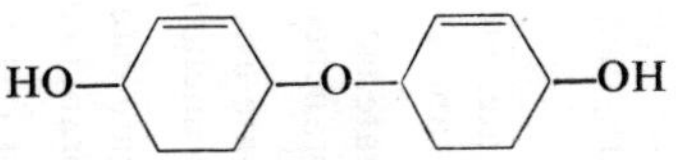

(XV)

TABLE 9. OXIDATIONS OF MISCELLANEOUS ORGANIC COMPOUNDS

Substrate	Boron compound	Wt. (%)	Other	Temp. (°C)	Pressure	Time (hr)	Gas, % O_2 in N_2	Flow rate (l./kg-hr)	% Conv.	Products and yields	Ref.
$B(OCH_2CH_2CH_3)_3$	H_3BO_3	14.3		165	10–12 atm	5	4	650	—	Propanediol	139
2-Ethylhexylborate	H_3BO_3	3	0.5% NH_4Br	160–175		1.75	5	650	—	Glycol (65%)	139
Borate ester of C_{13} branched alcohol	H_3BO_3	5		160–175		1.75	Air	650	15	Glycol (62%)	139
Borate ester of C_{13} branched alcohol	$(CH_3O)_3B$	10		170	15 atm	1.75	Air	—	—	Glycol (> 62%)	139
Borate ester of C_{13} branched alcohol	$[(CH_3)_3CO]_3B$	11.7		170		2	Air	—	35	Glycol (60%)	139
Borate ester of C_{13} branched alcohol	—	—		160–175		1.75	Air	650	—	No glycol	139
$(CH_3CH_2O)_3B$	—			150	10 kg/cm²	10	Air	300		Acetic acid (80%)	158
$(CH_3CH_2CH_2O)_3B$	—			170	10 kg/cm²	10	Air	200		Propionic acid (53%)	158
$(CH_3CH_2CH_2CH_2O)_3B$	—		0.1% cobalt naphthenate	150		10	Oxygen	50		Butyric acid (71%)+a small amount of n-butylbyrate	158

Substrate	Additive		Catalyst	Temp. (°C)	Pressure		Gas		Product	Ref.
$(CH_3CH_2CH_2CH_2O)_3B$	—		0.1% cobalt naphthenate	80		4	Oxygen	100	Butyric acid (> 40%)	158
$(CH_3CH_2CH_2CH_2CH_2O)_3B$	—		—	80		8	Oxygen	40	Valeric acid (69%)	158
CH_3CHO	1:1 boric acid: oxalic acid	1	2% manganese acetate	60	40 psig	~3–1/3	Oxygen		Acetic anhydride (30–40%)	207
$CH_3\overset{\text{O}}{\overset{\|}{C}}CH_2(CH_2)_5CH_3$	H_3BO_3	5		140–150		4	Air		Methyl hydroxyheptyl ketones	138
Stearic acid	H_3BO_3	3.2		170			8			138
Caprylic acid	H_3BO_3	5		140–150			Air	25	Hydroxycaprylic acids	138
Methyl stearate	H_3BO_3	3.2		170			8	25	Methyl esters of hydroxy stearic acids	138
Aliphatic nitriles (87% C_{14})	H_3BO_3	4.8		140–150			Air	30	Hydroxynitriles	138

F. *Oxidation of Miscellaneous Organic Compounds*

1. *Alcohols and Borate Esters*

Several examples of air oxidations of alcohols in the presence of boron compounds have appeared in the literature. The usual initial step in these reactions is conversion of the alcohols to their respective borate esters. Glycols have been obtained from liquid-phase air, or oxygen-in-nitrogen, oxidations of borates from n-propyl alcohol, 2-ethylhexyl alcohol and branched tridecyl alcohols (see Table 9).[137, 139] Boric acid, trimethyl borate and tri-t-butyl borate were used as the borate additives with comparable effectiveness in concentrations from 3 to 11 per cent.

In a Japanese patent[158] a process is described for the liquid-phase air oxidation of borate esters of primary alcohols to carboxylic acids (see Table 9). In some cases oil-soluble metal salts such as cobalt naphthenate were added as catalysts. The production of carboxylic acids by this method is unexpected, since the oxidation conditions are comparable to those used in normal hydrocarbon oxidation reactions such as described above.[137, 139] In these "normal" cases the product alcohols are converted to borate ester intermediates which are apparently stable to further oxidation. Three of the five examples claimed in the patent do not even use an added oxidation catalyst. One possible explanation is that longer reaction times are used in these conversions to carboxylic acids than are normally used in hydrocarbon oxidation processes. The times reported were 10, 4, 10, 8 and 10 hours, whereas the reaction period in the conversion of hydrocarbons to alcohols is usually 3–5 hours. The only other apparent difference is the inclusion of added borates in the form of boric acid or organic borates in the reactions noted in references 137 and 139.

Oxidation of alcohols to carbonyl compounds in the presence of phosphoric or boric acids has been disclosed in the patent literature.[57] No experimental details were given for the boric acid case.

Oxidations of several aromatic (benzyl) alcohols were discussed in Section III, D on aromatic oxidations.

2. *Carbonyl Compounds*

Oxidation of acetaldehyde with oxygen at 60°C in the presence of a 1:1 mixture of boric acid and oxalic acid plus 2 per cent manganese acetate resulted in a 30–40 per cent yield of acetic anhydride.[207]

Boric acid (0.1–0.15 per cent) has been shown to be a poor catalyst for the autoxidation of furfural.[60]

Air oxidation of methylheptyl ketone at 140–150°C in the presence of 5 per cent boric acid yielded a mixture of methyl hydroxyheptyl ketones.[138]

3. Carboxylic Acid Derivatives

Air oxidations of stearic and caprylic acids at 140–150°C in the presence of boric acid gave hydroxy acids; a related ester, methyl stearate, yielded methyl esters of hydroxystearic acids.[138] A mixture of aliphatic nitriles was air oxidized under the same conditions to hydroxynitriles.

IV. OXIDATIONS WITH OXIDIZING AGENTS OTHER THAN OXYGEN

Most reported organic oxidations in the presence of boron compounds involve molecular oxygen as the oxidizing agent. More recently, a number of reports on the effects of boron compounds on oxidations using peroxides and hydroperoxides have appeared, and this area appears to be one of increasing interest.

The liquid phase oxidation of a paraffinic hydrocarbon, decane, with hydroperoxides has been studied in the presence of boric acid and boric oxide.[199] Good yields of decanols were obtained at 130°C with n-octyl-hydroperoxide (see Table 10). The product contained essentially pure secondary decanols; only a trace of primary derivatives was noted. In an addendum to the patent it was shown that prior dehydration of boric acid at 160°C significantly increased the oxidation rate compared to boric acid.

The liquid phase oxidation of cycloalkanes containing six to twelve carbon atoms to cyclic alcohols and ketones has been described in a Rhone-Poulenc patent.[236] The oxidizing agents claimed include the general class of tri(organoperoxy)boranes [B(OOR)$_3$]. In the only reaction described in detail, cyclohexane was oxidized in the presence of cyclohexane hydroperoxide and ethyl borate. Initially ethanol was recovered as the cyclohexane–ethanol azeotrope, purportedly with the formation of tricyclohexylperoxyborane (equation (48)) before the desired reaction temperature was reached. However, triethyl borate still was present in

$$3 \text{ ROOH} + (\text{R}'\text{O})_3\text{B} \longrightarrow (\text{ROO})_3\text{B} + 3 \text{ R}'\text{OH} \qquad (48)$$

excess, and the oxidation took place in the presence of approximately 2.5 weight per cent tricyclohexylperoxyborane and approximately 4.0

TABLE 10. OXIDATIONS OF DECANE BY HYDROPEROXIDES

Boron compound		Hydroperoxide		Temp. (°C)	Time (hr)	Products and yields[a]
%	Formula	%	Formula			
2.2	H_3BO_3	4.9	$CH_3(CH_2)_7OOH$	130	5	70% decanols (mostly 2°), 143% total alcohols, octanols + decanols
4.4	H_3BO_3	4.9	$CH_3(CH_2)_7OOH$	130	5	190% total alcohols
2.2	H_3BO_3	9.8	$CH_3(CH_2)_7OOH$	130	5	120% total alcohols
2.2	H_3BO_3 (+5.5% decanones)	4.9	$CH_3(CH_2)_7OOH$	130	5	75% total alcohols
2.2	H_3BO_3 (+0.03% manganese as naphthenate)	4.9	$CH_3(CH_2)_7OOH$	130	5	75% total alcohols
5	B_2O_3	3	t-BuOOH	130	9	Decanols in good yield

[a] Yields are based on hydroperoxide added.

per cent triethyl borate after the theoretical quantity of ethanol was removed. The product mixture contained 1.31 mole of cyclohexanol and 0.11 mole of cyclohexanone per mole of cyclohexane hydroperoxide charge. The reaction temperature was 165°C.

A Rhone-Poulenc patent has described the oxidation of various linear and cyclic olefins to the corresponding epoxides by alkylperoxyboranes.[51] Reaction details are summarized in Table 11. The peroxyboranes are prepared by transesterification of alkyl borates (equation (48)). Good yields (usually 70–85 per cent) of epoxides are obtained based on the perborates used.

In another Rhone-Poulenc patent, oxidations of C_6–C_{12} cyclic olefins to epoxycycloalkanes have been reported.[50] In each case the oxidizing agent was cyclohexyl hydroperoxide and the boron-containing additive was trimethyl borate or triethyl borate. From the ratios of reactants used and the conditions described, it is probable that the actual oxidizing agents are peroxyboranes as noted in reference 51, and the reaction shown in equation (43) takes place *in situ*. In most cases the final reaction occurs in the presence of excess trialkyl borate. Reported yields of epoxycyclo-alkanes varied from 61 to 97 per cent based on the hydroperoxide added.

In reactions similar to those with peroxyboranes described above, oxida-tions of a number of olefins to glycols by hydroperoxides in the presence of various boron compounds have been reported.[204, 247] These reactions are summarized in Table 12. Either boric acid or boric oxide were included as the added boron compound in these reactions, and both cumene and cyclohexane hydroperoxides were used as oxidizing agents. In several of the reactions described the reactions were carried out in the presence of acetic acid.[204] In these cases acetates of the diols are formed as well as the corresponding borates. In all cases noted in Table 12 vicinal diols are formed rather than epoxides as observed when alkylperoxyboranes are used as oxidizing agents.

Rhone-Poulenc has described a series of oxidations of aromatic com-pounds to phenols by hydroperoxides and also by hydrogen peroxide in the presence of various boron compounds.[205-6] These reactions are outlined in detail in Table 13. The basic reaction with hydroperoxides is shown in equation (49). The yields based on added hydroperoxide or

$$\underset{\text{R}}{\bigcirc} + R'OOH \xrightarrow[\text{or } B(OCH_3)_3]{B_2O_3, H_3BO_3} \underset{\text{R}}{\bigcirc}-OH + R'OH \qquad (49)$$

TABLE 11. OXIDATIONS OF OLEFINS TO EPOXIDES BY ALKYLPEROXIBORANES

Substrate	Perborate	%	Temp. (°C)	Pressure	Time (hr)	% Conversion based on perborates	Products
Propylene	$\left(\mathrm{S}\text{—}\mathrm{OO}\right)_3\mathrm{B}$	19.1	80	40 atm	2	79	Propylene oxide
Propylene	$\left(\underset{\mathrm{CH_3}}{}\mathrm{C_6H_5\text{—}CHOO}\right)_3\mathrm{B}$	21.6	80	40 atm	2	70.6	Propylene oxide
Allyl chloride	$\left(\mathrm{S}\text{—}\mathrm{OO}\right)_3\mathrm{B}$	8.9	100	10 atm	2	38.4	Epichlorohydrin
Butadiene	$\left(\mathrm{S}\text{—}\mathrm{OO}\right)_3\mathrm{B}$	20.4	100		3	81.7	3,4-Epoxy-1-butene
Butadiene	$\left(\underset{\mathrm{CH_3}}{}\mathrm{C_6H_5\text{—}CHOO}\right)_3\mathrm{B}$ + trace nitrobenzene	20.4	100		3	51	3,4-Epoxy-1-butene
Cyclohexene	$\left(\mathrm{S}\text{—}\mathrm{OO}\right)_3\mathrm{B}$	8.8	84		1	84.5	Epoxycyclohexane
Cyclohexene	$\left(\underset{\mathrm{CH_3}}{\overset{\mathrm{CH_3}}{\mathrm{C_6H_5\text{—}COO}}}\right)_3\mathrm{B}$	8.8	84		5	67	Epoxycyclohexane

TABLE 12. OXIDATIONS OF OLEFINS TO GLYCOLS BY HYDROPEROXIDES IN THE PRESENCE OF BORON COMPOUNDS

Substrate	Boron compound		Hydroperoxide		Temp. (°C)	Time (hr)	% Conversion	Products	Ref.
	Wt. (%)	Formula	Wt. %	Formula					
Propylene	8.1	B_2O_3	13.5	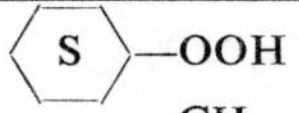 $\langle S\rangle$—OOH	110	2	—	73% propylene glycol based on hydroperoxide added	204
Cyclohexene	15.4	H_3BO_3	37.9	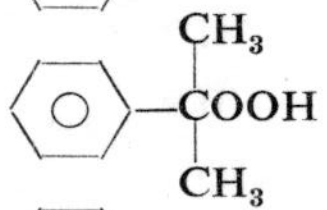 $\langle C_6H_5\rangle$—$C(CH_3)$—$COOH$	83–91	9.3	20	33% trans 1,2-dihydroxycyclohexane based on cyclohexene reacted	247
Cyclohexene	17.1	B_2O_3	14.0	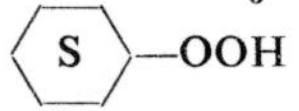 $\langle S\rangle$—OOH	88	1.33	—	65.3% cyclohexanediol based on hydroperoxide added	204
Cyclohexene	5.8	B_2O_3	5.0	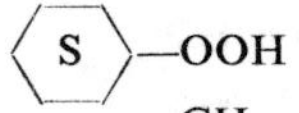 $\langle S\rangle$—OOH, CH_3	81	1.25	—	57% cyclohexanediol based on hydroperoxide added	204
Cyclohexene	4.2	B_2O_3	4.7	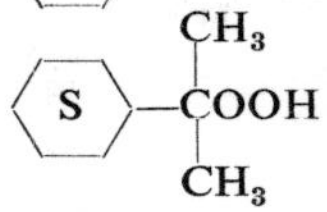 $\langle C_6H_5\rangle$—$C(CH_3)$—$COOH$	110	10	—	41% cyclohexanediol based on hydroperoxide added	204
2-Ethyl-1-hexene	27.8	H_3BO_3	63.3	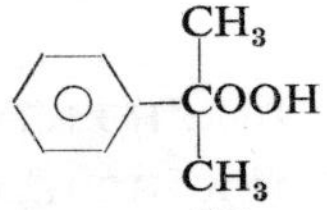 $\langle C_6H_5\rangle$—$C(CH_3)_2$—$COOH$	90	15	27	29% 2-ethyl-1,2-dihydroxyhexane based on reacted 2-ethyl-1-hexene	247
1-Dodecene	7.3	H_3BO_3	19.3	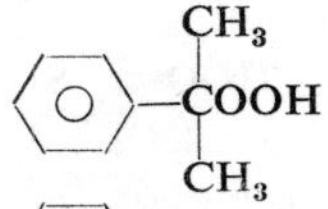 $\langle C_6H_5\rangle$—$C(CH_3)$—$COOH$	85–95	5	14	67.6% 1,2-dodecylglycol based on reacted 1-dodecene	247
Cyclododecene	5.5	B_2O_3	4.5	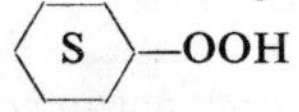 $\langle S\rangle$—OOH	110	1.5	—	64% cyclododecanediol based on hydroperoxide added	204
Cyclododecene	17.4	B_2O_3	17.0	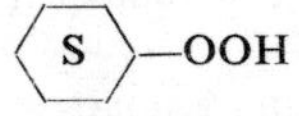 $\langle S\rangle$—OOH	110	1.5	—	79% cyclododecanediol based on hydroperoxide added	204

TABLE 13. OXIDATIONS OF AROMATIC COMPOUNDS BY HYDROPEROXIDES AND HYDROGEN PEROXIDE IN THE PRESENCE OF BORON COMPOUNDS

Substrate	Boron compound Formula	Wt. (%)	Oxidizing agent Wt. (%)	Oxidizing agent Formula	Temp. (°C)	Pressure	Time (hr)	% Conversion	Products	Ref.
Benzene	B_2O_3	2.4	2.0	C$_6$H$_5$–OOH	150	2 atm	2.5	31	Phenol	206
Benzene	B_2O_3	6.6	5.8	$C_6H_5CH_2OOH$	82		6	22.2	Phenol	206
Toluene	B_2O_3	2.3	2.1	$C_6H_5CH_2OOH$	109		3.5	61.2 based on –OOH	3 o-cresol/1 p-cresol	206
Toluene	B_2O_3	2.4	2.0	C$_6$H$_5$–OOH	108		3	59.2	3 o-cresol/1 p-cresol	206
Toluene	B_2O_3	6.1	6.1	C$_6$H$_5$–CH(CH$_3$)OOH	110		2	25	Cresols	206
Toluene	B_2O_3	1.9	2.0	C$_6$H$_5$–C(CH$_3$)$_2$OOH	150–165	5 atm	3.5	17	Cresols	206
Toluene	B_2O_3 + trace pyridine	1.9	2.0	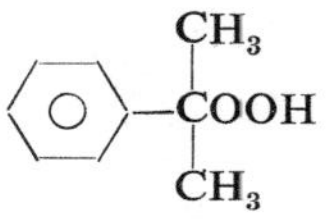 C$_6$H$_5$–C(CH$_3$)$_2$OOH	150–165	5 atm	3.5	35	Cresols	206
Toluene	B_2O_3 + trace pyridine	2.1	2.1	C$_6$H$_5$–CH(CH$_3$)OOH	110		6	49.3 / 15.5	o-Cresol / p-Cresol	206

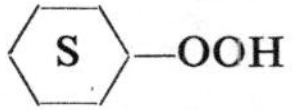

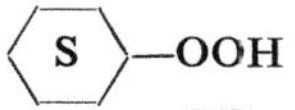

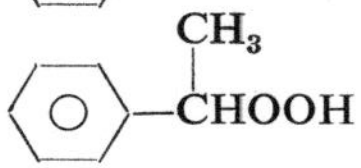

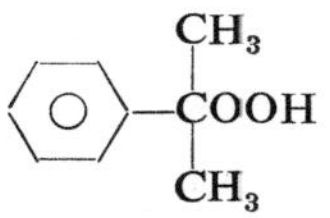

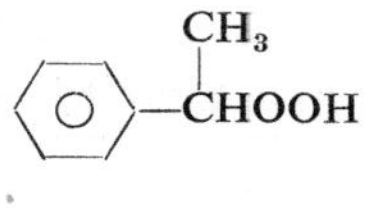

Substrate		Catalyst		Oxidant	Temp.		Yield	Products	Ref.
Toluene	2.3 +trace	B_2O_3 pyridine	2.0	⬡—CH_2OOH	110	4.17	48.6 / 17.1	o-Cresol / p-Cresol	206
Toluene	2.4 +trace	B_2O_3 pyridine	2.0	S—OOH	110	4	59.6	Cresols	206
Toluene	4.1 +trace	H_3BO_3 pyridine	2.0	⬡—CH_2OOH	109	3.75	60	3 o-cresol/1 p-cresol	206
Toluene	2.0 +trace	B_2O_3 piperazine	2.0	⬡—$\overset{CH_3}{CHOOH}$	110	11	51	3 o-cresol/2 p-cresol	206
Toluene	2.3 +trace	B_2O_3 triethyl-amine	2.3	⬡—$\overset{CH_3}{CHOOH}$	110	7.5	44.3	3 o-cresol/1 p-cresol	206
Toluene	2.8	B_2O_3	0.7	H_2O_2	90–100	2.0	47.5	o- and p-cresols	205
Chlorobenzene	2.3 +trace	B_2O_3 pyridine	2.0	⬡—CH_2OOH	110	13.5	20	1.25 o-chlorophenol/1 p-chlorophenol	206
m-Xylene	3.3	B_2O_3	1.5	H_2O_2	110	2.0	46	Xylenols	205
Phenol	53.2	$B(OCH_3)_3$	35.7	S—OOH	115	4	30	1.5 catechol/1 hydro-quinone	206
Anisole	2.2	B_2O_3	2.0	⬡—$\overset{CH_3}{CHOOH}$	110	2.5	76.5	3 guaiacol/1 **p-metho-xyphenol**	206
Anisole	2.4	B_2O_3	2.0	⬡—$\overset{CH_3}{CHOOH}$	110	2.5	82.5	Methoxyphenols	206

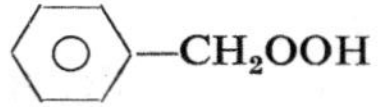

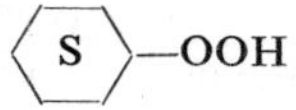

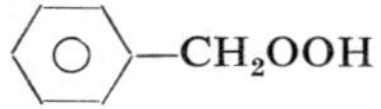

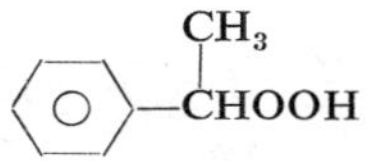

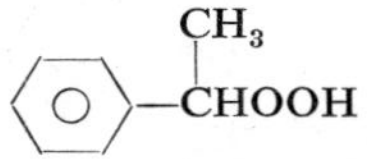

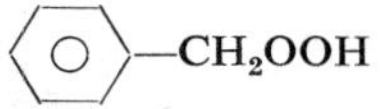

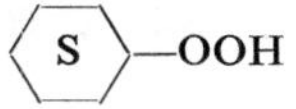

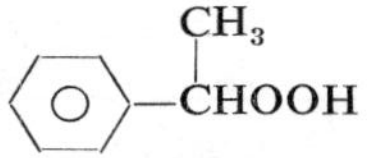

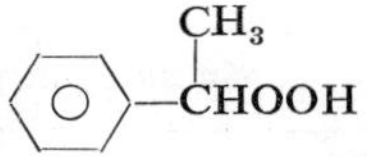

TABLE 13. (cont.)

Substrate	Boron compound		Oxidizing agent		Temp. (°C)	Pressure	Time (hr)	% Conversion	Products	Ref.
	Wt. (%)	Formula	Wt. (%)	Formula						
Anisole	14.8	$B(OCH_3)_3$	2.2	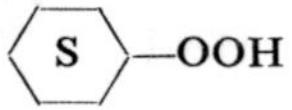 [thienyl]–OOH	130–140		1	58	2 guaiacol/1 p-methoxyphenol	206
Anisole	2.4	B_2O_3	0.625	H_2O_2 (2.3% in diethyl ether)	100		1.0	79	1.5 o-methoxyphenol/1 p-methoxyphenol	205
Anisole	3.0	$(HBO)_n$	0.625	H_2O_2	100		1.5	64.5 based on H_2O_2		205
Anisole	29	$B(OCH_3)_3$	0.625	H_2O_2	110–15		81			205
Anisole	1.9 + trace pyridine	B_2O_3	2.0	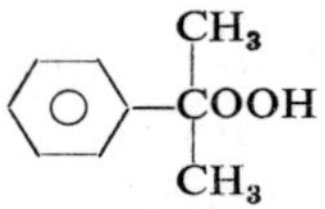 [phenyl]–C(CH₃)(CH₃)COOH	150		2.33	57	Methoxyphenols	206
Phenetole	2.3 + trace pyridine	B_2O_3	2.1	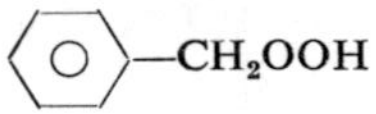 [phenyl]–CH_2OOH	116		4.16	59	3 o-ethoxyphenol/1 p-ethoxyphenol	206
Phenetole	2.75	B_2O_3	0.7	H_2O_2	70		1.5	63.5	Ethoxyphenols	205
Phenetole	2.75	B_2O_3	0.8	H_2O_2 (2.35% in ethyl acetate)	70		1.5	56.5	Ethoxyphenols	205
1,3-Dimethoxybenzene	2.4	B_2O_3	0.7	H_2O_2	100		1.5	89	Dimethoxyphenols	205

Substrate	Reagent		Peroxide					Product	Ref.
Allyloxybenzene	B$_2$O$_3$ + trace pyridine	7.6	S—OOH	6.4	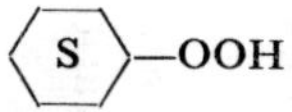 110	2		Allyloxyphenols	206
Trimethylsiloxybenzene	B(OCH$_3$)$_3$	16.3	S—OOH	1.8	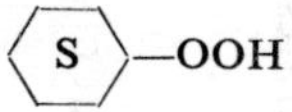 125–143	2.5	37	Diphenols	206
Trimethylsiloxybenzene	B(OCH$_3$)$_3$	49	S—OOH	7.1	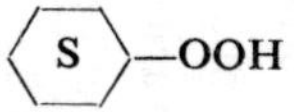 110	2.5	35	Equivalent amounts of ortho and para diphenols	206

hydrogen peroxide varied from 15 to 89 per cent. In most cases *ortho* phenols predominate over *meta* and *para* derivatives in the product mixture.

Organic peracids can be prepared by oxidation of carboxylic acids by hydrogen peroxide in the presence of boric oxide or metaboric acid.[253] The boric acid formed in the reaction by hydration does not lead to product decomposition, and unstable acids such as performic can be prepared by this method.

Oxidations of carbohydrates by mercuric chloride can be accomplished by adding borax followed by mercuric chloride to aqueous carbohydrate solutions.[228] Oxidation does not occur if borax and mercuric chloride are mixed previously, and it was postulated that oxidation is the result of formation of an activated mercuric oxide from borax and mercuric chloride. Complex carbohydrates are oxidized more easily than simple sugars.

Borates have been shown to inhibit the oxidation of sugars by the normal copper oxidizing solutions.[149, 153] Oxidations of glucose, saccharose and mannitol by potassium permanganate were also inhibited by boric acid.[249] Oxidation rates were reduced by three to four times. A similar reduction in oxidation rates of D-fructose, D-glucose and D-galactose has been noted using potassium ferricyanide.[119]

The effects of adding a variety of organic and inorganic borates on the oxidation of naphthalene in an aqueous fermentation mixture has been investigated.[283] This oxidation to salicylic acid by a variety of *pseudomonas* bacteria was improved by adding 0.2–0.6 weight per cent of various boron compounds.

In a series of publications the use of a boron trifluoride–peroxytrifluoroacetic acid mixture has been described.[52, 102–6, 268] This combination has been shown to be a very powerful reagent particularly for the oxidation of aromatics to phenols.

V. CATALYTIC OXIDATION REACTIONS

Most of the oxidation reactions described in this chapter involve the use of relatively large amounts of boron compounds, and the evidence which has accumulated indicates that the role of boron compounds is not catalytic in these processes. However, over the years a number of boron-containing catalysts have also been prepared and used in oxidative processes.

For example, air oxidation of ethylene has been accomplished at approximately 300°C over the same type of catalyst used for propylene oxidation.[213] The major product was acetic acid accompanied by a small amount of acetaldehyde.

Air oxidation of propylene has been carried out at 300°C over a catalyst prepared by calcination of a mixture of boric acid, phosphoric acid, molybdic oxide and silver nitrate on silica.[210] The final catalyst contained an equivalent of 3.5 per cent boric acid. In one pass a 17 per cent conversion of propylene was obtained with the production of acetone as the major component of the product mixture. Small amounts of acrolein, acetaldehyde and a significant quantity of carboxylic acids were also produced.

Activation of a molybdic oxide–arsenic oxide catalyst with boric oxide is reported to give increased yields of ethanol, acrolein and methanol in propylene oxidation.[85] The reactions described in the examples were carried out at 300–400°C, and propylene conversions were 10–15 per cent.

Propylene has also been oxidized to acrolein by molecular oxygen at 380–450°C over a molybdenum–cobalt–boron catalyst.[134] Acrolein yields were 52–60 per cent plus 25–35 per cent carbon monoxide and 13–15 per cent of other hydrocarbons.

A catalyst containing boric oxide, vanadium pentoxide and phosphorus pentoxide in the relative g-atom amounts of 0.1, 1.0 and 1.4, respectively, has been described.[84] Vapor phase oxidation of isobutylene gave >30 per cent yields of methylacrolein in the presence of this mixed oxide catalyst. A similar boric acid–phosphoric acid–metal oxide catalyst has been disclosed in another patent for the oxidation of simple olefins.[208] In a related patent, vapor phase oxidation of olefins to unsaturated aldehydes is claimed.[56] The catalyst used is a mixed bismuth oxide–molybdenum oxide promoted with boric and bismuth oxides. Boron is usually 0.5–1.0 per cent of the total catalyst.

A tungsten oxide catalyst containing boric oxide has been described in a patent but was not included in the specific examples given.[245] Using this type of catalyst, dimethyl ether or methanol containing dimethyl ether is oxidized preferentially to formaldehyde in air at 350–600°C.

Gas phase oxidation of acrolein to acrylic acid at 330–430°C with oxygen in the presence of steam has been reported using a catalyst containing boric oxide.[13] The complex catalyst also contains oxides of phosphorus, molybdenum, tin and antimony.

The air oxidation of a hydrocarbon fraction to carboxylic acids has been carried out at 160°C in the presence of 1 per cent boric acid and 1 per cent sodium chloride.[83] Metaboric acid was isolated and recycled. Boric acid was described as a catalyst in this reaction and is used as an alternative to the usual cobalt or manganese naphthenates.

Phthalic anhydride has been prepared in 70 per cent yield by air oxidation of o-xylene in the vapor phase over fused vanadium oxide (V_2O_5) containing 5 weight per cent boric acid or potassium monoborate.[162] A 150 : 1 molar ratio of air to o-xylene is passed at a rate of 400–600 pounds per hour over the catalyst at 420–430°C. Naphthalene is also claimed as a starting material but is not included in the examples or formal claims. Similar air oxidations of o-xylene and naphthalene in the presence of 0.3–0.5 per cent tri-n-butyl borate also gave good yields of phthalic anhydride.[163] Longer catalyst life is claimed as a major benefit of the boron additives.

Air oxidation of naphthalene at 430°C leads to low yields of 1,4-naphthoquinone in the presence of boric oxide on pumice or boric oxide plus vanadium pentoxide on pumice.[87] A product containing a high percentage of anthraquinone has been obtained by a related vapor-phase air oxidation of anthracene over a boron-containing catalyst at 400–430°C.[117] Relatively few examples of this type of vapor phase oxidation in the presence of boron compounds have been reported.

Air oxidation of toluene to benzyl alcohol in the presence of boric acid has been claimed in the introduction of a patent, but the examples and formal claims did not include the use of boron compounds.[288]

Air oxidation of cumene in the presence of 1.5 per cent potassium chlorate at 130°C gave good yields of cumene hydroperoxide with a variety of catalysts including metallic borates and perborates.[222]

In an early patent (1943) air oxidation of benzene or toluene to phenol was improved significantly by coating firebrick or silica bricks used as packing material with boric oxide.[167] Conversions of about 5 per cent were obtained at temperatures of approximately 800°C.

Hydroxybenzyl alcohols can be oxidized to aldehydes by passing air or oxygen through an alkaline solution in the presence of boric acid and a palladium catalyst at 15–40°C.[97] Further oxidation to acids is not observed. In one modification of this process, *o*-hydroxybenzyl alcohol was prepared from phenol, boric acid and formaldehyde and oxidized further to salicylaldehyde without isolation (equation (50)).

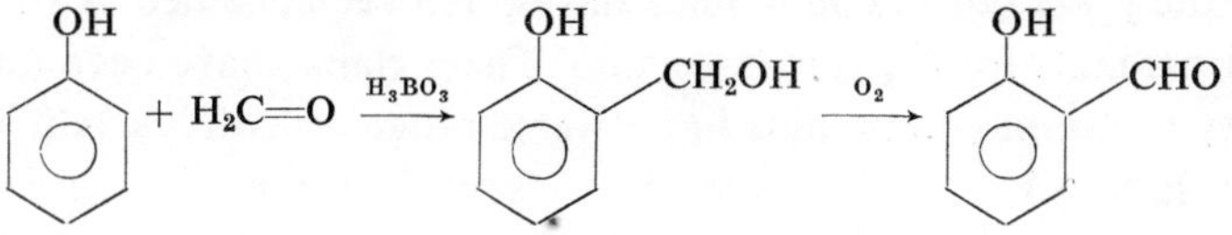

$$\text{(50)}$$

Baccaredda *et al.*[14] have noted a number of papers which discuss the advantages and disadvantages of borate-coated surfaces in the production of formaldehyde by oxidation of hydrocarbon fractions. Good yields of formaldehyde have been obtained when reactor walls and packing were coated with boric oxide or potassium tetraborate.

Boric acid has been used as a preferred catalyst for the vapor phase oxidation of pyridines by nitric acid.[242] For example, a 51 per cent yield of nicotinic acid was obtained from the oxidation of 2-methyl-5-ethyl-pyridine (1 part by weight) with 40 per cent nitric acid (20 parts by weight) at 230°C in the presence of 0.06 part by weight of boric acid (equation (51)). Isonicotinic acid was prepared in the same manner from γ-picoline or from 2,4,6-trimethylpyridine.

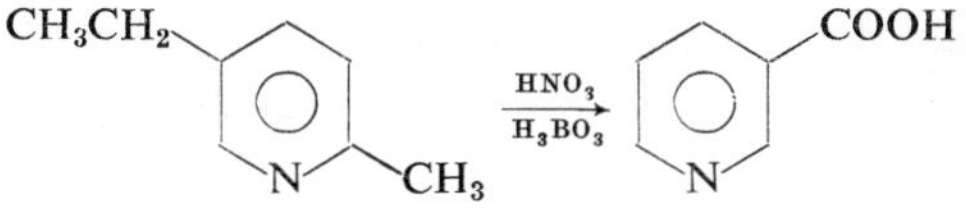

$$\text{(51)}$$

The air oxidation of methane in the presence of 0.2 per cent NO has been studied in the gas phase in a reactor coated with a series of metals and metallic salts.[254] Potassium tetraborate was apparently an effective coating material.

Coating the walls of glass reactors used in the partial oxidation of propane-butane mixtures at 270°C with boric oxide reduces the induction period significantly.[232]

7*

VI. BORON COMPOUNDS AS OXIDATION STABILIZERS OR ANTIOXIDANTS

A. *General*

Over the years boron compounds have often been claimed as stabilizers and antioxidants for organic molecules. These claims have been made for a variety of boron compounds in a diverse range of hydrocarbon systems and this subject is directly related to hydrocarbon oxidation. Elucidation of the mechanism of this antioxidant function would certainly help to clarify the role of boron compounds in other hydrocarbon oxidation reactions as noted in Section II. In some cases boron compounds are claimed as general "stabilizers" which can refer to stability towards heat, ultraviolet light or oxidative degradation or perhaps to combinations of these effects. An attempt has been made to limit this discussion to those reports dealing specifically with oxidative protection, although stability is poorly defined in some of the available published literature. This section has been divided into two parts covering antioxidants in monomers and in polymers; data from both are important for mechanistic considerations. Boron compounds which have been investigated as antioxidants are summarized in Tables 14 and 16.

Some of the information on antioxidants has been discussed in Section II on oxidation mechanisms. Antioxidation is also discussed in other sections; for example, it was observed in Section III, D that oxidation of aromatic hydrocarbons is inhibited in the presence of boron compounds.

B. *Monomers*

Oxidative stabilization of various hydrocarbon monomers by boron compounds has been reported. It is interesting that one of the earliest patents in this area claims the product of the air oxidation of a petroleum hydrocarbon distillate of molecular weight 235 in the presence of boric oxide as an antioxidant.[146] This product was described as a mixture of the borates formed from the alcohols produced, but the patent claims only the total fuel-borate mixture as an antioxidant.

Boron compounds which have been investigated as antioxidants in monomeric hydrocarbon systems are listed in Table 14. These applications include the petroleum oxidate described above and various secondary alkyl borates as synergists with alkylphenols as antioxidants in

TABLE 14. ANTIOXIDANTS AND STABILIZERS FOR HYDROCARBON MONOMERS

Boron compound	Substrate	Comments	Ref.
H_3BO_3	Kerosene	1.25 wt. % inhibited O_2 oxidation	90
H_3BO_3	Wool fat		286
$B(O\text{-}n\text{-}Bu)_3$	Kerosene	0.2 wt. % inhibited O_2 oxidation	90
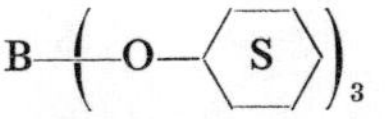	Mineral oil	Synergistic with aromatic amines	202
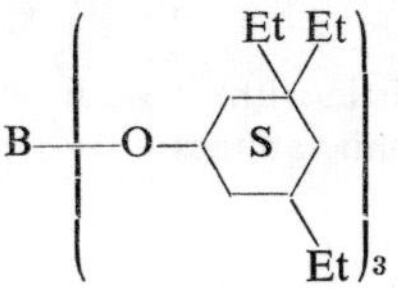	Mineral oil	Synergistic with aromatic amines	202
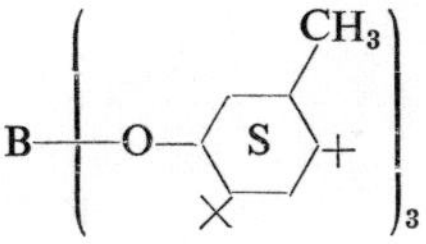	Mineral oil	Synergistic with aromatic amines	202
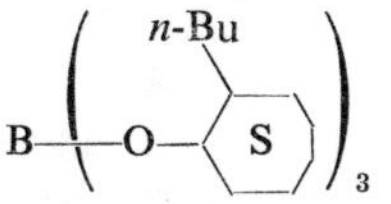	Mineral oil	Synergistic with aromatic amines	202
$RB(OH)_2$	Mineral oil	Synergistic with aromatic amines	203
$(RO)_3B$, where ROH is secondary alcohol	Hydrocarbon oil	Synergistic with alkylphenols	248
$B[OCH(CH_2CH_3)_2]_3$	Hydrocarbon oil	Synergistic with alkylphenols	248
	Hydrocarbon oil	Synergistic with alkylphenols	248
	Hydrocarbon oil	Synergistic with alkylphenols	248
	Hydrocarbon oil	Synergistic with alkylphenols	248

TABLE 14 *(cont.)*

Boron compound	Substrate	Comments	Ref.
$B{-\!\!\left[-OCH\left(CH_2CH\begin{smallmatrix}CH_3\\CH_3\end{smallmatrix}\right)_2\right]}_3$	Hydrocarbon oil	Synergistic with alkylphenols	248
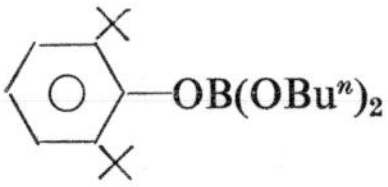 (dimethyl-substituted phenyl)—OB(OBu^n)$_2$	Mineral oil	Synergistic with aromatic amines	201
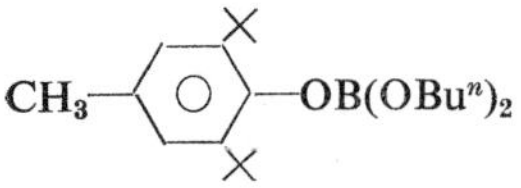 CH$_3$—(dimethyl-substituted phenyl)—OB(OBu^n)$_2$ (XXVIII)	White oil		46
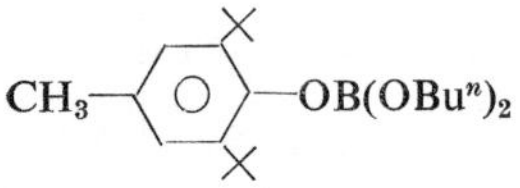 CH$_3$—(dimethyl-substituted phenyl)—OB(OBu^n)$_2$ (XXVIII)	Mineral oil	Synergistic with aromatic amines	201
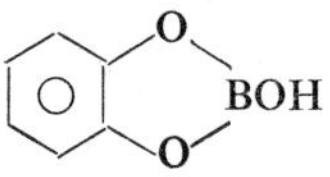 benzodioxaborole BOH (XXIV)	White oil		46
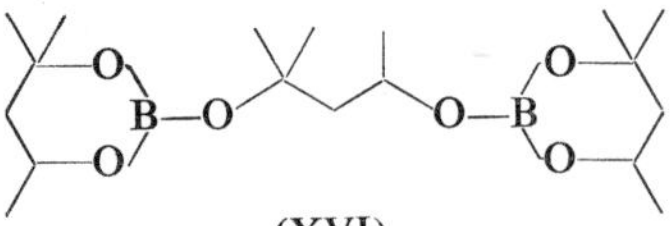 (XVI) diboron diester chain compound	White oil		46
PhB(OH)$_2$	White oil		46
RB(OH)$_2$	Purified naphthenic white oil	Effective as auxiliary peroxide reducing inhibitors	101
Boronic, borinic acids, salts and borates	Hydrocarbon oils	Used with other antioxidants, borates and minor compositions in patent which includes phosphorus, sulfur, aluminium, etc.	68
Alkylated arylborates, alkylated boronic and borinic acids, chlorinated wax alkylated phenolic borates	Lubricating oils		81
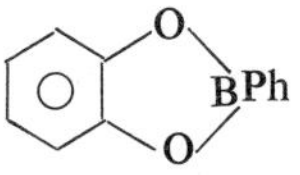 benzodioxaborole BPh (XXV)	White oil		46

TABLE 14 *(cont.)*

Boron compound	Substrate	Comments	Ref.
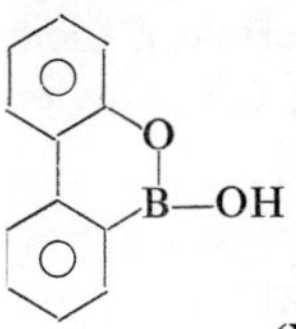 B—OH and derivatives (XIX-XXI)	White oils, hydrocarbon lubricants and fuels	—	45, 46, 159
$CaO_2B(CH_2)_{15}CH_3$ and other metal salts of boronic, borinic acids and related compounds	Hydrocarbon oils	—	64
Unknown borates obtained on oxidation in presence of B_2O_3–H_3BO_3	Hydrocarbon fuels Molecular weight 235	—	146–8, 196
Sulfurized oleyl borates	Lubricating oils	—	209
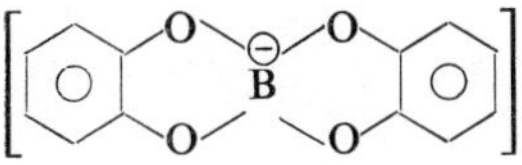 $H^{\oplus}$ (XXIX)	Lubricating oils	—	246
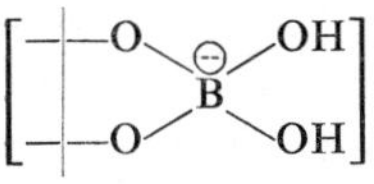 $H^{\oplus}$ (XXX)	Lubricating oils	—	246
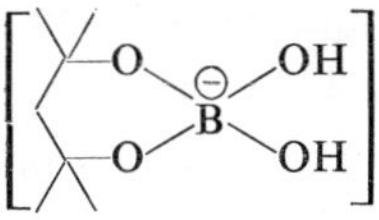 $H^{\oplus}$ (XXXI)	Lubricating oils	—	246
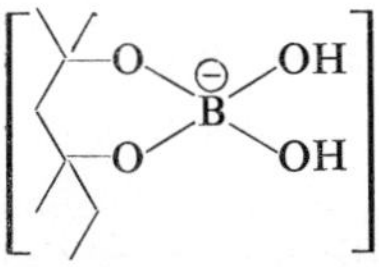 $H^{\oplus}$ (XXXII)	Lubricating oils	—	246

transformer type mineral oils.[248] In other patents a variety of boron compounds are claimed to stabilize hydrocarbon mineral oils either alone or in combination with known antioxidants.[64, 68, 81, 209]

In addition to this patent literature, several more detailed studies of boron compounds as antioxidants have been reported. Freidin studied the liquid phase air oxidation of a kerosene fraction at 170–3°C in the presence of boric acid and tri-n-butyl borate.[90] Boric acid led to complete inhibition of oxidation at a concentration of 1.25 weight per cent;

0.2 weight per cent of tri-n-butyl borate also inhibited the oxidation. The tri-n-butyl borate inhibition was easily overcome by adding manganese naphthenate (0.02 weight per cent as manganese). There may have been uromatic impurities present to account for this extremely strong inhibition.

Low[152] has discussed inhibition by boron compounds in terms of their mode of action. Normally antioxidants can be classified as (a) *primary* inhibitors which operate as chain breakers or chain terminators, and (b) *auxiliary* inhibitors which interrupt chain propagation by accelerating peroxide or hydroperoxide decomposition. Most primary inhibitors are phenols or aromatic amines. Auxiliary inhibitors include amines, sulfides, acids, bases, phosphorus acid esters and metal ions. A single inhibitor can operate by more than one type of mechanism. It was shown that several aromatic boron derivatives (XVI–XVIII) were basically ineffective as primary inhibitors in a radical-induced addition of a thiol to an

$Ph_2BOCH_2CH_2NHCH_3$

(XVI)   (XVII)†   (XVIII)

olefin, although XVI did show some activity, probably due to the amine groups. These compounds were not evaluated directly in oxidation tests, but their effectiveness as radical scavengers can be used to judge their potential antioxidant properties. The same laboratory found that borate esters synergize the oxidation inhibition of mineral oils by aromatic amines at 400°F.[201-3] These mineral oils apparently contain substantial amounts of aromatics. Formation of phenolic esters such as XVII or the corresponding phenol may be responsible for the inhibition by aromatic impurities of alkane oxidations in the presence of boron compounds[71, 114] (see Sections III, B and III, D above).

Compound XVII did not act synergistically with aromatic amines in the primary inhibition test.[152] Earlier studies quoted by the authors but not referenced showed that boric acid and borate esters did not catalyze hydroperoxide decomposition, and the authors could not explain the observed synergism with amines in oxidation studies. The earlier studies were probably in error, because borates have been shown to have significant effects on peroxide decomposition. See Section II, B above for a discussion of boron compounds as catalysts for hydroperoxide breakdown.

† This compound was incorrectly identified as $(PhO)_2BOH$ in reference 152 (private communication from the author).

Good antioxidant properties in white oil at 150°C and higher have been reported for the heterocyclic compound (XIX).[45, 46, 159] The oxidative half-lives for oxidations in the presence of XIX, its derivatives (XX and XXI) and other related boron and non-boron compounds (XXII–XXVIII) are shown in Table 15.[83]

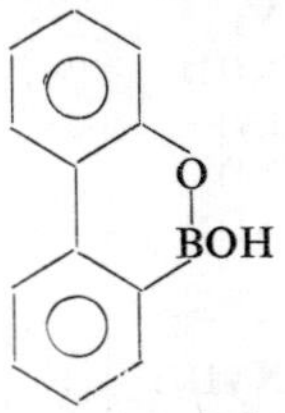

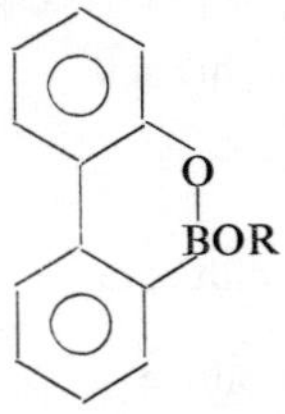

(XX), R = t-Bu
(XXI), R = CH₃

(XIX)

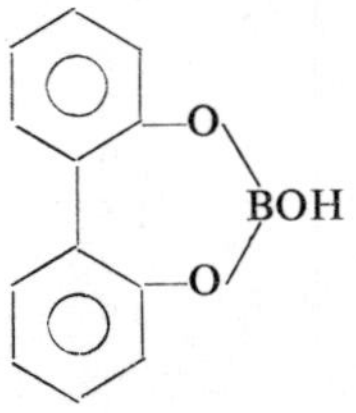

(XXII) 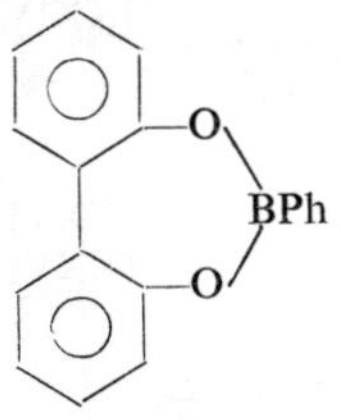   (XXIII) 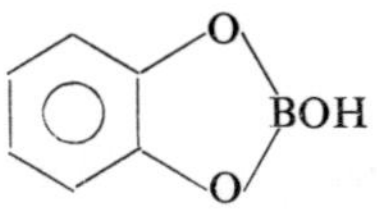   (XXIV) 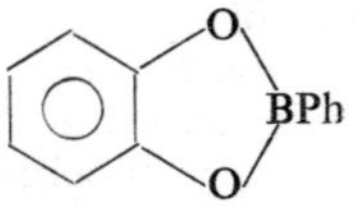   (XXV)

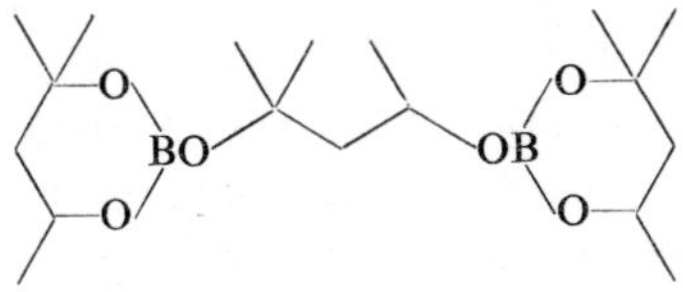

(XXVI) 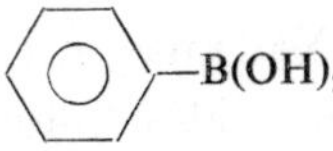   (XXVII)

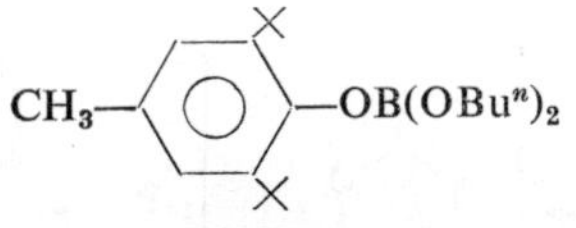

(XXVIII)

The striking activity of the boroxarophenanthrene derivatives compared to standard antioxidants such as N-phenyl-1-naphthylamine and zinc 0,0-di(4-methyl-2-pentyl)phosphorodithioate is somewhat unexpected. This is probably the first definitive work which shows a boron compound to have exceptional antioxidant properties in a hydrocarbon. In a study of the possible mechanism of this effect it was found that (XIX) is not a good radical acceptor and that the mechanism does not involve the B—O—H group since esters are also effective. The data suggest that

TABLE 15. COMPARISON OF BOROXAROPHENANTHRENE ANTIOXIDANTS
AND RELATED COMPOUNDS IN WHITE OIL

Boron compound (0.0051 mole/kg)	No.	$t_{0.5}$, hr (Absorption of 0.5 mole O_2/kg)
10-t-Butoxy-10,9-boroxarophenanthrene	(XX)	463
Product of 2,2'-diphenol and benzeneboronic acid	(XXIII)	395
10-Hydroxy-10,9-boroxarophenanthrene	(XIX)	354
10-Methoxy-10,9-boroxarophenanthrene	(XXI)	270
Product of 2,2'-diphenol and boric acid	(XXII)	230
Zinc 0,0-di(4-methyl-2-pentyl)phosphorodithioate		78
N-Phenyl-1-naphthylamine		64
2,6-Di-t-butyl-4-methylphenyl-di-n-butyl borate	(XXVIII)	38
2-Phenylbenzo-1,3-dioxa-2-borole	(XXV)	24
2,6-Di-t-butyl-4-methylphenol		18
2-Hydroxybenzo-1,3-dioxa-2-borole	(XXIV)	17
2,2'-Diphenol		5.7
Catechol		5.0
2-Hydroxylbiphenyl		2.0
Benzeneboronic acid	(XXVII)	1.8
Trihexylene glycol biborate	(XXVI)	0.9
None		0.9

borate catalysis of hydroperoxide decomposition plays an important role in the inhibitor mechanism and that secondary products are involved rather than (XIX) itself. It was postulated that (XIX) undergoes reaction with hydroperoxides followed by rearrangement (equation (52) to the seven-ring compounds. These intermediates could not be prepared in pure form but the measured inhibitory activities of the

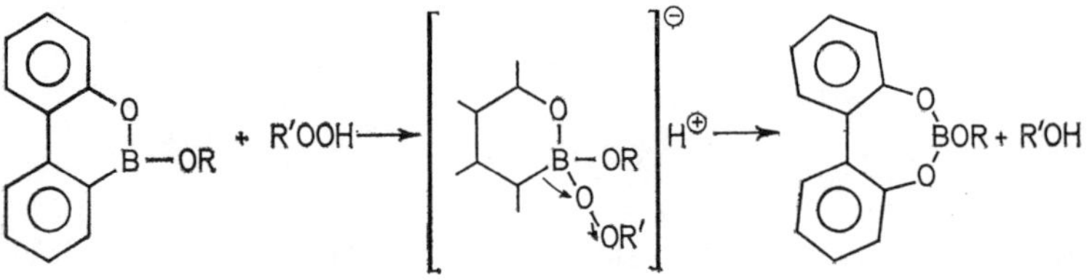

$$\text{(52)}$$

impure materials (XXII) and (XXIII) (see Table 14) confirm their unusual activity. The mechanism of the decomposition of hydroperoxides by these intermediates and other boron derivatives is discussed in Section II, B above.

In a paper presented at an American Chemical Society Meeting, Harle notes that alkyl boric acids are good auxiliary inhibitors for the oxidation of white oil at concentrations of 0.4–1.0 per cent.[101] A patent by Harle and Thomas discloses a number of borates as oxidation inhibitors in lubricating oils.[246] Compounds XXIX to XXXII were particularly effective in a series of antioxidant evaluation tests.

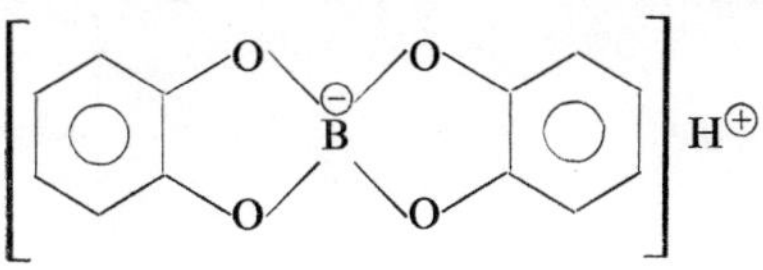

(XXIX) 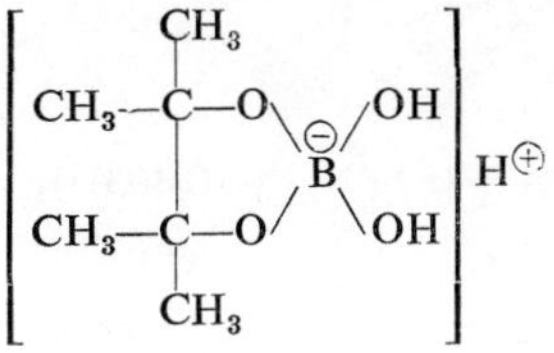　　　　　　　　　(XXX)

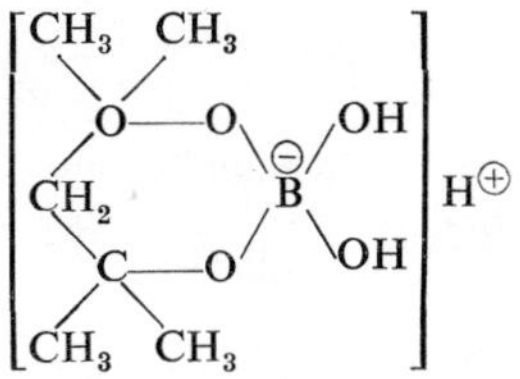

(XXXI) 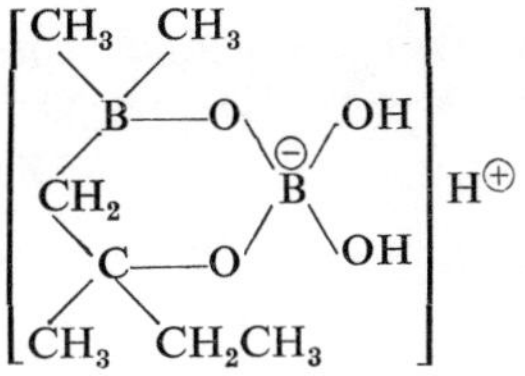　　　　　　　　　(XXXII)

In an isolated patent which is related to antioxidants in monomers, boric acid (2–3 per cent) is reported to be a more effective antioxidant for wool fat and wool fat mixtures than more conventional phenolic antioxidants.[286]

C. *Polymers*

A number of patents have described the use of boron compounds as antioxidants for polymers. Table 16 lists the various types of boron compounds which have been reported for this application. The earliest patent in this area was granted in 1931 and claimed aromatic borates (specifically β-naphthyl borate) as aging stabilizers in rubber.[54] In 1941 the use of phenolic borate esters and purportedly partial esters of *p*-hydroxydiphenylamine in natural and synthetic rubbers was revealed.[265] A complex stabilizing combination of a phenolic inhibitor, a sulfur-containing costabilizer and a borate ester has been patented for use in polyolefins.[100] Triisopropyl borate and tri-n-propyl borate were claimed specifically in amounts of 0.01–1 per cent, although a wide range of borates was included. Polyethylene and polypropylene are stabilized to oxidation by these combinations which are particularly effective when undesirable metallic impurities which could promote autoxidation are present. A study of boric acid esters as general polymer stabilizers has been noted.[287]

TABLE 16. BORATE STABILIZERS IN POLYMER SYSTEMS

Boron compound	Polymer system	Comments	Ref.
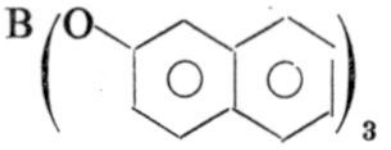 $B(O-C_{10}H_6)_3$ (dinaphthyl borate)	Rubber		54
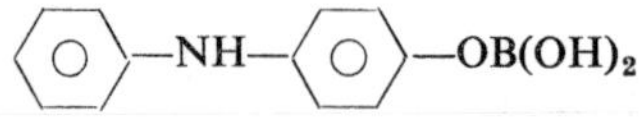 $C_6H_5{-}NH{-}C_6H_4{-}OB(OH)_2$	Rubber		191
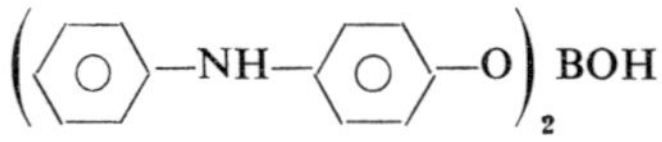 $(C_6H_5{-}NH{-}C_6H_4{-}O)_2 BOH$	Rubber		191
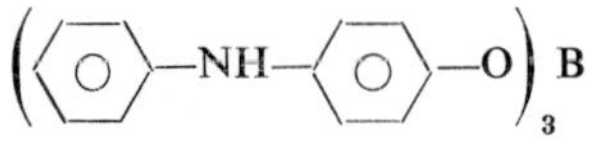 $(C_6H_5{-}NH{-}C_6H_4{-}O)_3 B$	Rubber		191
$B(OPr^n)_3$	Polyethylene, polypropylene	In combination with phenols and sulfides	44
$B(OPr^i)_3$	Polyethylene, polypropylene	In combination with phenols and sulfides	44
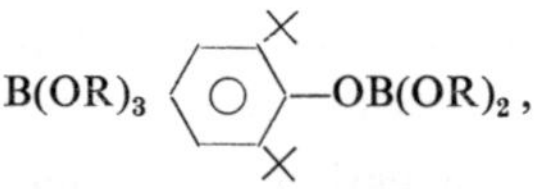 $B(OR)_3$, $X_2C_6H_3{-}OB(OR)_2$, glycol borates, $RB(OR')_2$ R_2BOR'	Polymers of mono alpha olefins including polyethylene, polypropylene		75, 226
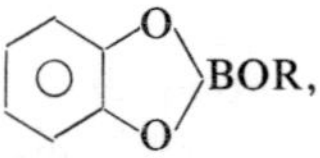 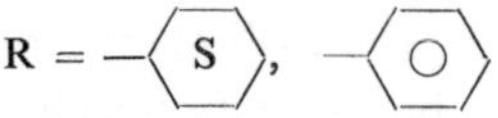 catechol-BOR, $R = -C_4H_3S$, $-C_6H_5$ (XXXIII), (XXXIV)	Polypropylene		211
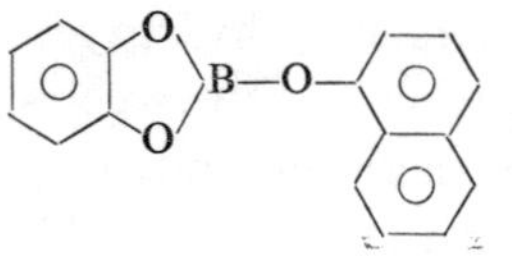 catechol-B—O—naphthyl (XXXV)	Polypropylene	Good oxidation inhibitor	211
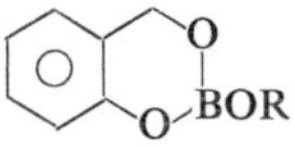  benzodioxaborine-BOR, $R = -C_4H_3S$, $-C_6H_5$ (XXXVII), (XXXVIII)	Polypropylene		212

TABLE 16. *(cont.)*

Boron compound	Polymer system	Comments	Ref.
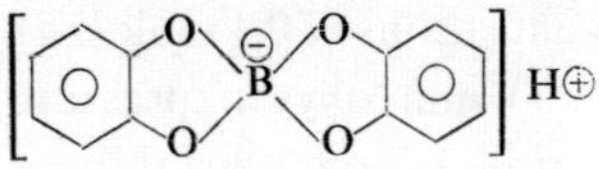 (XXIX)	Polypropylene		211
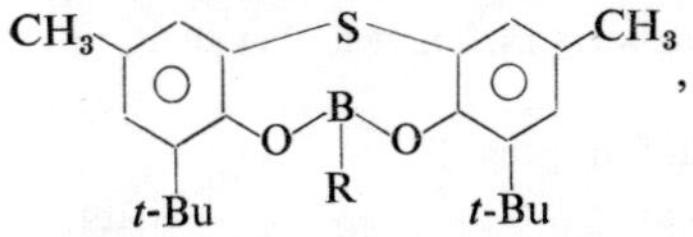 (XXXIX), (XL) R = OH, Ph	Polypropylene	Very good in- hibitor	212
H_3BO_3	Polyvinyl chloride		11, 37
H_3BO_3	Polyvinyl chloride	With added lead stearate	37
H_3BO_3	Polyvinyl chloride	With added barium hydroxide	195
Octyl borate	Polyvinyl chloride		63
$Na_2B_4O_7 \cdot 10 H_2O$	Polyvinyl chloride		66
KBO_2	Polyvinyl chloride		66
$Mg_2B_2O_6$	Polyvinyl chloride		66

Polypropylene stabilization has been the subject of a rather extensive study by Russian workers.[211-12] It was found that aliphatic derivatives such as tricyclohexyl borate increased the oxidative induction period of polypropylene insignificantly at 200–220°C.[211] Aromatic borates (XXIX and XXXIII to XXIX), on the other hand (at concentrations of 0.1 mole/kg), increased the induction period at 200°C and 300 mm oxygen pressure significantly (see Table 17). The induction period of the unstabilized

TABLE 17. AROMATIC BORATES AS ANTIOXIDANTS IN POLYPROPYLENE

Compound	Induction period, minutes		Ref.
	at 0.04 mole/kg	at 0.1 mole/kg	
Control	8–12		211
(XXXIII)	—	50	211
(XXXIV)	—	125	211
(XXXV)	—	345	211
(XXIX)	30	100	211
(XXXVI)	125	—	212
(XXXVII)	< 20	—	212
(XXXVIII)	125	—	212
(XXXIX)	950	—	212
(XL)	> 1200	—	212
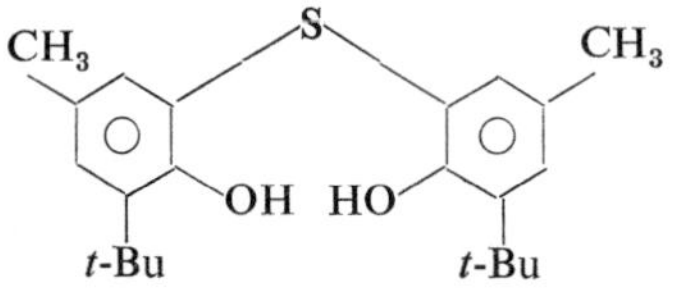	~ 300	—	212

polypropylene was 8–12 minutes and a mixture of the phenolic inhibitors, pyrocatechol and α-naphthol, raised this induction time to 125 minutes. The best borate studied (XXXV) resulted in an induction time of 345 minutes. Studies on inhibition versus borate concentration indicated that concentrations of approximately 0.04 to 0.1 mole/kg were necessary, although each material showed a slightly different time/concentration relationship. It is interesting that compounds XXIV and XXV, which are closely related to XXXIII and XXXIV were ineffective as antioxid-

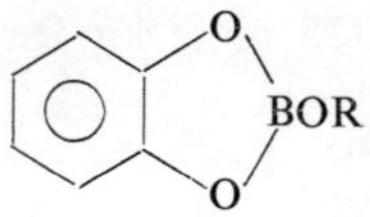

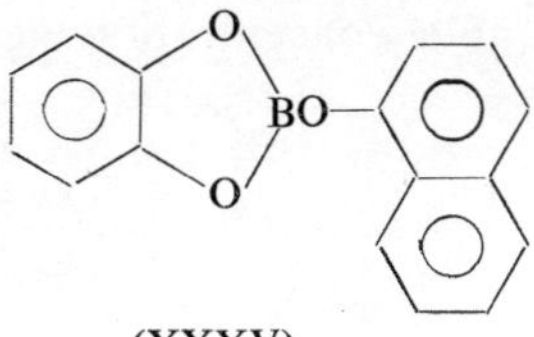

(XXXV)

(XXXIII), R = —⬡

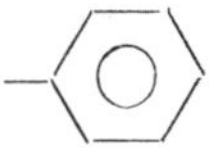

(XXXIV), R = —⟨ S ⟩

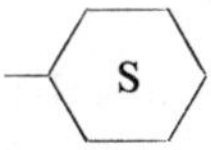

ants in monomeric paraffin oils, although the related phenanthrene derivatives (XIX) to (XXI) were very good.[46, 159]

In a continuing study of polypropylene oxidation inhibition at 200°C and 300 mm oxygen pressure, the following aromatic borates were also tested (Table 17).[170, 212] The cyclohexyl derivative (XXXVII) was relatively paren

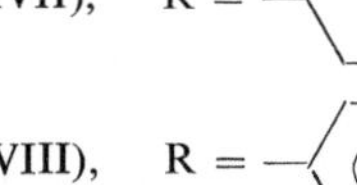

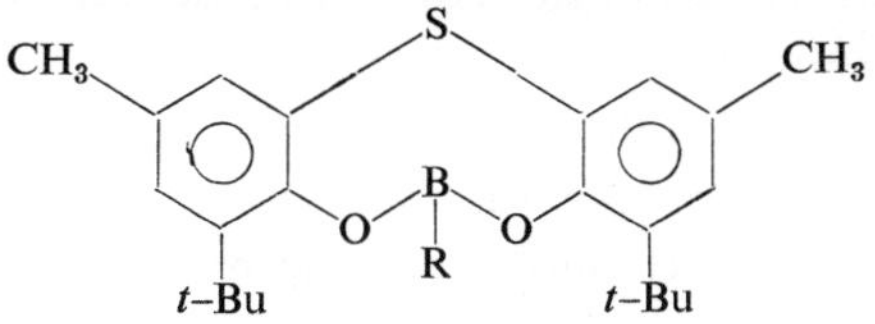

(XXXVI), R = H

(XXXVII), R = —⟨ S ⟩

(XXXVIII), R = —⬡

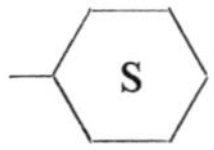

(XXXIX), R = OH

(XL), R = —⬡

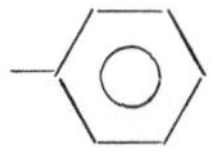

ineffective; compounds (XXXVI) and (XXXVIII) showed induction periods of approximately 125 minutes at 0.04 mole/kg compared to 8–12 minutes for unstabilized polypropylene. At the same concentration, the oxidative inhibition times for the sulfides (XXXIX) and (XL) were 950 minutes and over 1200 minutes, respectively. This is significantly better than any of the other borates tested. The thiobisphenol from which (XXXIX) and (XL) were derived, a known stabilizer, had an induction period of approximately 300 minutes at 0.04 mole/kg. The activities of (XXXIX) and

(XL) are postulated to arise from intramolecular synergism between the borate portion of the molecule and the sulfide, since sulfides are known oxidation inhibitors. Intermolecular synergism between systems of this type and the stable nitroxyl (XLI) has been observed. This effect was demonstrated at a concentration near 0.025 mole/kg for a mixture of

(XLI)

(XXXVIII) and (XLI), a known free radical inhibitor.

The extent of this synergism was shown by the fact that, at approximately 0.030 mole/kg, the inhibition time for (XXXVIII) was less than 20 minutes while that for 0.022 mole/kg of (XLI) was approximately 100 minutes. A mixture of the two materials at these concentrations required about 300 minutes before initiation of oxidation occurred. Similar synergistic effects were predicted for mixtures of borates and aromatic amines. The authors claimed that boron inhibits radical processes by radical termination because of its unfilled orbital.

In addition to the demonstrated superior antioxidant properties of these organic borates the following advantages over phosphates and other inhibitors were claimed: good thermal stability, low volatility, low toxicity and good compatibility with polymer systems.[211-12] It has also been reported that polymers of *alpha* monoolefins including polyethylene and polypropylene can reportedly be stabilized to thermal decomposition and to oxidation by addition of 0.1–5.0 per cent of a variety of borates, boronates and borinates.[75]

Stabilization of polyvinyl chloride by boric acid and inorganic borates has been claimed in several patents. A combination of 1 per cent boric acid added during the polymerization plus subsequent addition of 1 per cent lead stearate has been described as a heat stabilizer and antioxidant.[37] A mixture of 3–10 per cent boric acid and 1–3 per cent barium hydroxide has been claimed as a heat and light stabilizer; oxidation was not mentioned.[195] Borax ($Na_2B_4O_7 \cdot 10\ H_2O$), magnesium borate ($Mg_2B_2O_6$) and potassium metaborate (KBO_2) in 0.1–20 per cent concentrations have been claimed in a Czechoslovakian patent,[66] and the use of inorganic borates has also been claimed.[11] The addition of 10 per cent octyl borate has a stabilizing effect on plasticized polyvinyl chloride compositions.[63]

Velea *et al.* have found that borates of lead, calcium, barium and cadmium alone or in synergistic mixtures with known stabilizers are effective in stabilizing polyvinyl chloride.[256-9] In another case it was concluded on the basis of hydrogen chloride evolution and color change that boric acid inhibited the degradation of polyvinyl chloride.[11] Alkali borates were less effective.

An impedance to assessing the importance and role of stabilizers in these polyvinyl chloride systems is the lack of definition of the stabilizing action. It is difficult to determine from most of the literature whether the stabilizers are antioxidants, thermal stabilizers or light stabilizers. In a detailed paper, Velea *et al.* found that thermal degradation as measured by hydrochloric acid evolution was not affected significantly by boric acid.[256] However, aging as measured by color change was definitely inhibited by boric acid. On the basis of this observation the authors postulated that boric acid (or the metaboric acid formed) inhibited the oxidation of polyvinyl chloride. A rationalization of this stabilization was proposed in which the boric acid (as metaboric under the test conditions) directed the air oxidation to give largely hydroxyl groups on the PVC chain. These alcohol groups then would be esterified to give the oxidatively stable boric acid ester.[256] Metal borates, on the other hand, appear to act in two different ways: they can accept hydrochloric acid formed by thermal degradation, and the resulting boric acid (metaboric acid) can subsequently inhibit oxidative degradation.

Boric acid has been included as a color stabilizing antioxidant in polyvinyl chloride in a patent disclosure, although only silicic acid and silicon dioxide are noted specifically in the claims.[37]

An interesting note has appeared on the inhibition of oxidation of diamonds by boron.[233] Complete oxidation of synthetic and natural diamonds can be achieved readily at 800°C in oxygen. Addition of small amounts of boron reduces the observed oxidation rates dramatically (rates of weight loss were 70 to 100 times slower). It was postulated that diamond particles become coated with boric oxide during combustion and therefore are protected from further oxidative attack. In a related case, graphite oxidation can reportedly be inhibited by addi- tion of ammonium borate.[82]

VII. REFERENCES

1. ADDY, L. E., Brit. Pat. 967 822 (1964, to British Hydrocarbon Chemicals Ltd.); see also Belg. Pat. 623 468 (1963); Fr. Pat. 1 344 352 (1963); Ger. Pat. 1 249 253 (1967); Jap. Pat. 574/66.
2. ALAGY, J. *et al.*, Paper presented at the 7th World Petroleum Congress, Mexico City, April 1967.
3. ALAGY, J. and DEFOOR, F., Belg. Pat. 666 774 (1965, to Institut Français du Petrole des Carburants et Lubrifiants); see also Fr. Addn. 88 164 (1966).
4. ALAGY, J. and DEFOOR, F., Neth. Appl. 6 505 047 (1965, to Institut Français du Petrole des Carburants et Lubrifiants).
5. ALAGY, J. and DEFOOR, F., Fr. Pat. 1 442 272 (1966, to Institut Français du Petrole des Carburants et Lubrifiants).
6. ANON., *Chem. and Eng. News*, **44**, (23) 17 (1966).
7. ANON., *Chem. Week*, **96**, (21) 95 (1965).
8. ANON., *Chem. Week*, **96**, (24) 105 (1965).
9. ANON., *Chem. Week*, **100**, (4) 59 (1967).
10. ANON., *Chem. Week*, **102**, (4) 83 (1968).
11. ANON., *Plaste Kautschuk*, **5**, 231 (1960).
12. ARIMOTO, F. S., *J. Polymer Sci.*, Pt. A–1, **4**, 275 (1966).
13. Asahi Chemical, Brit. Pat. 1 094 328 (1967); see also Neth. Appl. 6 512 010 (1966).
14. BACCAREDDA, M., NENCETTI, G. F. and MAGAGNINI, P. L., *6th World Petroleum Congress*, Sect. IV, 451 (1963).
15. BARKER, R. S., Belg. Pat. 643 224 (1964, to Halcon International Inc.); see also Brit. Pat. 1 038 264 (1966); Fr. Pat. 1 422 043 (1965); Neth. Appl. 6 401 009 (1964).
16. BARKER, R. S., Belg. Pat. 659 334 (1965, to Halcon International Inc.); see also Brit. Pat. 1 059 575 (1967); Fr. Addn. 88 113 (1967).
17. BASHKIROV, A. N., *Khim. Nauka i Promy.*, **1**, 273 (1956).
18. BASHKIROV, A. N. *et al.*, *Neftekhimiya*, **1**, 527 (1961); *C.A.*, **57**, 684 (1962).
19. BASHKIROV, A. N. *et al.*, *Sintez i Svoistva Monomerov, Akad. Nauk SSSR, Inst. Neftekhim. Sinteza, Sb. Rabot 12-oi (Dvenadtsatoi) Konf. po Vysokomolekul. Soedin.*, 198 (1962); *C.A.*, **62**, 6405 (1965).
20. BASHKIROV, A. N. *et al.*, Russ. Pat. 106 914 (1957).
21. BASHKIROV, A. N. *et al.*, *6th World Petroleum Congress*, Sect. IV, 473 (1963).
22. BASHKIROV, A. N. *et al.*, Russ. Pat. 146 733 (1964).
23. BASHKIROV, A. N. *et al.*, *Petroleum Chem.* (Engl. transl.), **4**, 292 (1964).
24. BASHKIROV, A. N. and ALENTÉVA, Y. S., *Petroleum Chem.* (Engl. transl.), **4**, 204 (1964).
25. BASHKIROV, A. N. and CHERTKOV, Y. B., Russ. Pat. 106 872 (1957); *C.A.*, **52**, 3849 (1958).
26. BASHKIROV, A. N., CHERTKOV, Y. B. and ZIL'BERG, G. A., Russ. Pat. 105 932 (1957); *C.A.*, **51**, 16513 (1957).
27. BASHKIROV, A. N. and KAMZOLKIN, V. V., *Khim. Nauka i Promy.*, **4**, 607 (1959).
28. BASHKIROV, A. N. and KAMZOLKIN, V. V., *World Petrol. Congr. Proc. 5th, New York*, **4**, 175 (1959); *C. A.*, **57**, 13596 (1962).
29. BASHKIROV, A. N., KAMZOLKIN, V. V. and LODZIK, S. A., *Maslob. Zhir. Prom.*, **23**, 24 (1957).
30. BASHKIROV, A. N., KAMZOLKIN, V. V., SOKOVA, K. M. and ANDREEVA, T. P., *Proc. Acad. Sci. USSR, Chem. Tech. Sect.* (Engl. transl.), **118**, 1 (1958).
31. BASHKIROV, A. N., KAMZOLKIN, V. V., SOKOVA, K. M. and ANDREEVA, T. P., *Proc. Acad. Sci. USSR, Chem. Sect.* (Engl. transl.), **119**, 247 (1958).

32. BASHKIROV, A. N., KAMZOLKIN, V. V., SOKOVA, K. M. and ANDREEVA, T. P., *Okislenie Uglevodorodov z Zhidkoi Faze, Akad. Nauk SSSR, Inst. Khim. Fiz., Sb. Statei*, 159 (1959).
33. BASHKIROV, A. N. and KISTANOVA, A. I., *Proc. Acad. Sci. USSR, Chem. Sect.* (Engl. transl.), **131**, 301 (1960).
34. BASHKIROV, A. N. and KVASNEVSKAYA, K. S., *Proc. Acad. Sci. USSR, Chem. Sect.* (Engl. transl.), **173**, 305 (1967).
35. BASHKIROV, A. N. and PAL, S., *Proc. Acad. Sci. USSR, Chem. Sect.* (Engl. transl.), **128**, 875 (1959).
36. BASHKIROV, A. N., SHAIKHUTDINOV, E. M. and GILYAROVSKAYA, L. A., *Proc. Acad. Sci. USSR, Chem. Sect.* (Engl. transl.), **148**, 160 (1963).
37. BAUER, H. and HECKMAIER, J., U.S. Pat. 3 012 005 (1961, to Wacker-Chemie G. m. b. H.).
38. BECKER, M., U.S. Pat. 3 317 581 (1967, to Halcon International Inc.); see also Belg. Pat. 658 308 (1965); Ind. Pat. 91 064; Neth. Appl. 6 413 754 and 6 500 235 (1965); So. Af. Pat. 65/0223.
39. BECKER, M. M., Belg. Pat. 656 281 (1965, to Halcon International Inc.); See also Brit. Pat. 1 074 489 (1967); Fr. Pat. 1 434 164 (1966); Neth. Appl. 6 413 619 (1965).
40. BELLER, H. and SCHÜTTE, H., Ger. Pat. 652 541 (1937, to I. G. Farbenind. A.–G.).
41. BIGOT, J. A., Paper presented at the 7th World Petroleum Congress, Mexico City, April 1967.
42. BLANCHARD, H. S., *J. Am. Chem. Soc.*, **82**, 2014 (1960).
43. BOONSTRA, H. J. and ZWIETERING, P., *Chem. Ind. (London)*, 2039 (1966).
44. BROWN, D. E., POIROT, E. E. and SPEED, R. A., U.S. Pat. 3 242 135 (1966, to Esso Research and Engineering).
45. BRIDGER, R. F., McCABE, L. J. and WILLIAMS, A. L., U.S. Pat. 3 320 165 (1967, to Mobil Oil Corp.).
46. BRIDGER, R. F., WILLIAMS, A. L. and McCABE, L. J., *Ind. Eng. Chem. Prod. Res. and Develop.*, **5**, 226 (1966).
47. BROICH, F. and GRASEMANN, H., *Erdoel Kohle*, **18**, 360 (1965).
48. BROICH, F. and GRASEMANN, H., Brit. Pat. 1 089 252 (1967, to Chemische Werke Hüls, A.–G.); see also Fr. Pat. 1 423 382 (1966); Ger. Pat. 1 224 283 (1966).
49. BROTHMAN, A., Brit. Pat. 671 449 (1952).
50. BRUNIE, J. C. and CRENNE, N., Belg. Pat. 681 870 (1966, to Societe des Usines Chimiques Rhone-Poulenc); see also Fr. Pat. 1 445 653 (1966); So. Af. Pat. 66/3157.
51. BRUNIE, J. C. and CRENNE, N., Fr. Pat. 1 447 267 (1966, to Societe des Usines Chimiques Rhone-Poulenc); see also Ind. Pat. 105 530; So. Af. Pat. 66/3159.
52. BUEHLER, C. A. and HART, H., *J. Am. Chem. Soc.*, **85**, 2177 (1963).
53. BUKHSHTAB, Z. I., *Khim. i Tekhnol. Topliv i Masel*, **7**, No. 12, 8 (1962); *C.A.*, **58**, 6624 (1963).
54. BUNBURY, H. M., DAVIES, J. S. H. and NAUNTAN, W. J. S Brit. Pat. 363 483 (1931, to Imperial Chemical Industries Ltd.).
55. CAHN, R. P., Belg. Pat. 652 006 (1964, to Esso Research and Engineering); see also Brit. Pat. 1 064 167 (1967); Fr. Pat. 1 411 913 (1965); Neth. Appl. 6 409 181 (1965).
56. CALLAHAN, J. L., SZABO, J. J. and GERTISSER, B., Belg. Pat. 631 437 (1963, to Standard Oil Co. Ohio); see also Can. Pat. 764 468 (1967).
57. CAMPBELL, J. S. and RIDGWELL, S. A., Brit. Pat. 1 056 124 (1967, to Imperial Chemical Industries Ltd.).
58. Chemische Werke Hüls A.–G., Belg. Pat. 668 681 (1965); see also Fr. Pat. 1 445 874 (1966); Neth. Appl. 6 510 994 (1966).

59. Chemische Werke Hüls A.-G., Belg. Pat. 684 423 (1967).
60. CHERNYAEVA, G. N. and KHOLKIN, Y. I., *Issled. v. Obl. Khim. i Khim. Teknol. Drevesiny, Akad. Nauk SSSR, Sibirisk. Otd. Inst. Lesa i Drevesiny*, 38 (1963); *C.A.*, **60**, 14343 (1964).
61. CHERTKOV, Y. B., Russ. Pat. 106 355 (1957).
62. CHERTKOV, Y. B., *Zhur. Priklad. Khim.*, **32**, 363 (1959).
63. CHEVASSUS, F. and DE BROUTELLES, R., *The Stabilization of Polyvinyl Chloride*, St. Martins' Press Inc., New York, 1963, p. 31, 36.
64. CLAYTON, J. O. and FARRINGTON, B. B., U.S. Pat. 2 312 208 (1943, to Standard Oil Co. of Calif.).
65. COFFEE, E. C. J. and DAVIES, A. G., *J. Chem. Soc. (C)*, 1493 (1966); DAVIES, A. G. and ROBERTS, B. P., *Chem. Commun.*, 298 (1966).
66. Czech. Pat. 87 877 (1956).
67. DAVIES, A. G. and MOODIE, R. B., *J. Chem. Soc.*, 2372 (1958).
68. DENISON, G. H. and CONDIT, P. C., U.S. Pat. 2 346 155 (1944, to Standard Oil Co. of Calif.).
69. DIETRICH, W. and LUTHER, M., Ger. Pat. 581 238 (1933, to I. G. Farbenind. A.-G.).
70. DIMITROV, D. and PAPAZOVA, P., *Godishnik Khim. Tekhnol. Inst.*, **11**, 19 (1964); *C.A.*, **66**, 12702 (1967).
71. DIMITROV, D., PAPAZOVA, P. and PANAJOTOVA, E., *Compt. Rend. Acad. Bulg. Sci.* **19**, 125 (1966); *C.A.*, **65**, 3775 (1966).
72. DIMITROV, D. and STANKOVA, D., *Compt. Rend. Acad. Bulg. Sci.*, **17**, 33 (1964); *C.A.*, **61**, 9339 (1964).
73. DIMITROV, D., VOJNOVA, S. and PENCEV, N., *Compt, Rend. Acad. Bulg. Sci.*, **19**, 811 (1966).
74. DIMITROV, D., VOJNOVA, S. and STANEV, S., *Godishnik Khim. Tekhnol. Inst.*, **11**, 9 (1964); *C.A.*, **66**, 4643 (1967).
75. DOYLE, M. E. and JAFFE, G. S., U.S. Pat. 3 131 164 (1964, to Shell Oil Co.).
76. DROZDOVA, M. A. *et al.*, *Proc. Acad. Sci. USSR, Chem. Sect.* (Engl. transl.), **171**, 1061 (1966).
77. DROZDOVA, M. A., BASHKIROV, A. N. and KAMZOLKIN, V. V., *Proc. Acad. Sci. USSR, Chem. Sect.* (Engl. transl.), **169**, 779 (1966).
78. DUNCANSON, L. A. and MEE, A., Brit. Pat. 948 394 (1964, to Imperial Chemical Industries Ltd.).
79. E. I. du Pont de Nemours and Co., Brit. Pat. 716 820 (1954).
80. E. I. du Pont de Nemours and Co., Belg. Pat. 663 394 (1965); see also Neth. Appl. 6 505 838 (1965).
81. EBY, L. T. and MIKESKA, L. A., U.S. Pat. 2 462 616 (1949, to Standard Oil Development).
82. ELLIS, R. T. and JUEL, L. H., Fr. Pat. 1 454 593 (1966, to Great Lakes Carbon Corp.).
83. ENGELMANN, K., Ger. (East) Pat. 14 370 (1958); *C.A.*, **53**, 8668 (1959).
84. ETHERINGTON, R. W., U.S. Pat. 3 254 035 (1966, to Petro-Tex Chemical Corp.).
85. FALIZE, C. and GOBRON, G., Belg. Pat. 678 574 (1966, to Usines de Melle).
86. FEDER, J. B. and CARROLL, J. H., U.S. Pat. 3 239 552 (1966, to Halcon International Inc.); see also Belg. Pat. 634 343 (1963); Brit. Pat. 1 027 005 (1966); Can. Pat. 742 252 (1966); Fr. Pat. 1 384 576 (1965); So. Af. Pat. 63/2402.
87. FIERZ-DAVID, H. E., BLANGEY, L. and v. KRANNICHFELDT, W., *Helv. Chim. Acta*, **30**, 237 (1947); *C.A.*, **41**, 3088 (1947).
88. FOX, S. N. and COLTON, J. W., U.S. Pat. 3 109 864 (1963, to Halcon International Inc.); see also Brit Pat. 1 013 935 (1965).
89. FOX, S. N. and COLTON, J. W., U.S. Pat. 3 254 962 (1966, to Halcon International Inc.).

90. FREIDIN, B. G., *J. Applied Chem. USSR* (Engl. transl.), 1263 (1954).
91. GARDNER, C. and PRESCOTT, J. F., Brit. Pat. 1 054 053 (1967, to Imperial Chemical Industries).
92. GASH, V. W., U.S. Pat. 3 210 381 (1965, to Monsanto Co.); see also Brit. Pat. 1 038 182 (1966); Can. Pat. 727 122 (1966); Fr. Pat. 1 367 762 (1964); Jap. Pat. 17 908 (1966).
93. GERMAIN, J. and COGNION, J., *Chim. Ind. (Paris)*, **91**, 519 (1964).
94. GOLDEN, R. L. and MAZELLA, G., Fr. Pat. 1 469 227 (1967, to Halcon International Inc.); see also Neth. Appl. 6 516 567 (1966).
95. GOLDEN, R. L. and MAZZELLA, G., Belg. Pat. 675 370 (1966, to Halcon International Inc.); see also Fr. Pat. 1 466 717 (1967); Neth. Appl. 6 600 797 (1966); So. Af. Pat. 66/0499.
96. GREEN, R. C., Can. Pat. 749 991 (1967, to Esso Research and Engineering); see also Belg. Pat. 652 007 (1964); Brit. Pat. 1 070 800 (1967); Neth. Appl. 6 409 505 (1965).
97. GRENET, J. B. and MARCHAND, P., Fr. Pat. 1 337 243 (1963, to Rhone–Poulenc S.A.).
98. GROTEWOLD, J. and LISSI, E. A., *Chem. Commun.*, 21 (1965).
99. HABESHAW, J., WIRTH, M. M. and ADDY, L. E., Belg. Pat. 623 467 (1962, to British Hydrocarbon Chemicals Ltd.); see also Brit. Pat. 968 339 (1964); Fr. Pat. 1 337 002 (1963).
100. Halcon International Inc., Brit. Pat. 1 008 189 (1965); see also Neth. Appl. 6 410 476.
101. HARLE, O. L., *Am. Chem. Soc., Div. Petrol. Chem. Preprints*, **2**, No. 1, 51 (1957); *C.A.*, **55**, 6840 (1961).
102. HART, H. and BUEHLER, C. A., *J. Org. Chem.*, **29**, 2397 (1964).
103. HART, H., *et al.*, *J. Org. Chem.*, **30**, 331 (1965).
104. HART, H., COLLINS, P. M. and WARING, A. J., *J. Am. Chem. Soc.*, **88**, 1005 (1966).
105. HART, H. and LANGE, R. M., *J. Org. Chem.*, **31**, 3776 (1966).
106. HART, H. and LERNER, L. R., *J. Org. Chem.*, **32**, 2669 (1967).
107. HELBIG, J. E. *et al.*, U.S. Pat. 3 232 704 (1966, to Esso Research and Engineering).
108. HELLTHALER, T. and PETER, E., Ger. Pat. 552 886 (1934, to A. Riebeck'sche Montanwerke A.–G.).
109. HELLTHALER, T. and PETER, E., Ger. Pat. 564 196 (1934, to A. Riebeck'sche Montanwerke A.–G.); *C.A.*, **28**, 5832 (1934).
110. HELLTHALER, T. and PETER, E., U.S. Pat. 1 947 989 (1934, to A. Riebeck'sche Montanwerke A.–G.).
111. Imperial Chemical Industries Ltd., Brit. Pat. 1 085 892 (1967); see also Belg. Pat. 662 461 (1965); Fr. Pat. 1 431 459 (1965); Neth. Appl. 6 504 478 (1965).
112. Imperial Chemical Industries Ltd., Brit. Pat. 1 077 454 (1967); see also Fr. Pat. 1 405 545 (1965); Neth. Appl. 6 407 917 (1965).
113. Imperial Chemical Industries Ltd., Neth. Appl. 288 365 (1965).
114. ILLINGWORTH, G. E. and LESTER, G. W., Paper presented at the *154th National A.C.S. Meeting, Division of Petroleum Chemistry*, Chicago, Illinois, September 1967, p. 161.
115. Institut Français du Petrole des Carburants et Lubrifiants, Neth. Pat. 6 510 337 (1966); see also Belg. Pat. 668 290 (1966); Fr. Pat. 1 448 752 (1966).
116. Institut Français du Petrole des Carburants et Lubrifiants, Fr. Pats. 1 497 516, 1 497 519, 1 497 522, 1 497 525, 1 497 540, 1 498 351 (1967).
117. JOHNSON, J. Y., Brit. Pat. 199 886 (1923, to Badische Anilin and Soda Fabrik).
118. JOHNSON, J. Y., Brit. Pat. 353 047 (1931, to I. G. Farbenind. A.–G.).
119. JONES, M. M. and LAMBERT, D. G., *J. Inorg. Nucl. Chem.*, **29**, 579 (1967).

120. JUBB, A. H., Brit. Pat. 1 035 626 (1966, to Imperial Chemical Industries Ltd.); see also Belg. Pat. 634 236 (1963); Fr. Pat. 1 361 232 (1964).
121. JUBB, A. H., Brit. Pat. 1 035 625 (1966, to Imperial Chemical Industries Ltd.); see also Belg. Pat. 634 235 (1963); Fr. Pat. 1 378 051 (1964).
122. JUBB, A. H., Brit. Pat. 1 051 874 (1966, to Imperial Chemical Industries Ltd.).
123. KAMZOLKIN, V. V. et al., Neftekhimiya, 4, 599 (1964); C.A., 61, 13209 (1964).
124. KAMZOLKIN, V. V., BASHKIROV, A. N., KAMZOLKINA, E. V. and LODZIK, S. A., Neftekhimiya, 2, 750 (1962).
125. KAMZOLKIN, V. V., BASHKIROV, A. N. and LODZIK, S. A., Neftekhimiya, 1, 411 (1961); C.A., 57, 10997 (1962).
126. KAMZOLKIN, V. V., BASHKIROV, A. N. and LODZIK, S. A., Neftekhimiya, 1, 260 (1961).
127. KAMZOLKIN, V. V., BASHKIROV, A. N., SOKOVA, K. M. and ANDREEVA, T. P., Proc. Acad. Sci. USSR, Chem. Sect. (Engl. transl.), 128, 857 (1960).
128. KAMZOLKIN, V. V., BASHKIROV, A. N., SOKOVA, K. M. and ANDREEVA, T. P., Trudy Inst. Nefti, Akad. Nauk. SSSR, 14, 65 (1960); C.A., 56, 4608 (1962).
129. KAMZOLKIN, V. V., BASHKIROV, A. N., SOKOVA, K. M. and ANDREEVA, T. P., Neftekhimiya, 4, 96 (1964).
130. KAPKIN, V. D., BASHKIROV, A. N. and GROZHAN, M. M., Neftekhimiya, 7, 97 (1967); C.A., 66, 116207 (1967).
131. KAWAI, S. and NOBORI, H., J. Soc. Chem. Ind. Japan, 46, 765 (1943); C.A., 42, 6737 (1948).
132. KAWAI, S. and NOBORI, H., J. Soc. Chem. Ind. Japan, 46, 768 (1943); C.A., 42, 6737 (1948).
133. KAWAI, S. and NOBORI, H., J. Soc. Chem. Ind. Japan, 46, 895 (1943); C.A., 42, 6737 (1948).
134. KERNOS, Y. D. and MOLDAVSKII, B. L., Zhur. Priklad. Khim., 33, 2593 (1960); C.A., 55, 8831 (1961).
135. KHANNA, M. L., J. Sci. Ind. Res. (India), 20D, 255 (1961); C.A., 56, 1678 (1962).
136. KIRSHENBAUM, I., BARTLETT, J. H. and HILL, R. M., U.S. Pat. 3 082 192 (1963, to Esso Research and Engineering Co.).
137. KIRSHENBAUM, I., BARTLETT, J. H. and HILL, R. M., U.S. Pat. 3 202 694 (1965, to Esso Research and Engineering Co.).
138. KIRSHENBAUM, I., BARTLETT, J. H. and HILL, R. M., U.S. Pat. 3 214 449 (1965, to Esso Research and Engineering Co.).
139. KIRSHENBAUM, I. et al., Belg. Pat. 598 331 (1961, to Esso Research and Engineering Co.); see also Brit. Pat. 944 110 (1963).
140. KIRSHENBAUM, I., HILL, R. M. and BARTLETT, J. H., U.S. Pat. 3 301 887 (1967, to Esso Research and Engineering Co.).
141. KIRSHENBAUM, I., MUESSIG, C. W. and HILL, R. M., U.S. Pat. 3 042 661 (1962, to Esso Research and Engineering Co.).
142. KROPF, H., Angew. Chem. Intern. Ed. Engl., 5, 646 (1966).
143. KRUGLOVA, N. V. and FREIDLINA, R. K., Bull. Acad. Sci., USSR, Div. Chem. Sci. (Engl. transl.), 2008 (1965).
144. KURASOV, V. I., Vses. Soveshch. Sin. Zhirozamen, Poverkhnostnoaktiv Veshchestvam Moyushch Sredstvam, 3rd Shebekino, 89 (1965); C.A., 66, 30234 (1967).
145. LAMBRIS, G. and GERDES, J., Brennstoff-Chem., 22, 125; 139 (1941).
146. LAWRENCE, F. I. L., SMITH, R. K. and POHORILLA, M. J., U.S. Pat. 2 721 121 (1955, to Kendall Refining Co.).
147. LAWRENCE, F. I. L., SMITH, R. K. and POHORILLA, M. J., U.S. Pat. 2 721 180 (1955, to Kendall Refining Co.).
148. LAWRENCE, F. I. L., SMITH, R. K. and POHORILLA, M. J., U.S. Pat. 2 721 181 (1955, to Kendall Refining Co.).

149. LEVY, M. and DOISY, E. A., *J. Biol. Chem.*, **77**, 733 (1928); *C.A.*, **22**, 2925 (1928).
150. LI, HAI, LI, SHENG-CHING, CHIN, WEN-CHU and LU, P'EI-CHANG, *Jan Liao Hsueh Pao*, **4**, 253 (1959); *C.A.*, **54**, 24337 (1960).
151. LAIO, C. W., HOOK, E. O. and BUSH, C., U.S. Pat. 3 076 013 (1963, to Standard Oil Co., Ohio).
152. LOW, H., *Ind. Eng. Chem. (Prod. Res. and Dev.)*, **5**, 80 (1966).
153. LUNDIN, H. *Biochem. Z.*, **207**, 107 (1929); *C.A.*, **23**, 3240 (1929).
154. LUTHER, M. and DIETRICH, W., U.S. Pat. 1 931 501 (1934, to I. G. Farbenind. A.-G.).
155. MARCELL, R. L., U.S. Pat. 3 317 614 (1967, to Halcon International Inc.); see also Belg. Pat. 626 227 (1963); Brit. Pat. 1 013 839 (1965); Jap. Pat. 19889/66.
156. MARCELL, R. L., Belg. Pat. 627 494 (1963, to Halcon International Inc.); Brit. Pat. 1 026 971 (1966); Fr. Pat. 1 384 586 (1965).
157. MARCELL, R. L., U.S. Pat. 3 275 695 (1966, to Halcon International Inc.); see also Belg. Pat. 641 112 (1964); Brit. Pat. 1 068 948 (1967); Fr. Addn. 84 984 (1967).
158. MATSUURA, T., SUZUKI, Y. and FUKAZAWA, K., Jap. Pat. 19431/1964 (1964, to Mitsubishi Kasei Kogyo Co.).
159. MCCABE, L. J. and BRIDGER, R. F., U.S. Pat. 3 287 270 (1966, to Mobil Oil Corp.).
160. MCCLURE, J. D. and WILLIAMS, P. H., *J. Org. Chem.*, **27**, 24 (1962).
161. MCNAMARA, L. S. and COHEN, C. A., U.S. Pat. 3 238 238 (1966, to Esso Research and Engineering Co.).
162. MIHAIL, R. *et al.*, Brit. Pat. 1 011 678 (1965, to Ministerul Industriei Petrolului Si Chimiei); see also Ger. Pat. 1 205 520 (1965).
163. Ministerul Industriei Petrolului Si Chimiei, Brit. Pat. 1 035 105 (1966); see also Fr. Addn. 83 565 (1964).
164. MISONO, A., UCHIDA, Y. and YAMADA, K., *Bull. Chem. Soc. Japan*, **39**, 2458 (1966).
165. Mitsubishi Chemical Ind. Ltd., Jap. Pat. 11121/62 (1962).
166. MONROE, G. C. and PATTON, L. W., Belg. Pat. 666 793 (1966, to E. I. du Pont de Nemours and Co.); see also Brit. Pat. 1 061 877 (1967); Can. Pat. 767 158 (1967); Fr. Pat. 1 440 441 (1966); Neth. Appl. 6 509 034 (1966).
167. MOYER, W. W., U.S. Pat. 2 328 920 (1943, to Solvay Process Co.).
168. Societe des Usines Chimiques Rhone-Poulene, Belg. Pat. 679, 792; see also Fr. Pat. 1, 458, 153; Neth. Appl. 6, 604, 935.
169. Nagase Kao Soap Co., Jap. Pat. 161 763 (1944); *C.A.*, **43**, 2219 (1949).
170. NEIMAN, M. B., *Russ. Chem. Rev.* (Engl. Transl.), **33**, 13 (1964).
171. NEISWENDER, D. D., Can. Pat. 739 013 (1966, to Socony Mobil Oil Co.); see also Brit. Pat. 976 975 (1964).
172. NELSON, D. O. *et al.*, U.S. Pat. 3 336 390 (1967, to Esso Research and Engineering).
173. NEW, R. G. A. *et al.*, Brit. Pat. 972 656 (1964, to Imperial Chemical Industries Ltd.); see also Fr. Pat. 1 351 666 (1964).
174. NEW, R. G. A., Brit. Pat. 1 073 522 (1967, to Imperial Chemical Industries Ltd.).
175. NEW, R. G. A., Brit. Pat. 1 076 843 (1967, to Imperial Chemical Industries Ltd.).
176. NEWSOM, H. C., U.S. Borax Research Corporation, unpublished results.
177. NIES, N. P. and CAMPBELL, G. W., "Inorganic Boron-Oxygen Chemistry", Chapter 3 in *Boron, Metallo-boron Compounds and Boranes*, R. M. ADAMS, Ed., Interscience, New York, N.Y., 1964.
178. NOBORI, H., *J. Soc. Chem. Ind. Japan*, **47**, 16 (1944); *C.A.*, **42**, 6738 (1948).
179. NOBORI, H., *J. Soc. Chem. Ind. Japan*, **47**, 595 (1944); *C.A.*, **42**, 6738 (1948).
180. NOBORI, H. and KAWAI, S., *J. Soc. Chem. Ind. Japan*, **46**, 1208 (1943); *C.A.*, **42**, 6737 (1948).

181. NOBORI, H. and KAWAI, S., *J. Soc. Chem. Ind. Japan*, **46**, 1209 (1943); *C.A.*, **42**, 6737 (1948).
182. NOBORI, H., KAWAI, S., NOGUCHI, M. and NAKAJIMA, K., *J. Soc. Chem. Ind. Japan*, **46**, 1252 (1943); *C.A.*, **42**, 6737 (1948).
183. NOBORI, H., NOGUCHI, M. and NAKAJIMA, H., *J. Soc. Chem. Ind. Japan*, **47**, 473 (1944); *C.A.*, **42**, 6738 (1948).
184. NOBORI, H., NOGUCHI, M. and NAKAJIMA, H., *J. Soc. Chem. Ind. Japan*, **47**, 475 (1944); *C.A.*, **42**, 6738 (1948).
185. NOBORI, H. and YONESHIRO, A., *J. Soc. Chem. Ind. Japan*, **48**, 79 (1945); *C.A.*, **42**, 6738 (1948).
186. NOVAK, F. I. *et al.*, *Neftekhimiya*, **7**, 248 (1967); *C.A.*, **67**, 107900 (1967).
187. OLENBERG, H., U.S. Pat. 3 324 186 (1967, to Halcon International Inc.); see also Belg. Pat. 640 732 (1963); Brit. Pat. 1 068 948 (1967); Fr. Pat. 1 397 197 (1965); Ind. Pat. 85 864, 90 930; Jap. Pat. 16782/66; So. Af. Pat. 63/5656.
188. OLENBERG, H., Brit. Pat. 1 040 115 (1966, to Halcon International Inc.).
189. OLENBERG, H. and FEDER, J. B., Brit. Pat. 1 044 824 (1966, to Halcon International Inc.).
190. OLENBERG, H. and KANTROWITZ, L., U.S. Pat. 3 240 820 (1966, to Halcon International Inc.); see also Brit. Pat. 969 148 (1964); Fr. Pat. 1 384 585 (1965).
191. PAUL, P. T., U.S. Pat. 2 259 175 (1941, to U.S. Rubber).
192. PEDAYAS, V. M., *Zhur. Prikl. Khim.*, **37**, 2692 (1964).
193. PEDAYAS, V. M., Russ. Pat. 181 075 (1966); *C. A.*, **65**, 8760 (1966).
194. PEDAYAS, V. M., LYUBARSKAYA, I. I. and MIROSHNICHENKO, V. A., *Maslob. Zhir. Prom.*, **30**, 30 (1964).
195. PIACENTE, A. N., U.S. Pat. 3 053 802 (1962, to Congoleum-Nairn, Inc.).
196. POHORILLA, M. J. and SMITH, R. K., U.S. Pat. 2 815 325 (1957, to Kendall Refining Co.).
197. Produits Chimiques Pechiney Saint-Gobain, Fr. Pat. 1 379 747 (1964).
198. Produits Chimiques Pechiney Saint-Gobain, Fr. Pat. 1 379 783 (1964).
199. QUIN, D. C., Brit. Pat. 892 723 (1962, to Distillers Co. Ltd.).
200. RAKSZAWSKI, J. F. and PARKER, W. E., *Carbon*, **2**, 53 (1964); *C.A.*, **61**, 14175 (1964).
201. REYNOLDS, W. W. and LOW, H., U.S. Pat. 3 007 873 (1961, to Shell Oil Co.).
202. REYNOLDS, W. W. and LOW, H., U.S. Pat. 3 014 869 (1961, to Shell Oil Co.).
203. REYNOLDS, W. W. and LOW, H., U.S. Pat. 3 014 870 (1961, to Shell Oil Co.).
204. RHONE-POULENC, S. A., Belg. Pat. 676 960 (1966); see also Fr. Pat. 1 469 464 (1967).
205. RHONE-POULENC, S. A., Belg. Pat. 683 399 (1966); see also Fr. Pat. 1 453 355 (1966).
206. RHONE-POULENC, S. A., Brit. Pat. 1 029 419 (1966); see also Belg. Pat. 656 082 (1965); Fr. Pat. 86 646 (1966); Ind. Pat. 96 644; Jap. Pat. 10942/67.
207. RIESSER, G. H. and SMITH, R. F., U.S. Pat. 3 281 462 (1966, to Shell Oil Co.); see also Brit. Pat. 1 012 556 (1965); Fr. Pat. 1 405 273 (1965); Neth. Appl. 6 405 626 (1964).
208. ROELEN, O. and ROTTIG, W., U.S. Pat. 3 255 238 (1966, to Ruhrchemie Aktiengesellschaft); see also Belg. Pat. 620 834 (1963); Can. Pat. 726 220 (1966).
209. ROGERS, D. T. and McNAB, J. G., U.S. Pat. 2 526 506 (1950, to Standard Oil Development Co.).
210. ROTTIG, W., Ger. Pat. 1 216 863 (1966, to Ruhrchemie Aktiengesellschaft); see also Fr. Pat. 1 397 639 (1965).
211. ROZANTSEV, E. G., KRINITSKAYA, L. A. and ROZYNOV, B. V., *Plasticheskie Massy*, **46** (1963); *C.A.*, **60**, 4302 (1964).
212. ROZANTSEV, E. G., KRINITSKAYA, L. A. and TROITSKAYA, L. S., *Khim. Prom.*, **20** (1964).

213. Ruhrchemie Aktiengesellschaft, Brit. Pat. 1 052 085 (1966).
214. RUSSELL, G. A., in *Peroxide Reaction Mechanisms*, J. O. EDWARDS, Ed., Interscience, New York, N.Y., 1962.
215. RUSSELL, J. L., Belg. Pat. 661 761 (1965, to Halcon International Inc.); see also Brit. Pat. 1 049 846 (1966); Fr. Pat. 1 455 077 (1966); Ind. Pat. 98 222; Neth. Appl. 6 503 884 (1965).
216. RUSSELL, J. L., Belg. Pat. 668 480 (1966, to Halcon International Inc.); see also Fr. Pat. 1 459 896 (1966); Neth. Appl. 6 510 612 (1966).
217. RUSSELL, J. L. and BECKER, M., Belg. Pat. 633 933 (1963, to Halcon International Inc.); see also Brit. Pat. 1 036 206 (1966); Can. Pat. 762 517 (1967); Fr. Pat. 1 397 141 (1965); So. Af. Pat. 63/2860.
218. RUSSELL, J. L. and BECKER, M., Can. Pat. 751 925 (1967, to Halcon International Inc.); see also Brit. Pat. 1 069 733 (1967); Fr. Pat. 1 447 962 (1966).
219. RUSSELL, J. L. and FEDER, J. B., Belg. Pat. 668 664 (1966, to Halcon International Inc.); see also Brit. Pat. 1 092 924 (1967); Fr. Pat. 1 462 269 (1966); Ger. Pat. 1 238 886 (1967); Neth. Appl. 6 510 957 (1966); So. Af. Pat. 65/4784.
220. RUSSELL, J. L. and OLENBERG, H., Ger. Pat. 1 182 228 (1964, to Halcon International Inc.); see also Belg. Pat. 635 301 (1963); Fr. Pat. 1 359 084 (1964).
221. RUSSELL, J. L. and OLENBERG, H., Brit. Pat. 1 017 214 (1966, to Halcon International Inc.).
222. SAFUE, H. *et al.*, Jap. Pat. 5920 (1957, to Tokai Electrode Manufacturing Co.).
223. SCIPIONI, A., *Ann. Chim. Applicata*, **39**, 311 (1949); *C.A.*, **46**, 720 (1952).
224. SCIPIONI, A., *Ann. Chim. Applicata*, **40**, 135 (1950); *C.A.*, **46**, 720 (1952).
225. SCIPIONI, A., *Ann. Chim. Applicata*, **40**, 143 (1950); *C.A.*, **46**, 720 (1952).
226. SCHOOTEN, J. V. and WIJGA, P. W. O., *Soc. Chem. Ind. (London)*, Monograph 13, 432 (1961).
227. SELWITZ, C. M., U.S. Pat. 2 969 380 (1961, to Gulf Research and Development Corp.).
228. SEN, B. N., *Gazz. Chim. Ital.*, **78**, 422 (1948); *C.A.*, **43**, 1729 (1949).
229. Shell Internationale Research Maatschappij N. V., Neth. Appl. 6 612 501 (1967).
230. Shell Internationale Research Maatschappij N. V., Neth. Appl. 6 612 575 (1967); see also Belg. Pat. 686 459 (1967); Brit. Pat. 1 088 810 (1967); Fr. Pat. 1 491 194 (1967).
231. SHIMAN, A. M. *et al.*, *Vses. Soveshch. po Sintetich. Zhirozamenitelyam, Poverkhnostnoaktivn. Veshchestvam i Moyushchim. Sredstvam, 3rd, Sb., Shebekino*, 197 (1965); *C.A.*, **66**, 4083 (1967).
232. SHOIKHET, P. A., TROTSENKO, M. A. and POLYAKOV, M. V., *Doklady Akad. Nauk S.S.S.R.*, **89**, 519 (1953); *C.A.*, **48**, 10327 (1954).
233. SIMONS, E. L. and CANNON, P., *Nature*, **210**, 90 (1966).
234. SMITH, P. and MEE, A., Brit. Pat. 939 534 (1963, to Imperial Chemical Industries, Ltd.).
235. Société anon. d'Innovations Chimique Sinnova on Sidac, Fr. Pat. 1 166 679 (1958); *C.A.*, **55**, 378 (1961).
236. Société des Usines Chimiques Rhone-Poulenc, Fr. Pat. 1 466 528 (1967); see also Belg. Pat. 690 989 (1967); Neth. Appl. 6 617 026 (1967).
237. SPIELMAN, M., *Am. Inst. Chem. Engrs. J.*, **10**, 496 (1964).
238. STAMICARBON, N. V., U.S. Pat. 3 287 423 (1966); see also Belg. Pat. 626 256 (1963); Brit. Pat. 970 450 (1964); Fr. Pat. 1 346 563 (1963).
239. STARKS, C. M. and KENNEDY, E. F., U.S. Pat. 3 346 614 (1967, to Continental Oil Co.).
240. STEEMAN, J. W. M. and VON DEN HOFF, J. P. H. U.S. Pat. 3 350 465 (1967, to Stamicarbon, N. V.); see also Belg. Pat. 627 084 (1963); Brit. Pat. 1 025 443 (1966); Fr. Pat. 1 346 607 (1963); Ger. Pat. 1 246 723 (1967).

241. STEINBERG, H., *Organoboron Chemistry*, Vol. 1, Interscience, New York, N.Y., 1964.
242. STIRNEMANN, E., MOSER, W. and PERL, S., U.S. Pat. 2 996 511 (1961, to Lonza Electric and Chemical Works, Ltd.); see also Brit. Pat. 757 958 (1956).
243. SUGERMAN, G., Belg. Pat. 667 462 (1966, to Halcon International Inc.); see also Fr. Pat. 1 463 011 (1967); Neth. Appl. 6 509 857 (1966).
244. SUGERMAN, G. and WINNECK, C. N., Belg. Pat. 668 432 (1966, to Halcon International Inc.).
245. TADENUMA, H., MURAKAMI, T. and MITSUSHIMA, H., Ger. Pat. 1 239 285 (1967, to Akita Petrochemicals Co. Ltd.).
246. THOMAS, J. R. and HARLE, O. L., U.S. Pat. 2 795 548 (1957, to California Research Corp.).
247. TOLAND, W. G., U.S. Pat. 3 251 888 (1966, to Chevron Research Co.).
248. TRAUTMAN, C. E., U.S. Pat. 2 813 830 (1957, to Gulf Research and Development Co.).
249. TRONOV, B. V. and SHEROVA, M. I., *Materialy XI Nauchnoi Konf. Prof.-Prepod. Sostova Biolog. Fakulteta, Kirgizskii Gos. Univ. Frunze*, 110 (1962).
250. TSYSKOVSKII, V. K. and SHCHEGLOVA, F. T., *Khim. Prom*, 325 (1961); *C.A.*, **55**, 27010 (1961).
251. TWIGG, G. H., *Angew. Chem. Intern. Ed., Engl.*, **4**, 886 (1965).
252. TYUTYUNNIKOV, B. N. and BUKHSHTAB, Z. I., *Izv. Vysshikh Uchebn. Zavedenii. Pishchevaya Tekhnol.*, 59 (1963).
253. UGINE-KUHLMANN, Belg. Pat. 694 748 (1967); see also Fr. Pat. 1 492 059 (1967); Neth. Appl. 6 702 996 (1967).
254. URIZKO, V. I. and POLYAKOV, M. V., *Ukr, Khim. Zh.*, **24**, 177 (1958); *C.A.*, **52**, 19899 (1958).
255. VAN MOURIK, J. and VAN BEVEREN, J., unpublished results quoted in reference 248.
256. VELEA, I. *et al.*, *Materiale Plastice (Rumania)*, **2**, 199 (1965).
257. VELEA, I. *et al.*, *Rev. Chim. (Bucharest)*, **14**, 435 (1963).
258. VELEA, I. *et al.*, Brit. Pat. 1 066 214 (1967, to Ministerul Industriei Petrolului Si Chimiei).
259. VELEA, I. *et al.*, *Materiale Plastice (Rumania)* 3, 9 (1966); *C.A.*, **65**, 15594 (1966).
260. VESELOV, V. V. and PEDAYAS, V. M., *Tr. Nauchn.-Issled. Inst. Sintetich Zhirozamenitelei i Moyushchikh Sredstv.*, 25 (1961).
261. VESELOV, V. V., PEDAYAS, V. M. and ALEKSEYENKO, E. V., *Tr. Nauchn.-Issled. Inst. Sintetich. Shirozan. i Moyuschikh Sredsty*, 33 (1961).
262. VESELOV, V. V., PEDAYAS, V. M., KASHCHEEVA, E. D. and TERESHCHENKO, E. E., *Tr. Nauchn.-Issled Inst. Sintetich. Zhirozamenitelei i Moyushchikh Sredstv.*, 39 (1961); *C.A.*, **59**, 15097 (1963).
263. VESELOV, V. V., PEDAYAS, V. M. and LYUBARSKAYA, I. I., *Tr. Nauchn.-Issled. Inst. Sintetich. Zhirozamenitelei i Moyushchikh Sredstv.*, 30 (1961).
264. VESELOV, V. V., PEDAYAS, V. M. and TERESHCHENKO, E. E., *Tr. Nauchn.-Issled. Inst. Sintetich. Zhirozamenitelei i Moyushchikh Sredstv.*, 45 (1961); *C.A.*, **59**, 3757 (1963).
265. VESELOV, V. V., SIPEEVA, Z. V. and BRUSENINA, L. I., *Tr. Nauchn.-Issled. Inst. Sintetich. Zhirozamenitelei i Moyushchikh Sredstv.*, 78 (1961).
266. VESELOV, V. V., SIPEEVA, Z. V. and BRUSENINA, L. I., *Tr. Nauchn.-Issled. Inst. Sintetich. Zhirozamenitelei i Moyushchikh Sredstv.*, 5 (1961).
267. WALLING, C., *Free Radicals in Solution*, John Wiley & Sons, Inc., New York, N. Y., 1957, p. 421.
268. WARING, A. J. and HART, H., *J. Am. Chem. Soc.*, **86**, 1454 (1964).
269. WENZEL, W., Ger. Pat. 884 043 (1953, to Badische Anilin and Soda-Fabrik).

270. WINNICK, C. N., Brit. Pat. 996 791 (1965, to Halcon International Inc.); see also Ger. Pat. 1 158 963 (1963).
271. WINNICK, C. N., Brit. Pat. 996 792 (1965, to Halcon International Inc.); see also Fr. Pat. 1 359 064 (1964).
272. Winnick, C. N., Can. Pat. 716 097 (1965, to Halcon International Inc.); see also Brit. Pat. 997 348 (1965); Fr. Pat. 1 356 221 (1964).
273. WINNICK, C. N., Can. Pat. 751 931 (1967, to Halcon International Inc.); see also Belg. Pat. 635 117 (1963); Brit. Pat. 1 002 083 (1965); Fr. Pat. 1 356 222 (1964).
274. WINNICK, C. N., U.S. Pat. 3 243 449 (1966, to Halcon International Inc.); see also Belg. Pat. 626 496 (1963); Brit. Pat. 1 031 661 (1966); Fr. Addn. 85 255 (1965); Ind. Pat. 97234.
275. WINNICK, C. N., Can. Pat. 695 850 (1964, to Halcon International Inc.); see also Belg. Pat. 633 934 (1963); Brit. Pat. 1 008 314 (1965); Fr. Pat. 1 352 116 (1964).
276. WINNICK, C. N., Belg. Pat. 651 625 (1965, to Halcon International Inc.); see also Brit. Pat. 1 058 155 (1967); Ind. Pat. 88 201; So. Af. Pat. 63/2401.
277. WINNICK, C. N. and SCHLOSSMAN, I., Belg. Pat. 685 981 (1967, to Halcon International Inc.); see also Fr. Pat. 1 495 078 (1967); Neth. Appl. 6 612 133 (1967).
278. WOODS, W. G. and WHITEN, J. G., Paper presented at the *154th National A.C.S. Meeting, Chicago, Illinois*, September 1967; see Division of Organic Coatings and Plastics Chemistry preprints, p. 210.
279. YAMAZAKI, K. and NAGATA, H., *Rept. Inst. Sci. and Technol. Univ. Tokyo*, **4,** 11 (1950); *C.A.*, **45,** 6026 (1951).
280. YANG, TAO-SHENG, TIEN, Y-L and CHANG, F-Y, *Jan Liao Hsüch Pao*, **4,** 210 (1959); *C.A.*, **54,** 24336 (1960).
281. YOULE, P. V., GARDNER, C. and NEW, R. G. A., Brit. Pat. 1 035 624 (1966, to Imperial Chemical Industries Ltd.); see also Fr. Pat. 1 345 832 (1963).
282. ZAHHARKIN, L. I. *et al.*, Russ. Pat. 145 579 (1963).
283. ZAJIC, J. E. and DUNLAP, W. J., U.S. Pat. 3 274 074 (1966, to Kerr-McGee Oil Industries).
284. ZAMYSHLYAEVA, A. M. and ZOLOTAREV, N. S., *Tr. Nauchn.-Issled. Inst. Sintetichn. Zhirozamenitelei i Moyushchikh Sredstv.*, 63 (1960); *C.A.*, **60,** 7040 (1964).
285. ZAMYSHLYAEVA, A. and ZOLOTAREV, N. S., *Tr. Nauchn.-Issled. Inst. Sintetichn. Zhirozamenitelei i Moyushchikh Sredstv.*, 59 (1960); *C.A.*, **60,** 7040 (1964).
286. ZATHURECKY, L. *et al.*, *Cesk. Farm.*, **14,** 2 (1965).
287. ZOLOTOREVA, K. A., TARANENKO, A. S. and MIKHAILOV, V. V., *Sintez i Issled. Effektivh. Stabilizatorov dlya Polimern. Materialov. Sb.*, *Voronezh*, 33 (1964); *C.A.*, **65,** 18768 (1966).
288. ZUMSTEIN, F. and KOENIGSBERGER, R., Ger. Pat. 1 224 291 (1966, to Rhone-Poulenc S. A.); see also Fr. Pat. 1 366 078 (1964).

2

NEIGHBORING-GROUP EFFECTS OF BORON IN ORGANOBORON COMPOUNDS

by DONALD S. MATTESON

Department of Chemistry, Washington State University, Pullman, Washington 99163

CONTENTS

I. INTRODUCTION

The neighboring-group effects to be discussed all depend on the ability of the vacant orbital of the trivalent boron atom to accept electrons. Two classes of interactions are possible. One is π-bonding between the vacant p-orbital of boron and a filled p-orbital on an adjacent atom. The other involves σ-bonding of the acidic boron atom to a nearby atom having an unshared electron pair, with interactions between boron and a nucleophilic atom both bonded to the same carbon being of primary interest. Both classes of interaction may occur simultaneously. This review is primarily concerned with the influence of these two types of interactions on reaction mechanisms.

The utility of boron to the organic chemist as a tool for providing insight into structure and reaction mechanisms rests on its close relationship to carbon. For example, trialkylboranes, R_3B, are isoelectronic with carbonium ions, R_3C^+, and boronic acids, $RB(OH)_2$, are isoelectronic with carboxylic acids, RCO_2H. Boron is the only metalloidal element which can participate in strong p-π-bonding to carbon, without the theoretical complication of d-π-bonding. The close similarities frequently allow the organic chemist to view the boron atom as a modified carbon atom, though the carborane chemist might say that carbon is just boron with an extra nuclear proton.

II. THEORETICAL DESCRIPTION OF CARBON–BORON π-BONDING

A. *Resonance Description*

The qualitative character of carbon–boron π-bonding can be clearly and simply illustrated by means of resonance structures. Boron is less electronegative than carbon and therefore is electron-donating, except that the vacant p-orbital can act only as a π-electron acceptor. This may be illustrated with vinyldimethylborane, which is isoelectronic with the vinyldimethylcarbonium ion.

$$CH_2{=}CH{-}B(CH_3)_2 \longleftrightarrow \overset{+}{C}H_2{-}CH{=}\overset{-}{B}(CH_3)_2$$

$$CH_2{=}CH{-}\overset{+}{C}(CH_3)_2 \longleftrightarrow \overset{+}{C}H_2{-}CH{=}C(CH_3)_2$$

However, more chemical intuition about the boron compound may be gained by comparing it with vinyldimethylamine, which shows the same pattern of opposing inductive and resonance effects but with opposite direction to those of the boron compound.

$$CH_2{=}CH{-}N(CH_3)_2 \longleftrightarrow \overset{-}{C}H_2{-}CH{=}\overset{+}{N}(CH_3)_2$$

Just as the net chemical effect of the nitrogen atom in enamines or aniline is that of electron donation, the effect of boron in vinylboron and arylboron compounds should be electron withdrawal.

The foregoing interpretation is essentially the same as that presented over 30 years ago by Branch and co-workers.[2, 5, 85] Measurements of ionization constants of substituted benzeneboronic acids and of dihydroxyboronosubstituted benzoic acids confirmed the chemical significance of the resonance effect. A Hammett treatment[32] of the data of Bettman, Branch and Yabroff[2] for p-fluoro- and p-ethoxybenzeneboronic acid shows good correlation with normal σ values, but not with σ^+.[8] Thus, the decrease in resonance interaction when the boronic acid is converted to its anion is about the same fraction of the total interaction as when a carboxylic acid is converted to its anion.

Branch's group formulated boronic acid ionization as loss of a proton, but this is now believed to be incorrect. Instead, the boron atom acts as a Lewis acid and adds hydroxide ion.

$$OH^- + R—B(OH)_2 \rightleftharpoons R—B(OH)_3^-$$

The finding by Bettman, Branch and Yabroff[2] that the acidity of o-nitro-benzeneboronic acid is anomalously low is readily rationalized by the modern view on steric grounds. This interpretation was introduced by McDaniel and Brown,[59a] and is now widely accepted, e.g. in mechanistic studies of deboronation by Kuivila and co-workers.[35, 36] The ionization of boronic acids is thus electronically analogous to the mechanism of base-catalyzed ester hydrolysis, which is a classical example of a reaction that correlates well in the Hammett treatment.[32]

B. Molecular Orbital Calculations

1. Vinylboron Compounds

(a) *Introduction.* Simple molecular orbital calculations are useful for interpretation of the chemistry of vinylboron compounds. The qualitative predictions are much the same as those based on resonance structures or just organic chemists' intuition. However, the molecular orbital method does provide semiquantitative correlations of various chemical and physical properties, including some that cannot be handled at all by resonance. Also, the author once risked publishing some general predictions of the chemistry of vinylboronic esters based on molecular orbital calculations,[38] and these predictions survived subsequent experimental tests reasonably well.[40, 54, 58]

A detailed understanding of the Hückel molecular orbital method is not necessary for the discussion which follows. However, readers wishing to clarify certain points, or to obtain explicit directions for solving the equations, may refer to a standard text on the subject such as that by Streitwieser.[80]

The energy units used in Hückel calculations are the Coulomb integral on carbon, α_C, and the carbon–carbon bond integral, β_{CC}. The Coulomb integral α_C is in principle the energy released when a free electron is placed in a vacant p-orbital, as in the reaction $CH_3^+ + e^- \longrightarrow CH_3^{\cdot}$. The bond integral β_{CC} is the additional energy released when the electron is delocalized into an adjacent p-orbital, or half the π-bond energy of ethylene. Both α_C and β_{CC} vary somewhat with structure but are treated as constants in the simple calculations. In order to introduce a heteroatom such as boron into the calculations, reasonable values must be assigned to the boron–carbon bond integral, β_{CB}, and the Coulomb integral, α_B.

The value of β_{BC} depends on orbital overlap, which ought to be about the same as for a carbon–nitrogen bond, or near $0.9\beta_{CC}$. In vinylboron compounds, the boron–carbon single bond is compared with a carbon–carbon double bond, which requires a correction factor of about 0.8, leading to $\beta_{BC} = 0.7\beta_{CC}$.

Since boron is less electronegative than carbon, α_B must be smaller than α_C, or if α_B is expressed as $\alpha_C + h_B\beta_{CC}$ in the customary fashion for heteroatoms,[80] h_B is negative. As a first rough guess, $h_B = -h_N$, the corresponding parameter for nitrogen, which suggests that rounded off to a whole number $h_B = -1$. In practice, the best value of h_B is the one which gives the best correlation of experimental data, so that attempts to evaluate α_B tend to involve somewhat circular reasoning. However, adjustment of α_B to give the best correlation of one set of data should yield a value which works well for other boron compounds and other data. The value $h_B = -1$, which initially seemed about right for describing a vinylboronic ester,[38] also gives good results with vinylboranes.[22]

(b) *Vinylboranes.* Because of their inherent theoretical simplicity, dimethylvinylborane (I), methyldivinylborane (II), and trivinylborane (III) will

$(CH_3)_2BCH{=}CH_2$ $CH_3B(CH{=}CH_2)_2$ $B(CH{=}CH_2)_3$
     (I)   (II)   (III)

be discussed first. Good and Ritter[22] used the parameters $\alpha_B = \alpha_C - 1\beta_{CC}$ and $\beta_{BC} = 0.7\beta_{CC}$ to calculate the energy levels and wave functions listed (with corrections and revisions) in Table 1. The derived delocalization energies, ultraviolet absorption energies, π-electron densities, and π-bond orders are summarized in Table 2.

The ultraviolet spectra of the vinylboranes correlate well with the calculated differences in energy between the highest occupied level and the lowest unoccupied level. For ethylene the energy level difference is 2β and the observed λ_{max} is at 163 mμ. Assuming a linear energy correlation, the calculated λ_{max} for dimethylvinylborane (I) is 192 mμ, that for methyldivinylborane (II) is 225 mμ, and for trivinylborane (III), 238 mμ. The observed values of λ_{max} in the gas phase are, respectively, 196, 220 and 234 mμ.[22]

TABLE 1. VINYLBORANE ENERGY LEVELS AND WAVE FUNCTION COEFFICIENTS

Compound	Energy level	Electrons in level	Coefficients						
			B	C_1	C_2	C_1'	C_2'	C_1''	C_2''
$Me_2BC_2H_3$ (I)	$\alpha-1.545\beta$	0	(not useful and not reported)						
	$\alpha-0.577\beta$	0	0.638	0.385	−0.668	—	—	—	—
	$\alpha+1.122\beta$	2	0.239	0.725	0.646	—	—	—	—
$MeB(C_2H_3)_2$ (II)	$\alpha-1.793\beta$	0							
	$\alpha-1.000\beta$	0							
	$\alpha-0.449\beta$	0	0.593	0.233	−0.519	0.233	0.519	—	—
	$\alpha+1.000\beta$	2	0	0.500	0.500	−0.500	−0.500	—	—
	$\alpha+1.242\beta$	2	0.323	0.521	0.419	0.512	0.419	—	—
$B(C_2H_3)_3$ (III)	$\alpha-1.989\beta$	0							
	$\alpha-1.000\beta$	0 ⎫	(degenerate pair due to symmetry)						
	$\alpha-1.000\beta$	0 ⎭							
	$\alpha-0.370\beta$	0	0.555	0.167	−0.450	0.167	−0.450	0.167	−0.450
	$\alpha+1.000\beta$	2	0	0.500	0.500	−0.500	−0.500	0	0
	$\alpha+1.000\beta$	2	0	0.289	0.289	0.289	0.289	−0.577	−0.577
	$\alpha+1.359\beta$	2	0.383	0.430	0.316	0.430	0.316	0.430	0.316

TABLE 2. CALCULATED VINYLBORANE PROPERTIES

Compound	Delocalization energy	U.v. abs. energy	π-electron density			π-bond order	
			B	C_1	C_2	$B-C_1$	C_1-C_2
$Me_2BC_2H_3$	0.244β	1.699β	0.11	1.05	0.84	0.35	0.94
$MeB(C_2H_3)_2$	0.484β	1.449β	0.21	1.04	0.85	0.34	0.94
$B(C_2H_3)_3$	0.718β	1.370β	0.29	1.04	0.87	0.33	0.94

The vinylborane delocalization energies represent the calculated difference in π-bond energy between the boron compound and a corresponding number of isolated ethylene molecules. For calculating delocalization energies, a suitable value for β is 16.5 kcal/mole. The calculated delocalization energy for dimethylvinylborane (I) is 4.0 kcal/mole, that for methyldivinylborane (II) is 8.0, and for trivinylborane (III), 11.8. The relative slowness of disproportionation of the methylvinylboranes is consistent with stabilization energies of this magnitude.[22]

The change in calculated π-electron density on boron is very nearly linear at 0.1 electron per vinyl group. This correlates well with the ^{11}B n.m.r. spectra of trimethylborane, the methylvinylboranes, and trivinylborane, which show a progression of 5 parts per million toward higher field with each additional vinyl group. Since the inductive effect of the vinyl group is electron-withdrawing relative to methyl, the net electron-donating effect must be due to π-electron donation.[22]

The carbon–carbon π-bond order for all three of the vinylboranes is calculated to be close to that of butadiene. The infrared stretching frequencies of 1597 cm^{-1} for butadiene and 1603–5 cm^{-1} for the vinylboranes are in reasonable agreement with the calculation.[22]

All simple Hückel calculations, including those in the preceding paragraphs, greatly underestimate ultraviolet transition energies, overestimate delocalization energies, and overestimate electric charges on atoms. The u.v. spectral correlation amounts to assigning β, the "variable constant", a value of 60 kcal/mole, which is about twice the value it would have according to the definition used when it was fed into the equations at the beginning. For calculating delocalization energies, β must be reduced to 15–20 kcal/mole. Linear correlations of calculated with observed quantities are the best results that can be expected with such an approximate method of calculation.

(c) *Vinylboronic ester.* Attention will now be turned to dibutyl vinylboronate (IV).† Molecular orbital calculations are complicated by the oxygen

$$CH_2=CH-B(OC_4H_9)_2$$
$$(IV)$$

atoms adjacent to boron. Values must be assigned to the parameters α_O and β_{BO}, which together with β_B and β_{BC} add up to four rather arbitrary assumptions in one calculation. Fortunately, the calculated properties do not depend very strongly on the parameters assigned to oxygen. In fact, except for the π-electron density of boron, the calculated properties of the C–C–B unit are about the same whether the oxygen atoms are included or omitted. Thus, the calculation for dimethylvinylborane is a good first approximation for dibutyl vinylboronate.

For the first set of calculations on vinylboronic acid or its equivalent, dibutyl vinylboronate (IV), h_O was assigned the value $+3$, or $\alpha_O = \alpha_C + 3\beta_{CC}$, and β_{BO} was assumed equal to β_{CC}.[38] An auxiliary inductive parameter[80]

---

† This was originally named "dibutyl ethyleneboronate" in conformity with *Chemical Abstracts* nomenclature, but "dibutyl vinylboronate" seems clearer.

(the fifth "fudge factor"!) equal to $0.1h_B$ was added to α_C for the carbon adjacent to boron. The value $0.7\beta_{CC}$ was assigned to β_{BC}. Then h_B was given values ranging from -0.5 to -2.0 and the calculations were carried out. Since the most significant experiment for correlation with the calculations involved addition of a free radical to dibutyl vinylboronate, a quantity called the radical stabilization energy (related to the atom localization energy) was calculated. This was defined as the difference in energy between the reactions

$$R^{\cdot} + CH_2\!\!=\!\!CH_2 \longrightarrow RCH_2CH_2^{\cdot}$$

and

$$R^{\cdot} + CH_2\!\!=\!\!CH\!\!-\!\!B(OBu)_2 \longrightarrow R\!\!-\!\!CH_2\!\!-\!\!\dot{C}H\!\!-\!\!B(OBu)_2$$

Another customary index of reactivity toward radicals, the free valence, was also calculated. $\left(\text{This is defined as } 3+\sqrt{3} \text{ minus the total } \sigma\text{- and cal-}\right.$ culated π-bond orders around a given carbon atom.$\left.\right)$ The calculated results are summarized in Table 3.

TABLE 3. CALCULATED PROPERTIES OF VINYLBORONIC ACID

h_B	Radical stabilization	Free valence at C_2	π-electron density	
			C_1	C_2
-0.5	0.129β	0.787	1.021	0.879
-1.0	0.098β	0.770	0.997	0.932
-1.5	0.076β	0.759	0.973	0.977
-2.0	0.057β	0.754	0.946	1.015

In the radical-catalyzed addition of carbon tetrachloride to dibutyl vinylboronate, the intermediate radical (V) shows a strong preference for reaction with dibutyl vinylboronate rather than carbon tetrachloride.[38]

$Cl_3C^{\cdot} + CH_2\!\!=\!\!CH\!\!-\!\!B(OBu)_2$
    (IV)

$\longrightarrow Cl_3CCH_2\dot{C}HB(OBu)_2$
    (V)

$\xrightarrow[k_1]{CCl_4} Cl_3CCH_2CHB(OBu)_2 + Cl_3C^{\cdot}$
$\quad\quad\quad\quad\quad\quad\quad\quad\quad |$
$\quad\quad\quad\quad\quad\quad\quad\quad\quad Cl$

$\xrightarrow[k_2]{IV} Cl_3CCH_2CHB(OBu)_2$
$\quad\quad\quad\quad\quad\quad |$
$\quad\quad\quad\quad\quad\quad CH_2\dot{C}HB(OBu)_2$
$\quad\quad\quad\quad\quad\quad\quad\quad (VI)$

9*

The chain transfer constant C_1, defined as k_1/k_2, is about 3.3×10^{-3}. The corresponding constant for styrene is also low, 6×10^{-4}. These contrast with that for propylene, which is approximately 1. Electron-withdrawing substituents also lead to low transfer constants with carbon tetrachloride, but the inductive effect of the dibutoxyboryl group was expected to be electron-donating. It was therefore concluded that the delocalization of the unpaired electron in the radical (V) from carbon to boron must be substantial. The delocalization energy would be lost in the reaction of (V) with carbon tetrachloride, but the reaction of (V) with dibutyl vinylboronate would lead to a similarly stabilized radical (VI) and thus be favored. It was suggested that the delocalization energy was a substantial fraction of that of a benzyl radical.

One factor was partially overlooked by the author at that time. Delocalization of the odd electron from carbon to boron leaves a partial positive charge on the carbon. This charge would help reduce the reactivity of the radical towards carbon tetrachloride. Also, the inductive effect of the dibutoxyboryl group is not as electron-donating as was originally thought. (See Section III, B and IV, B, 2 of this chapter.) The benzyl radical derived from styrene would differ in having no charge due to π-electron delocalization, and the low transfer constant with carbon tetrachloride would be due almost solely to delocalization energy with a small contribution from the inductive effect of the phenyl group.

For comparison with published calculations on styrene and benzyl radical with all β_{CC}'s assumed equal, vinylboronic acid was recalculated with β_{BC} set at $0.9\beta_{CC}$ and h_B at -1.0. Free valences at the terminal carbon were calculated to be 0.821 in styrene, 0.796 in vinylboronic acid, and 0.732 in ethylene. Radical stabilization energies were 0.296β for styrene and 0.128β for vinylboronic acid relative to zero for ethylene. From these and other results, the author concluded that h_B should not be set more negative than -1 for qualitative correlation of calculated and observed properties.[33] It may be noted that with any reasonable parameters for boron, the radical stabilization energy of vinylboronic acid is much less than that of styrene. This is consistent with the polarity considerations outlined in the preceding paragraph, but weakens the argument for setting a lower limit on h_B.

Although the calculated electron density shifts from being greater at C_1, adjacent to boron, to being greater at C_2 as the value of h_B is reduced to -1.5 and below in Table 3, this is a secondary effect resulting from use of the auxiliary inductive parameter, which is a questionable quantity. Inasmuch as the experimental evidence seemed to point to a value of h_B no lower than -1.0, predictions of the chemistry of dibutyl vinylboronate were made assuming that C_2 would bear some positive charge and C_1 would be essentially neutral. It was suggested that electrophiles should preferentially attack C_1. Nucleophiles, if any could be found that did not react exclusively at boron, should attack C_2. The positive character of C_2 and the lowered carbon–carbon π-bond order suggested that dibutyl vinylboronate should act as a dienophile.[38]

In fact, dibutyl vinylboronate did turn out to be a moderately active dienophile[58] and to show considerable activity in the electronically related cycloaddition of diazo compounds.[40] The nucleophilic carbon atom of the diazo compound was shown to attack the terminal carbon of the vinylboron compound, as predicted. (See Section V, A, 2.) In order to provide better correlation with other types of compounds, the molecular orbital calculations were repeated with parameters close to those suggested by Streitwieser. Only minor changes from the previous calculations resulted. Ether-type oxygen, which contributes two electrons to the π-system, was assigned the parameters $h_{\ddot{O}} = 2.0$ and $\beta_{B-O} = 0.6\beta_{C=C}$. For comparison purposes, acrylic acid was calculated with the carbonyl oxygen parameters $h_{\ddot{O}} = 1.0$ and $\beta_{C=O} = \beta_{C=C}$. Other parameters used included $\beta_{C-O} = \beta_{C-C} = 0.8\beta_{C=C}$, $\beta_{B-C} = 0.7\beta_{C=C}$, carbonyl carbon $h_{C_0} = 0.2$, and carbon adjacent to boron $h_{C_1} = 0.1h_B$. The value of h_B was varied from -1.0 to -0.5. Some calculated properties of acrylic acid and vinylboronic acid are compared in Table 4.

TABLE 4. CALCULATED PROPERTIES OF ACRYLIC AND VINYLBORONIC ACID

			Energies (units of β)				
h_{C_0} or h_B	Charge at C_2	Free valence at C_2	Lowest unfilled orbital	u.v. abs.	Delocalization	Radical stabilization, C_2	Anion stabilization, C_2
$+0.20$[a]	$+0.146$	0.817	0.428	1.428	0.323[b]	0.202	0.70
-0.50	$+0.147$	0.804	0.487	1.569	0.212	0.163	0.50
-0.75	$+0.115$	0.790	0.599	1.658	0.160	0.140	0.41
-1.00	$+0.087$	0.777	0.692	1.731	0.113	0.121	0.33

[a] This line refers to acrylic acid, the others to vinylboronic acid.
[b] Values in this column are total π-bond energy minus the sum of the π-bond energy of ethylene and the carboxyl or dihydroxyborono group separately.

The rate of reaction of ethyl diazoacetate with ethyl acrylate is about 3.5 times that with dibutyl vinylboronate at 25°C. Diphenyldiazomethane favors ethyl acrylate by a factor of 6. The rates of reaction of the diazo compounds with a simple olefin, 1-octene, are too small to measure accurately but must be no greater than 1/500 to 1/1000 those with vinyl-boronic ester. Conjugated olefins such as styrene also react relatively slowly.[40] From these results, it appears that the properties of the double bond in the vinylboronic ester resemble those of the acrylic ester more closely than they resemble those of a simple or conjugated olefin. Thus, the chosen range of h_B, -1.0 to -0.5, yields qualitatively appropriate calculated properties.[40]

More accurate correlation of the properties of vinylboronic and acrylic esters would require knowledge of the appropriate reactivity index and its relation to the rates of reaction of diazo compounds. The author has suggested that the energy of the lowest unoccupied molecular orbital of the dienophile may be the best simple choice.[40] This is based on the assumption that the transition state resembles a π-complex in structure and energy. Other things being equal, the closer in energy the lowest unoccupied orbital of the π-electron acceptor and the highest occupied orbital of the π-electron donor, the stronger will be the bonding between them. Thus, a low-lying unfilled π-orbital is associated with high reactivity toward dienes and diazo compounds. From the comparisons in Table 4, $h_B = -0.5$ yields a value (0.487β) suitably close to that for acrylic acid (0.428β) and well below those for conjugated olefins $(0.7\text{--}0.8\beta)$ or ethylene (1.0β). With $h_B = -0.75$, the energy of the lowest unfilled orbital (0.599β) is a bit on the high side.

Other possible reactivity indices include the calculated positive charge and free valence at C_2. The charge at C_2 should be smaller in the vinyl-boron compound than in acrylic acid, and the result with $h_B = -0.5$, which yields equal charges, is inappropriate. The charge seems reasonable when h_B is -0.75, too low when h_B is -1.0. The previously mentioned radical reactions point toward values of h_B in the -0.5 to -1.0 range. The ultraviolet spectrum of the cyclic vinylboronic ester of 2-methyl-pentane-2,4-diol shows a maximum at 198.5 mμ,[84] which is consistent with the calculated u.v. absorption energy of 1.66β when h_B is near -0.75.

From the foregoing discussion, it is apparent that good correlations of calculated and observed properties of vinylboronic ester are obtained with h_B in the range -0.5 to -1.0, but that neither the available data nor the inherent approximations in the calculations would justify any attempt to specify h_B more narrowly than this. Energy levels and wave functions for vinylboronic acid, $CH_2{=}CH{-}B(OH)_2$, and the system $C{-}B(OH)_2$

(cation, radical, or anion) resulting from reaction at the terminal carbon are listed in Table 5. The calculations with $h_B = -0.5$, -0.75, and -1.0 (previously unpublished data obtained by the author) are given, and results for intermediate values of h_B may be interpolated.

TABLE 5. ENERGY LEVELS AND WAVE FUNCTIONS FOR C—C—BO$_2$ and C—BO$_2$

Compound and h_B	Energy level, n in $\alpha + n\beta$	Coefficient of				
		O_1	O_2	B	C_1	C_2
C—BO$_2$ h_B -0.5	-1.16633	-0.15656	-0.15656	$+0.82619$	-0.51807	—
	$+0.33616$	-0.16913	-0.16913	$+0.46901$	$+0.85019$	—
	$+2.00000$	$+0.70711$	-0.70711	0	0	—
	$+2.28017$	$+0.66849$	$+0.66849$	$+0.31215$	$+0.09377$	—
C—C—BO$_2$ h_B -0.5	-1.42991	$+0.10773$	$+0.10773$	-0.61586	$+0.63345$	-0.44300
	-0.48669	-0.16675	-0.16675	$+0.69108$	$+0.29899$	-0.61433
	$+1.08165$	-0.13612	-0.13612	$+0.20835$	$+0.70411$	$+0.65096$
	$+2.00000$	$+0.70711$	-0.70711	0	0	0
	$+2.28496$	$+0.66488$	$+0.66488$	-0.31577	$+0.11650$	$+0.50599$
C—BO$_2$ h_B -0.75	-1.34944	-0.15328	-0.15328	$+0.85565$	-0.46997	—
	$+0.26705$	-0.14861	-0.14861	$+0.42923$	$+0.87840$	—
	$+2.00000$	$+0.70711$	-0.70711	0	0	—
	$+2.25739$	$+0.67411$	$+0.67411$	$+0.28918$	$+0.08679$	—
C—C—BO$_2$ h_B -0.75	-1.54676	$+0.11737$	$+0.11737$	-0.69380	$+0.58850$	-0.38047
	-0.59910	-0.14538	-0.14538	$+0.62976$	$+0.38498$	-0.64260
	$+1.05932$	-1.12185	-0.12185	$+0.19104$	$+0.70269$	$+0.66335$
	$+2.00000$	$+0.70711$	-0.70711	0	0	0
	$+2.26154$	$+0.67100$	$+0.67100$	$+0.29249$	$+0.10808$	$+0.04779$
C—BO$_2$ h_B -1.0	-1.54284	-0.14895	-0.14895	$+0.87952$	-0.42670	—
	$+0.20507$	-0.13123	-0.13123	$+0.39258$	$+0.90080$	—
	$+2.00000$	$+0.70711$	-0.70711	0	0	—
	$+2.23777$	$+0.67867$	$+0.67867$	$+0.26894$	$+0.08053$	—
C—C—BO$_2$ h_B -1.0	-1.68765	$+0.12401$	$+0.12401$	-0.76216	$+0.53613$	-0.31768
	-0.69234	-0.12492	-0.12492	$+0.56056$	$+0.46053$	-0.66518
	$+1.03861$	-0.10975	-0.10975	$+0.17585$	$+0.70027$	$+0.67424$
	$+2.00000$	$+0.70711$	-0.70711	0	0	0
	$+2.24138$	$+0.67600$	$+0.67600$	$+0.27195$	$+0.10045$	$+0.04481$

From the coefficients listed in Table 5, the following molecular diagrams may be constructed for vinylboronic acid.

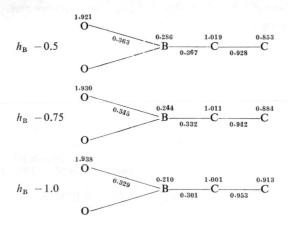

The numbers over the atoms are calculated π-electron densities, and the numbers under the bonds are π-bond orders. For comparison, the molecular diagram for acrylic acid is shown.

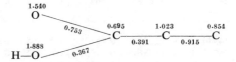

The kinds of changes in electron density which occur when a reagent attacks the terminal carbon of a vinylboronic ester are indicated by the following molecular diagrams, all based on $h_B = -0.75$.

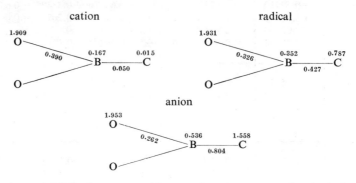

Having established a reasonable set of parameters and calculations for dibutyl vinylboronate, further predictions of its chemistry may be made. The first of these is that anions are unlikely to add to the double bond.

Although the charge density at the terminal carbon atom is perhaps nearly as positive as that of ethyl acrylate, the anion stabilization energies are not comparable. Dibutyl vinylboronate with $h_B = -0.75$ has a calculated anion stabilization of 0.41β, only slightly greater than that of butadiene, which is 0.36β when the simplest model is used, and considerably less than the 0.70β calculated for ethyl acrylate. As can be seen from the molecular diagram for the boron anion, the negative charge is carried entirely by the boron and carbon atoms. However, the anion from acrylic ester shows a calculated negative charge of 0.859 on the carbonyl oxygen, an energetically more favorable atom for it.

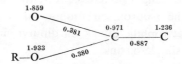

It is not only the relative difficulty of forming the anion from vinyl-boronic ester which determines its chemistry, but the relative ease of attack of anions at the boron atom. Only 0.64β of π-bond energy is sacrificed by adding an anion to boron, instead of the 1.59β lost when the terminal carbon is attacked. Although the carbon–boron σ-bond formed by attack at boron is probably weaker than a carbon–carbon σ-bond, the 0.95β difference in π-bond energy, nearly half a bond worth, should more than compensate. In contrast, acrylic ester sacrifices 2.02β of π-bond energy when the anion is added to the carbonyl carbon, only 1.30β when the terminal carbon is attacked.

Before calculating the hopelessness of it, the author tried to achieve conjugate addition to a boronic ester by sterically hindering attack at boron. However, di-t-butyl vinylboronate and mesitylmagnesium bromide gave a good yield of B-mesityl-B-vinyl-B-alkoxyborane, with no evidence for conjugate addition.[49] The generation of boron-substituted carbanions or α-boronalkylmetallic compounds by another route will be discussed in Section V, B of this chapter.

Electrophilic cations should attack the double bond of dibutyl vinyl-boronate, the question being which carbon is most susceptible. On the basis of electron densities, it was originally predicted that the α-carbon would be attacked preferentially,[38] but π-bond energies of the carbonium ion product favor attack at the β-carbon by a scant 0.03β, in accord with the calculated electron density of 0.015 at the carbonium ion center donated by the dialkoxyborono group. The inductive effect of boron should also help stabilize an adjacent carbonium ion. Thus, if the electrophilic addition is highly exothermic, or if it involves prior attack at boron as does hydrogen peroxide deboronation, the α carbon should be attacked, but if bonding to boron is not involved and the transition state occurs relatively late along the reaction coordinate, β attack should be favored. It should also be noted that addition of an electrophile to the thermodynamically more favorable β carbon costs 2.13β of π-bond energy, which is 0.13β less favorable than addition to ethylene. This slight deactivating effect of the π-bond system is opposed by the activating inductive effect of boron, and the net result cannot be predicted. Several types of experimental results, summarized in Section IV of this chapter, indicate that electrophiles may show slight preference for either the α- or the β-carbon, that formation of the α-boronocarbonium ion is somewhat facilitated, and that rates of electrophilic attack on dibutyl vinylboronate are comparable to those on simple olefins.

(d) *Vinylborinic esters.* Molecular orbital calculations were carried out on butyl divinylborinate, or B,B-divinyl-B-butoxyborane (VII),[39] and on B-vinyl-B-phenyl-B-butoxyborane (VIII)[49] with the older set of parameters, $h_B = -1$, $h_O = +3$, and $\beta_{BO} = \beta_{CC}$. For the phenyl group, β_{CC} was set at $0.84\beta_{CC}$ for the vinyl group.

$$(CH_2{=}CH)_2BOC_4H_9 \qquad\qquad C_6H_5{-}\underset{\underset{\displaystyle OC_4H_9}{|}}{B}{-}CH{=}CH_2$$

(VII)   (VIII)

The calculated results suggest that the vinyl groups in these compounds should resemble that in dibutyl vinylboronate (IV) rather closely. The calculated π-electron densities at the terminal carbon are for (IV), 0.932, for (VII), 0.920, and for (VIII), 0.920. The free valences are for (IV), 0.770, for (VII), 0.780, and for (VIII), 0.780. Radical stabilization energies are for (IV), 0.098β, for (VII), 0.139β, and for (VIII), 0.141β. Appropriate introduction of an auxiliary inductive parameter for the boron atom in the boronic ester (IV), which changes h_B to perhaps -0.8 or -0.7, would decrease the electron density at C_2 to 0.91, increase the free valence to 0.78, and increase the radical stabilization to $0.11-0.12\beta$, which moves the calculated properties even closer together.

The gross chemical properties of the vinylborinic esters are quite different from those of dibutyl vinylboronate inasmuch as the borinic esters are far more susceptible to oxygen attack, oxygen-initiated polymerization, and acid-catalyzed deboronation. With suitable precautions, radical additions to the borinic ester vinyl groups can be carried out and the results seem qualitatively similar to those obtained with dibutyl vinylboronate.

2. Heteroaromatic Boron Compounds

M. J. S. Dewar, the originator and principal developer of this field, has reviewed this topic in Volume 1 of this series.[17] The only aspects to be covered here are recent developments in the application of molecular orbital theory to these compounds.

Dewar and Rogers[20] have studied π-complexes of tetracyanoethylene with aromatic hydrocarbons and borazaro compounds. The energy of the electronic charge-transfer transition, estimated from the position of the absorption maximum in the visible spectrum, gives a linear correlation with the calculated energy of the highest occupied molecular orbital for a series of hydrocarbons. For a series of four borazaro compounds, parameters for nitrogen and boron were estimated by an approximate perturbation method and gave good agreement between calculated and observed spectra. Assuming to begin with that $h_B = -h_N$, the estimated value of h_B was -1.59.

Since Dewar and Rogers' value of h_B seemed low to the author, the perturbation calculations were checked by performing the actual Hückel calculations on a computer.[40] It turned out that the perturbation treatment, a linear extrapolation, had been carried beyond the region of linearity, and the Hückel energies were not very close. Changing h_B from -1.59 to -0.9 had little effect, and it appeared that any reasonable parameter for boron would yield about three times as much scatter in the energy correlations of the borazaro compounds as in that of the hydrocarbons. However, changing h_N from 1.59 was not tried, and it is possible that a "good" correlation might be obtained by modifying this and other parameters in a reasonable range. Another uncertainty is whether Dewar and Rogers adequately eliminated the possibility that some of the supposed π-complexes of borazaro compounds instead might have been tricyanovinyl dyes, which can form under mild conditions with aromatic amines and have visible spectra at least superficially similar to those of the π-complexes.[60]

Dewar, Gleicher and Robinson[19] have recently synthesized 10,9-borazaronaphthalene (IX). From the proton n.m.r. spectrum and theoretical considerations they concluded that the π-electron density was highest

(IX)

at carbon 2. Hückel calculations suggested that the π-electron charge on carbon 4 should also be negative, but the n.m.r. chemical shift evidence strongly indicates a positive contribution to the charge from the π electrons. The electron-donating effect of boron on the σ-electron framework causes an upfield shift of the absorption of the proton on the adjacent carbon, but in this case the shift is reduced from its usual value by the π-electron contribution. The qualitatively correct π-electron distribution was given by the split p-orbital method of self-consistent field calculation described by Chung and Dewar.[13] It might be noted that the use of the auxiliary inductive parameter as suggested by Streitwieser[80] would also correct the Hückel calculation in the right direction. Either approach involves some arbitrary assumptions. The self-consistent field approach is sounder theoretically, but the auxiliary inductive parameter is much easier for the ordinary chemist to use.

3. Borazines

A review of molecular orbital calculations on borazines is beyond the scope of a chapter on organoboron compounds. However, two relevant recent articles will be cited.

Brown and McCormack[6] have correlated ionization potentials of substituted borazines with Hückel calculations. It was assumed that $h_B = -h_N$ and β was adjusted to an appropriate value for the bond order. The parameters $h_B = -0.922$ and $\beta_{BN} = 0.888\beta_{CC}$ were calculated.

Chalvet, Daudel and Kaufman[12] have reported self-consistent field calculations on borazine. In their best model, the calculated π-electron density on boron was 0.478 and on nitrogen was 1.522. This π-electron charge distribution is opposed by a σ-electron charge distribution of similar magnitude in the opposite direction, resulting in net charges close to zero on the boron and nitrogen atoms. The calculated results correlated well with the ionization potential and ultraviolet spectrum.

4. The Diazadiboretidine Question

Among the most surprising conclusions in organoboron chemistry in the last few years are the reports of four different research groups[10, 21, 37, 64] that certain compounds containing the grouping R—BN—R', where R and R' are bulky, exist not as trimers (borazines) but as dimers, postulated to have the 1,3-diaza-2,4-diboretidine structure (X).

(X)

However, what appeared to be the most thoroughly documented example of structure $X^{(10)}$ has, on X-ray and 32-megacycle ^{11}B n.m.r. examination, been shown to have the five membered ring structure (XI) instead (personal communication from J. Casanova).

(XI)

The remaining postulated examples of structure (X) must be viewed with considerable skepticism, inasmuch as the structure is isoelectronic with the notoriously unstable cyclobutadiene. However, simple Hückel calculations predict a substantial delocalization energy for the four-membered ring, compared to localized B-N π-bonds. If the parameters $h_N = -h_B = 1.0\beta_{BN}$ are used, the calculated delocalization energy is 0.8β, in sharp contrast to the zero value found for cyclobutadiene, and there is no orbital degeneracy. This calculated result is practically insensitive to the parameter h in any reasonable range. Moreover, if the ring is square there must be considerable boron–boron π-bonding diagonally across the ring, since the boron atoms are considerably larger than the nitrogens. Slater orbitals suggest that β_{BB} should be about $0.75\beta_{BN}$. Inclusion of this normally neglected interaction increases the delocalization energy to 1.3β. The bonding in this structural model is analogous to that in the carborane $HC(BH)_3CH$, except that the symmetry is different. (Carboranes are reviewed in this volume by Williams.[82])

Hoffmann[29, 30] has carried out extended Hückel calculations on diaza-diboretidine, $(HNBH)_2$, which included σ as well as π interactions. His calculated binding energy for square $(HNBH)_2$, 351.375 eV, is only 4 kcal (0.174 eV) poorer than two-thirds that of $(HNBH)_3$ (borazine), 527.323 eV. In contrast, one mole of $(HNBH)_2$ is more stable than a half mole of the cyclooctatetraene analog, $(HNBH)_4$, by 18.5 kcal. However, Hoffmann's method of calculation does not yield strain energies in small rings,[28] and the energy found for $(HNBH)_2$ is probably grossly overoptimistic for this reason.

From the foregoing discussion, it is apparent that Hückel-type calculations suggest that the four-membered ring structure (X) might be stable if the substituents R can provide sufficient steric hindrance or other influence to prevent further regrouping to the normally more stable borazine system. Claims for these compounds cannot be dismissed as inherently fantastic on theoretical grounds, contrary to one's first intuitive impulse. However, Hückel calculations are more likely to be overly optimistic in predicting stabilities than overly pessimistic. The possibility exists in this case that the calculations are seriously and qualitatively erroneous, and that the $(HNBH)_2$ system is almost as difficult to form as cyclobutadiene. Even if the calculations are "right", the ring strain energy may outweigh the π-electron delocalization. Indeed, Casanova's work provides a clue in this direction. The five-membered ring structure (XI) is formed by a ring contraction during an alkyl migration from boron to electron-deficient carbon. (See Section IV, B, 4(c).) Repetition of this process would convert the five-membered ring (XI) to the four-membered ring (X) (with $R = CEt_3$, $R' = C_6H_5$). The stability of structure (XI) under the conditions of its formation (300°) as well as toward vigorous acid or base treatment[10] implies that the four-membered ring (X) is relatively very unstable, too much so even to serve as a transient intermediate. The author thus views all claims for the four-membered ring structures (X) as highly questionable, unless and until someone provides a proof of structure by X-ray crystallography.[†]

5. Conclusions

It is clear that boron participates strongly in π-bonding to carbon. The vacant p-orbital of boron accepts a chemically significant amount of electron density from an adjacent carbon π-bond system. Simple molecular orbital calculations provide a useful interpretive framework for correlations and predictions of properties of unsaturated organoboron compounds.

The appropriate value for the boron–carbon bond integral, β_{BC}, is only slightly less than β_{CC} for a carbon–carbon bond of similar bond order. In vinylboron compounds, the appropriate value for β_{B-C} is about $0.7\beta_{C=C}$. The best value for the Coulomb integral on boron, α_B, in vinylboronic acid appears to be about $\alpha_C - 0.75\beta_{C=C}$, that is, $h_B = -0.75$. For vinylboranes, $h_B = -1.0$ has given good results. Different values of h_B are expected because of the different electronegativities of the oxygen and carbon atoms adjacent to boron.

The suggested parameters for boron are intended to be useful with the other heteroatom parameters in Streitwieser's "recommended" list.[80] One discrepancy has so far been overlooked. It is usually assumed that $\alpha_{\ddot{N}} - \alpha_{\dot{C}} = \alpha_{\dot{C}} - \alpha_B$.[6, 12, 20] The dots over each atomic symbol represent the number of π-electrons contributed by that atom, and it should be noted that $\alpha_{\ddot{N}}$ is not the same as $\alpha_{\dot{N}}$.[80] This assumption leads to the value $h_B = -1.5$.

[†] Note added in proof. X-Ray evidence for one of the four–membered ring structures has been reported recently.[25a]

However, this disagreement with the author's assignment of $h_B = -0.75$ in vinylboronic acid is not nearly as large as it looks. The -1.5 value of h_B becomes about -1.35 when converted from aromatic β_{CC} to olefinic $\beta_{C=C}$ units, and if the auxiliary inductive parameter is $0.1h_{\ddot{o}}$, h_B becomes -0.95. In view of the numerous gross approximations involved, there is hardly any discrepancy left. Furthermore, there is no good reason for assuming h_B to be exactly equal to $-h_{\ddot{N}}$. It seems possible that the nearly empty p-orbital on boron could have a somewhat higher electron affinity than linear extrapolation from carbon and nitrogen would suggest.

It should finally be noted that π-electron acceptance and σ-electron donation by boron tend to make the atom more spherically symmetrical. (The angular dependences of the p-orbitals are sine and cosine functions. Since $\sin^2\theta + \cos^2\theta = 1$, equal electron densities in each of the three p-orbitals would result in spherical symmetry for the total electron density around the atom.) Spherical symmetry is associated with closed-shell structures, since it allows maximum attraction of electrons by the nucleus with minimum electron repulsion. The tendency of boron and nitrogen to acquire as nearly a spherical electron distribution as other factors will allow is most clearly illustrated by the self-consistent field calculations on borazine.[12]

III. BORON-SUBSTITUTED FREE RADICALS

A. Radical Additions to Vinylboron and Ethynylboron Compounds

1. Dibutyl Vinylboronate

The radical-catalyzed addition of carbon tetrachloride to dibutyl vinylboronate has been described and interpreted in molecular orbital terms in Section II, B, 1(c) of this chapter. The theoretically interesting low value of the transfer constant is synthetically frustrating, but a number of more active reagents do add efficiently to dibutyl vinylboronate. The products from these radical additions opened up the new field of organofunctional boronic esters, which with the exception of a limited series of aromatic compounds, had been inaccessible by previous synthetic methods.

Bromotrichloromethane adds efficiently to dibutyl vinylboronate (IV) to yield dibutyl 1-bromo-3,3,3-trichloropropane-1-boronate (XII).[38]

$$Cl_3C^\bullet + CH_2{=}CH{-}B(OC_4H_9)_2 \longrightarrow Cl_3CCH_2\dot{C}H{-}B(OC_4H_9)_2$$
$$(IV)$$

$$Cl_3CCH_2\dot{C}H{-}B(OC_4H_9)_2 + CCl_3Br \longrightarrow Cl_3CCH_2\underset{\underset{Br}{|}}{C}H{-}B(OC_4H_9)_2 + Cl_3C^\bullet$$
$$(XII)$$

A radical initiator such as azobisisobutyronitrile or light leads to formation of trichloromethyl radicals, which presumably catalyze the addition by the typical radical chain propagation mechanism shown. A methyl group at either the α or β position in the vinylboron compound (dibutyl 1-propene-1-boronate or propene-2-boronate) does not seriously interfere with the bromotrichloromethane addition, though there is some reduction in yield. Other halomethanes which have been added to dibutyl vinylboronate include carbon tetrabromide and bromoform.[42]

An unusual reaction of dibutyl vinylboronate is the radical-catalyzed addition of bromomalononitrile to yield dibutyl 1-bromo-3,3-dicyano-propane-1-boronate (XIII). The yield is 85 per cent.[52]

$$(NC)_2CHBr + CH_2{=}CH{-}B(OC_4H_9)_2 \longrightarrow (NC)_2CHCH_2\underset{\underset{Br}{|}}{CH}{-}B(OC_4H_9)_2$$

(XIII)

It was certainly fortunate that we first tried this reaction with the boron compound and not with some cheaper "model system" such as ethyl acrylate, acrylonitrile, 1-octene, or vinyl acetate, none of which gave any isolable adduct with bromomalononitrile. Dibutyl bromomalonate also adds to dibutyl vinylboronate to yield 40 per cent of dibutyl 1-bromo-3,3-dicarbobutoxypropane-1-boronate (XIV).

$$(C_4H_9O_2C)_2CHCH_2CHBrB(OC_4H_9)_2$$

(XIV)

The other known radical additions to dibutyl vinylboronate may be generalized as X—H additions. For example, hydrogen bromide adds efficiently

$$X^{\bullet} + CH_2{=}CH{-}B(OC_4H_9)_2 \longrightarrow X{-}CH_2\overset{\bullet}{C}H{-}BOC_4H_9)_2$$

$$X{-}H + X{-}CH_2\overset{\bullet}{C}H{-}B(OC_4H_9)_2 \longrightarrow X{-}CH_2CH_2{-}B(OC_4H_9)_2 + X^{\bullet}$$

under ultraviolet irradiation at 70°C to yield dibutyl 2-bromopropaneboronate.[46] Mercaptans and thiol acids also add efficiently to dibutyl vinylboronate.[38, 55] A complete list of these adducts is available in an earlier review.[42] An interesting variant on this type of reaction is the addition of sodium bisulfite to yield the dihydroxyboronoethanesulfonate anion (XV).[55]

$$(HO)_2BCH_2CH_2SO_3^-$$

(XV)

Dibutyl vinylboronate polymerizes readily and is best stored at 0°C with a suitable inhibitor such as 0·01 per cent of phenothiazine. The butoxy groups in the polymer are rapidly hydrolyzed by atmospheric moisture, resulting in a crumbly white solid.

2. *2-Vinyl-4,4,6-trimethyl-1,3,2-dioxaborinane*

Woods, Bengelsdorf, and Hunter[84] have synthesized this boronic ester, which might also be named 2-methylpentane-2,4-diyl vinylboronate (XVI). This compound is resistant to oxygen-initiated polymerization, which was attributed

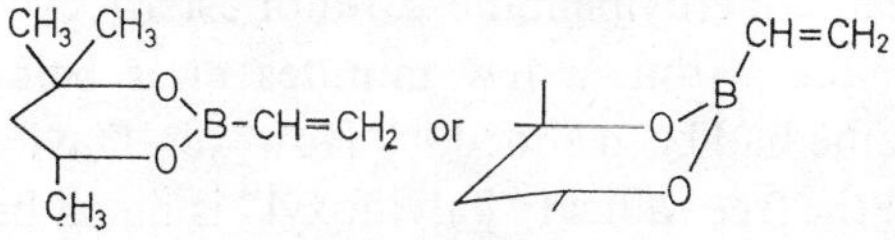

(XVI)

to steric inhibition of attack of oxygen on the boron atom. Such attack would force either the oxygen or the vinyl group into an unfavorable 1,3-diaxial interaction with a methyl group. Azobisisobutyronitrile initiates polymerization of the vinylboronic ester (XVI) in benzene to a thermoplastic polymer of low molecular weight. Films of the polymer showed considerable stability in moist air. Copolymers of (XVI) with stryrene, methyl methacrylate, and acrylonitrile have also been prepared.

Woods and Bengelsdorf[83] added hexanethiol to the double bond of the boronic ester (XV) at −70°C under ultraviolet irradiation. Since they only obtained a 51 per cent yield under conditions where the author once claimed a 93 per cent yield for the analogous addition of hexanethiol to dibutyl vinylboronate,[38] Woods and Bengelsdorf concluded that the cyclic ester (XV) was less reactive than the butyl ester. This conclusion is doubtful because the author only ran his reaction once and the temperature control was dubious (unpublished recollections). Subsequently, addition of ethanethiol to dibutyl vinylboronate was found to be more efficient at higher temperatures, 25°C having been recommended.[39]

3. *N-Methyldiethanolamine Vinylboronate*

This compound, named tetrahydro-2-vinyl-6-methyl-6-aza-1,3,2-dioxaborocine (XVII) by Woods, Bengelsdorf and Hunter,[84] failed to polymerize

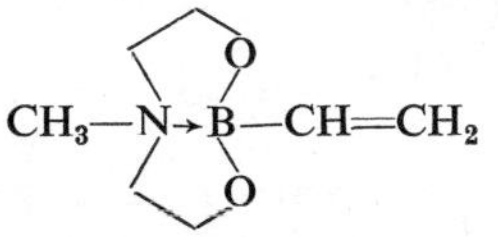

(XVII)

when heated with azobisisobutyronitrile. Addition of a radical to the terminal carbon would be less facile than normal for a vinylboronic ester, since the boron atom in (XVII) does not have a vacant p-orbital to delocalize the odd electron. Perhaps more important is the bulky group surrounding the boron atom, which would hinder addition of the adjacent radical to a double bond. No attempt to add mercaptans or other active reagents to (XVII) has been reported.

4. Polymerizable Cyclic Vinylboron Compounds

In contrast to the 2-methylpentane-2,4-diol ester (XVI), catechyl vinylboronate polymerized within a few minutes after being distilled, even though phenothiazine had been added.[52] However, Davies and Roberts[16] have reported that the free radical "galvinoxyl" is much better than phenothiazine for preventing autoxidation of boronic acids, and the possibility of stabilizing reactive vinylboron compounds has not been fully explored.

Vinylboronic acid itself, $CH_2{=}CH{-}B(OH)_2$, polymerizes readily but can be isolated without difficulty if it is kept a little moist and is inhibited with phenothiazine.[38] Vinylboronic anhydride (XVIII) was evidently isolated once.

$$CH_2{=}CH{-}B \overset{\overset{\displaystyle CH{=}CH_2}{\overset{\displaystyle |}{\overset{\displaystyle B}{\diagup \diagdown}}}}{\underset{\displaystyle O}{\overset{\displaystyle O \qquad O}{}}} B{-}CH{=}CH_2$$

(XVIII)

The mobile liquid polymerized explosively after a short time in the refrigerator.[40a] Traces of this anhydride may be responsible for the rapid, exothermic oxidative decomposition of vinylboronic acid samples when exposed to the air after a little too thorough drying.

In summary, the vinylboronic acid derivatives which polymerize uncontrollably are those in which attack of oxygen or a peroxy radical at the boron atom should be sterically facilitated, and those which do not polymerize in air are those in which such attack is sterically hindered. This is in accord with the autoxidation mechanism suggested by Davies and Roberts.[16] The polymerization might be initiated either by further breakdown of the organoperoxyborane $ROOB(OH)_2$ to free radicals or by attack of some of the chain-carrying radicals in the autoxidation sequence on the vinylboron compound or both.

Initiation:

$$RB(OH)_2 + O_2 \longrightarrow \left[R{-}\overset{\displaystyle \cdot}{\underset{\displaystyle O{-}O\cdot}{B}}(OH)_2 \right] \longrightarrow R\cdot + (HO)_2B{-}O{-}O\cdot \,(?)$$

Propagation:

$$R^{\cdot} + O_2 \longrightarrow R-O-O^{\cdot}$$

$$R-O-O^{\cdot} + R-B(OH)_2 \longrightarrow \left[\begin{array}{c} R-\overset{\cdot}{B}(OH)_2 \\ | \\ O-O-R \end{array} \right] \longrightarrow R^{\cdot} + R-O-O-B(OH)_2$$

5. *B-Trivinyl-N-triphenylborazine*

Seyferth and co-workers have studied radical additions of a variety of reagents of the X—H type to the vinyl groups of *B-trivinyl-N-triphenyl-borazine* (XIX). Hydrogen bromide bubbled through a benzene solution of the vinylborazine (XIX) at room temperature with benzoyl peroxide as the

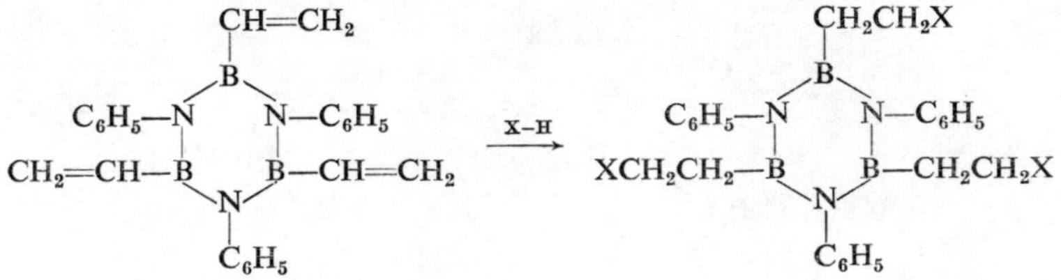

(XIX)

initiator gave a good yield of *B*-tris(2-bromoethyl)-*N*-triphenylborazine. If ether was used as solvent instead of benzene, the borazine ring was cleaved to yield anilinium bromide.[78] Benzenethiol[73] and diphenylphosphine[76] also added readily to the vinylborazine (XVIII) in the presence of peroxide initiators. Triphenyltin hydride added without an initiator, and triethyltin hydride added in the presence of azobisisobutyronitrile.[77]

Attempted radical-catalyzed additions of trichlorosilane and methyl-dichlorosilane to the vinylborazine (XIX) failed. However, the desired adducts were obtained when chloroplatinic acid was used as the catalyst.[77]

Bromotrichloromethane and carbon tetrabromide have also been added to the vinylborazine (XIX) in the usual manner with benzoyl peroxide as the initiator. Carbon tetrachloride failed to add.[78]

Pellon, Deichert and Thomas[73] have reported the polymerization of the trivinylborazine (XIX).

6. *Vinylborinic Esters*

Butyl divinylborinate or *B*-butoxy-*B,B*-divinylborane (VII) polymerizes rapidly if solutions of it are concentrated.[39] However, dilute solutions of this compound have been prepared in benzene and treated with hydrogen sulfide at 80°C with azobisisobutyronitrile as the initiator. The heterocyclic adduct 4-butoxy-1-thia-4-boracyclohexane (XX) resulted in 12 per cent yield, together with considerable polymeric residue. Transesterification of the butoxy compound (XIX) with N,N-dimethylaminoethanol yielded the stable, chelated B-(dimethylaminoethoxy)-B,B-divinylborane (XXI). This was converted to the heterocyclic 4-dimethylaminoethyl-1-thia-4-boracyclohexane (XXII) in 36 per cent yield on treatment with hydrogen sulfide and the radical initiator. Though more azo compound was required to initiate the addition of hydrogen sulfide to the dimethylaminoethanol ester (XXI), little polymer was formed.

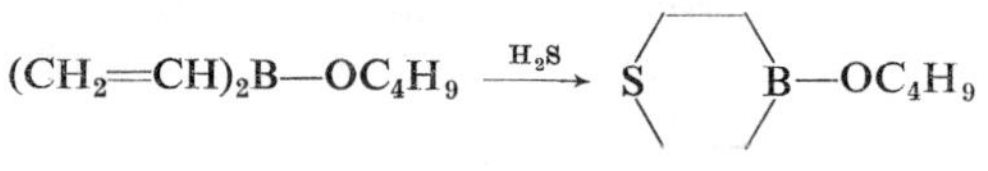

$$(CH_2{=}CH)_2B{-}OC_4H_9 \xrightarrow{H_2S} \quad S \qquad B{-}OC_4H_9$$

    (VII)   (XX)

$$(CH_2{=}CH)_2B \overset{N(CH_3)_2}{\underset{O}{\diagup}} \xrightarrow{H_2S} S \qquad B \overset{N(CH_3)_2}{\underset{O}{\diagup}}$$

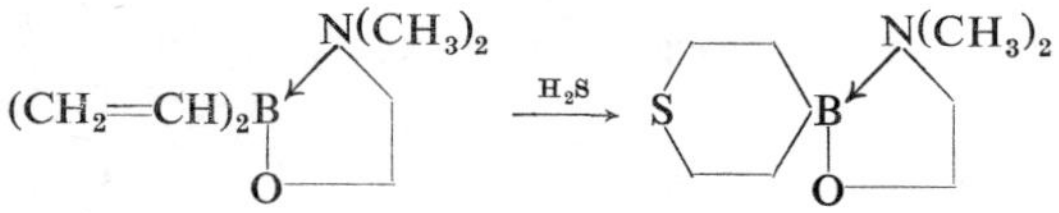

    (XXI)   (XXII)

    The behavior of the two borinic esters (XIX) and (XXI) is consistent with other vinylboron chemistry. B-Butoxy-B,B-divinylborane (XIX) should be far more susceptible to attack by oxygen than would a boronic ester, which accounts for its polymerization. Chelation in the N,N-dimethylaminoethoxy derivative (XXI) should stabilize the boron against peroxide attack and should also make it more difficult for radicals to attack the terminal carbons of the vinyl groups. Although the author suggested that the polymeric by-product from the reaction of the butoxy compound (XIX) with hydrogen sulfide reflected a low transfer constant, consistent with the theoretically expected high degree of delocalization of the odd electron to boron in the intermediate,[39] the experimental evidence is questionable. It is possible that the polymerization merely occurred faster than the hydrogen sulfide could be delivered to the reaction mixture.

    Butyl B-phenyl-B-vinylborinate or B-butoxy-B-phenyl-B-vinylborane (VIII) can be isolated without undue difficulty.[49] Vinyl polymerization occurred readily. A sample which was protected from air and stored several

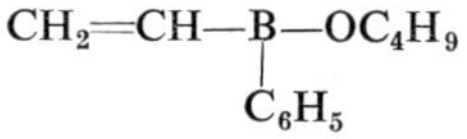

$$CH_2{=}CH{-}\underset{\underset{C_6H_5}{|}}{B}{-}OC_4H_9$$

(VIII)

months with phenothiazine at $-15°C$ appeared to disproportionate into boranes and boronic esters, since on redistillation a forerun was obtained which ignited spontaneously on exposure to air. Both polymerization and disproportionation were prevented when an *ortho* methyl group was incorporated into the aryl substituent. *B*-Butoxy-*B*-*o*-tolyl-*B*-vinylborane and *B*-butoxy-*B*-(2,5-dimethylphenyl)-*B*-vinylborane were much easier to handle than the phenyl compound, and *B*-butoxy-*B*-mesityl-*B*-vinylborane failed to polymerize even when heated with azobisisobutyronitrile.[49]

Three of the *B*-butoxy-*B*-aryl-*B*-vinylboranes, the phenyl, 2,5-dimethylphenyl, and mesityl compounds, were reacted with bromotrichloromethane at room temperature under ultraviolet irradiation to yield the 1 : 1 adducts of the general formula (XXIII). The phenyl compound gave good results only when freshly distilled. *B*-Butoxy-*B*-mesityl-*B*-vinylborane, in spite

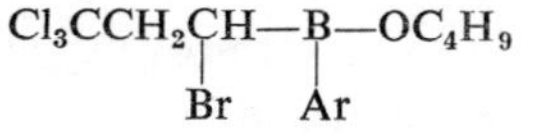

$$Cl_3CCH_2\underset{Br}{\underset{|}{C}}H—\underset{Ar}{\underset{|}{B}}—OC_4H_9$$
$$(XXIII)$$

of its reluctance to polymerize, added bromotrichloromethane in 75 per cent yield.

7. Dibutyl Acetyleneboronate

Polymerization of dibutyl acetyleneboronate (XXIV) has not been observed, and moderately active radical reagents such as carbon tetrachloride do not react. Active reagents such as bromotrichloromethane do react readily, and dibutyl 1-bromo-3,3,3-trichloropropene-1-boronate (XXVI) has been obtained in 90 per cent yield.[51] Likewise, the addition of benzenethiol proceeds smoothly to yield dibutyl 2-phenylthioetheneboronate (XXVI). Other thiols, including thiolacetic acid and hexanethiol, add similarly.

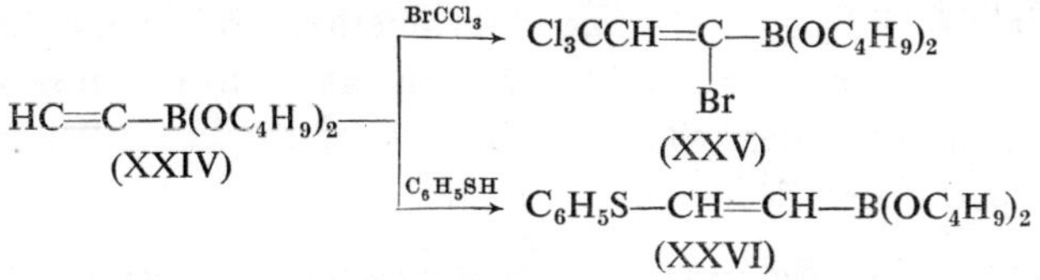

$$HC{\equiv}C—B(OC_4H_9)_2$$
$$(XXIV)$$

$$\xrightarrow{BrCCl_3} Cl_3CCH{=}\underset{Br}{\underset{|}{C}}—B(OC_4H_9)_2$$
$$(XXV)$$

$$\xrightarrow{C_6H_5SH} C_6H_5S—CH{=}CH—B(OC_4H_9)_2$$
$$(XXVI)$$

Hydrogen bromide addition to the acetylenic compound (XXIV) yields dibutyl 2-bromovinylboronate. Bromine addition to the triple bond also proceeds by a radical mechanism, the reaction occurring at $25°C$ only when the reactants are irradiated. The product is dibutyl 1,2-dibromoetheneboronate, $BrCH{=}CBrB(OC_4H_9)_2$. The stereochemistry of these radical additions has not been investigated, though the reaction conditions used would in general be conducive to equilibration of *cis* and *trans* isomers.

B. *Radical Substitutions on Methylboron Compounds*

1. *Trimethylborane*

Reaction of chlorine with trimethylborane in the gas phase at −95°C yields chloromethyldimethylborane, $ClCH_2B(CH_3)_2$.[74] It seems likely that this is a typical free-radical chlorination, although it does not appear that this mechanistic question has been examined.

2. *Di-t-butyl Methaneboronate*

In the hope of finding a useful synthesis of a halomethaneboronic ester, the author studied the halogenation of methaneboronic acid derivatives.[41] Methaneboronic anhydride (XXVII) proved inert to *N*-bromosuccinimide or bromine. Photoinitiated reaction with t-butyl hypochlorite yielded methyl chloride and an unstable residue which appeared to be tri-t-butoxyboroxine (XXVIII). The postulated mechanism, which may or may not be concerted, involves addition of a t-butoxy radical to a boron atom and

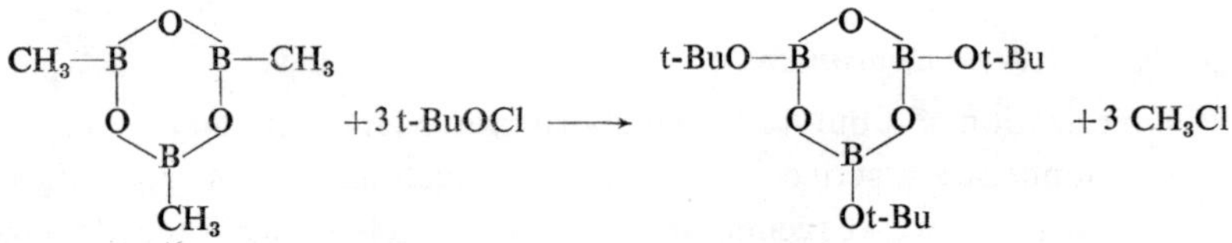

elimination of a methyl radical. The methyl radical would react with t-butyl hypochlorite to produce methyl chloride and another t-butoxy radical.

Irradiation of t-butyl hypochlorite and di-t-butyl methaneboronate (XXIX) at 0°C did produce some chloromethylboron compound, which was isolated in 9 per cent yield after conversion to the di-n-butyl ester (XXX). The problem was that chlorination of the *B*-methyl group was barely favored energetically (by a factor of 1.1–1.5) over attack on the

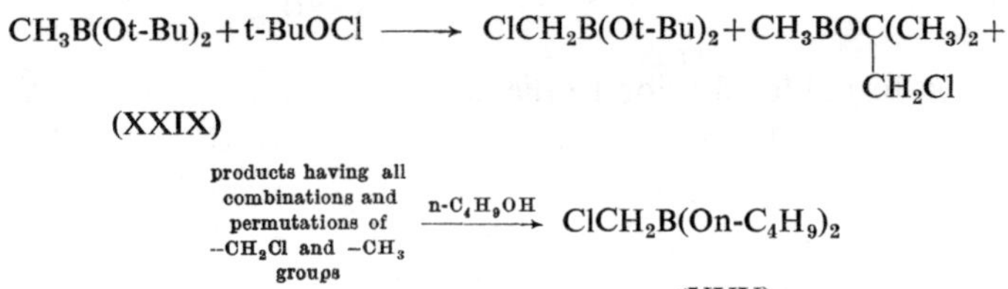

C-methyl groups, which outnumber the former nine to one in an equimolar mixture of the reactants. Some attack of t-butoxy radicals on the boron atoms also occurred in spite of steric hindrance, resulting in carbon–boron bond cleavage. Thus, it would not be possible to increase the yield to a practical level by adding more chlorinating agent, since the product would be degraded. A different synthetic route to a halomethaneboronic ester, based on the reaction of iodomethylmercuric iodide with boron tribromide, has been found to be practical.[44]

The inertness of the *B*-methyl group to radical attack was neither anticipated nor satisfactorily explained.[44] The author later made the mistaken suggestion that hydrogen atom transfer to the t-butoxy radical was so exothermic that the transition state was reached before enough carbon–hydrogen bond breaking had occurred to allow delocalization of the bonding electrons into the vacant *p*-orbital of the boron atom.[42] The trouble with this suggestion is that t-butoxy radicals are not sufficiently avid hydrogen atom abstractors to be indiscriminate, and the inertness of the *B*-methyl group toward less active radicals is not consistent with low activation energies.

It seems more likely that the transition state for attack of the t-butoxy radical on the *B*-methyl group does involve a considerable degree of carbon–hydrogen bond breaking and some consequent delocalization of the bonding electrons into the vacant *p*-orbital on the boron atom. This competes with the oxidizing character of the t-butoxy radical.

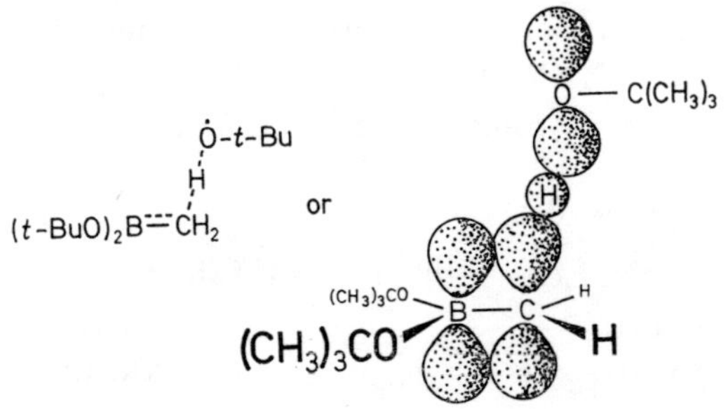

(XXXI)

The electronegative oxygen of the radical seizes the greater share of the electrons from the breaking carbon–hydrogen bond in the transition state (XXXI). To whatever extent the boron atom can delocalize electrons toward itself, the energy gain must be nearly nullified by the reduction in availability of these electrons to the oxygen atom [Structure XXXI].

If the dialkoxyboryl group showed a strong electron-donating inductive effect toward carbon, the π-bonding effects in the hydrogen atom abstraction would not matter and the methyl group would be activated toward attack by the t-butoxy radical. Evidently the dibutoxyboryl group is weakly electron donating or possibly weakly electron withdrawing. The proton n.m.r. absorption of the B-methyl group in methaneboronic acid derivatives appears near $\tau 9.8$, implying inductive electron donation by the boron atom. However, the dialkoxyboryl group must direct the positive end of its dipole toward the B-methyl group, a field effect which might influence the chemical reactivity more than it does the n.m.r. spectrum. Both this radical reaction and the ease of generating a carbonium ion adjacent to the dialkoxyboryl group (see Section IV, B, 2) imply that the dialkoxyboryl group is less electronegative than a hydrogen atom but more electronegative than a methyl group, though the inductive effect of the latter cannot be untangled from hyperconjugative stabilization of transition states.

IV. BORON-SUBSTITUTED ELECTROPHILIC CENTERS

A. *Electrophilic Additions to Vinylboron Compounds*

1. *Hydrogen Halide Additions*

Polar addition of hydrogen bromide to unsaturated boronic esters has been reported by Matteson and Liedtke.[45] Addition of dibutyl *trans*-2-butene-2-boronate (XXXII) to excess liquid hydrogen bromide yielded 61 per cent of dibutyl 2-bromobutane-2-boronate (XXXIII).

$$
\begin{array}{ccc}
\underset{\text{H}}{\overset{\text{CH}_3}{\diagdown}}\text{C}=\text{C}\underset{\text{CH}_3}{\overset{\text{B(OC}_4\text{H}_9)_2}{\diagup}} & + \text{HBr} \longrightarrow & \text{CH}_3\text{CH}_2-\underset{\underset{\text{Br} \quad \text{CH}_3}{}}{\overset{\text{B(OC}_4\text{H}_9)_2}{\diagup}}\text{C} \\
(\text{XXXII}) & & (\text{XXXIII})
\end{array}
$$

Similarly, dibutyl propene-2-boronate yielded dibutyl 2-bromopropane-2-boronate. However, the directing influence of a methyl group is stronger than that of the dibutoxyboryl group, as shown by the formation of dibutyl 2-bromopropane-1-boronate, $CH_3CHBrCH_3B(OBu)_2$, from dibutyl 1-propeneboronate, $CH_3CH=CHB(OBu)_2$.

Dibutyl vinylboronate (IV) did not react with liquid hydrogen bromide at atmospheric pressure, but the author and Schaumberg[52] found that the addition occurred in a bomb at 0° and yielded a mixture of isomers. More conveniently, liquid hydrogen iodide added to the vinyl compound (IV) at atmospheric pressure to give a mixture containing about 60 per cent dibutyl 1-iodoethaneboronate (XXXIV) and 40 per cent dibutyl 2-iodoethaneboronate (XXXV).

$$CH_2{=}CH{-}B(OC_4H_9)_2 + HI \longrightarrow CH_3\underset{\underset{I}{|}}{C}H{-}B(OC_4H_9)_2 + ICH_2CH_2B(OC_4H_9)_2$$

(IV)   (XXXIV)   (XXXV)

The transition states for hydrogen iodide addition to dibutyl vinyl-boronate (IV) probably have polar four-center character and may be represented by the ion-pair structures (XXXVI) and (XXXVII). Most but

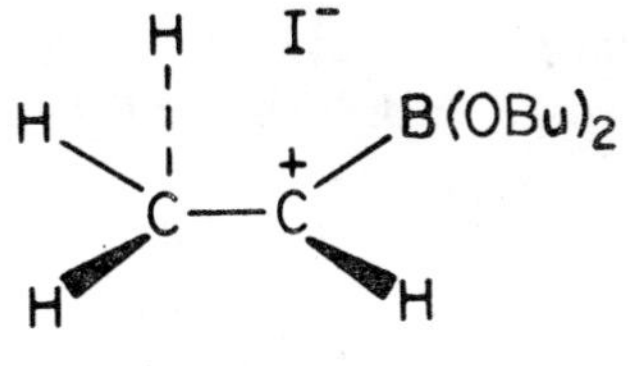

not all of the π-bonding in the original vinylboron compound (IV) has been lost in these transition states, making the directing influence of the remaining π-bonding uncertain. (See Section II, B of this chapter.) If the inductive effect of the dibutoxyboryl group is slightly more electron dona-ting than that of a hydrogen atom (see Section III, B), this effect would favor the α-cation (XXXVI). The α-borono substituent also appears to facilitate the approach of nucleophiles (see Section IV, B), an effect which would also tend to favor the α-cation (XXXVI). These transition state models are based on the work of Dewar and Fahey,[18] who have studied the stereochemistry of additions of hydrogen halides to olefins.

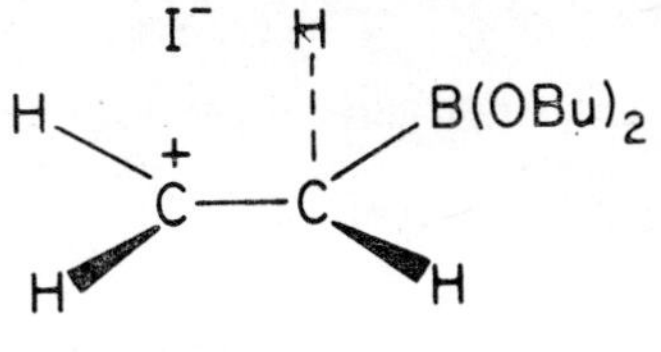

Although kinetic control of these hydrogen halide additions has not been proved, it is highly unlikely that equilibration of the α- and β-isomers is taking place. When elimination reactions of the β-haloalkaneboronic esters occur, boron halide and not hydrogen halide is evolved. (See Section IV, B.) Thus, the α-isomer cannot be formed from the β-isomer. If the β-isomer were formed irreversibly from the α-isomer, the relative amounts of dibutyl 1- and 2-iodoethaneboronate would vary in different prepara-tive reactions due to different reaction times. Such variation has never been observed, even though the reaction has been run a number of times and by different operators.

2. Bromination

Bromine adds rapidly to dibutyl vinylboronate at $-70°C$ to yield dibutyl 1,2-dibromoethaneboronate, $BrCH_2CHBrB(OBu)_2$.[51, 61] Stereospecific *trans* addition of bromine to dibutyl *cis*- and *trans*-2-butene-2-boronate has been reported by Matteson and Liedtke.[47]

3. Hydroboration

(a) *Dibutyl vinylboronate.* The hydroboration of dibutyl vinylboronate (IV) was first reported by Mikhailov and Aronovich[62] and studied independently by Matteson and Shdo.[53, 54] Addition of the boronic ester (IV) to a solution of diborane in tetrahydrofuran gave a mixture of boranes. Treatment with butanol resulted in evolution of hydrogen and conversion to a mixture of tetrabutyl ethane-1,1-diboronate (XXXVIII)

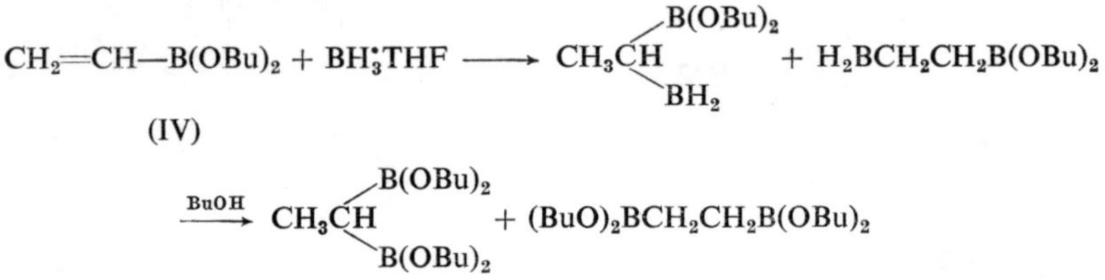

$$CH_2{=}CH{-}B(OBu)_2 + BH_3^\cdot THF \longrightarrow CH_3\overset{\displaystyle B(OBu)_2}{\underset{\displaystyle BH_2}{\overset{|}{\underset{|}{C}}H}} + H_2BCH_2CH_2B(OBu)_2$$

(IV)

$$\xrightarrow{\text{BuOH}} CH_3\overset{\displaystyle B(OBu)_2}{\underset{\displaystyle B(OBu)_2}{\overset{|}{\underset{|}{C}}H}} + (BuO)_2BCH_2CH_2B(OBu)_2$$

        (XXXVIII)   (XXXIX)

and ethane-1,2-diboronate (XXIX) in an overall yield of 44 per cent. The yield was kept low by the formation of adducts of two and three molecules of boronic ester (IV) to one of BH_3. These were isolated in analytically pure form though as mixtures of isomers by Matteson and Shdo.[54]

Hydrolysis of the mixture of tetrabutyl esters (XXXVIII) and (XXXIX) yielded 15 per cent of insoluble ethane-1,2-diboronic acid and 67 per cent of the relatively soluble ethane-1,1-diboronic acid. Thus, the electrophilic boron atom preferentially attacks the electron-rich α carbon atom of the vinylboronic ester (IV). Brown and Zweifel[9] have postulated that hydroborations proceed through four-center transition states in order to account for their *cis* stereochemistry. This type of transi-

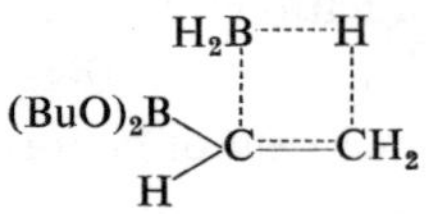

$$(BuO)_2B\underset{\displaystyle H}{\overset{\displaystyle H_2B\cdots\cdots H}{\diagdown C\!\!\!-\!\!-\!\!-CH_2}}$$

(XL)

tion state (XL) is closely related to that postulated for hydrogen iodide addition (XXXVII), except that it may be less polar than the latter and may preserve more of the π-bond character of the vinylboron system.

The predominant isomer from the hydroboration is that which would be predicted on the basis of π-electron densities in the vinylboronic ester (IV).

The quantitative value of the isomer ratio for the addition of BH_3 to dibutyl vinylboronate is indeterminate. The ratio of four or five parts 1,1-isomer to one part 1,2-isomer implied by the isolation of the boronic acids may be too high, since the intermediate 1,2-borane, $H_2BCH_2CH_2B(OBu)_2$, would condense more rapidly than the 1,1-isomer with a second molecule of vinylboronic ester for steric reasons and thus be lost from the route to ethanediboronic acids.

Finally, it should be pointed out that the difference between isomer ratios in hydroboration and hydrogen halide addition is small in thermodynamic terms, about 1 kcal/mole or less.

(b) *Dibutyl syreneboronate.* The rather closely balanced isomer ratio in the products of hydroboration of dibutyl vinylboronate can be reversed by the steric interference of an α-substituent on the vinyl group. Pasto, Chow and Arora[67] have found that reaction of dibutyl α-styreneboronate with excess diborane in tetrahydrofuran followed by treatment with ethylene glycol yields a mixture containing 90–95 per cent ethanediyl phenylethane-1,2-diboronate (XLI) and 5–10 per cent of its 1,1-isomer. Dibutyl β-styreneboronate, in which the steric and electronic effects work together, yields ethanediyl phenylethane-2,2-diboronate (XLII) containing less than 5 per cent of the 1,2-isomer (XLI).

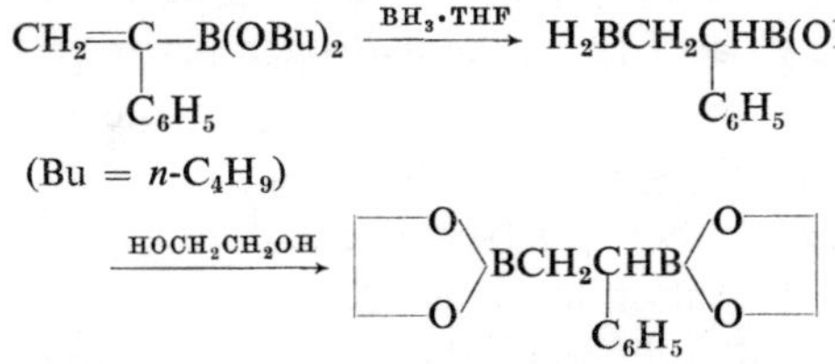

$$CH_2\!\!=\!\!\underset{\underset{C_6H_5}{|}}{C}\!\!-\!\!B(OBu)_2 \xrightarrow{BH_3\cdot THF} H_2BCH_2\underset{\underset{C_6H_5}{|}}{C}HB(OBu)_2$$

$$(Bu = n\text{-}C_4H_9)$$

$$\xrightarrow{HOCH_2CH_2OH}$$

(XLI)

$$C_6H_5CH\!\!=\!\!CHB(OBu)_2 \longrightarrow C_6H_5CH_2\underset{\underset{BH_2}{\diagdown}}{\overset{\diagup B(OBu)_2}{CH}} \longrightarrow C_6H_5CH_2CH$$

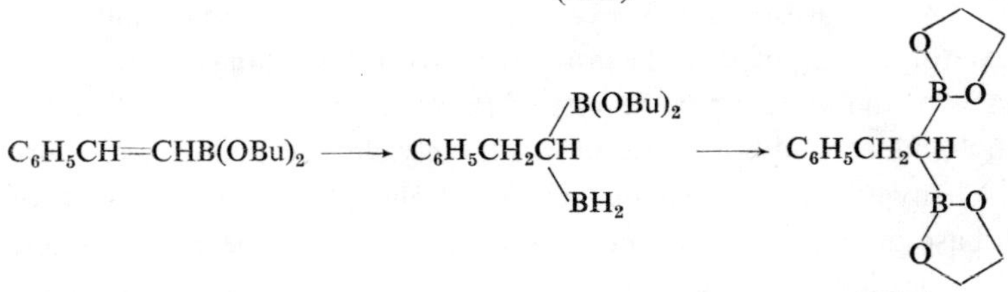

(XLII)

(c) *Dihydroboration of acetylenes.* Zweifel and Arzoumanian[87] have studied the dihydroboration of 1-hexyne in considerable detail. With diborane in tetrahydrofuran, 1-hexyne yielded a mixture containing approximately 85 per cent 1,1-diborylhexanes (XLIII), 10 per cent 1,2-diborylhexanes (XLIV), and 5 per cent 2,2-diborylhexanes (XLV). Hydroboration with dicyclohexylborane, $(C_6H_{11})_2BH$, increased the proportion of 1,1-diborylhexane (XLIII, R = cyclohexyl) to 96 per cent.

$$C_4H_9C{\equiv}CH \xrightarrow{R_2BH} \begin{cases} C_4H_9CH{=}CHBR_2 \xrightarrow{R_2BH} \begin{cases} C_4H_9CH_2\underset{\diagdown BR_2}{\overset{\diagup BR_2}{C}}H & \text{(XLIII)} \\ C_4H_9\underset{\underset{BR_2}{|}}{CH}{-}CH_2BR_2 & \text{(XLIV)} \end{cases} \\ C_4H_9\underset{\underset{BR_2}{|}}{C}{=}CH_2 \xrightarrow{R_2BH} C_4H_9\underset{\underset{BR_2}{|}}{\overset{\overset{BR_2}{|}}{C}}{-}CH_3 & \text{(XLV)} \end{cases}$$

(R = H or alkyl)

Interpretation of the results is complicated because the boranes themselves are not isolable. Treatment of the 1,2-diborylhexanes (XLIV) with alkaline hydrogen peroxide gives good yields of 1,2-hexanediol. However, the 1,1-diborylhexanes (XLIII) are in part cleaved by base to 1-borylhexanes, $C_4H_9CH_2CH_2BR_2$, and may also be converted in part to various hydroxyboron compounds (boronic and borinic acids) which are cleaved much more slowly by base. Thus, the hydrogen peroxide treatment yields mixtures of 1-hexanol and hexaldehyde. Similarly, the 2,2-diborylhexanes (XLV) lead to 2-hexanol as well as 2-hexanone.

Pasto[65] has also hydroborated 1-hexyne and concluded that 30 per cent of the product was 1,2-diborylhexanes (XLIV). Because Pasto's reaction conditions were somewhat different, there is no reason to expect the isomer ratios to be the same. However, Pasto concluded that some of the 1-hexanol and all of the 2-hexanol arose from 1,2-diborylhexanes (XLIV). In view of the more recent results of Pasto, Chow and Arora,[67] in which phenylethane-1,2-diboronic acid was largely cleaved by hydrogen peroxide to 1-phenylethanol but yielded mostly 2-phenylethanol when hydrolyzed with base prior to the hydrogen peroxide treatment, the earlier conclusion seems questionable. Zweifel and Arzoumanian[87] appear to have good evidence that they have avoided fragmentation and cleavage reactions during the hydrogen peroxide deboronation step, perhaps because they used higher concentrations of both base and peroxide.

Pasto[65] found that hydroboration of diphenylacetylene in tetrahydrofuran followed by oxidation with alkaline hydrogen peroxide yielded 31 per cent of d,l-hydrobenzoin, 37 per cent of deoxybenzoin, and 11 per cent of 1,2-diphenylethanol. From the boron hydride consumption, about 40 per cent of the reaction stopped at the borylstilbene (monohydroboration) stage, the source of most of the deoxybenzoin, under the conditions chosen. From these results it is clear that the major dihydroboration products are 1,2-diboryl-1,2-diphenylethanes, which are the source of the d,l-hydrobenzoin. This direction of hydroboration can be rationalized on steric grounds. Similarly, Pasto found evidence for considerable amounts of 1,2-diboryl products from phenylacetylene and from 3-hexyne. These results cannot now be interpreted quantitatively because of the probability that cleavages during the hydrogen peroxide treatment led to the major portion of the monoalcohols. However, quantitative data are not meaningful anyway unless the exact nature of the hydroborating agent is known. Both Pasto[65] and Zweifel and Arzoumanian[87] used limited amounts of diborane. As the reaction proceeds, diborane is converted to various alkenyl- and alkylboranes, which then become the effective hydroborating agents.

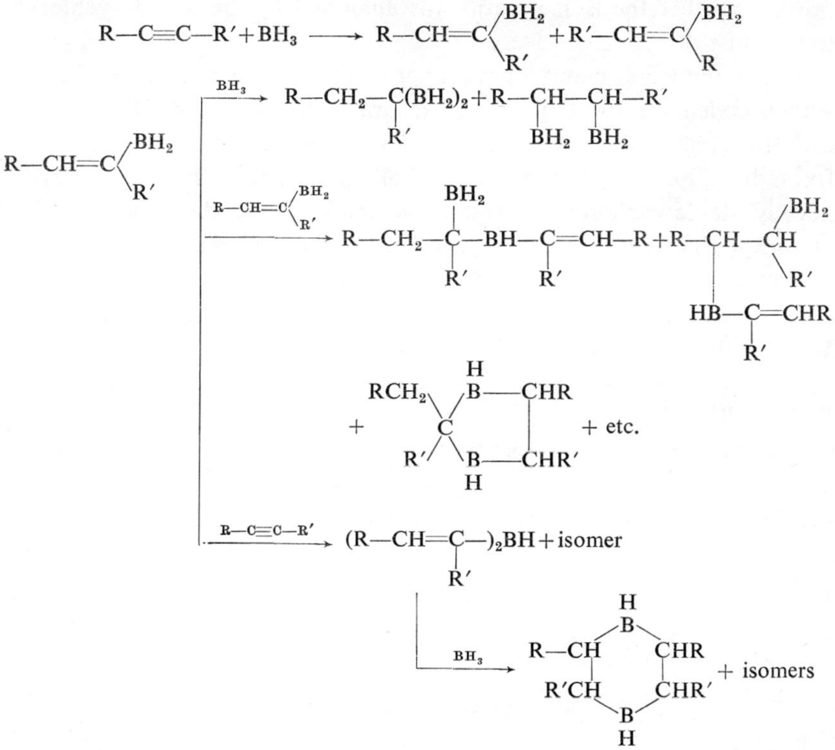

Only a few of the initial steps of the bewilderingly complex array of conceivable and probably functioning routes to the ultimate polymeric products of the class R_2BH or R_3B are illustrated. In general, each different reaction path will produce a different ratio of vicinal and geminal dihydroboration products. Therefore, none of these data can supply any meaningful quantitative information about the isomer ratios produced in the "hydroboration of alkenylboranes".

On the other hand, the hydroborations with dicyclohexylborane and 2,3-dimethyl-2-butylborane reported by Zweifel and Arzoumanian are quantitatively significant but governed by steric rather than electronic factors.

The dihydroboration of acetylene itself in tetrahydrofuran has been described by Zakharkin and Kovredov.[86] The polymeric product, $(C_2H_4B)_n$, was cleaved by boron trichloride at 180–200°C to yield approximately equal amounts of 1,1- and 1,2-bis(dichloroboryl)ethane, $CH_3CH(BCl_2)_2$ and $Cl_2BCH_2CH_2BCl_2$. It seems likely that the relatively large amount of 1,2-isomer arises from sterically hindered addition of dialkylborane to alkenylborane during the later stages of the hydroboration. It was not proved whether the isomer ratio was unaltered by the boron trichloride treatment.

Köster and Rötermund[34] have reported that the reaction of ethylborane with acetylene at 200°C yields a carborane, $CH_3C(BEt)_3CCH_3$. Related carboranes have also been obtained in low but synthetically useful yields from dialkylpropynylboranes and dialkylboranes.[33] The thermodynamically stable carboranes result from cracking and disproportionation of the original hydroboration products, and these reactions provide no information whatever about the original isomer ratio in the hydroboration of the alkenylborane intermediates. Carboranes are reviewed by R. E. Williams elsewhere in this volume.[82]

4. Dichlorocarbene

The addition of dichlorocarbene to a vinylboronic ester to form a cyclopropane[83] is discussed together with other types of reactions of dichlorocarbene and boranes in Section IV, B, 4 (d).

B. *Displacements and Boron-to-Carbon Migrations*

1. *Diazoalkanes and Boron Compounds*

The earliest examples of migration of nucleophiles from an electron-rich boron atom to an electron-deficient carbon atom are provided by the reactions of diazo compounds with boron halides or boronic esters, BX_3. Diazomethane is

$$BX_3 + CH_2N_2 \longrightarrow X_3\bar{B}\!-\!CH_2\!-\!\overset{+}{N_2} \xrightarrow{-N_2} X_2BCH_2X \xrightarrow{CH_2N_2} X_2\bar{B}\!-\!CH_2\!-\!\overset{+}{N_2}$$
$$\underset{CH_2X}{\big|}$$

$$\xrightarrow{-N_2} X_2B\!-\!CH_2CH_2X \xrightarrow[\;]{(n-2)CH_2N_2} X_2B\!-\!(CH_2)_n\!-\!X$$

polymerized to polymethylene, evidently by a mechanism which involves migration of the *B*-alkyl chain from boron to the electron-deficient methylene carbon atom each time another molecule of diazomethane complexes with the boron atom. Bawn and Ledwith[1] have reviewed this general topic in Volume 1, to which the reader is referred for further details. Diazo compounds also add to the double bond of unsaturated boronic esters, a topic which is reviewed in Section V, A, 2 of this chapter.

Diarylboron azides, $Ar_2B\!-\!N_3$, are somewhat analogous to the diazomethaneboron trifluoride system. Paetzold and co-workers[63, 64] have reported that these azides lose nitrogen at about 200°C with aryl migration from boron to nitrogen. On the basis of molecular weight measurements, the product was believed to be the dimer of Ar—BN—Ar containing a four-membered ring (see Section II, B, 4, structure (X), but the author considers this conclusion doubtful for the reasons set forth in Section II, B, 4.

2. *α-Haloalkaneboronic Esters*

(a) *General considerations.* α-Haloalkaneboronic esters are stable compounds suitable for mechanistic studies. Because of their isoelectronic relationship to α-halo carboxylic esters, the mechanisms of their reactions are relevant to "pure" organic chemistry.

Nucleophiles (X^-) displace bromide ion from α-bromoalkaneboronic esters extremely rapidly compared to their rates of reaction with ordinary alkyl bromides. There is extensive evidence for direct involvement of the boron atom in the transition state (XLVI), which is illustrated with both conventional and qualitative molecular orbital structural formulas.

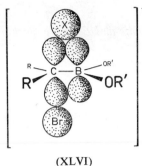

(XLVI)

Nucleophiles ranging in basicity from iodide ion or water to butoxide ion and various carbanion sources have been tested. The bond orders of the partial bonds in the transition state (XLVI) must vary over a wide range, but there is no evidence for any transition state that cannot be represented accurately as a variation of (XLVI). Basic nucleophiles add to the boron atom to form a metastable complex ion, $R_2CBrBX(OR)_2^-$, prior to the transition state. Nonbasic nucleophiles such as iodide ion probably do not. It should be noted that the transition state structure (XLVI) places one electron pair in a three-membered ring, which is a stable electronic configuration, and that (XLVI) closely resembles all transition states (or intermediates) for migration of a nucleophile from one electrophilic atom to another adjacent one. The carbon–boron π-bonding in (XLVI) is probably significant if the nucleophile X^- is weakly basic, but weak when a strongly basic nucleophile is used.

(b) *Strongly basic nucleophiles.* The first α-haloalkaneboronic ester to become available was dibutyl 1-bromo-3,3,3-trichloropropane-1-boronate (XII).[38] (See Section III, A, 1.) Matteson and Mah[48] have found that displacement of bromide ion from (XII) by a variety of nucleophilic reagents is greatly accelerated by the neighboring boron atom. Strong bases add to the boron atom, then displace bromide ion as they migrate to the α-carbon. Butoxide ion yields 1-butoxy-3,3,3-trichloropropane-1-boronate (XLVII).

It is remarkable that the bromoalkaneboronic ester (XII) undergoes displacement reactions rather than dehydrobromination, which is the characteristic mode of reaction of 1-bromo-3,3,3-trichloropropane and 1-bromo-1-carbethoxy-3,3,3-trichloropropane. Even potassium t-butoxide in t-butyl alcohol converted the bromoalkaneboronic ester (XII) to the n-butoxyboronic ester (XLVII) (after re-esterification with n-butyl alcohol). We were able to dehydrobrominate (XII) to dibutyl 3,3,3-trichloropropene-1-boronate, $Cl_3CCH=CHB(OBu)_2$, only by using t-butylamine in a nonhydroxylic medium.[50] Steric hindrance makes this amine very weakly basic toward boron.

Evidence for the proposed mechanism of displacement of bromide ion from (XII) is provided by the reaction with sodium butyl mercaptide in butanol. Although the expected sulfide is the major product, a small

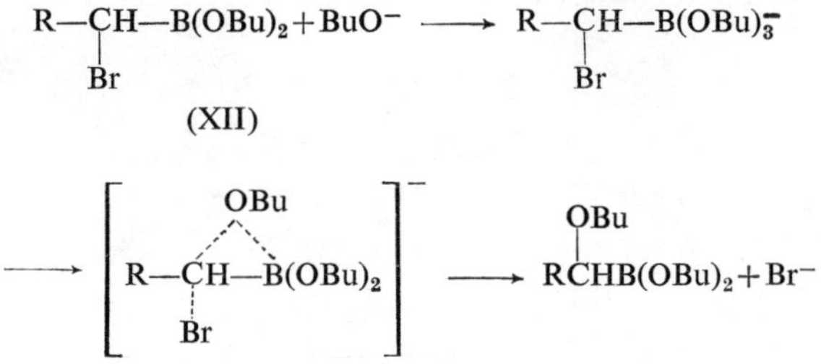

$$R-\underset{\underset{Br}{|}}{C}H-B(OBu)_2 + BuO^- \longrightarrow R-\underset{\underset{Br}{|}}{C}H-B(OBu)_3^-$$

(XII)

$$\longrightarrow \left[R-\underset{\underset{Br}{\overset{OBu}{|}}}{C}H-\overset{\frown}{B}(OBu)_2 \right]^- \longrightarrow RCHB(OBu)_2 + Br^-$$

(XLVII)

$$R = Cl_3CCH_2^-$$
$$Bu = n\text{-}C_4H_9$$

amount of the butoxy derivative (XLVII) is also formed.[48] This result is unexpected because mercaptide ions normally react faster than alkoxide ions in displacement reactions, and because the concentration of the weakly basic mercaptide ion must be much greater than that of the strongly basic butoxide ion, just as the concentration of HS^- far exceeds that of OH^- when alkali is added to an excess of aqueous hydrogen sulfide. However, boron bonds relatively weakly to sulfur, and the concentration of anionic adducts of the type $R-B(OBu)_3^-$ may exceed that of $R-B(OBu)_2SBu^-$. If the group R contains two alkyl substituents on the α-carbon, bromide ion departure evidently requires little nucleophilic assistance and sulfur loses its advantage over oxygen. Thus, sodium thiophenolate in butanol reacts with dibutyl 2-bromopropane-2-boronate (XLVIII) to yield almost exclusively the product of displacement of bromide by butoxide, dibutyl 2-butoxypropane-2-boronate (XLIX).[52]

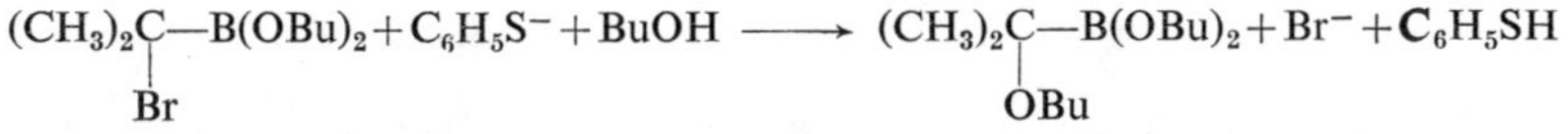

$$(CH_3)_2\underset{\underset{Br}{|}}{C}-B(OBu)_2 + C_6H_5S^- + BuOH \longrightarrow (CH_3)_2\underset{\underset{OBu}{|}}{C}-B(OBu)_2 + Br^- + C_6H_5SH$$

(XLVIII)   (XLIX)

It has been shown[48, 52] that the α-haloalkaneboronic esters do not solvolyze under the foregoing reaction conditions.

The most conclusive evidence for the general mechanism of displacement reactions of bromoalkaneboronic esters is provided by the behavior of (XII) with Grignard reagents. Phenylmagnesium bromide phenylates the boron atom of (XII) at $-70°C$ to yield the anion (L). Acidification of (L) below $-40°C$ removes a butoxide ion and yields *B*-phenyl-*B*-1-bromo-3,3,3-tri-chloro-1-propyl-*B*-butoxyborane (LI). However, allowing a solution of the anion (L) to warm to room temperature leads to phenyl migration from boron to carbon, resulting in formation of dibutyl 1-phenyl-3,3,3-tri-chloropropane-1-boronate (LII).

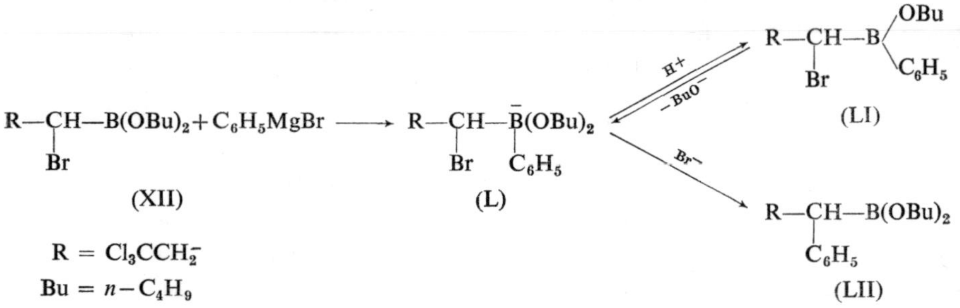

Similar reactions have been observed with alkyl Grignard reagents.[48]

Proof of the structure of the bromoalkylborinic ester (LI) was provided by an alternate synthesis from *B*-vinyl-*B*-phenyl-*B*-butoxyborane (VIII), Section III, A, 6). Rearrangement of the borinic ester (LI) to the boronic ester (LII) was effected by treatment with sodium butoxide or even by brief shaking of an ethereal solution of (LI) with sodium bicarbonate.

Direct displacement of bromide ion by butoxide ion in borinic esters such as (LI) does not occur. Ethylmagnesium bromide and the bromoboronic ester (XII) yield only dibutyl 1,1,1-trichloropentane-3-boronate (LIII) from bromide displacement. The isomeric α-butoxy *B*-ethyl borinic ester (LIV) has been synthesized from the α-butoxy boronic ester (XLVII) and ethylmagnesium bromide. From bond energies, the boronic ester (LIII) is more stable than its borinic ester isomer (LIV) by 30–40 kcal/mole.

$$Cl_3CCH_2\text{---}\underset{\underset{C_2H_5}{|}}{C}H\text{---}B(OBu)_2 \qquad Cl_3CCH_2\text{---}\underset{\underset{BuO}{|}}{C}H\text{---}\underset{\underset{C_2H_5}{|}}{B}\text{---}OBu$$

$$\text{(LIII)} \qquad\qquad\qquad\qquad \text{(LIV)}$$

Since displacement of bromide directly by alkoxide would be exothermic, the combined displacement and rearrangement process for converting bromoborinic esters such as (LI) to boronic esters such as (LII) must be very exothermic.[48]

The highly exothermic nature of the rearrangement process precludes the possibility of any intermediate zwitterion (LV).

$$R-\underset{\underset{Br}{|}}{C}H-\underset{\underset{C_6H_5}{|}}{B}-OBu+BuO^- \xrightarrow{\quad\times\quad} Br^- + R-\overset{+}{C}H-\overset{-}{B}(OBu)_2$$
$$\underset{(LV)}{}$$

The zwitterion (LV) would have to be at an energy minimum to have finite existence, and this is hardly possible when the steep downhill energy slope for rearrangement is considered.[48] The active participation of aryl and alkyl neighboring groups in solvolysis reactions (nonclassical carbonium ions) occurs with far less than 30 kcal/mole driving the rearrangement. Aryl participation occurs in open-chain systems even when the energies of the rearranged and unrearranged products are equal.[14] To unequivocally rule out (LV) as an intermediate it would be necessary to prove the stereospecificity of the reaction, which has not yet been done because of the anticipated considerable experimental difficulty in resolving a bromoalkylboron compound and the small likelihood of any surprising result.

We have recently found a useful synthetic application for the rearrangement of α-haloalkylborinic esters (work with G. Srivastava, to be published). A series of 1-arylethane-1-boronic esters was desired for a mechanistic study, but the synthesis of these compounds by the conventional route encounters difficulties.[43] The addition of hydrogen iodide to B-vinyl-B-aryl-B-butoxyboranes followed by treatment with sodium butoxide has proved to be a useful route to dibutyl 1-arylethaneboronates (LVI).

$$CH_2{=}CH-\underset{\underset{Ar}{|}}{B}-OBu+HI \longrightarrow CH_3\underset{\underset{I}{|}}{C}H-\underset{\underset{Ar}{|}}{B}-OBu \xrightarrow{NaOBu} CH_3\underset{\underset{Ar}{|}}{C}H-B(OBu)_2$$
$$\text{(LVI)}$$

Our first attempts to react diethyl sodiomalonate and related carbanions in alcoholic media with α-haloalkaneboronic esters failed.[48] Evidently the tendency of the neighboring boron atom to bond to oxygen favored either reaction with alkoxide ion or oxygen alkylation of the sodiomalonate. However, Matteson and Schaumberg[52] found that treatment of dibutyl 1-bromo-3,3-dicyanopropane-1-boronate (XIII) with pyridine or sodium butoxide yields the cyclic alkylation product, dibutyl 2,2-dicyanocyclopropaneboronate (LVII).

$$(C_4H_9O)_2B\underset{\underset{Br}{|}}{C}HCH_2CH(CN)_2 \xrightarrow{base} (C_4H_9O)_2B-\underset{}{\overset{}{<}}\overset{CN}{\underset{}{{-}CN}}$$
$$\text{(XIII)}\qquad\qquad\qquad\qquad\text{(LVII)}$$

Matteson and Cheng[44] have found that dibutyl iodomethaneboronate (LVIII) alkylates malononitrile in the presence of potassium t-butoxide to yield dibutyl 2,2-dicyanoethaneboronate (LIX). Dibutyl iodomethaneboronate also

$$ICH_2B(OC_4H_9)_2 + CH_2(CN)_2 \xrightarrow{\text{KOt-Bu}} (C_4H_9O)_2B-CH_2CH(CN)_2$$
$$\text{(LVIII)} \qquad\qquad\qquad \text{(LIX)}$$

alkylates sodiomalonic ester, amines, and some other nucleophiles that have failed to yield alkylation products with other α-haloalkaneboronic esters. This halomethaneboronic ester is prepared from iodomethylmercuric iodide and boron tribromide followed by esterification with butanol.

Dibutyl 1-iodoethaneboronate, $CH_3CHIB(OBu)_2$, recently has been found to alkylate malonitrile under the same conditions used with the iodomethyl compound (LVIII), though the yields are only moderate (unpublished work with J. Ebbert).

(c) *Weakly basic nucleophiles.* Matteson and Schaumberg[52] have compared the rates of reaction of dibutyl 2-bromopropane-2-boronate (XLVIII) and dibutyl 1-bromoethaneboronate (LX) with that of allyl bromide with sodium iodide in acetone. The relative rates at 22–24°C were (XLVIII),

$$\underset{\substack{|\\ \text{Br}}}{CH_3CH}-B(OBu)_2 + NaI \longrightarrow \underset{\substack{|\\ \text{I}}}{CH_3CH}-B(OBu)_2 + NaBr$$
$$\text{(LX)} \qquad\qquad\qquad\qquad \text{(XXXIV)}$$

0.419; allyl bromide, 1.0; and (LX), 1.44. Comparison with literature values for the reaction of alkyl bromides gave the values isopropyl bromide, 1.0; ethyl bromide, 40; dibutyl 2-bromopropane-2-boronate (XLVIII), 1600; allyl bromide, 4000; and dibutyl 1-bromoethaneboronate (LX), 6000. Thus, the dibutoxyboryl group greatly accelerates the displacement of bromide by iodide and reduces the sensitivity of the reaction to steric hindrance by the second methyl group in the 2-propane compound (XLVIII). These results are in accord with the previously postulated general transition state structure (XLVI). The vacant p-orbital on the boron atom may interact directly with either the attacking iodide ion or, perhaps more likely, the departing bromide ion. (If the attacking and leaving nucleophiles were identical, e.g. both bromide ion, the direct interaction would have to be with the attacking species half the time and the departing species the other half, or else both equally all the time. This is required by the principle of microscopic reversibility. Entropy should favor the first alternative.)

Surprisingly, a β-trichloromethyl substituent on the bromoalkaneboronic ester does not retard the rate of displacement of bromide by iodide. Dibutyl 1-bromo-3,3,3-trichloropropane-1-boronate (XII) reacts with iodide ion at the same rate, within experimental error, as does dibutyl 1-bromoethaneboronate (LX). In contrast, ethyl 1-bromo-3,3,3-trichloropropane-1-carboxylate reacts only 0.2–0.3 times as fast as n-butyl bromide with sodium iodide or, as a very rough estimate, at about 1/1000 the rate of an α-bromo carboxylic ester without the trichloromethyl substituent.[48] Thus, the boronic ester reacts hundreds of times faster than the carboxylic ester in this case. Perhaps the retardation due to the electron-withdrawing trichloromethyl group is partially relieved by electron donation from the dibutoxyboryl group but enhanced by the electronegative carbethoxy group. Ordinarily, the carbethoxy group activates the bromine toward displacement more than does the dibutoxyboryl group.

The dibutoxyboryl group also activates a β-halogen toward displacement. Matteson and Liedtke[46] found that dibutyl 2-bromoethaneboronate, $BrCH_2CH_2B(OBu)_2$, reacts about one-fifth as fast as allyl bromide with sodium iodide, or several times faster than a simple alkyl bromide. It was suggested that this activating effect is due to the inductive effect of the boron atom, though it is also possible that the boron atom bonds weakly to the attacking or leaving halide ion, much as postulated when the displacement site is α to the boron. Such an interaction would be more like what is normally thought of as solvation rather than covalent bond formation.

Matteson and Schaumberg[52] have studied the solvolysis of dibutyl 2-bromopropane-2-boronate (XLVIII) and dipropyl 1-bromoethaneboronate, $CH_3CHBrB(OC_3H_7)_2$, in aqueous ethanol. The actual species undergoing solvolysis are thought to be the boronic acids. Solvolysis of dibutyl 2-bromopropane-2-boronate in 50 per cent ethanol was found to be an efficient method for preparing 2-hydroxypropane-2-boronic acid (LXI).

$$(CH_3)_2\underset{\underset{Br}{|}}{C}\!-\!B(OH)_2 + H_2O \longrightarrow (CH_3)_2\underset{\underset{OH}{|}}{C}\!-\!B(OH)_2 + H^+ + Br^-$$

$$(LXI)$$

It was found to be impossible to measure the solvolysis rates by titration of the hydrogen ion liberated, since the bromoalkaneboronic acids are attacked extremely rapidly by base in the useful endpoint range above pH 4. However, measurement of the liberated acid with a pH meter yielded good pseudo-first-order kinetics. For dibutyl 2-bromopropane-2-boronate (XLVIII) in 50 per cent ethanol, the rate constant is 3.86

$(\pm 0.2) \times 10^{-4} sec^{-1}$, ΔH^* is 17.5 kcal/mole, and ΔS^* is -15.5 cal/deg-mole. In the range 40–70 per cent ethanol, the rate constants varied according to the Grunwald–Winstein equation[23] with a slope m of 0.68 (± 0.13). To solvolyze dipropyl 1-bromoethaneboronate at a measurable rate, it was necessary to go to 90 per cent water, where k is 0.858×10^{-4} sec^{-1}, ΔH^* is 17.7 cal/mole and ΔS^* is -17.8 cal/deg-mole.

For mechanistic interpretation, it is useful to compare the foregoing results with the behavior of alkyl halides. Extrapolated to 50 per cent ethanol, the relative orders of magnitude of the rates are isopropyl bromide, 1; dipropyl 1-bromoethaneboronate, 1; dibutyl 2-bromopropane-2-boronate, 100; t-butyl bromide, 10,000. The α-bromoalkaneboronic acids do not appear to yield carbonium ion intermediates, but must involve a tightly bound water molecule in the transition state (LXII), which is another vari-

$$(CH_3)_2\underset{\underset{Br}{|}}{C} \cdots\cdots \underset{\overset{|}{O}}{\overset{H \diagdown \diagup H}{}} \cdots\cdots B(OH)_2$$

(LXII)

ation of the postulated general transition state structure (XLVI) for boron participation. The Grunwald–Winstein slope m for 2-bromopropane-2-boronic acid, 0.68, is between that for ethyl bromide, 0.34, and t-butyl chloride, 1.0, suggesting an intermediate polarity between concerted displacement and carbonium ion formation. Phenacyl bromide solvolysis, which should be similar in mechanism to that of the boronic acid but should have more concerted and less carbonium ion character, shows a Grunwald–Winstein slope m of 0.2.[69] The entropy of activation, -15.5 cal/deg-mole, for 2-bromopropane-2-boronic acid is far more negative than that for t-butyl chloride, -1.4, in 50 per cent ethanol. Less is known about 1-bromoethaneboronic acid, but its entropy of activation is 30 units below that of t-butyl chloride in 10 per cent ethanol, which seems sufficient evidence to exclude a carbonium ion intermediate.

Similar attack of water on a boron atom occurs in the solvolytic decomposition of 2-bromoethaneboronic acid, $BrCH_2CH_2B(OH)_2$, to bromide ion, ethylene, and boric acid. Matteson and Liedtke[47] found that the rate is very sensitive to the basicity of the solvent, the Grunwald–Winstein slope m in aqueous ethanol is 0.40, ΔH^* in 90 per cent ethanol is 12 kcal/mole, and ΔS^* is -28 cal/deg-mole. The isotope effect k_H/k_D is 1.7, comparing ordinary 79 per cent ethanol with fully 0-deuterated solvent. The kinetics rule out the intermediacy of any anion such as $BrCH_2CH_2B(OH)_3^-$

and the relatively small value of the isotope effect indicates that proton loss does not occur from the intermediate, $BrCH_2CH_2B(OH)_2 \cdot H_2O$, in the transition state. The data suggest the same kind of interaction between water and the boron in the transition state for this elimination reaction as in the transition states for displacement of bromide from α-bromoalkaneboronic acids, except that the water is somewhat more tightly bound in the particular elimination which was studied.

Treatment of dibutyl 2-bromoethane-2-boronate (XLVIII) with thiourea followed by hydrolysis yields 2-(S-thioureido)propane-2-boronic acid hydrobromide (LXIII). This displacement by the sulfur nucleophile contrasts

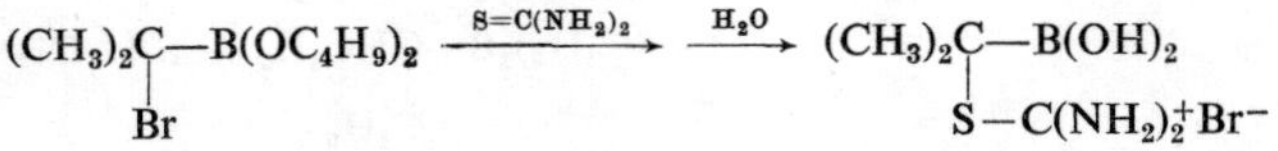

$$(CH_3)_2\underset{\underset{Br}{|}}{C}\!\!-\!\!B(OC_4H_9)_2 \xrightarrow{\text{S=C(NH}_2)_2} \xrightarrow{\text{H}_2\text{O}} (CH_3)_2\underset{\underset{S-C(NH_2)_2^+Br^-}{|}}{C}\!\!-\!\!B(OH)_2$$

(XLVIII)   (LXIII)

with the previously mentioned behavior of (XLVIII) toward the more basic thiophenoxide in butanol, which yielded only the product of butoxide displacement, dibutyl 2-butoxypropane-2-boronate (XLIX).[52]

3. A γ-Trichloropropeneboronic Ester

Matteson and Mah[50] prepared dibutyl 3,3,3-trichloropropene-1-boronate (LXIV) by dehydrobromination of dibutyl 1-bromo-3,3,3-trichloropropane-1-boronate (XII) with t-butylamine. This trichloropropeneboronic ester (LXIV) readily undergoes S_N2' displacements of a chloride ion by butoxide ion to yield dibutyl 1,1-dichloro-3-butoxypropene-3-boronate (LXV) or by phenylmagnesium bromide to yield dibutyl 1,1-dichloro-3-phenylpropene-3-boronate (LXVI). Attempts to obtain a borinic ester,

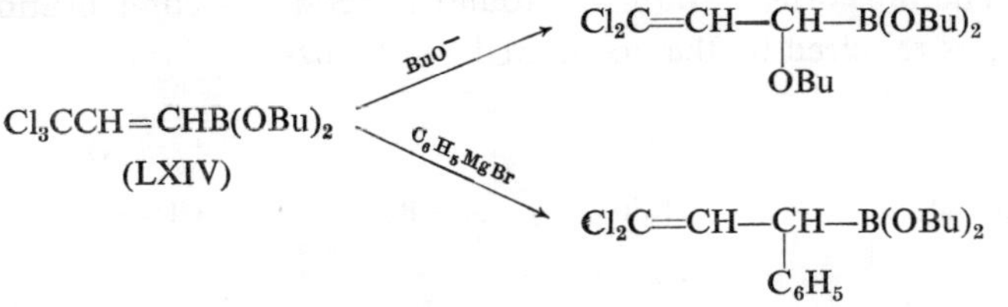

$Cl_3CCH\!=\!CHB(OBu)_2$ $\xrightarrow{BuO^-}$ $Cl_2C\!=\!CH\!-\!\underset{\underset{OBu}{|}}{C}H\!-\!B(OBu)_2$ (LXV)

(LXIV)

$\xrightarrow{C_6H_5MgBr}$ $Cl_2C\!=\!CH\!-\!\underset{\underset{C_6H_5}{|}}{C}H\!-\!B(OBu)_2$ (LXVI)

$Cl_3CCH\!=\!CHB(OBu)C_6H_5$, by acidification of the cold reaction mixture from (LXIV) and phenylmagnesium bromide were not successful.

4. Trialkylborane Derivatives

(a) *Tetraalkylboron anions.* Jaeger and Hesse[31] have reported that treatment of lithium tetrabutylboron with benzyl chloride in ether at 110–120° results in an oxidative rearrangement to *B*-4-octyl-*B*,*B*-dibutyl-borane (LXVII), with toluene and lithium chloride as the other products.

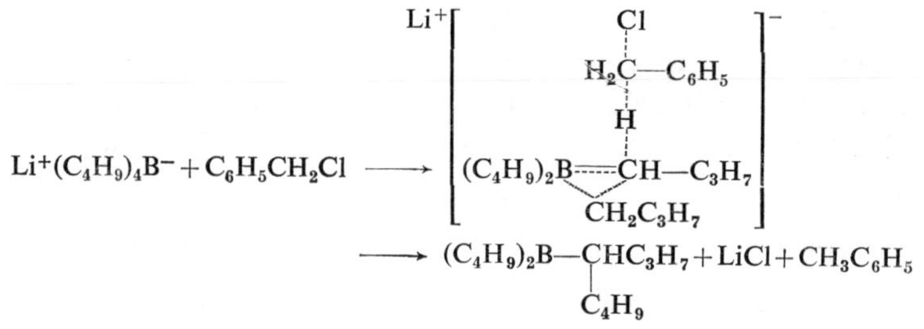

$$Li^+(C_4H_9)_4B^- + C_6H_5CH_2Cl \longrightarrow Li^+\begin{bmatrix} & Cl & \\ & H_2C-C_6H_5 & \\ & H & \\ (C_4H_9)_2B\text{-----}CH-C_3H_7 \\ & CH_2C_3H_7 & \end{bmatrix}^-$$

$$\longrightarrow (C_4H_9)_2B-\underset{\underset{C_4H_9}{|}}{C}HC_3H_7 + LiCl + CH_3C_6H_5$$

(LXVII)

Some displacement of chloride from benzyl chloride by butyl anion transfer also occurs, yielding tributylborane and n-pentylbenzene. The author has suggested his own transition state for the reaction, which is probably a concerted rearrangement and displacement somewhat related to reactions of α-haloalkaneboronic esters and the general transition state (XLVI) previously described. For the same reasons mentioned before in connection with reactions of α-haloalkaneboronic esters, it is unlikely that the intermediate "boron ylid", $Bu_3B^--CH^{\pm}-C_3H_7$, suggested by Jaeger and Hesse[31] has any finite existence, though it is a useful mnemonic device for remembering the outcome of this and related rearrangements.

Tufariello and Lee[81] have found that dimethyloxosulfonium methylide adds to trialkylboranes to yield the postulated zwitterion (LXVIII). One of the alkyl groups then migrates from boron to displace dimethyl sulfoxide from carbon. The norbornyl group was found to retain its configuration in this process, as required by the postulated mechanism.

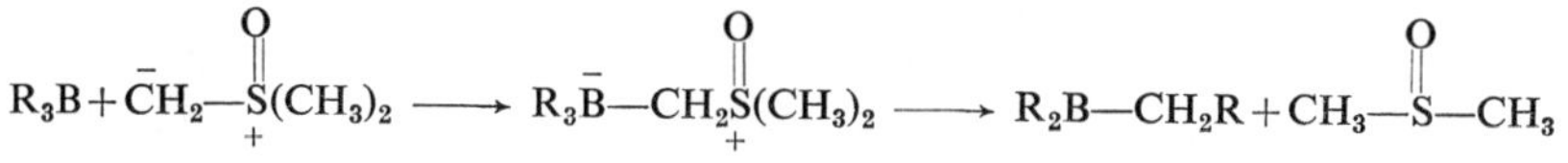

$$R_3B + \overset{-}{C}H_2-\overset{\overset{O}{\|}}{\underset{+}{S}}(CH_3)_2 \longrightarrow R_3\overset{-}{B}-CH_2\overset{\overset{O}{\|}}{\underset{+}{S}}(CH_3)_2 \longrightarrow R_2B-CH_2R + CH_3-\overset{\overset{O}{\|}}{S}-CH_3$$

(LXVIII)

Zweifel, Arzoumanian and Whitney[88] have recently found that treatment of *B*,*B*-dicyclohexyl-*B*-*trans*-hexenylborane with sodium hydroxide and iodine results in attack of I^+ on the double bond followed by migration of a cyclohexyl group from boron to carbon. The initially formed

β-iodoalkylboron compound undergoes rapid deboronoiodination to yield *cis*-1-hexenylcyclohexane.

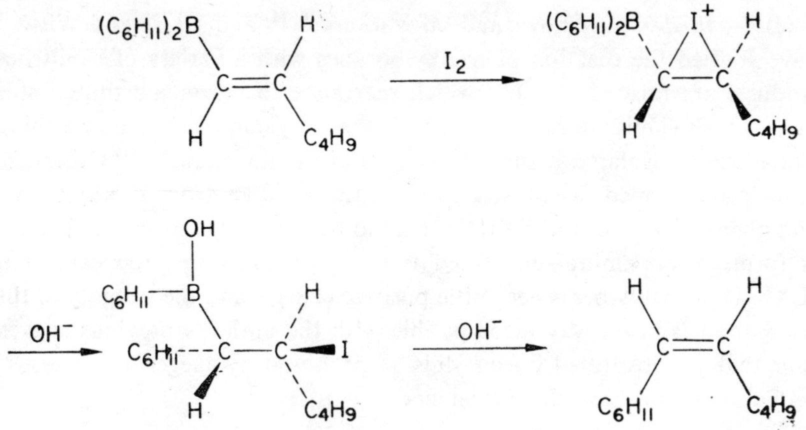

(b) *Carbon monoxide.* Hillman[27] has studied the reaction of various trialkylboranes with carbon monoxide under pressure. The alkyl groups migrate in stepwise fashion from the boron atom to the carbon monoxide carbon atom. For example, triethylborane and carbon monoxide yield 2,3,3,5,6,6-hexaethyl-1,4-dioxa-2,5-diboracyclohexane (LXIX), which rearranges in water at 140° to triethylmethaneboronic acid (LXX).

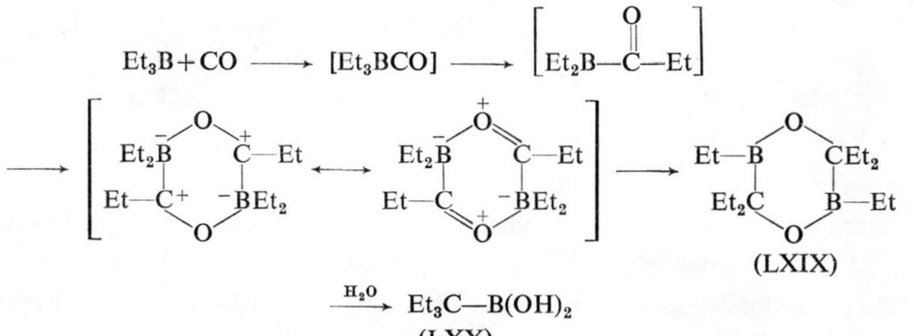

The first postulated product of ethyl migration, Et_2BCOEt, is not stable and probably dimerizes. Although this dimer is not isolable, the analogous structures from isonitriles and boranes have recently been prepared (next paragraph). Hillman has postulated that the first intermediate, Et_2BCOEt, is trapped by aldehydes, which yield 1,3-dioxa-4-boracyclopentanes (LXXI).

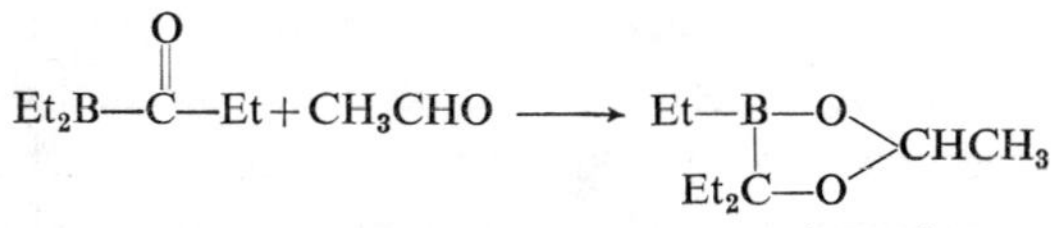

Hillman's mechanisms are reasonable rationalizations of the formation of the observed products, though there is no other kind of evidence to support them.

(c) *Isonitriles.* Casanova and co-workers[10, 11] and Hesse and Witte[26] have studied the reaction of trialkylboranes with a variety of isonitriles. Adducts are formed at $-60°$ which rearrange to boron-substituted aldimines by alkyl migration from boron to the isocyanide carbon on warming. These can be isolated if the substituents are bulky enough.[11] Otherwise, dimers are formed, as illustrated with the product from triethylborane and phenyl isocyanide (LXXII). It should be noted that structure (LXXII) is formally an amino-substituted boron ylid. However, the carbon in (LXXII) probably bears very little positive charge, and the stability of this compound is in no way incompatible with the author's previous suggestion that unsubstituted boron ylids as proposed by Jaeger and Hesse[31] probably do not have finite existence.

$$C_6H_5\text{—}NC + BEt_3 \xrightarrow{-60°} C_6H_5NCBEt_3 \xrightarrow{warm} C_6H_5N{=}\underset{Et}{C}\text{—}BEt_2 \longrightarrow$$

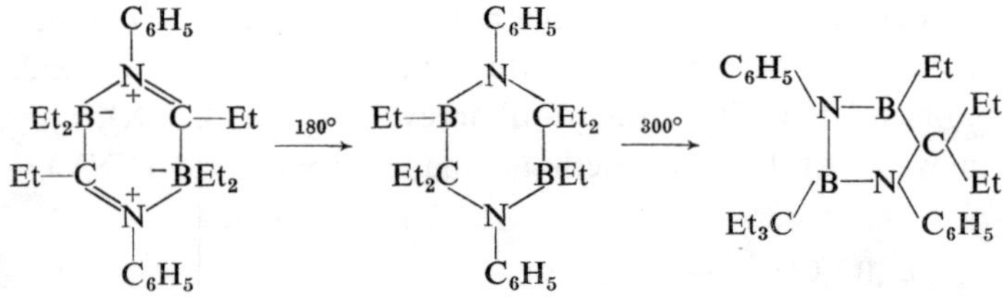

(LXXII)   (LXXIII)   (LXXIV)

At 180° the second stage of ethyl migration occurs, converting the aldimine dimer (LXXII) to the cyclic structure (LXXIII), which undergoes the third and final stage of ethyl migration to the former isocyanide carbon at 300°. However, only one of the two *B*-ethyl groups migrates, contracting the six-membered ring to the very stable five-membered cyclic structure (LXXIV). (See Section II, B, 4 for a discussion of the possible instability of the four-membered ring that would arise if the second *B*-ethyl group migrated.)

(d) *Dichlorocarbene.* Dichlorocarbene, $:CCl_2$, bears some electronic analogy to carbon monoxide and isonitriles. The unshared electron pair of the "divalent" carbon can lead to coordination with boranes, leaving the carbon atom electron deficient and permitting boron–carbon alkyl migration. Seyferth and Prokai[75] have used the thermal decomposition of phenyl(bromodichloromethyl)mercury as a dichlorocarbene source in the presence of tributylborane. The major product isolated was 4-nonene

(58 per cent *cis*, 42 per cent *trans*), believed to be formed by way of the dichlorocarbene insertion product, *B,B*-dibutyl-*B*-1,1-dichloropentyl-borane (LXXV). The formation of (LXXV) is written in two steps, which is not unreasonable though there is no evidence on this question. The next step is a migration of chloride from carbon to boron with butyl migration from boron to carbon. Finally, the second chloride migrates to boron and a hydride shifts from a β-carbon to the α-carbon to yield the 4-nonene.

$$Bu_3B + C_6H_5HgCCl_2Br \xrightarrow[C_6H_6]{70°} Bu_3\bar{B}-\overset{+}{C}Cl_2 \longrightarrow Bu_2B-CCl_2Bu \longrightarrow \underset{Bu}{\overset{Cl}{Bu-B-CClBu}}$$
$$+ C_6H_5HgBr$$
$$(LXXV)$$

$$[BuBCl_2 + :CBu_2] \text{ or } \left[\underset{Cl}{\overset{Cl}{Bu-B}} = \underset{Bu}{\overset{H}{C}} - CH - C_3H_7 \right] \longrightarrow BuBCl_2 + BuCH = CHC_3H_7$$

Although Seyferth and Prokai[75] write a carbene intermediate, :CBu₂, as a precursor to 4-nonene, this is probably not meant to represent a free species. They have proved that butylchlorocarbene, Bu—C̈—Cl, is not an intermediate in the preceding step, since a mixture of tripropylborane and tributylborane with the dichlorocarbene source yields only the products of intramolecular reactions, 3-heptene and 4-nonene, and none of the octenes that would result from intermolecular reactions of a free carbene intermediate. A dialkylcarbene intermediate should be more difficult than an alkylchlorocarbene to generate, assuming comparable energy terms in the other products, dialkyl-, chloro- and alkyldichloroboranes.

Seyferth and Prokai obtained evidence that the phenyl analog of (LXXV), $(C_6H_5)_2BCCl_2C_6H_5$, is stable in solution. Transfer of phenylchlorocarbene, C_6H_5—C̈—Cl, to cyclohexene to yield 7-chloro-7-phenylnorcarane was observed. However, this result in no way implies that the carbene unit transferred is ever a free carbene intermediate. For example, the reaction of iodomethylzinc iodide with olefins to form cyclopropanes has been conclusively shown not to involve free carbene, :CH₂.[79] Another product formed indirectly from the α,α-dichlorobenzylborane was trichlorostyrene, $Cl_2C=CClC_6H_5$, evidently formed by way of a deborochlorination of $(C_6H_5)_2B—CCl_2—CCl_2—C_6H_5$.

Dimethyl benzeneboronate, $C_6H_5B(OCH_3)_2$, reacted with the dichlorocarbene source to yield products derived both from phenyl and from methoxy migration from boron to carbon.[75] In view of the very large energy difference in favor of phenyl migration,[48] this seems to indicate that the boronic ester—dichlorocarbene complex is exceedingly unstable and migration is exothermic and indiscriminate. A reasonable alternative explanation is that addition of dichlorocarbene to the boron atom is concerted with phenyl or methoxy migration from boron to carbon, or in other words, that this is a carbene insertion into the boron–carbon or boron–oxygen bond.

Seyferth and Prokai[75] have also suggested that methoxychlorocarbene, CH_3O—C—Cl, splits out easily from the grouping B—CCl_2OCH_3, but the sole evidence for this is failure of an "expected" formation of 7-chloro-7-methoxynorcarbene, which the author finds unconvincing. It would not be surprising if methoxychloromethylboron compounds such as (LXXVI) failed to persist simply because phenyl migration to carbon with displacement of chloride to boron is very much facilitated by the methoxy group, and the author could write several speculative mechanisms to account for the products, one of which is illustrated.

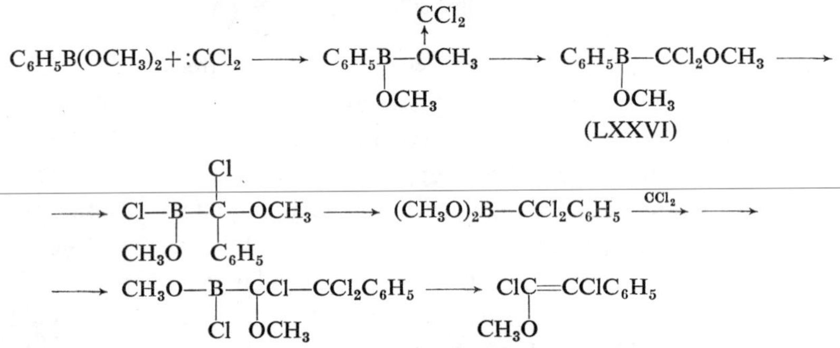

$$C_6H_5B(OCH_3)_2 + :CCl_2 \longrightarrow C_6H_5\underset{\underset{OCH_3}{|}}{\overset{\overset{CCl_2}{\uparrow}}{B}}-OCH_3 \longrightarrow C_6H_5\underset{\underset{OCH_3}{|}}{B}-CCl_2OCH_3 \longrightarrow$$

(LXXVI)

$$\longrightarrow Cl-\underset{\underset{CH_3O}{|}}{B}-\underset{\underset{C_6H_5}{|}}{\overset{\overset{Cl}{|}}{C}}-OCH_3 \longrightarrow (CH_3O)_2B-CCl_2C_6H_5 \xrightarrow{CCl_2} \longrightarrow$$

$$\longrightarrow CH_3O-\underset{\underset{Cl}{|}}{B}-\underset{\underset{OCH_3}{|}}{C}Cl-CCl_2C_6H_5 \longrightarrow \underset{\underset{CH_3O}{|}}{Cl}C{=}CClC_6H_5$$

Woods and Bengelsdorf[83] added dichlorocarbene to the double bond of the cyclic vinylboronic ester (XVI), using phenyltrichloromethylmercury as the carbene source. The 2,2-dichlorocyclopropaneboronic ester (LXXVII) was obtained in 60 per cent yield and its structure unequivocally established by infrared and n.m.r. spectra. Treatment of the cyclopropyl compound (LXXVII) with potassium t-butoxide yielded a small amount (4–7 per cent) of 1-chloroallene. No evidence was found for insertion of dichlorocarbene into the carbon–boron bond, which would yield

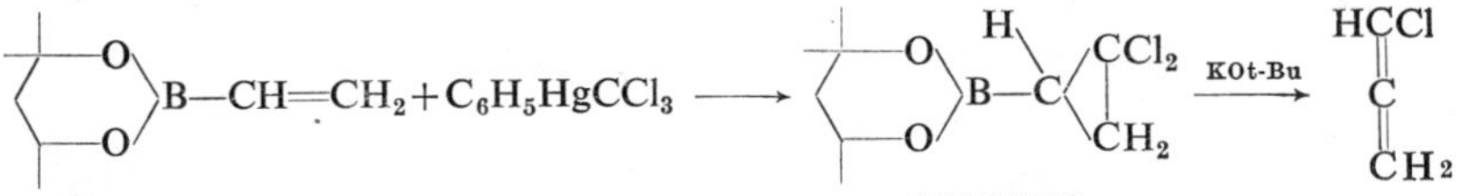

(XVI)   (LXXVII)

$H_2C{=}CHCCl_2BO_2C_6H_{12}$, though the possibility of a small amount was not excluded.

5. Hydroboration of Substituted Olefins

(a) *Haloalkylboranes.* Hawthorne and Dupont[25] treated vinyl chloride in dimethyl ether at $-75°C$ with diborane. The product, believed to be tris(2-chloroethyl)-borane, $(ClCH_2CH_2)_3B$, decomposed very exothermically when the solution warmed toward room temperature and the dimethyl ether evaporated. Hawthorne and Dupont succeeded in isolating a small amount of the dimethyl ether complex of chloroethylboron dichloride, $ClCH_2CH_2BCl_2 \cdot (CH_3)_2O$. On treatment with water, this compound rapidly decomposed to boric acid, dimethyl ether, ethylene, and hydrochloric acid, the last two being obtained in quantitative yield.

Hawthorne and Dupont[25] also hydroborated allyl chloride in ether at room temperature. A moderate yield of tris(3-chloropropyl)-borane (LXXVIII) was obtained, together with some bis(3-chloropropyl)-boron chloride and 3-chloropropylboron dichloride. On treatment with aqueous sodium hydroxide, these compounds were converted to cyclopropane in high yields. Hawthorne[24] has developed this method into a useful general synthesis of cyclopropanes.

$$CH_2{=}CH{-}CH_2{-}Cl + B_2H_6 \longrightarrow (ClCH_2CH_2CH_2)_3B \xrightarrow{\text{OH}^-} Cl^- + \triangle + \text{borate}$$
$$\text{(LXXVIII)}$$

Several other eliminations of boron halide from β-haloalkylboranes have been reported.[4, 7, 15] No attempt to collect a complete list of references or to review this type of reaction in detail will be made here.

Pasto and Snyder[71] have reported that hydroboration of 4-t-butyl-1-chlorocyclohexene followed by deboronation with hydrogen peroxide yielded a mixture of alcohols containing 65 per cent *trans-* and 15 per cent *cis*-4-t-butylcyclohexanol, with the remainder 3-t-butylcyclohexanols. Since the hydroboration of 4-t-butylcyclohexene yields a 50:50 mixture of 3- and 4-t-butylcyclohexanols, Pasto and Snyder concluded that no more than 40 per cent of the t-butylcyclohexanol mixture from the chlorocyclohexene could have arisen by way of t-butylcyclohexene. In other words, 60 per cent of the hydroboration yielded 1-boryl-1-chloro-4-t-butylcyclohexene (LXXIX), which underwent migration of chloride from carbon to boron and hydride from boron to carbon, called "α-transfer", ultimately resulting in 4-t-butylcyclohexanol. The other 40 per cent of the hydroboration yielded 2-boryl-1-chloro-4-t-butylcyclohexene (LXXX), which eliminated boron halide to give 4-t-butylcyclohexene, which in turn was hydroborated and then oxidized to a 50:50 mixture of 3- and 4-t-butylcyclohexanols.

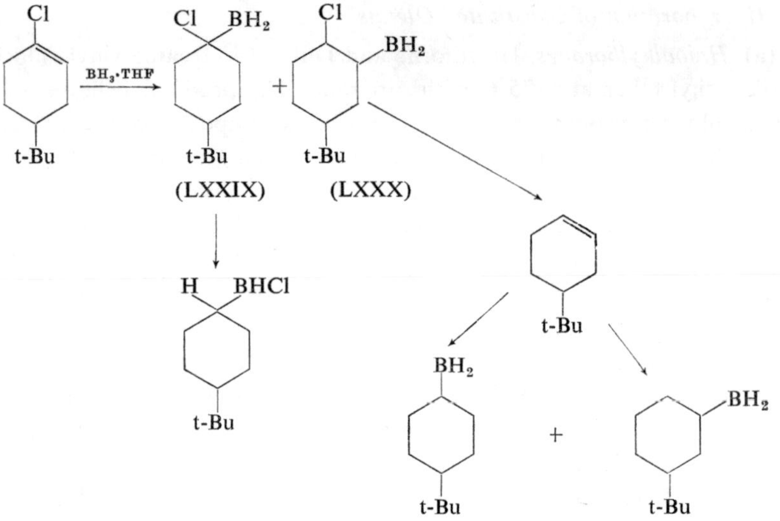

The postulated chloride–hydride interchange which converts the α-chloro-alkylborane (LXXIX) to the alkylboron chloride is probably mechanistically related to the nucleophilic displacement reactions of α-halo-alkaneboronic esters, described in Section IV, B, 2. Pasto[71] also has claimed evidence that the displacement of chloride by hydride occurs with more than 90 per cent inversion, as would be expected in a nucleophilic displacement, though the details were not published at that time. The elimination of boron chloride from the β-chloroalkaneboronic ester (LXXX) is certainly the expected behavior and has close analogy in the boronic ester series, where the β-haloalkylboron compounds have been isolated and the mechanisms of deboronobromination studied.[46, 47]

Pasto and Snyder[71] have ruled out carbene formation as a mode of decomposition of α-haloalkylboron hydrides in tetrahydrofuran. The most convincing evidence is from the hydroboration of 2-chloro-2-butene. If the 2-boryl-2-chlorobutane decomposed to methylethylcarbene, C_2H_5—$\ddot{C}$—CH_3, rearrangement of the carbene to a mixture of 1- and 2-butene followed by hydroboration of the 1-butene to 1-butanol would result, but 1-butanol was proved absent. It should be noted that the hydroboration conditions, which include boron–hydrogen bonds and a nucleophilic ether as solvent, are quite different from the conditions where Seyferth and Prokai[75] observed that α-chloroalkylboranes served as a carbene source in the absence of boron–hydrogen bonds with benzene as the solvent.

Pasto and Snyder[71] also studied the hydroboration of β-bromostyrene. After hydrogen peroxide treatment, the major product was 2-phenyletha- nol. No 1-phenylethanol was found. This seems to indicate that the ori- ginal hydroboration product was almost entirely 2-boryl-2-bromo-1- phenylethane (LXXXI) and not more than 3–4 per cent 1-boryl-2-bromo-1- phenylethane (LXXXII), since if the latter were formed it would deboro- nobrominate to styrene, which would yield about 17 per cent of 1-phenyl- ethanol under the reaction conditions. Pasto and Balasubramaniyan[66] showed that elimination of bromoborane, $BrBH_2$, from the β-bromoalkyl- borane (LXXXII) followed by recombination of the bromoborane with the styrene to yield the 2-phenylethylboron compound, $C_6H_5CH_2CH_2BHBr$, did not occur, since haloboranes are slower hydroborating agents than borane itself and give some 1-phenylethanol after the reaction sequence anyway.

$$C_6H_5CH{=}CHBr + BH_3{\cdot}THF \longrightarrow C_6H_5CH_2\underset{\underset{Br}{|}}{C}H{-}BH_2 + C_6H_5\underset{\underset{BH_2}{|}}{C}H{-}CH_2Br$$

$$\text{(LXXXI)} \qquad\qquad \text{(LXXXII)}$$

The α-bromoalkylboron hydride (LXXXI) would rearrange to the alkyl- boron bromide, $C_6H_5CH_2CH_2BHBr$, which would deboronate to 2-phe- nylethanol with hydrogen peroxide. The isomer (LXXXII) must have been present at least in small amount, since some styrene oxide was found in the product mixture, but the apparent very low yield of this isomer and total absence of 1-phenylethanol is anomalous. For both steric and electronic reasons, the bromine atom should tend to direct the boron atom β (yielding LXXXII), analogous to the behavior of the ethoxy group (next paragraph). Speculation on the possible mechanisms of hydrobora- tion is outside the scope of this review, and the chain of reasoning which indicates the absence of the β-bromoalkylboron compound (LXXXII) involves too many steps to allow the author to reach any conclusions about possible rearrangements and neighboring group effects in either bromoalkylborane isomer, (LXXXI) or (LXXXII). Rapid future devel- opment of this field can be expected, though it will probably be very difficult to get the kind of kinetic information needed to really define the details of the hydroboration and rearrangement mechanisms.

(b) *Alkoxyalkylboranes and alkylthioalkylboranes.* Pasto and Cumbo[68] have shown that hydroboration of β-ethoxystyrene yields mainly 1-boryl- 2-ethoxy-1-phenylethane (LXXXIII), which gives a 60 per cent yield of 2-ethoxy-1-phenylethanol if treated immediately with alkaline hydrogen peroxide. In less than an hour at room temperature, this β-ethoxyalkyl-boron compound (LXXXIII) undergoes deoxyborination to styrene, which hydroborates to the usual mixture of 1- and 2-borylphenylethanes.

$$C_6H_5CH\!=\!CH\!-\!OEt + BH_3 \cdot THF \longrightarrow C_6H_5CH\!-\!CH_2OEt \xrightarrow{H_2O_2} C_6H_5CH\!-\!CH_2OEt$$

$$BH_2 \qquad\qquad\qquad\quad OH$$

$$(LXXXIII) \qquad C_6H_5CH\!=\!CH_2 + H_2BOEt$$

Pasto and Cumbo concluded that some of the β-ethoxyalkylboron compound (LXXXIII) underwent transfer of the β-ethoxy group from carbon to boron with concurrent migration of hydrogen from boron to carbon, a "β-transfer" reaction. This is because more 1-phenylethanol than 2-phenylethanol was obtained when the hydroboration was only allowed to proceed for a short time at 0°C, though the yields of both were small. However, it is not apparent to the author why "β-transfer" should initially be faster than β-elimination, but that β-elimination clearly becomes the dominant mode of decomposition of (LXXXIII) as the reaction proceeds. Another anomaly is that some 2-ethoxy-1-phenyl-ethane, $C_6H_5CH_2CH_2OEt$, is formed in all cases, and the 2–8 per cent yields bear no consistent relation to the reaction conditions. It seems possible that some of these minor by-products result from the deboronation with hydrogen peroxide and not by rearrangement of the boranes themselves, as has been found recently to be the case when diborylalkane intermediates are involved.[67]

The stereochemistry of elimination of boron alkoxide from the β-ethoxy-alkylboron compound (LXXXIII) has been established by Pasto and Snyder.[72] Deuterioboration of cis-β-ethoxystyrene with dipropyldeuterio-borane, Pr_2BD, presumably proceeds in the usual cis fashion to yield the deuteriated 1-phenyl-2-ethoxyethylborane (LXXXIV). On cis deoxy-borination, which occurred without added catalyst in the course of an hour, (LXXXIV) yielded trans-β-deuteriostyrene. Acid (boron trifluoride) or base (butyllithium) caused trans deoxyborination to yield cis-β-deu-teriostyrene.

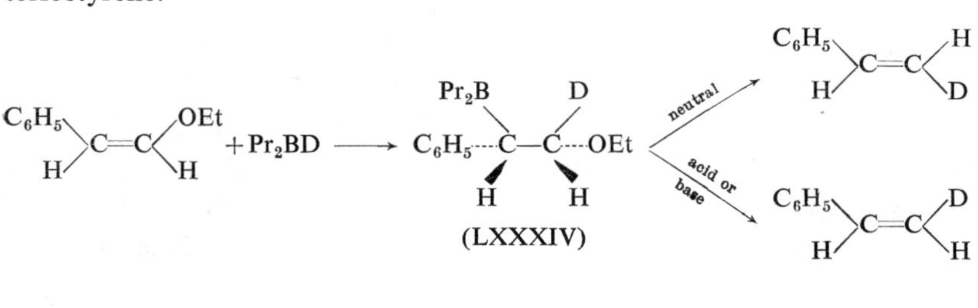

Pasto and Cumbo,[68] Miesel[70] and Snyder[72] studied the hydroboration of a variety of other oxygen, sulfur and nitrogen substituted olefins. Evidence for other β-eliminations resembling the one just described was obtained. It was also found that product ratios occur which can be attributed to "α-transfer" and "β-transfer" reactions, in which the alkoxy or alkylthio group migrates from carbon to boron and hydrogen (or, to a small extent, alkyl) goes the other way. Interpretation is complicated by the fact that some of the boranes undergo proto-deboronation on treatment with base, and by the present uncertainty about the specificity of the hydrogen peroxide deboronation.[67]

Because of the extreme complexity of the data obtained by Pasto's group, and because of the imminent likelihood of new data which could render any present conclusions obsolete, the author will not attempt to present a detailed evaluation of this work on the so-called "transfer reactions". Because boron facilitates nucleophilic displacement at the α-carbon in well-characterized systems,[48, 52] Pasto's postulated "α-transfer" reactions seem reasonable. The postulated "β-transfer" seems questionable on purely intuitive grounds, since it is difficult to see how such a process could compete with the very facile β-elimination reaction. The evidence for "β-transfer" rests largely on semiquantitative yields of rather minor products from a four-step process: hydroboration, rearrangement, basic hydrolysis and oxidation with hydrogen peroxide. There seems to be little hope of clearly defining exactly what happens and with what rate law in each of these four steps and in each of several simultaneously competing reactions, yet without such evidence any mechanisms that can be written are pure speculation.

V. BORON-SUBSTITUTED NUCLEOPHILIC CENTERS

A. Nucleophilic Additions to Unsaturated Boron Compounds

1. Diels–Alder Reactions

The only known nucleophilic additions to the double bond of unsaturated boron compounds are those of the four-center type, in which the π-bond system of the boron compound remains intact though perturbed in the transition state. The Diels–Alder reaction falls in this category.

As expected on theoretical grounds,[38] dibutyl vinylboronate is a moderately active dienophile. Matteson and Waldbillig[58] found that reaction of dibutyl vinylboronate (IV) with cyclopentadiene at about 100° yielded a mixture containing 60 per cent dibutyl *exo*-5-norbornene-2-boronate (LXXXV) and 40 per cent of the *endo* isomer (LXXXVI). When the *tert*-butyl ester was used, the proportion of *exo*

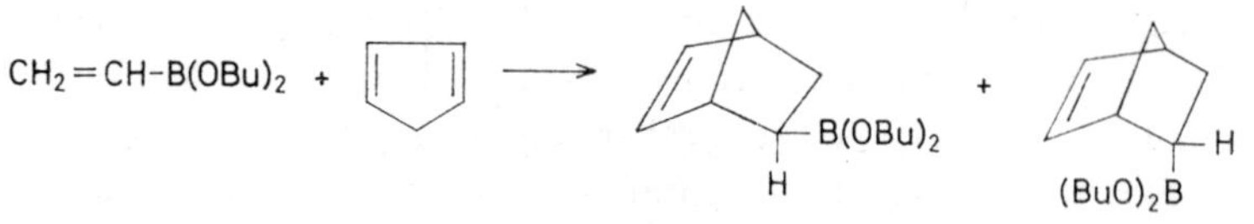

(LXXXV)   (LXXXVI)

isomer increased to 73 per cent. Vinylboronic acid decreased the *exo* to 40 per cent. It was found possible to separate the pure *endo* isomer by fractional crystallization of the free boronic acid, and the pure *exo* isomer by way of the *o*-phenylenediamine derivative. These compounds have yielded particularly interesting results in studies of the mechanism and stereochemistry of electrophilic displacement of boron at saturated carbon,[56, 59] a topic outside the scope of this review.

Dibutyl vinylboronate has also yielded Diels–Alder adducts with isoprene,[58] perchlorocyclopentadiene[58] and cyclohexadiene.[57] Dibutyl acetyleneboronate is a less reactive dienophile than the vinyl compound, but does react with cyclopentadiene to give a rather low yield of dibutyl norbornadiene-2-boronate (LXXXVII).

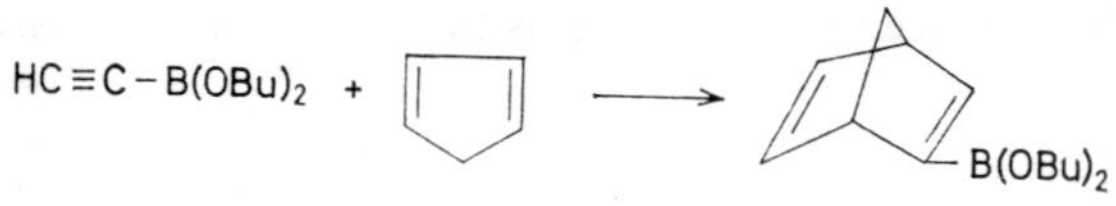

(LXXXVII)

The transition state for the Diels–Alder reaction is thought to resemble that for 1,3-dipolar additions, described in the next section. The boron atom confers dienophilic properties on the vinyl group by π-electron withdrawal, as discussed in Section II, B, 1(c).

2. 1,2-*Dipolar Additions*

The author[40] has studied the reaction of dibutyl vinylboronate and related compounds with ethyl diazoacetate. The first step is the addition of the diazo compound to the vinylboronic ester (IV) to yield an unstable boron-substituted pyrazoline (LXXXVIII). The boron atom is then transferred from carbon to nitrogen, and ethanolysis yields 5-carbethoxy-2-pyrazoline (LXXXIX).

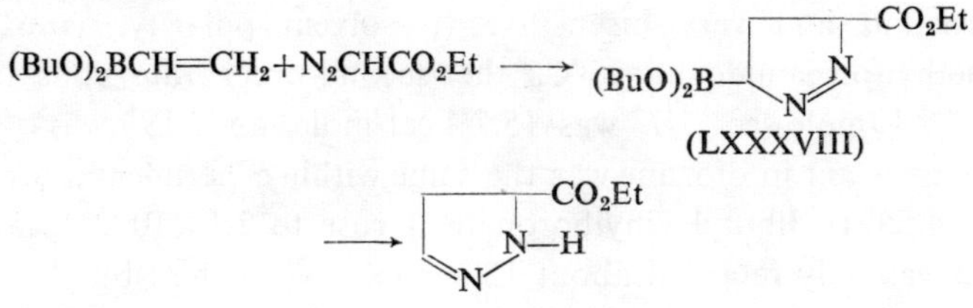
$(BuO)_2BCH{=}CH_2 + N_2CHCO_2Et$
CO_2Et
$(BuO)_2B$
N
N
CO_2Et
N—H
N

(LXXXVIII)

(LXXXIX)

It may be noted that this is not the same as the conjugated pyrazoline normally obtained from ethyl acrylate and diazomethane.

Although the intermediate (LXXXVIII) was unstable, Woods and Bengelsdorf[83] found that the analogous 2-methyl-2,4-pentanediyl ester underwent proton tautomerization instead of boron migration to yield a boronopyrazoline assigned the structure (XC) on the basis of infrared evidence, which seemed to indicate that the double bond was not in the expected position in conjugation with the carbethoxy group. Compound (XC) could not be obtained in analytically pure condition.

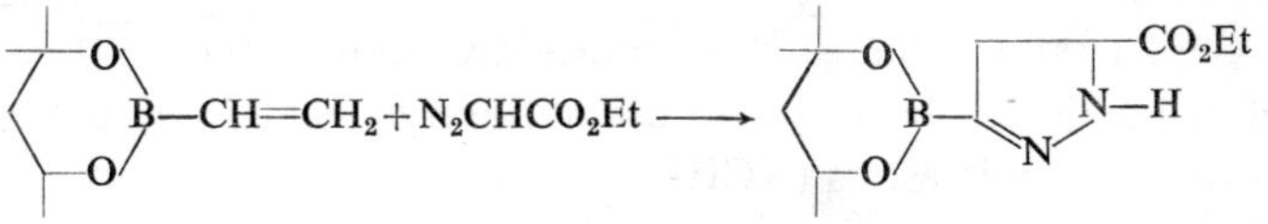
$B{-}CH{=}CH_2 + N_2CHCO_2Et$
CO_2Et
N—H
N

(XC)

Further evidence for the course of the addition was obtained by reacting dibutyl acetyleneboronate with ethyl diazoacetate, which yielded 5-carbethoxypyrazole-2-boronic acid (XCI) after hydrolysis.[40]

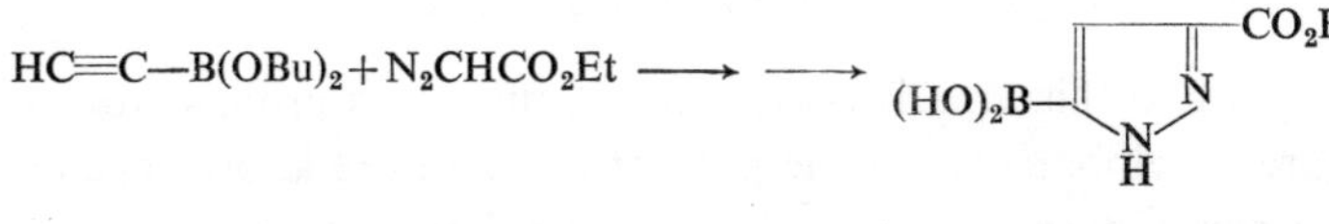
$HC{\equiv}C{-}B(OBu)_2 + N_2CHCO_2Et$
CO_2Et
$(HO)_2B$
N
N
H

(XCI)

The direction of addition is of particular interest, since molecular orbital calculations (Section II, B, 1 (c)) suggest that the most nucleophilic atom of the ethyl diazoacetate, the α-carbon, should attack the β-carbon of the unsaturated boronic ester. This was confirmed in the case of the acetylenic boronic ester by hydrogen peroxide deboronation of the pyrazoleboronic ester (XCI) to the known 3-carbethoxy-5-pyrazolone. Dibutyl 1-propene-1-boronate and propene-2-boronate also yielded carbethoxypyrazolines with the methyl groups in the right locations, as shown by oxidation to the known pyrazoles.[40]

The author[40] followed the kinetics of the reaction of ethyl diazoacetate with dibutyl vinylboronate and found the rate to be first-order in each reactant and very insensitive to solvent polarity, as expected. In dimethylformamide at 25°C, the second-order rate constant was 1.50×10^{-5} l./mole-sec, ΔH^* was 15.7 kcal/mole, and ΔS^* was -28 e.u. The rate constant in dioxane was the same within experimental error, and in neat (4.55 M) dibutyl vinylboronate it rose to 2.1×10^{-5} l./mole-sec. The rate was only retarded about 15 per cent when the t-butyl ester was substituted for the *n*-butyl, but was reduced by a factor of 20 by an α-methyl substituent on the vinyl group and a factor of over 100 by a β-metyl substituent. These data are all consistent with the transition state structure (XCII).

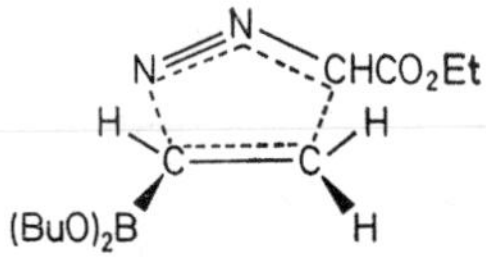

(XCII)

The rate of reaction of ethyl diazoacetate with vinyltriethoxysilane is only $\frac{1}{20}$ that with dibutyl vinylboronate. Other rate comparisons and their correlation with molecular orbital calculations have been noted in Section II, B, 1(c).

Bianchi, Cogoli and Grünanger[3] have recently reported the 1,3-dipolar addition of aromatic nitrile oxides to dibutyl acetyleneboronate to yield 3-arylisoxazole-5-boronic esters (XCIII).

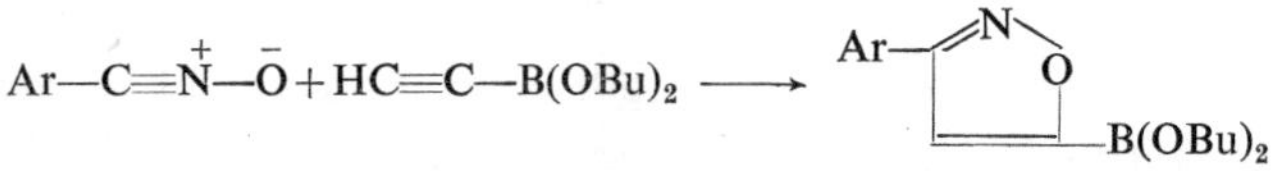

(XCIII)

B. *Electrophilic Displacements Assisted by Boron*

The literature in this area is rather scanty, but holds promise for interesting future developments. At the present time, all the known chemistry is based on *gem*-diboron compounds and their derivatives.

Matteson and Shdo[54] found that ethane-1,1-diboronic acid reacts readily with mercuric chloride and sodium hydroxide at 0°C to yield ethylidenedimercuric chloride (XCIV). This is unusual behavior for an alkaneboronic acid, since it is usually very difficult to replace boron by mercury. The chlo-

$$CH_3CH(B(OH)_2)_2 + 2\,HgCl_2 + 2\,NaOH \longrightarrow CH_3CH(HgCl)_2 + 2\,NaCl + 2\,B(OH)_3$$

(XCIV)

romercuric group also must accelerate the displacement. No estimate of the degree of rate acceleration is possible at present, except that several orders of magnitude are involved. Ethane-1,1-diboronic acid was also observed to decompose at a moderate rate in basic deuterium oxide, and the n.m.r. spectrum suggested that deuteriodeboronation to 1-deuterioethaneboronic acid, $CH_3CHDB(OH)_2$, was occurring.

Zweifel and Arzoumanian[87] have shown that 1,1-diborylalkanes, $RCH(BR_2)_2$, are readily cleaved by sodium hydroxide to the alkylborane, RCH_2BR_2. Reaction of the dihydroboration product from 1-pentyne with methyllithium has yielded 1-lithio-1-borylpentanes (XCV), which react with ethyl bromide to yield 3-heptylboranes (XCVI), which are converted to 3-heptanol by alkaline hydrogen peroxide.

$$C_3H_7C\!\equiv\!CH + BH_3 \longrightarrow BuCH(BR_2)_2 \xrightarrow{CH_3Li} BuCH\!\!\begin{smallmatrix}Li\\[-2pt]\\BR_2\end{smallmatrix}$$

(XCV)

$$\xrightarrow{EtBr} BuCH\!\!\begin{smallmatrix}Et\\[-2pt]\\BR_2\end{smallmatrix} \xrightarrow{H_2O_2} BuCHEt\!\!\begin{smallmatrix}\\[-2pt]\\OH\end{smallmatrix}$$

(XCVI)

Pasto[65] has also obtained evidence that 1,1-diboryl compounds undergo rapid protodeboronation in aqueous base. Pasto's results also suggest that protodeboronation is accelerated by a boron atom in the β-position. Unfortunately, the recent results of Pasto, Chow and Arora[67] showing that much replacement of boron by hydrogen can occur during the hydrogen peroxide deboronation of phenylethane-1,2-diboronic esters make it unnecessary to invoke β-boron participation in the hydrolysis step to explain the observed products. However, the hypothesis that a β-boron atom can stabilize a carbanion through three-membered ring formation is attractive and logical, and it will not be surprising if the effect is confirmed by future experiments.

The author would like to point out that the acceleration of electrophilic displacement at the carbon atom adjacent to boron implies a good deal of carbanion character in the transition state, which is stabilized by electron delocalization to the assisting boron atom, illustrated by the transition state (XCVII). The α-lithioalkylborane (XCV) probably also has a considerable degree of carbon–boron double bond character. It seems possible that a free intermediate anion containing a carbon–boron double bond (XCVIII) might have transient existence if generated from a 1,1-diboronic ester and alkoxide in a polar, inert solvent. Although structures such as (XCVIII) are likely to polymerize, it might be possible to trap the species with a suitable reagent, much in the manner that benzyne and similar active species have been trapped. This seems a promising direction for future research.

$$CH_3CH = \bar{B}(OR)_2 \longleftrightarrow CH_3\bar{C}H - B(OR)_2$$

XCVII   XCVIII

VI. REFERENCES

1. BAWN, C. E. H. and LEDWITH, A., *Progress in Boron Chemistry*, 1, 345 (1964).
2. BETTMAN, B., BRANCH, G. E. K. and YABROFF, D. L., *J. Am. Chem. Soc.*, 56, 1865 (1934).
3. BIANCHI, G., COGOLI, A. and GRÜNANGER, P., *J. Organometal. Chem.*, 6, 598 (1966).
4. BINGER, P. and KÖSTER, R., *Tetrahedron Letters*, 156 (1961).
5. BRANCH, G. E. K., YABROFF, D. L. and BETTMAN, B., *J. Am. Chem. Soc.*, 56, 937 (1934).
6. BROWN, D. A. and McCORMACK, C. G., *Theoret. Chim. Acta* (Berlin), 6, 350 (1966).
7. BROWN, H. C. and COPE, O. J., *J. Am. Chem. Soc.*, 86, 1801 (1964).
8. BROWN, H. C. and OKAMOTO, Y., *J. Am. Chem. Soc.*, 80, 4979 (1958).
9. BROWN, H. C. and ZWEIFEL, G., *J. Am. Chem. Soc.*, 82, 4708 (1960).
10. CASANOVA, J., KIEFER, H., KUWADA, D. and BOULTON, A., *Tetrahedron Letters*, 703 (1965).
11. CASANOVA, J. and SCHUSTER, R. E., *Tetrahedron Letters*, 405 (1964).
12. CHALVET, O., DAUDEL, R. and KAUFMAN, J. J., *J. Am. Chem. Soc.*, 87, 399 (1965).
13. CHUNG, A. L. H. and DEWAR, M. J. S., *J. Chem. Phys.*, 42, 756 (1965).
14. CRAM, D. J., *J. Am. Chem. Soc.*, 71, 3863 (1949).
15. CRISTOL, S. J., PARUNGO, F. P. and PLORDE, D. E., *J. Am. Chem. Soc.*, 87, 2870 (1965).
16. DAVIES, A. G. and ROBERTS, B. P., *J. Chem. Soc.* (B), 17 (1967).
17. DEWAR, M. J. S., *Progress in Boron Chemistry*, 1, 235 (1964).
18. DEWAR, M. J. S. and FAHEY, R. C., *J. Am. Chem. Soc.*, 85, 2248 (1963).
19. DEWAR, M. J. S., GLEICHER, G. J. and ROBINSON, B. P., *J. Am. Chem. Soc.*, 86, 5698 (1964).
20. DEWAR, M. J. S. and ROGERS, H., *J. Am. Chem. Soc.*, 84, 395 (1962).
21. GEYMEYER, P., ROCHOW, E. G. and WANNAGAT, U., *Angew. Chem.*, 76, 499 (1964).
22. GOOD, C. D. and RITTER, D. M., *J. Am. Chem. Soc.*, 84, 1162 (1962).
23. GRUNWALD, E. and WINSTEIN, S., *J. Am. Chem. Soc.*, 70, 846 (1948).
24. HAWTHORNE, M. F., *J. Am. Chem. Soc.*, 82, 1886 (1960).

25. HAWTHORNE, M. F. and DUPONT, J. A., *J. Am. Chem. Soc.*, **80**, 5830 (1958).
25a. HESS, H., *Angew. Chem.* (Internat. Ed. in English), **6**, 975 (1967).
26. HESSE, G. and WITTE, H., *Ann.*, **687**, 1 (1965).
27. HILLMAN, M. E. D., *J. Am. Chem. Soc.*, **84**, 4715 (1962); *ibid.*, **85**, 982, 1626 (1963).
28. HOFFMANN, R., *J. Chem. Phys.*, **39**, 1397 (1963).
29. HOFFMANN, R., *Boron–Nitrogen Chemistry*, Advances in Chemistry Series, No. 42, K. NIEDENZU, Ed., American Chemical Society, Washington, D.C., 1964, p. 78.
30. HOFFMANN, R., *J. Chem. Phys.*, **40**, 2474 (1964).
31. JAEGER, H. and HESSE, G., *Ber.*, **95**, 345 (1962).
32. JAFFÉ, H. H., *Chem. Rev.*, **53**, 191 (1953).
33. KÖSTER, R., HORSTSCHÄFER, H. J. and BINGER, P., *Angew. Chem.* (Internat. Ed. in English), **5**, 730 (1966).
34. KÖSTER, R. and RÖTERMUND, G. W., *Tetrahedron Letters*, 1667 (1964).
35. KUIVILA, H. G. and ARMOUR, A. G., *J. Am. Chem. Soc.*, **79**, 5659 (1957).
36. KUIVILA, H. G. and MULLER, T. C., *J. Am. Chem. Soc.*, **84**, 377 (1962).
37. LAPPERT, M. F., and MAJUMDAR, M. K., *Proc. Chem. Soc.*, 88 (1963); *Boron–Nitrogen Chemistry*, Advances in Chemistry Series, No. 42, K. NIEDENZU., Ed., American Chemical Society, Washington, D. C., 1964, p. 208.
38. MATTESON, D. S., *J. Am. Chem. Soc.*, **82**, 4228 (1960).
39. MATTESON, D. S., *J. Org. Chem.*, **27**, 275 (1962).
40. MATTESON, D. S., *J. Org. Chem.*, **27**, 4293 (1962).
40a. MATTESON, D. S., *J. Org. Chem.*, **27**, 3712 (1962).
41. MATTESON, D. S., *J. Org. Chem.*, **29**, 3399 (1964).
42. MATTESON, D. S., *Organometal. Chem. Rev.*, **1**, 1 (1966).
43. MATTESON, D. S. and BOWIE, R. A., *J. Am. Chem. Soc.*, **87**, 2587 (1965).
44. MATTESON, D. S. and CHENG, T. C., *J. Organometal. Chem.*, **6**, 100 (1966).
45. MATTESON, D. S. and LIEDTKE, J. D., *Chem. Ind.* (London), 1241 (1963).
46. MATTESON, D. S. and LIEDTKE, J. D., *J. Org. Chem.*, **28**, 1924 (1963).
47. MATTESON, D. S. and LIEDTKE, J. D., *J. Am. Chem. Soc.*, **87**, 1526 (1965).
48. MATTESON, D. S. and MAH, R. W. H., *J. Am. Chem. Soc.*, **85**, 2599 (1963).
49. MATTESON, D. S. and MAH, R. W. H., *J. Org. Chem.*, **28**, 2171 (1963).
50. MATTESON, D. S. and MAH, R. W. H., *J. Org. Chem.*, **28**, 2174 (1963).
51. MATTESON, D. S. and PEACOCK, K., *J. Org. Chem.*, **28**, 369 (1963).
52. MATTESON, D. S. and SCHAUMBERG, G. D., *J. Org. Chem.*, **31**, 726 (1966).
53. MATTESON, D. S. and SHDO, J. G., *J. Am. Chem. Soc.*, **85**, 2684 (1963).
54. MATTESON, D. S. and SHDO, J. G., *J. Org. Chem.*, **29**, 2742 (1964).
55. MATTESON, D. S., SOLOWAY, A. H., TOMLINSON, D. W., CAMPBELL, J. D. and NIXON, G. A., *J. Med. Chem.*, **7**, 640 (1964).
56. MATTESON, D. S. and TALBOT, M. L., *J. Am. Chem. Soc.*, **89**, 1119 (1967).
57. MATTESON, D. S. and TALBOT, M. L., *J. Am. Chem. Soc.*, **89**, 1123 (1967).
58. MATTESON, D. S. and WALDBILLIG, J. O., *J. Org. Chem.*, **28**, 366 (1963).
59. MATTESON, D. S. and WALDBILLIG, J. O., *J. Am. Chem. Soc.*, **86**, 3778 (1964).
59a. MCDANIEL, D. H. and BROWN, H. C., *J. Am. Chem. Soc.*, **77**, 3756 (1955).
60. MCKUSICK, B. C., HECKERT, R. E., CAIRNS, T. L., COFFMAN, D. D. and MOWER, H. F., *J. Am. Chem. Soc.*, **80**, 2806 (1958).
61. MIKHAILOV, B. M. and ARONOVICH, P. M., *Izv. Akad. Nauk SSSR, Otd. Khim. Nauk*, 927 (1961).
62. MIKHAILOV, B. M. and ARONOVICH, P. M., *Izv. Akad. Nauk SSSR, Otd. Khim. Nauk*, 1233 (1963).
63. PAETZOLD, P. I. and HABEREDER, P. P., *J. Organometal. Chem.*, **7**, 61 (1967).
64. PAETZOLD, P. I., HABEREDER, P. P. and MÜLLBAUER, R., *J. Organometal. Chem.*, **7**, 51 (1967).
65. PASTO, D. J., *J. Am. Chem. Soc.*, **86**, 3039 (1964).

66. PASTO, D. J. and BALASUBRAMANIYAN, P., *J. Am. Chem. Soc.*, **89,** 295 (1967).
67. PASTO, D. J., CHOW, J. and ARORA, S. K., *Tetrahedron Letters,* 723 (1967).
68. PASTO, D. J. and CUMBO, C. C., *J. Am. Chem. Soc.*, **86,** 4343 (1964).
69. PASTO, D. J., GARVES, K. and SERVE, M. P., *J. Org. Chem.*, **32,** 774 (1967).
70. PASTO, D. J. and MIESEL, J. L., *J. Am. Chem. Soc.*, **85,** 2118 (1963).
71. PASTO, D. J. and SNYDER, R., *J. Org. Chem.*, **31,** 2773 (1966).
72. PASTO, D. J. and SNYDER, R., *J. Org. Chem.*, **31,** 2777 (1966).
73. PELLON, J., DEICHERT, W. G. and THOMAS, W. M., *J. Polymer Sci.*, **55,** 153 (1961).
74. SCHAEFFER, R. and TODD, L. J., *J. Am. Chem. Soc.*, **87,** 488 (1965).
75. SEYFERTH, D. and PROKAI, B., *J. Am. Chem. Soc.*, **88,** 1834 (1966).
76. SEYFERTH, D., SATO, Y. and TAKAMIZAWA, M., *J. Organometal. Chem.*, **2,** 367 (1964).
77. SEYFERTH, D. and TAKAMIZAWA, M., *Inorg. Chem.*, **2,** 731 (1963).
78. SEYFERTH, D. and TAKAMIZAWA, M., *J. Org. Chem.*, **28,** 1142 (1963).
79. SIMMONS, H. E., BLANCHARD, E. P. and SMITH, R. D., *J. Am. Chem. Soc.*, **86,** 1347 (1964).
80. STREITWIESER, A., *Molecular Orbital Theory for Organic Chemists*, John Wiley Sons, Inc., New York, N.Y., 1961.
81. TUFARIELLO, J. J. and LEE, L. T. C., *J. Am. Chem. Soc.*, **88,** 4757 (1966).
82. WILLIAMS, R. E., *Progress in Boron Chemistry*, **2,** 37 (1969).
83. WOODS, W. G. and BENGELSDORF, I. S., *J. Org. Chem.*, **31,** 2769 (1966).
84. WOODS, W. G., BENGELSDORF, I. S. and HUNTER, D. L., *J. Org. Chem.*, **31,** 2766 (1966).
85. YABROFF, D. L., BRANCH, G. E. K. and BETTMAN, B., *J. Am. Chem. Soc.*, **56,** 1850 (1934).
86. ZAKHARKIN, L. I. and KOVREDOV, A. I., *Izv. Akad. Nauk SSSR, Ser. Khim.*, 393 (1964).
87. ZWEIFEL, G. and ARZOUMANIAN, H., *J. Am. Chem. Soc.*, **89,** 291 (1967).
88. ZWEIFEL, G., ARZOUMANIAN, H. and WHITNEY, C. C., *J. Am. Chem. Soc.*, **89,** 3652 (1967).

3

THERMOCHEMISTRY OF BORON COMPOUNDS

by Arthur Finch and P. J. Gardner

Royal Holloway College, University of London, Englefield Green, England

CONTENTS

I. INTRODUCTION

The aims of modern thermochemistry, in a very general sense, are excellently reviewed by Mortimer and Springall.[1] Detailed descriptions of the techniques and apparatus utilized to achieve these aims are contained in the two volumes[2] of *Experimental Thermochemistry*. It remains to discuss briefly the limitations of conventional techniques when applied to boron thermochemistry and, conversely, to indicate those methods that are particularly suitable.

Several comprehensive reviews of both inorganic and organic aspects of boron chemistry have appeared[3-6] recently and, without exception, there is very little emphasis placed on the thermochemistry or thermodynamic properties of boron compounds. This situation exists partly because of the scarcity of reliable data for boron compounds, compared with those for, say, carbon/hydrogen/oxygen compounds. The classical method for the determination of heats of formation of compounds is to monitor thermally their combustion in excess oxygen under pressure, the reaction proceeding at constant volume in a steel vessel ("static bomb calorimetry"). From the viewpoint of the combustion thermochemist utilizing this technique, boron is most definitely not a "well-behaved" element, i.e. combustions of boron-containing compounds are not "clean". This means that it is difficult to define the final thermodynamic state due to incomplete combustion and undesirable side-reactions. For example, most of the boron is oxidized to B_2O_3 which may partially hydrate and, in addition, small amounts of residue are usually found. This residue contains not only soot but B_4C and elemental boron. Initial admixture of the sample with combustion promoters such as hydrocarbon oils does not mitigate the problem.

Recent work[30] using a rotating-bomb appears to have overcome some of these drawbacks. In this method, the bomb contains a solvent capable of dissolving the expected combustion products; such a solution may be defined precisely in thermodynamic terms. To effect dissolution the bomb is rotated slowly end-over-end and axially after firing the sample, ensuring intimate mixing and dissolution of the combustion products. The technique as applied to boron compounds was developed at the Bartlesville Petroleum Research Center and involves using a fluorine-containing combustion promoter (vinylidene fluoride polymer) and initially charging the bomb with an HF solution. Under these conditions, no solid products remain and the final thermodynamic state may be accurately defined in terms of $HBF_4 \cdot xHF \cdot yH_2O$.

Another technique which is particularly promising is that of fluorine bomb calorimetry. In this case, the oxidant is fluorine, the combustion is complete and the final thermodynamic state readily specified. This method has been applied recently to borides, which are generally rather intractable compounds.

The technique which undoubtedly offers most scope to the imagination of the thermochemist is reaction calorimetry. Here, any reaction of the boron compound which is established as quantitative, is thermally monitored at constant pressure at temperatures around ambient. Many types of reaction are used, e.g. hydrolysis, hydroboronation, aqueous oxidation and reduction, etc., the reaction commonly, but not necessarily, taking place in solution.

Equilibrium studies at elevated temperatures have been applied to many systems, e.g. the oxides and halides, particularly in attempts to establish $\Delta H_f^\circ[\text{B, g}]$ unequivocally. For example, in a reaction such as:

$$0.25\text{B}_4\text{C(c)} \rightleftharpoons \text{B(g)} + 0.25\text{C(c)} \qquad (1)$$

at ca. 2500°K, two estimates of ΔH_{298}° are usually made. The "second law" value comes from the integrated van't Hoff isochore and the "third law" figure from ΔG_T° (in this particular case simply $-RT \ln p_e$, where p_e is the equilibrium pressure), and

$$\Delta H_{298}^\circ = \Delta G_T^\circ - T\left[\Delta\left(\frac{G_T^\circ - H_{298}^\circ}{T}\right)\right]. \qquad (2)$$

In this review, we shall be concerned almost exclusively with standard enthalpies of formation (ΔH_f°) which will be quoted at 298.15°K unless stated otherwise. However, for many of the simple species described, comprehensive statistical thermodynamic functions are available. The best source of the latter data is the JANAF compilation.[24, 67] References from the literature up to mid-1967 are included (but the review is critical rather than comprehensive) and compounds mentioned will include both inorganic species (e.g. oxides, borax, etc.) and organic compounds (e.g. trialkyl borates). We specifically exclude the following: the boron hydrides, and derivatives, carboranes and donor–acceptor adducts. We shall include certain trigonal compounds containing the B—H bond and species containing the borohydride ion (BH_4^-).

The bonding in boron compounds is that of a typical first-row element. All three-covalent compounds exhibit trigonal geometry, the system as such being a Lewis acid because of the vacant 2_{p_z} orbital. In the presence of Lewis bases, adducts are readily formed, the geometry being approximately tetrahedral.

II. THERMOCHEMICAL DATA

A. *Reference State*

Boron is known to exist in several allotropic modifications[4] including at least three tetragonal forms, two rhombohedral and one hexagonal. The detailed structures of these forms are not certain and the existence of other allotropes is likely. It is difficult to specify the thermodynamically most stable modification at 25°C and one atmosphere, as the transformations appear to be kinetically controlled and generally irreversible. The reference state is normally quoted as "crystal" and the β-rhombohedral modification is a suitable choice, being the subject of careful heat capacity studies. It is probable that this form is thermodynamically stable above 1100°C, the α-rhombohedral being stable at lower temperatures. It is, however, not uncommon for an unstable allotrope to be selected as the reference state, phosphorus being a further example. Many pyrolytic decomposition reactions of boron compounds have been studied thermochemically[7] and, under these conditions, the boron is produced in an amorphous form. Various estimates of ΔH_f°[B, am.] have been made and the currently accepted figure is based on

(i) the pyrolytic decomposition of diborane:

$$B_2H_6(g) \longrightarrow 2\,B(am.) + 3\,H_2(g) \qquad (3)$$

and (ii) a weighted average of the values for ΔH_f°[B$_2$H$_6$, g] from reactions not involving elemental boron. The accepted value[8] for ΔH_f°[B, am.] is +0.8 kcal g atom^{-1} but the error in this figure is probably at least ±0.5 kcal.

The current position regarding the sublimation energy of boron is far from satisfactory. Values for the function ΔH_f°[B, g] at 298.15°K spread over ca. 30 kcal. The majority of data is based on vapor pressure measurements from effusion or free-evaporation methods in the temperature range 1600–2500°K. The data obtained from such methods are subject to two principal uncertainties. First, the error in the measurement of the temperature, and second, the high reactivity of boron at elevated temperatures, which makes the selection of a material for an inert container virtually impossible (i.e. the thermodynamic state of the condensed phase cannot be defined). This latter objection was first circumvented by Robson and Gillies,[9] who performed vapor pressure measurements on the dissociation of B$_4$C in a graphite crucible. They first established the composition of the condensed phase in equilibrium with graphite at the temperature of the experiment and subsequently recorded the decomposition pressures of

$$0.25B_4C(c) \rightleftharpoons 0.25C(c) + B(g) \qquad (4)$$

using a Knudsen effusion technique. This equilibrium was subsequently investigated by Hildenbrand[10] using a torsion effusion method. Both these sets of workers ignore the small concentrations of $BC_2(g)$, $B_2(g)$ and $B_2C(g)$ which exist in equilibrium with $B(g)$ at the temperature of the experiment. A disadvantage experienced by a vapor pressure study of B_4C is that to translate the results into $\Delta H_f^\circ[B, g]$ a knowledge of the heat of formation of B_4C is necessary and this is still subject to an appreciable uncertainty[67] (ca. ± 2 kcal). Some accumulated data are given in Table 1.

TABLE 1. THE SUBLIMATION ENTHALPY OF BORON†

Investigator	Process or reaction		ΔH_f° [B, g]298.15°K
Searcy and Myers (1957)[12]	B(c)	$\rightleftharpoons$ B(g)	139.1 ± 2.5
Schissel and Williams (1959)[13]	B(c)	$\rightleftharpoons$ B(g)	129 ± 5
Akishin et al. (1959)[14]	B(c)	$\rightleftharpoons$ B(g)	131.2 ± 5
Priselkov et al. (1960)[15]	B(c)	$\rightleftharpoons$ B(g)	101 ± 2
Chupka (1961)[16]	B(c)	$\rightleftharpoons$ B(g)	129.9
Verhaegen and Drowart (1962)[17]	B(am)	$\rightleftharpoons$ B(g)	129.9 ± 5
ditto	$0.25B_4C(c) \rightleftharpoons B(g) + 0.25C(c)$		132.1 ± 5‡
Alcock and Grieveson (1962)[18]	B(c)	$\rightleftharpoons$ B(g)	140
Schissel and Trulson (1962)[20]	B(c)	$\rightleftharpoons$ B(g)	132.8 ± 1.4
Paule and Margrave (1963)[19]	B(c)	$\rightleftharpoons$ B(g)	136.4 ± 1.0
Hildenbrand and Hall (1964)[10]	$0.25B_4C(c) \rightleftharpoons B(g) + 0.25C(c)$		136.2 ± 1.3‡
Hildenbrand (1964)[22]	B(c)	$\rightleftharpoons$ B(g)	136.1 ± 0.3
Robson and Gillies (1964)[9]	$0.25B_4C(c) \rightleftharpoons B(g) + 0.25C(c)$		135.4 ± 1.3‡
Linevsky et al. (1964)[21]	B(c)	$\rightleftharpoons$ B(g)	131.4
Burns et al. (1967)[23]	B(c)	$\rightleftharpoons$ B(g)	135.0 ± 3.6

† The majority of the data has been recalculated by the compilers of the JANAF tables, using consistent free-energy functions.
‡ Based on $\Delta H_f^\circ[B_4C, c] = -9.2 \pm 2.4^{[67]}$ kcalmole^{-1}.

The value recommended by JANAF thermochemical tables,[24] obtained by a weighted average of the data in Table 1 (excluding Priselkov's, Alcock et al. and Burns' data), is $\Delta H_f^\circ[B, g] = 132.8 \pm 4.0$ kcalg atom^{-1}.

A recent reappraisal of the problem[23] suggests that Paule and Margrave's contention that an evaporation coefficient for B(c) of 0.15 may explain the divergence between the free evaporation studies and the Knudsen effusion studies is probably in error. Redetermination of the evaporation coefficient (α_v) yields $0.96 \leqslant \alpha_v(B, c) \leqslant 1.0$, i.e. boron exhibits the normal behavior of crystalline materials evaporating as mono-mers. The figure of 135.0±3.6 kcal derived from these electron impact studies of Burns *et al.* does not suggest any drastic revision of the JANAF mean datum. The figure 132.8±4.0 kcal is therefore a recommended value.

B. *The Oxyacids*

Metaboric acid (HBO_2), which may be obtained by dehydration of orthoboric acid (H_3BO_3) at ca. 150°C, exists in at least three crystalline forms[4]: HBO_2 (I) cubic, HBO_2 (II) monoclinic and HBO_2 (III) ortho-rhombic or $HBO_2(\alpha)$. From a thermochemical study of the reactions:

$$HBO_2(c) + H_2O(liq) = H_3BO_3(soln) \qquad (5)$$

$$HBO_2(c) + NaOH(soln) = [NaBO_2 + H_2O] (soln) \qquad (6)$$

Kilday *et al.*[25, 26] derive $\Delta H_f^\circ[HBO_2, I, c] = -192.3±0.2$ kcalmole^{-1}. These workers also report[26] $\Delta H_f^\circ[HBO_2, c, II] = -190.0±0.5$ and $\Delta H_f^\circ[HBO_2, c, III] = -188.7±0.4$ kcalmole^{-1}. These data have been re-calculated using the recommended value for $\Delta H_f^\circ[H_3BO_3, c]$ (see below). A slightly different value was obtained by Russian workers[27] who investigated:

$$HBO_2(c, I) + 501 H_2O(liq) = H_3BO_3 500H_2O \qquad (7)$$

giving $\Delta H_f^\circ[HBO_2, II, c] = -189.8±0.2$ kcalmole^{-1}. and $\Delta H_f^\circ[HBO_2, III, c] = -188.5±0.2$ kcalmole^{-1}. The equilibrium pressures for the re-action (at ca. 1200°K):

$$\tfrac{1}{2}B_2O_3(liq) + \tfrac{1}{2}H_2O(g) \rightleftharpoons HBO_2(g) \qquad (8)$$

have been extensively studied, and a weighted average[24] of $\Delta H_f^\circ[HBO_2, g]$ is $-134.4±1.0$ kcalmole^{-1}. The gas phase structure has been proposed as H—O—B≡O and spectroscopic constants are available.

Trimeric metaboric acid (I) has been studied in the gas phase[28, 29] from the equilibria:

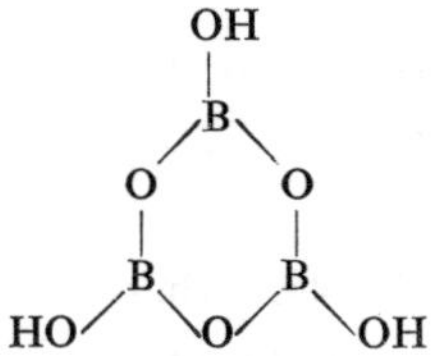

I

$$H_3BO_3(l) + B_2O_3(l) \rightleftharpoons (HOBO)_3(g) \qquad (9)$$

or

$$1.5 B_2O_3(l) + 1.5 H_2O(g) \rightleftharpoons (HOBO)_3(g) \qquad (10)$$

using mass-spectrometric detection or by thermal transpiration methods. The "best" value for the standard heat of formation is $\Delta H_f^\circ[HOBO]_3, g] = -543±3$ kcalmole^{-1}.

The enthalpy of formation of orthoboric acid, a "key" compound in boron thermochemistry, has been the subject of many careful investigations. One of the most recent is that of Good et al.[30] whose used a rotating-bomb to combust B (β-rhombohedral) to a final thermodynamic state of fluoroboric acid in excess aqueous HF. These workers then generated the same final thermodynamic state by dissolving crystalline boric acid in aqueous HF. Their final result, $\Delta H_f^\circ[H_3BO_3, c] = -261.5 \pm 0.2$ kcalmole^{-1}, differs marginally from the recommended value[8] of -261.7 kcalmole^{-1}. The enthalpy of transition, $\Delta H(c \rightarrow g)$ has been recalculated from van Stackelberg's[31] data as 24.3 ± 0.6 kcalmole^{-1}.

For the solution reaction:

$$H_3BO_3(c) + 1000\ H_2O(liq) \longrightarrow H_3BO_3 1000\ H_2O(soln) \qquad (11)$$

Evans et al.[32] report $+5.24 \pm 0.05$ kcalmole^{-1} which is in only reasonable agreement with Fasolino's datum[33] for $H_3BO_3 504 H_2O$ of $+5.45 \pm 0.01$. The heat of dilution[34] of $H_3BO_3 504 H_2O$ to $H_3BO_3 1000 H_2O$ is less than 10 cal. and does not account for the divergence in the data. Hence we recommend a mean value for $\Delta H_f^\circ[H_3BO_3 nH_2O]$ where $500 < n < 20{,}000$ of -256.35 ± 0.2 kcalmole^{-1}.

Hypoboric (or sub-boric) acid, $B_2(OH)_4$, has been studied in a reaction calorimeter, utilizing its reduction in the presence of the silver ion,[35]

$$B_2(OH)_4 + 2\ Ag^+ + 2\ H_2O = 2\ Ag + 2\ H_3BO_3 + 2\ H^+ \qquad (12)$$

The value reported is $\Delta H_f^\circ[B_2(OH)_4, c] = -335.6 \pm 0.8$ kcalmole^{-1}, using updated ancillary data.

C. Oxides

Boric oxide (B_2O_3), obtained from the dehydration of boric acid at temperatures in excess of 1000°C, exists in two distinct crystalline forms (normal hexagonal and high-temperature, high-pressure monoclinic), in addition to an amorphous form.[4] The melting point of the hexagonal form is ca. 450°C and the resulting viscous liquid is stable even at temperatures far in excess of 1000°C. Values for the standard enthalpy of formation of B_2O_3, generally obtained by direct combination of the elements or hydration to boric acid, vary over a range[24] of ca. 15 kcal. A recent appraisal of the available data recommends[30]

$$\Delta H_f^\circ[B_2O_3, c] = -304.6 \pm 0.4 \text{ kcalmole}^{-1}$$
$$\Delta H_f^\circ[B_2O_3, am] = -300.2 \pm 0.4 \text{ kcalmole}^{-1}$$

(corrected to $\Delta H_f^\circ[H_3BO_3, c] = -261.7$ kcalmole^{-1}). The enthalpies of transition[24] are $\Delta H(c \rightarrow l) = 4.4$ and $\Delta H(l \rightarrow g) = 100.1$ kcalmole^{-1}. The liquid (or glassy) state is very close thermodynamically to the amorphous state.

Boron monoxide (BO) has been reported in the gas and in at least two crystalline forms[4]. The dimeric form B_2O_2 is known in the gas and the equilibria between boron and boric oxide or carbon and boric oxide have been studied,[24] yielding $\Delta H_f^\circ[B_2O_2, g] = -109 \pm 2$ kcalmole^{-1}. The corresponding datum for BO(g) is within the range 3 ± 10 kcalmole^{-1}. The establishment of reliable thermodynamic data for boron monoxide is potentially important because this compound provides a convenient chemical route[36] for the synthesis of compounds containing the B—B bond:

e.g. $\quad\quad 6/n\,(BO)_n + 4\,BCl_3 = 3\,B_2Cl_4 + 2\,B_2O_3$ at 200°C. $\quad\quad$ (13)

D. *Boron Halides*

Some recent data for BF$_3$ are given in Table 2.

TABLE 2. SOME THERMODYNAMIC DATA FOR BF$_3$

Investigator	$\Delta H_f^\circ[BF_3, g]$
Circ. 500[37]	−265.2
Up-dated Circ. 500[38]	−271.7
Gross et al.[40]	−270.8
Gross et al. (from B(am.))[40]	−271.6 (−269.8)†
Hubbard et al. (1961)[39, 41]	−269.88 ± 0.24
Hubbard et al. (1966)[42]	−271.65 ± 0.22

† From B(c) using ΔH_f° B(am.) = 0.8 kcal mole^{-1}.

Re-examination of the boron sample used in Hubbard's 1961 determination revealed the presence of small concentrations of impurities.[42] Gunn's recommended[8] figure (−269.7 kcalmole^{-1}) is based on a thermochemical cycle involving $\Delta H_f^\circ[HFnH_2O]$. As this latter datum is not unequivocally established, the recommended value for BF$_3$ is that of Hubbard (1966), i.e. −271.65 ± 0.22 kcalmole^{-1}. Boron trifluoride, unlike the other boron trihalides, does not hydrolyze quantitatively in water to boric acid and the corresponding hydrogen halide. The products of aqueous BF_3 hydrolysis depend to some extent on the final concentration, but the principal reaction is probably:

$$BF_3 + (n+1)H_2O = [H^+ + BF_3OH]nH_2O \qquad (14)$$

In dilute HF solutions (from 0.3 to 0.6 molal) the heat of reaction[43] appears independent of the variation in HF concentration. At higher HF/BF_3 ratios, the main reaction is undoubtedly:

$$BF_3 + HF + nH_2O = [H^+ + BF_4^-]nH_2O \qquad (15)$$

and thermochemical data for this reaction are given by Gunn.[8]

The hydrolysis reaction:

$$BCl_3(liq) + (n+3)H_2O(liq) = [H_3BO_3 + 3\ HCl]nH_2O \qquad (16)$$

has been studied independently several times.

Investigator	n (in equation (16))	$\Delta H_f^\circ BCl_3$ (liq)
Skinner and Smith (1953)[44]	5000	-101.8 ± 1.1
Gunn and Green (1960)[45]	2000	-102.3 ± 0.2
Fasolino (1965)[46]	590	-101.5 ± 0.3

The data in the above table were recalculated from the boric acid figure recommended earlier and using HCl and H_2O data from ref. 38. In addition to this work, values for $\Delta H_f^\circ[BCl_3, liq]$ and $\Delta H_f^\circ[BCl_3, g]$ have been reported,[130] utilizing the combustion of amorphous boron in chlorine gas. The results, corrected to $\Delta H_f^\circ B(am.) = 0.8 \pm 0.5$, are $\Delta H_f^\circ[BCl_3, liq] = -101.7 \pm 0.8$ and $\Delta H_f^\circ[BCl_3, g] = -95.8 \pm 0.9$ kcalmole^{-1}.

A "best" value for $\Delta H_f^\circ[BCl_3, liq]$ is -102.0 ± 0.3 kcalmole^{-1}. A value for $\Delta H_f^\circ[BCl_3, g]$ of -96.3 ± 0.3 kcalmole^{-1} is obtained by combining the above datum with $\Delta H(BCl_3, 1 \to g)$[47] $= 5.7$ kcalmole^{-1}. The standard enthalpy of formation in the gas phase has been independently checked by a direct study of the reaction[48]

$$B(am.) + 3/2\ Cl_2(g) = BCl_3(g) \qquad (17)$$

which yields $\Delta H_f^\circ[BCl_3, g] = -96.7 \pm 0.6$ kcalmole^{-1} based on $\Delta H_f^\circ[B, am.] = 0.8 \pm 0.5$ kcalmole^{-1}.

The hydrolysis of boron tribromide (liquid) has been studied by Skinner and Smith.[49] Using the following ancillary data, ΔH_f°[HBr 3000H$_2$O][50] $= -29.05\pm0.09$, dilution data for HBr from ref. 38 and boric acid figure mentioned earlier, we derive the result ΔH_f°[BBr$_3$, liq] $= -57.4\pm0.5$ kcalmole^{-1}, which differs slightly from that (-57.0 ± 0.2) recommended in the JANAF compilation. Using Stock and Kuss' value[51] for ΔH[BBr$_3$, liq $\to$ g] of 8.2 kcalmole^{-1}, ΔH_f°[BBr$_3$, g] $= -49.2 \pm0.5$ kcalmole^{-1}.

Boron triiodide has been studied[52] by a hydrolysis procedure similar to that of boron tribromide, and the standard enthalpy of formation for BI$_3$(c), up-dated using ΔH_f°[HI 50.3H$_2$O][53] $= -13.56\pm0.03$, and dilution data from ref. 38, is -11.3 ± 0.3 kcalmole^{-1}. An estimate of this datum has also been made by Tiensuu[54] who studied quantitatively the yield of the reaction

$$3 \text{ LiI(c)} + \text{BBr}_3(\text{liq}) \leftrightharpoons 3 \text{ LiBr(c)} + \text{BI}_3(\text{liq}) \text{ at } 220°C. \qquad (18)$$

Using various simplifying assumptions and known statistical thermodynamic functions for BI$_3$(g), Tiensuu derived ΔH_f°[BI$_3$, c] $= -10.1\pm1.0$, which is in good agreement with the datum of Finch et al.[35] Tiensuu also estimated ΔH(BI$_3$, c $\to$ g) as 15.5 ± 1.5 kcalmole^{-1} from which ΔH_f°[BI$_3$, g] is 4.2 ± 1.7 kcalmole^{-1}, a far more precise figure than that given in ref. 24 of 17.0 ± 12 kcalmole^{-1}, which was based on some fairly drastic assumptions.

There have been two independent investigations[55, 56] of the equilibrium:

$$\text{BF}_3(\text{g}) + \text{BCl}_3(\text{g}) \leftrightharpoons \text{BCl}_2\text{F(g)} + \text{BClF}_2(\text{g}) \qquad (19)$$

yielding ΔH(reaction, 298.15) $= 1.6(\pm0.5)$ kcalmole^{-1}. This fixes the composite term $[\Delta H_f^\circ(\text{BCl}_2\text{F, g}) + \Delta H_f^\circ(\text{BClF}_2, \text{g})]$ fairly precisely and a reasonable division of this term on the basis of bond energies from BF$_3$ and BCl$_3$ gives individual values within ±5 kcal. Making the further assumption that disproportionation reactions of the type:

$$(3-n)\text{BBr}_3(\text{g}) + n\text{BX}_3(\text{g}) \rightleftharpoons 3\text{BBr}_{3-n}\text{X}_n(\text{g}) \qquad (20)$$

where X $=$ F or Cl and $n = 1,2$ are thermoneutral[52] one can estimate[24] the standard enthalpies of the mixed halides of boron.

Of the sub-halides of boron, only B$_2$F$_4$ and B$_2$Cl$_4$ are stable[57] at room temperatures. The reaction between B$_2$Cl$_4$ (liq) and chlorine gas to give BCl$_3$(g) was studied by Gunn et al.[58] and this yields ΔH_f°[B$_2$Cl$_4$, liq] $= -124.6\pm2.0$ kcalmole^{-1}. Combination of this figure with

TABLE 3. THE MIXED BORON HALIDES

Compound	$\Delta H_f^\circ(g)298.15$ (in kcalmole^{-1})	Method of estimation
BBrCl$_2$	-80.5 ± 10	thermoneutrality
BBrF$_2$	-196 ± 10	thermoneutrality
BBr$_2$Cl	-65 ± 10	thermoneutrality
BBr$_2$F	-123 ± 10	thermoneutrality
BClF$_2$	-211.6 ± 5	refs. 55, 56
BCl$_2$F	-154 ± 5	refs. 55, 56

Schlesinger's[59] $\Delta H(1 \rightarrow g) = 8.03$ kcalmole^{-1}, gives $\Delta H_f^\circ[B_2Cl_4, g] = -116.6\pm2.0$ kcal/mole^{-1}. Study of the reaction:

$$B_2F_4(g) + Cl_2(g) = 0.92\ BF_3(g) + 0.46\ BF_2Cl(g) + 0.32\ BFCl_2(g) + 0.30\ BCl_3(g) \quad (21)$$

by Gunn et al.[60] yielded $\Delta H_f^\circ[B_2F_4, g] = -343.4\pm3.0$ kcalmole^{-1}. Very approximate estimates[35] of the standard heats of formation are available for the remaining sub-halides, viz:

$$\Delta H_f^\circ[B_2Br_4, g] \sim -50 \quad \text{and} \quad \Delta H_f^\circ[B_2I_4, g] \sim +20 \text{ kcalmole}^{-1}.$$

Brief mention will be made of the trimeric oxyhalides (BOF)$_3$ and (BOCl)$_3$. Both were[61, 62] studied via the equilibrium:

$$BX_3(g) + B_2O_3(liq) \rightleftarrows (BOX)_3(g) \quad (22)$$

yielding $\Delta H_f^\circ[(BOCl)_3, g] = -390\pm2$kcalmole^{-1} and $\Delta H_f^\circ[(BOF)_3, g] = -565\pm1.0$ kcalmole^{-1}. In addition, Magee[63] gives $\Delta H_f^\circ[(BOF)_3, c] = -586.5\pm3.0$ kcalmole^{-1}. However, there is some doubt that (BOF)$_3$ is capable of crystalline existence at 25°C.

E. Borohydrides

The thermodynamic data that are available for the borohydrides are the result of careful solution calorimetric studies. The lithium and sodium salts were studied by Davis et al.[64] and in both cases the reaction chosen was:

$$MBH_4(c) + 1.25\ HCl\ 253\ H_2O = [MCl + H_3BO_3 + 0.25\ HCl]\ 250\ H_2O + 4\ H_2(g) \quad (23)$$

where M is Li or Na. They utilized a bomb calorimeter with the borohydride contained in a specially designed ampoule. As the system is one of constant volume, the experimentally measured quantity is ΔU. Further, the assumption was made that the heat of mixing of the products was zero. Using $\Delta H_f^\circ[\text{NaCl, c}]^{(24)} = -98.26 \pm 0.08$, $\Delta H_f^\circ[\text{LiCl, c}]^{(24)} = -97.58 \pm 0.27$, $\Delta H_f^\circ[\text{H}_2\text{O, liq}]^{(38)} = -68.315$, enthalpies of dilution for NaCl and LiCl from Parker,[65] hydrochloric acid data from ref. 38, and $\Delta H_f^\circ[\text{H}_3\text{BO}_3 n\text{H}_2\text{O}] = -256.35 \pm 0.2$, we derive $\Delta H_f^\circ[\text{NaBH}_4, \text{c}] = -45.0 \pm 0.3$ and $\Delta H_f^\circ[\text{LiBH}_4, \text{c}] = -45.8 \pm 0.4$ kcalmole^{-1}.

These workers[75] also used a "bomb" solution calorimeter to study the hydrolysis of aluminium borohydride. In this case, and when conventional oxygen combustion was attempted, undefineable products were generated and incomplete reaction was suspected. Finally, they utilized three sets of reaction conditions:

(a) reaction in water vapor in equilibrium with a little water,
(b) reaction in small excess of dilute aqueous ammonia, and
(c) reaction in moist oxygen.

In each case, the reaction was "damped" with 3 to 5 atmospheres of argon. They assumed the following thermochemical reactions, based on careful analysis of reaction products. For reaction conditions (a) and (b):

$$\text{Al(BH}_4)_3(\text{liq}) + 12\text{H}_2\text{O(liq)} = \text{Al(OH)}_3(\text{am.}) + 3\text{H}_3\text{BO}_3(\text{c}) + 12\text{H}_2(\text{g}) \quad (24)$$

and for (c):

$$\text{Al(BH}_4)_3(\text{liq}) + 6\text{O}_2(\text{g}) = \tfrac{1}{2}\text{Al}_2\text{O}_3(\text{cryst. corundum}) + \tfrac{3}{2}\text{B}_2\text{O}_3(\text{c}) + 6\text{H}_2\text{O(liq)} \quad (25)$$

For (a) and (b) using $\Delta H_f^\circ[\text{Al(OH)}_3, \text{am.}]^{(38)} = -305$ kcalmole^{-1}, boric acid *(loc. cit.)* we derive $\Delta H_f^\circ[\text{Al(BH}_4)_3, \text{liq}] = -74.1 + 1.3$ kcalmole^{-1}. For (c), using $\Delta H_f^\circ[\text{Al}_2\text{O}_3, \alpha, \text{corundum}]^{(24)} = -400.4 \pm 0.3$ and $\Delta H_f^\circ[\text{B}_2\text{O}_3, \text{c}] = -304.6$ kcalmole^{-1} we derive $\Delta H_f^\circ[\text{Al(BH}_4)_3, \text{liq}] = -77.9 \pm 4.0)$ kcalmole^{-1}. However, if the B_2O_3 is assumed to be produced in an amorphous form, this figure becomes -71.3 kcalmole^{-1}. We recommend a mean of the two sets, assuming crystalline B_2O_3 is produced, i.e. -76.0 kcalmole^{-1}. Schlesinger *et al.*[83] report $\Delta H(1 \rightarrow \text{g}) = 7.6$ kcalmole^{-1} for $\text{Al(BH}_4)_3$.

A similar reaction, studied this time in a conventional solution calorimeter, was used by Johnson *et al.*[66]: The reduced equation is:

$$\text{KBH}_4(\text{c}) + \text{HCl(g)} + 3\text{H}_2\text{O(liq)} = \text{KCl(c)} + \text{H}_3\text{BO}_3(\text{c}) + 4\text{H}_2(\text{g}). \quad (26)$$

Using $\Delta H_f^\circ[\text{KCl, c}]^{(67)} = -104.37 \pm 0.06$, $\Delta H_f^\circ[\text{HCl, g}]^{(38)} = -22.063 \pm 0.05$, and the boric acid data mentioned previously, $\Delta H_f^\circ[\text{KBH}_4, \text{c}] = -54.4 \pm 0.5$ kcalmole^{-1}.

Under this section, we include mention of the solution calorimetric work of Bills and Cotton[68] to derive thermochemical data for $KBF_4(c)$. They utilized the reaction:

$$H_3BO_3 + 4HF + KNO_3 = KBF_4 + HNO_3 + 3H_2O \qquad (27)$$

and from their results, the JANAF compilation reports -449.7 kcalmole^{-1} for the standard enthalpy of formation. This figure depends, of course, on the choice of data for $\Delta H_f^\circ[\text{HF } nH_2O]$. This is, currently (1967), a controversial datum, the following figures for $\Delta H_f^\circ[\text{HF } 20H_2O]$ being a selection: -77.40,[69] -75.63,[37] -76.28[38] kcalmole^{-1}. A "concensus" value, recommended by Cox,[70] is -76.95 ± 0.2 kcalmole^{-1}, and using this figure insert, with $\Delta H_f^\circ[\text{KNO}_3, c] = -118.2$[24] kcalmole^{-1} and the boric acid figure (loc. cit.), we obtain $\Delta H_f^\circ[\text{KBF}_4, c] = -451.6 \pm 1.4$ kcalmole^{-1}.

F. *Inorganic Borates*

The bulk of the thermodynamic data for the oxo-borates derives from work by Shartsis *et al.*[71, 72]

The crystal structure of oxo-borates can be described in terms of aggregates (cyclic or linear) of trigonal BO_3 or tetrahedral BO_4 units via shared oxygen atoms. This situation leads to a wide range of anions whose stoichiometric composition gives little insight into their molecular structure. Many borates can be formulated in terms of $M_2^IOnB_2O_3mH_2O$ and $M^{II}OnB_2O_3mH_2O$ where M^I and M^{II} are uni- and divalent metals respectively and n and m are usually integrals. Anhydrous borates may often be synthesized by fusion of a metal oxide with the appropriate mole ratio of B_2O_3. The hydrated salt may subsequently be synthesized by simply crystallizing from water. The structures and interrelations of such systems are excellently reviewed by Nies and Campbell.[73] Shartsis *et al.* measured the heats of solution of mixtures of alkali oxides and B_2O_3 in various ratios in 2N nitric acid. From a reappraisal[24] of their results it is possible to deduce the enthalpy changes accompanying the reactions:

$$M_2^IO(c) + nB_2O_3(c) \longrightarrow M_2^IOnB_2O_3(gl.) \qquad (28)$$

and

$$M^{II}O(c) + nB_2O_3(c) \longrightarrow M^{II}OnB_2O_3(gl.) \qquad (29)$$

They examined cases where $n = 1$, 2 or 3 and where M^I is Li, Na or K, and where M^{II} is Pb. Combination of these results with appropriate ancillary data allows the tentative proposal of standard enthalpies of formation of borates of the type M^IBO_2, $M^{II}B_2O_4$, $M_2^IB_4O_7$, $M^{II}B_4O_7$, $M_2^IB_6O_{10}$,

$M^{II}B_6O_{10}$, $M_2^IB_8O_{13}$ and $M^{II}B_{10}O_{17}$. In the particular case of lithium meta-borate, Shartsis' work was repeated by Sinke[74] who studied the reaction:

$$LiBO_2(c) + HNO_3 \; 112 \; H_2O(aq) + H_2O(liq) \longrightarrow [LiNO_3 + H_3BO_3] \; 112 \; H_2O \quad (30)$$

Using the ancillary data outlined in ref. 24 and the boric acid data recommended here $\Delta H_f^\circ[LiBO_2, c] = -243.4 \pm 0.3$ kcalmole^{-1}. The available data for borates are collected in Table 4.

TABLE 4. THE OXO-BORATES

Compound	ΔH_f°(c) kcalmole^{-1}	Source
$LiBO_2$	-246.0 ± 0.8	Shartsis[71, 72] recalculated by JANAF
$LiBO_2$	-243.4 ± 0.3	Sinke[74]
$NaBO_2$	-234 ± 2	Shartsis[71, 72] recalculated by JANAF
PbB_2O_4	-372 ± 1.5	Shartsis[71, 72] recalculated by JANAF
$Li_2B_4O_7$	-803.6 ± 1.5	Shartsis[71, 72] recalculated by JANAF
$Na_2B_4O_7$	-783.2 ± 2.0	Shartsis[71, 72] recalculated by JANAF
$K_2B_4O_7$	-796.9	Shartsis[71, 72] recalculated by JANAF
PbB_4O_7	-683 ± 1.5	Shartsis[71, 72] recalculated by JANAF
$Li_2B_6O_{10}$	-1113.7 ± 1.4	Shartsis[71, 72] recalculated by JANAF
$Na_2B_6O_{10}$	-1094.7 ± 2.2	Shartsis[71, 72] recalculated by JANAF
$K_2B_6O_{10}$	-1107.4 ± 2.4	Shartsis[71, 72] recalculated by JANAF
PbB_6O_{10}	-1003 ± 2.0	Shartsis[71, 72] recalculated by JANAF
$Li_2B_8O_{13}$	-1413.6 ± 1.6	Shartsis[71, 72] recalculated by JANAF
$K_2B_8O_{13}$	-1420.9 ± 1.4	Shartsis[71, 72] recalculated by JANAF
$Pb_2B_{10}O_{17}$	-1694 ± 3	Shartsis[71, 72] recalculated by JANAF

G. Borides

In general, borides are hard refractory materials and chemically inert.[76] The thermodynamic data currently available derive from high-temperature equilibrium studies of the decomposition into elements and, latterly, from fluorine bomb calorimetry, a technique to which these compounds are ideally suited. Wright and Walsh[77] have studied the equilibria (at ca. 1100°K):

$$MgB_4(c) \rightleftharpoons Mg(g) + 4 \; B(c) \quad (31)$$

and
$$2 \; MgB_2(c) \rightleftharpoons Mg(g) + MgB_4(c) \quad (32)$$

and Schissel and Trulson[20] examined:

$$TiB(c) \rightleftharpoons Ti(g) + B(g) \quad (33)$$

at considerably higher temperatures (ca. 2300°K), both workers deriving approximate data for the borides at 298.15°K. For TiB_2, the thermodynamic position is very uncertain. For example, the following figures for $\Delta H_f^\circ[\text{TiB}_2, \text{c}]$ have been reported: -66.9, < -51, -62.9, -66.85 ± 2.7, and -48.2 ± 5 kcalmole^{-1}. The JANAF compilation,[67] from which the above figures were taken, recommend $\Delta H_f^\circ[\text{TiB}_2, \text{c}] = -66.8 \pm 4.0$ kcalmole^{-1}. A recent value[82] obtained by conventional oxygen bomb calorimetry for $\text{TiB}_{2.02 \pm 0.01}$ is -77.6 ± 0.9 kcalmole^{-1}. Johnson et al.[78] have applied fluorine-bomb techniques to zirconium and hafnium borides. Some of the available data are collected in Table 5.

TABLE 5. ENTHALPIES OF FORMATION OF SOME BORIDES

Compound	$\Delta H_f^\circ(\text{c})$ kcalmole^{-1}	Ref.
MgB_2	-22.0 ± 2.0	77
MgB_4	-25.1 ± 2.0	77
TiB	-38.3 ± 9	20
TiB_2	-66.8 ± 4	67
$\text{TiB}_{2.02 \pm 0.01}$	$-77.6 \pm .9$	82
$\text{ZrB}_{1.993}$	-77.9 ± 1.5	78
$\text{HfB}_{2.003}$	-78.6 ± 2.1	78
VB_2	-24	79
CrB_2	-31	80
NbB_2	-36	80
TaB_2	-51.7	81

H. Boron–Nitrogen Compounds

Boron nitride has been the subject of several investigations, both equilibrium studies on

$$\text{BN(c)} \rightleftharpoons \text{B(c)} + \tfrac{1}{2}\text{N}_2(\text{g}) \tag{34}$$

and by combustion calorimetry. Some values are collected in Table 6.

TABLE 6. $\Delta H_f^\circ \text{BN(c)}$

Investigator	Method	$\Delta H_f^\circ \text{BN(c)}$
Gross[84]	F bomb-calorimetry	-60.7 ± 0.6
Sinke[85]	combustion of B(c) and BN(c) in NF$_3$	-60.1 ± 2.0
Gal'chenko[86]	nitrogen combustion	-60.3 ± 0.6
Dworkin[87]	oxygen bomb calorimetry	-59.9 ± 0.6
Hildenbrand[88]	equilibrium (see above)	-59.4 ± 0.7
Dreger[89]	equilibrium (see above)	-60.3 ± 2.0
Hubbard[90]	F bomb-calorimetry	-59.97 ± 0.37

The concurrence of these data is good and a mean figure $\Delta H_f^\circ[\text{BN, c}]$ $= -60.1 \pm 0.4$ kcalmole^{-1} is proposed. Brief mention will be made here of a formally similar molecule, boron phosphide.

Gal'chenko et al. report thermochemical data[91] for the reaction:

$$BP(c) + 4\ Cl_2(g) = BCl_3PCl_5(c) \tag{35}$$

and combination of their data with $\Delta H_f^\circ[\text{PCl}_5 \cdot \text{BCl}_3, \text{c}] = -251.5 \pm 0.9^{(92)}$ kcalmole^{-1} yields $\Delta H_f^\circ[\text{BP, c}] = -59.3 \pm 1.0$ kcalmole^{-1}. This is in wide divergence to a figure of -19 kcalmole^{-1} reported in ref. 38.

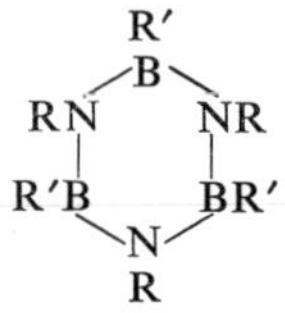

(II)

Compounds of general formula (II) (variously called borazoles or bora-zines) have not been studied extensively. Some of the available data are collected in Table 7.

TABLE 7. ENTHALPIES OF FORMATION OF SOME BORAZOLES (II)

R	R'	ΔH_f°	$\Delta H(c \rightarrow g)$	$\Delta H(l \rightarrow g)$	$\Delta H(c \rightarrow l)$
Me	H	-232.3 ± 0.1(liq)[93]	—	9.3[97]	—
H	H	-204.9 ± 0.3(liq)[93]	—	7.0[96]	—
H	H	-129.7 ± 3.0(liq)†[94]	—	7.0[96]	—
H	Cl	-256.1 ± 0.2(c)[93]	16.85 ± 0.09[98]	11.86 ± 0.04[98]	5.0 ± 0.1[98]
H	Cl	-255.15(c)± 1.0†[95]	17.2[38]	—	—
H	Br	—	20.6 ± 0.1[98]	11.2 ± 1.3[98]	9.4 ± 1.4[98]
H	F	—	15.07 ± 0.02[98]	—	—

† Recalculated.

The divergence between the two data for borazole (II : R=R'=H) is very large.

Skinner and Smith[99] have studied the solution thermochemistry of the series $(NMe_2)_{3-n} BCl_n$ (where $n = 0,1,2$) via the acid hydrolysis reac-tion:

e.g. $$B(NMe_2)_2Cl + 3\ H_2O + HCl = H_3BO_3 + 2\ \overset{+}{N}H_2Me_2\overset{-}{Cl} \tag{36}$$

The recalculated data are:

Compound	$\Delta H_f^{\circ}(l)$	$\Delta H(l \rightarrow g)$
$(NMe_2)BCl_2$	-104.8 ± 1.1 kcalmole^{-1}	8.9 ± 0.3
$(NMe_2)_2BCl$	-90.1 ± 1.1 kcalmole^{-1}	10.0 ± 0.5
$(NMe_2)_3B$	-69.9 ± 0.5 kcalmole^{-1}	11.2 ± 0.2

Conventional oxygen bomb calorimetry[100] was used to determine the standard enthalpies of formation of some dibutylaminoboranes. Corrections were made for unburned material;

$$\Delta H_f^{\circ}[Bu_2BNH_2, liq] = -95.2 \pm 0.8 \text{ kcalmole}^{-1}.$$
$$\Delta H_f^{\circ}[Bu_2BNHBu, liq] = -108.1 \pm 1.3 \text{ kcalmole}^{-1}$$

I. Trialkyl and Triarylboranes

Methods to determine the standard enthalpies of formation of compounds of the type R_3B have been based on three reactions:

(i) conventional oxygen combustion,
(ii) hydroboronation in non-aqueous solvents,
(iii) aqueous oxidative hydrolysis.

As mentioned in the introduction, static bomb methods using oxygen as oxidant are unsatisfactory for boron compounds because of the difficulty in defining the final thermodynamic state. When static oxygen bomb techniques are utilized it is mandatory to estimate the extent of combustion (e.g. by assessing the CO_2 produced if the original compound contains carbon) and to analyze the bomb residues to estimate the thermal effect of side reactions or simply correct for incomplete combustion. This procedure was followed by Gal'chenko.[104] Skinner et al.[103, 107] have exploited the hydroboronation reaction first documented by H. C. Brown et al.[116]:

$$B_2H_6 + 6 (RCH\!\!=\!\!CH_2) = 2 B(CH_2CH_2R)_3 \qquad (37)$$

The reactions were performed in "monoglyme" or "diglyme" using a specially designed reaction calorimeter to introduce the diborane into the system. The disadvantage of this method is that it requires accurate enthalpy of formation data for the reactants. These are available, of course, for the olefins, but $\Delta H_f^{\circ}[B_2H_6, g]$ is still equivocal. In a recent review,

Skinner[109] used 9.2 ± 0.5 kcalmole^{-1} but the JANAF compilation recommends rather more pessimistic error limits, i.e. 9.8 ± 4.0 kcalmole^{-1}. In his paper on "key" data for boron compounds, Gunn[8] recommends 8.6, and the up-dated[38] NBS Circular 500, $+8.5$ kcalmole^{-1}.

Aqueous oxidative hydrolysis[108] has been utilized to determine the standard enthalpies of formation of triphenylborane and tricyclohexylborane in a conventional reaction calorimeter:

$$R_3B + 3H_2O_2 = 3\,ROH + H_3BO_3 \tag{38}$$

The ancillary data required for this reaction are well documented. Some of the currently available data are collected in Table 8.

TABLE 8. ENTHALPIES OF FORMATION FOR R_3B IN KCALMOLE^{-1}

Compound	Method	$\Delta H_f^{\circ \dagger}$	$\Delta H(c \to g)$	$\Delta H(l \to g)$	Ref.§
$(CH_3)_3B(liq)$	(i)	-34.3 ± 5.5		4.83	101
$(CH_3)_3B(liq)$	(i)	-34.3 ± 5.5			102
$(C_2H_5)_3B(liq)$	(i)	-46.7 ± 3.7		8.8 ± 0.2	101
$(C_2H_5)_3B(liq)$	(ii)	$-43.1 \pm 1.2\ddagger$			103
$(C_3H_7)_3B(liq)$	(i)	-66.7 ± 3.1			104
$(i\text{-}C_3H_7)_3B(liq)$	(i)	-70.4 ± 2.8			104
$(C_4H_9)_3B(liq)$	(i)	-82.7 ± 2.5		13.8	101 : 105
$(i\text{-}C_4H_9)_3B(liq)$	(i)	-91.9 ± 1.3			104
$(i\text{-}C_4H_9)_3B(liq)$	(ii)	-80.6		13.8 ± 0.5	103
$(s\text{-}C_4H_9)_3B(liq)$	(i)	-73 ± 6		14.3 ± 0.5	106 : 103
$(i\text{-}C_5H_{11})_3B(liq)$	(i)	-108.6 ± 1.6			104
$(C_6H_5)_3B(c)$	(iii)	$+11.4 \pm 0.9$	~ 4	15.4 ± 0.2	108
$(c\text{-}C_6H_{11})_3B(c)$	(iii)	-115.5 ± 1.5			108
$(C_6H_{13})_3B(liq)$	(ii)	$-115.6 \pm 2.2\ddagger$		21.2 ± 0.5	107 : 103
$(C_7H_{15})_3B(liq)$	(ii)	$-133.3 \pm 1.8\ddagger$		24.4 ± 0.5	107 : 103
$(C_8H_{17})_3B(liq)$	(ii)	$-151.8 \pm 1.8\ddagger$		27.6 ± 0.5	107 : 103

$\dagger$ All data converted to $\Delta H_f^{\circ}[H_3BO_3, c] = -261.7 \pm 0.2$ kcalmole^{-1}.

$\ddagger$ Assuming $\Delta H_f^{\circ}[B_2H_6, g] = 9.2 \pm 0.5$ kcalmole^{-1}.

§ Second reference is to the enthalpy of transition data when the source is not the first reference.

Skinner[109] finds his value[105] for $\Delta H(Bu_3B, l \to g)$ rather low and prefers the estimate 14.8 ± 0.5 kcalmole^{-1}. With this proviso, one can interpolate readily some missing data from Table 8 using a graph of $\Delta H_f^{\circ}(R_3B)$ vs. number of C atoms in R.

$$\Delta H_f^{\circ}[(C_5H_{11})_3B, liq] = -100 \pm 3 \text{ kcalmole}^{-1}$$
$$\Delta H[(C_5H_{11})_3B, liq \longrightarrow g] = 18.3 \pm 0.7 \text{ kcalmole}^{-1}$$
$$\Delta H[(C_3H_7)_3B, liq \longrightarrow g] = 11.9 \pm 0.7 \text{ kcalmole}^{-1}$$

J. *Boronic and Borinic Acids and their Anhydrides*

Little data are available for boronic acids (III), borinic acids (IV) and their respective anhydrides (V, VI)

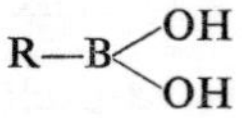

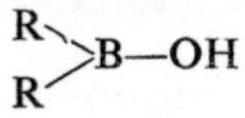

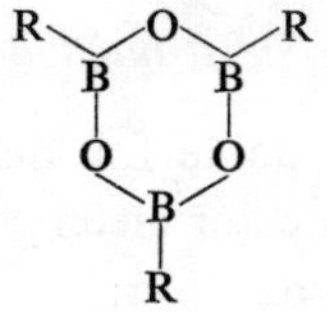

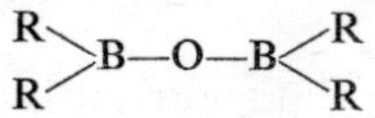

(III)   (IV)   (V)   (VI)

Skinner[105] has studied the hydrolysis of $(n\text{-}C_4H_9)_2BX$(liq) where $X = Cl, Br, I$:

$$(C_4H_9)_2BX(liq) + (n+1)H_2O(liq) = (C_4H_9)_2BOH(liq) + HXnH_2O \qquad (39)$$

The standard enthalpies of formation of the dibutylboron halides ($X = Br, I$) are known from the study of independent reactions. This gives $\Delta H_f^\circ[\text{Bu}_2\text{BOH, liq}] = -145.4 \pm 3.0$ kcalmole^{-1} for $X = Br$ in the above reaction and -146.2 ± 3.0 kcalmole^{-1} for $X = I$. A mean value of -145.8 ± 3.0 is selected. The ancillary data for this calculation were: $\Delta H_f^\circ[\text{HBr 300 0H}_2\text{O}]^{(50)} = -29.05 \pm 0.09$ kcalmole^{-1}, $\Delta H_f^\circ[\text{HI} \cdot 50\,\text{H}_2\text{O}]^{(53)} = -13.56 \pm 0.3$, $\Delta H_f^\circ[\text{H}_2\text{O, liq}]^{(38)} = -68.315$ kcalmole^{-1} and dilution enthalpies from ref. 38.

Phenylboronic acid, diphenylborinic acid and their anhydrides have been investigated thermochemically[111] via their oxidative hydrolyses:

$$(\text{ØBO})_3 + 3\,H_2O_2 + 3\,H_2O = 3\,\text{ØOH} + 3\,H_3BO_3 \qquad (40)$$
$$(\text{Ø}_2\text{B})_2\text{O} + 4\,H_2O_2 + H_2O = 4\,\text{ØOH} + 2\,H_3BO_3 \qquad (41)$$

The experiments were performed in a conventional reaction calorimeter in a basic medium (pH $\sim$ 13). The following results were obtained:

$$\Delta H_f^\circ[(\text{ØBO})_3, c] = -301.8 \pm 2.3 \text{ kcalmole}^{-1}$$
$$\Delta H_f^\circ[\text{ØB(OH)}_2, c] = -172.2 \pm 0.5 \text{ kcalmole}^{-1}$$
$$\Delta H_f^\circ[(\text{Ø}_2\text{B})_2\text{O}, c] = -84.4 \pm 2.0 \text{ kcalmole}^{-1}$$

Diphenylborinic acid is an unstable, intractable oil[10] and was not investigated directly. Instead $\Delta H_f^\circ[\text{Ø}_2\text{BOH}$, hypothetical crystalline state] ~ -78 kcalmole^{-1} was estimated. Further work[112] on the hydrolysis of diphenylboron chloride yielding $(\Delta H_f^\circ[\text{Ø}_2\text{BCl, liq}] - \Delta H_f^\circ[\text{Ø}_2\text{BOH, oil}])$ and the oxidative hydrolysis of the same compound yielding $\Delta H_f^\circ[\text{Ø}_2\text{BCl, liq}]$ gave:

$$\Delta H_f^\circ[\text{Ø}_2\text{BOH, oil}] = -77.5 \pm 1.8 \text{ kcalmole}^{-1}.$$

Dworkin and van Artsdalen[113] have recorded the enthalpies of hydrolysis of butylboronic anhydride and the enthalpy of solution of butylboronic acid as -11.6 ± 0.2 and 2.7 ± 0.1 kcalmole^{-1} respectively.

K. *Substituted Boron Halides*

The data currently available for these compounds derive exclusively from work using reaction calorimeters.

Skinner[105] studied the reaction:

$$(n-C_4H_9)_3B(liq)+HX(g) = (C_4H_9)_2BX(liq)+C_4H_{10}(g) \qquad (X=Br, I) \qquad (42)$$

in an adiabatic reaction calorimeter at 56.5°C with the reactants and products dispersed in an inert solvent. Estimation of the extent of reaction was effected by both measuring the volume of butane produced and analyzing the dibutylboron halide. Using $\Delta H_f^\circ[(C_4H_9)_3B, liq] = -82.5\pm2.5$ and $\Delta H_f^\circ[HI, g]^{(38)} = 6.33$ kcalmole^{-1}, we derive:

$$\Delta H_f^\circ[(C_4H_9)_2BBr, liq] = -84.2\pm1.4 \text{ kcalmole}^{-1}$$

and
$$\Delta H_f^\circ[(C_4H_9)_2BI, liq] = -66.7\pm1.3 \text{ kcalmole}^{-1}$$

Skinner also measured the enthalpy of hydrolysis of dibutylboron chloride, and using the figure for dibutylborinic acid mentioned previously, we have $\Delta H_f^\circ[(C_4H_9)_2BCl, liq] = -98.8\pm3$ kcalmole^{-1}. The latent heats[105] of vaporization ($\Delta H(l \rightarrow g)$) of dibutylboron chloride, bromide and iodide are 11.5 ± 0.2, 12.1 ± 0.2 and 13.0 ± 0.5 (est.) kcalmole^{-1} respectively.

The phenylboron bromides and chlorides have been studied[108] via their hydrolyses and, in the case of diphenylboron chloride, via its oxidative hydrolysis. This latter reaction in conjunction with the aqueous hydrolysis of the same compound was used to establish the enthalpy of formation of phenylborinic acid. Both phenylboron dibromide and diphenylboron chloride have melting points close to 25°C ($28-29°$, $29-30°$ resp.). The chloride could be supercooled to 25°C so the thermodynamic data for this compound refers to the metastable liquid at 25°C. The bromide would not supercool and the experiments were performed at 30°C and the difference between the enthalpies of formation at 25° and 30° assumed to be within experimental error. The following data were obtained:

$$\Delta H_f^\circ[(C_6H_5)_2BCl, liq] = -31.8\pm0.8 \text{ kcalmole}^{-1}$$
$$\Delta H_f^\circ[(C_6H_5)_2BBr, liq] = -16.1\pm0.9 \text{ kcalmole}^{-1}$$
$$\Delta H_f^\circ[C_6H_5BCl_2, liq] = -71.8\pm0.3 \text{ kcalmole}^{-1}$$
$$\Delta H_f^\circ[C_6H_5BBr_2, liq] = -41.5\pm0.3 \text{ kcalmole}^{-1}$$

The enthalpies of vaporization of these compounds, in the above order, are 9.9 ± 0.5, 14.4 ± 0.5, 8.05 ± 0.2, 10.5 ± 0.5 kcalmole^{-1}.

Hydrolysis reactions in conventional solution calorimeters have been used to study $(EtO)_2BCl$,[114] $(MeO)_2BCl$[115] and $EtOBCl_2$.[114] These compounds are rather unstable and disproportionate on heating. It is pertinent that the N.B.S. sample of $(MeO)_2BCl$ was purified by fractional crystallization, and even then had an estimated purity of only ca. 95 per cent.

Recalculating Skinner's data we obtain:

$$\Delta H_f^\circ[(C_2H_5O)_2\,BCl,\ liq] = -205.6\pm1.2 \text{ kcalmole}^{-1}$$
$$\Delta H_f^\circ[(C_2H_5O)\,BCl_2,\ liq] = -157.6\pm1.2 \text{ kcalmole}^{-1}$$

and from Johnson's data:

$$\Delta H_f^\circ[(CH_3O)_2BCl,\ liq] = -186.4\pm0.5 \text{ kcalmole}^{-1}$$

Vapor pressure data of Wiberg and Sütterlin[117] give the enthalpies of vaporization of these compounds in the above order as 9.3 ± 0.2, 8.4 ± 0.2 and 8.2 ± 0.3 kcalmole^{-1}. Johnson et al.[115] estimate a value for $\Delta H_f^\circ[CH_3OBCl_2,\ g]$ of ca. -141 kcalmole^{-1}.

The thermochemistry of some 1,3,2-dioxaborole and borolane cyclic systems (VII–X) has been studied[121]

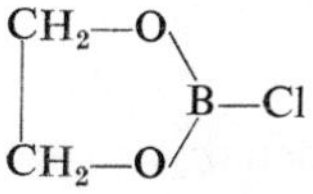

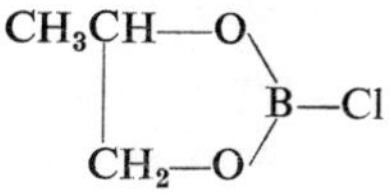

via their hydrolyses in an attempt to evaluate the strain energy in the heterocyclic ring. The standard enthalpies of formation of (VII)–(X) (all (liquid are -180.4 ± 1.0, -190.2 ± 1.0, -199.0 ± 1.7, and -181.8 ± 0.7 kcalmole^{-1}: This final figure was recalculated using a new value[122] for $\Delta H_f^\circ[HO(CH_2)_3OH,\ liq] = -110.7\pm0.4$ kcalmole^{-1}. The latent heat of vaporization[123] of (X) is 13.7 ± 0.5 kcalmole^{-1}.

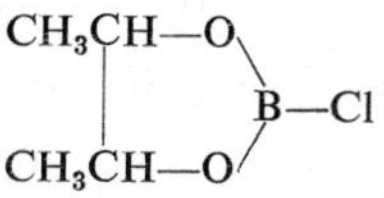

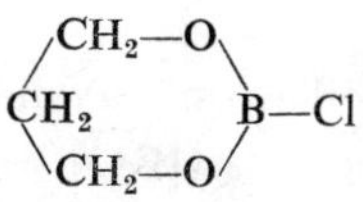

L. Alkyloxyboranes

Two dialkoxyboranes $((RO)_2BH)$ have been studied. Skinner[118] studied the iso-propoxy derivative and Cooper[119] the methoxy derivative. Skinner investigated two reactions with diborane, both yielding the same product:

$$B_2H_6+4\,(CH_3)_2CO = 2\,[CH_3CH(CH_3)O]_2BH \tag{43}$$

and $$B_2H_6+4\,CH_3CH(CH_3)OH = 2\,[CH_3CH(CH_3)O]_2BH+4\,H_2 \tag{44}$$

The reactions were performed either by passing diborane into the pure second reactant or into a hexane solution of it. The latter reaction is very much slower than the former, reaction times of ca. 30 minutes being observed. Notwithstanding, the spread in the experimental results for this second experiment was smaller

$$\Delta H_f^\circ[(i-PrO)_2BH, liq] \text{ (reaction (43))} = -182.8 \pm 1.3 \text{ kcalmole}^{-1}$$
$$\Delta H_f^\circ[(i-PrO)_2BH, liq] \text{ (reaction (44))} = -183.7 \pm 0.8 \text{ kcalmole}^{-1}$$

Both these data depend upon data for diborane, and Skinner's choice was $\Delta H_f^\circ[B_2H_6, g] = +9.2 \pm 0.5 \text{ kcal/mole}^{-1}$. This figure has been commented upon previously (see Section II, I).

Cooper and Masi[119] used a hydrolysis reaction for their studies,

$$HB(OCH_3)_2(liq) + (n+3)H_2O(liq) = H_2(g) + [H_3BO_3 + 2 CH_3OH]nH_2O \qquad (45)$$

The compound is somewhat unstable with respect to disproportionation into trimethyl borate and diborane so samples are stored at $-80°C$ prior to calorimetry

$$\Delta H_f^\circ[HB(OCH_3)_2, liq] = -144.7 \pm 1.5 \text{ kcalmole}^{-1}$$

Burg[120] reports 6.14 kcalmole^{-1} for the latent heat of vaporization for this compound.

M. *Organic Borates*

The only available thermochemical data on this important class of compounds ($[RO]_3B$) are those of Skinner[124] who studied their hydrolyses:

$$B(OR)_3(liq) + (n+3)H_2O(liq) = [H_3BO_3 + 3 ROH]nH_2O \qquad (46)$$

Using the following ancillary data we have recalculated Skinner's results: $\Delta H_f^\circ[MeOH, liq]^{[125]} = -57.02 \pm 0.05$, $\Delta H_f^\circ[EtOH, liq]^{[125]} = -66.36$, $\Delta H_f^\circ[n\text{-}PrOH, liq]^{[126]} = -72.31 \pm 1.0$ and $\Delta H_f^\circ[n\text{-}BuOH, liq]^{[127]} = -78.49 \pm 0.2 \text{ kcalmole}^{-1}$.

TABLE 9. ENTHALPIES OF FORMATION OF SOME BORATES

R	$\Delta H_f^\circ[(RO)_3B, liq]$	$\Delta H(liq \longrightarrow g)$
Me	-223.5 ± 0.3	$8.3(\pm 0.2)$
Et	-250.9 ± 0.3	$10.5(\pm 0.2)$
n–Pr	-268.9 ± 1.2	11.8 ± 1.0
n–Bu	-287.6 ± 0.3	12.5 ± 1.0

N. *Thioboron Compounds*

Compared with boron–oxygen chemistry, the literature of thioboron chemistry is extremely scant—only 2.3 per cent of Steinberg's comprehensive treatise[3] is devoted to such compounds. The only thermodynamic datum for B_2S_3 is a very approximate figure for the enthalpy of formation[128] in the gas phase of -11 (± 10) kcalmole^{-1}, deriving from a mass-spectroscopic study of the volatile species above $ZnS(c)$ and $B(c)$ at 700–900°C.

The thioborates, $(RS)_3B$, have been studied[129] via their hydrolyses:

$$(RS)_3B(liq) + (n+3)H_2O(liq) = 3RSH(s.s.) + H_3BO_3 n H_2O \qquad (47)$$

The experiments were performed in saturated solutions of the appropriate thiol so that the thiol produced by hydrolysis is present in its standard state (s.s.). The available data are collected below:

TABLE 10. ENTHALPIES OF FORMATION OF THIOBORATES

R	$\Delta H_f^\circ (RS)_3B(liq)$	$\Delta H(liq \rightarrow g)$
CH_3	-50.2 ± 0.6	12.9 ± 0.2
C_2H_5	-82.7 ± 0.5	14.6 ± 0.5
$n-C_3H_7$	-100.7 ± 0.6	20.8 ± 0.5
$n-C_4H_9$	-117.0 ± 0.9	22.9 ± 0.5
$n-C_5H_{11}$	-135.6 ± 1.2	25.0 ± 0.5
C_6H_5	$+15.4 \pm 0.9$(cryst.)	26.0 ± 1.5(est.)

Some cyclic boron–sulfur compounds have also been investigated.[123]

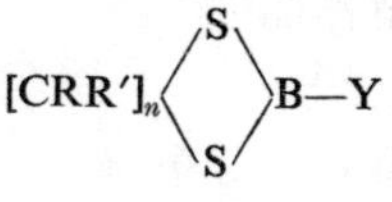

$$[CRR']_n \underset{\diagdown S \diagup}{\overset{\diagup S \diagdown}{\Big\langle}} B-Y \qquad
\begin{array}{l} n = 2,3 \\ R, R' = CH_3 \text{ or } H \\ Y = Cl, \varnothing \end{array}$$

(XI)

The hydrolyses were studied and lead to different products dependent on the nature of Y

$$\overline{S\,RS\,B}Cl + 3\,H_2O = R(SH)_2 + H_3BO_3 + HCl \qquad (48)$$

$$\overline{S\,RS\,B}\,\varnothing + 2\,H_2O = R(SH)_2 + \varnothing B(OH)_2 \qquad (49)$$

The dithiols produced by hydrolysis were relatively slow to dissolve compared with the other hydrolysis products. Careful analysis of the temperature/time profile during the reaction period yielded the enthalpy changes for both of the following reactions:

e.g. $\overline{S\,RSB}\,Cl(liq)+(n+3)H_2O = [R(SH)_2+H_3BO_3+HCl]nH_2O$ (50)

$\overline{S\,RSB}\,Cl(liq)+(n+3)H_2O = R(SH)_2(liq)+[H_3BO_3+HCl]nH_2O$ (51)

Enthalpy changes deduced for the former reaction were rather more accurate. The data are collected in Table 11.

TABLE 11. SOME 1,3,2-DITHIABOROLES AND BOROLANES

Compound	$\Delta H_f^\circ(liq)$ kcalmole^{-1}	$\Delta H(liq \rightarrow g)$
$\overline{S\,CH_2CH_2CH_2S\,B}\,Cl$	-73.0 ± 0.7	13.1 ± 0.5 (est.)
$\overline{S\,CH_2CH_2S\,B}\,Cl$	-66.1 ± 0.7	11.6 ± 0.5 (est.)
$\overline{S\,CH(CH_3)CH_2S\,B}\,Cl$	-72.2 ± 0.7	
$\overline{S\,CH_2CH_2CH_2S\,B}\,\varnothing$	-37.0 ± 1.2	
$\overline{S\,CH_2CH_2S\,B}\,\varnothing$	-32.0 ± 1.1	

III. COLLATED STANDARD ENTHALPIES OF FORMATION

TABLE 12. SELECTED RECOMMENDED STANDARD ENTHALPIES OF FORMATION at 298.15°K IN KCALMOLE^{-1}

Compound	ΔH°	Compound	ΔH_f°
B(amorphous)	0.8 ± 0.5	BBr$_3$(liq)	-57.4 ± 0.5
B(gas)	132.8 ± 4.0	BBr$_3$(g)	-49.2 ± 0.5
HBO$_2$(c, I)	-192.3 ± 0.2	BI$_3$(c)	-11.3 ± 0.3
HBO$_2$(c, II)	-190.0 ± 0.5	BI$_3$(g)	4.2 ± 1.5
HBO$_2$(c, III)	-188.7 ± 0.4	B$_2$Cl$_4$(liq)	-124.6 ± 2.0
HBO$_2$(g)	-134.4 ± 1.0	B$_2$Cl$_4$(g)	-116.6 ± 2.0
(HOBO)$_3$(g)	-543 ± 3	B$_2$F$_4$(g)	-343.4 ± 3.0
H$_3$BO$_3$(c)	-261.7 ± 0.2	(BOCl)$_3$(g)	-390 ± 2
H$_3$BO$_3$(g)	-237.4 ± 0.7	(BOF)$_3$(g)	-565.3 ± 1.0
H$_3$BO$_3$nH$_2$O		NaBH$_4$(c)	-45.0 ± 0.3
$500 < n < 20,000$ }	-256.35 ± 0.2	LiBH$_4$(c)	-45.9 ± 0.4

TABLE 12. (cont.)

Compound	ΔH_f°	Compound	ΔH_f°
$B_2(OH)_4(c)$	-335.6 ± 0.8	$KBH_4(c)$	-54.4 ± 0.5
$B_2O_3(c)$	-304.6 ± 0.4	$KBF_4(c)$	-451.6 ± 1.4
$B_2O_3(amorphous)$	-300.2 ± 0.4	$Al(BH_4)$ (liq)	-76.0 ± 4.0
$B_2O_3(liq)$	-300.2 ± 0.4	$Al(BH_4)$ (g)	-68.4 ± 4.0
$B_2O_3(g)$	-200.1 ± 0.5		
$B_2O_2(g)$	-109 ± 2	$LiBO_2(c)$	-243.4 ± 0.3
		$NaBO_2(c)$	-234 ± 2
$BF_3(g)$	-271.65 ± 0.22	$BN(c)$	-60.1 ± 0.4
$BCl_3(liq)$	-102.0 ± 0.3	$BP(c)$	-59.3 ± 1.0
$BCl_3(g)$	-96.3 ± 0.3	$(NMe_2)_3B(liq)$	-69.9 ± 0.5
		$(NMe_2)_3B(g)$	-58.7 ± 0.6
$(CH_3)_3B(liq)$	-34.3 ± 5.5	$(C_4H_9O)_3B(liq)$	-287.6 ± 0.3
		$(C_4H_9O)_3B(g)$	-275.1 ± 1.0
$(CH_3)_3B(g)$	-29.5 ± 5.5	$(CH_3S)_3B(liq)$	-50.2 ± 0.6
$(C_2H_5)_3B(liq)$	-46.7 ± 3.7	$(CH_3S)_3B(g)$	-37.3 ± 0.7
$(C_2H_5)_3B(g)$	-37.9 ± 3.7	$(C_2H_5S)_3B(liq)$	-82.7 ± 0.5
$(C_3H_7)_3B(liq)$	-66.7 ± 3.1	$(C_2H_5S)_3B(g)$	-68.1 ± 0.8
$(C_3H_7)_3B(g)$	-54.8 ± 3.2	$(C_3H_7S)_3B(liq)$	-100.7 ± 0.6
$(C_4H_9)_3B(liq)$	-82.7 ± 2.5	$(C_3H_7S)_3B(g)$	-79.9 ± 0.8
$(C_4H_9)_3B(g)$	-67.9 ± 2.5	$(C_4H_9S)_3B(liq)$	-117.0 ± 0.9
$(C_5H_{11})_3B(liq)$	-100 ± 3	$(C_4H_9S)_3B(g)$	-94.1 ± 1.0
$(C_5H_{11})_3B(g)$	-81.7 ± 3.1	$(C_5H_{11}S)_3B(liq)$	-135.6 ± 1.2
$(C_6H_5)_3B(c)$	11.4 ± 0.9	$(C_5H_{11}S)_3B(g)$	-110.6 ± 1.3
$(C_6H_5)_3B(g)$	30.8 ± 2.0	$(C_6H_5S)_3B(c)$	$+15.4 \pm 0.9$
$(c-C_6H_{11})_3B(c)$	-115.5 ± 1.5		
$(C_4H_9)_2BOH(liq)$	-145.8 ± 3.0		
$(ØBO)_3(c)$	-301.8 ± 2.3		
$ØB(OH)_2(c)$	-172.2 ± 0.5		
$(Ø_2B)_2O(c)$	-84.4 ± 2.0		
$Ø_2BOH(oil)$	-77.5 ± 1.8		
$(CH_3O)_3B(liq)$	-223.5 ± 0.3		
$(CH_3O)_3B(g)$	-215.2 ± 0.4		
$(C_2H_5O)_3B(liq)$	-250.9 ± 0.3		
$(C_2H_5O)_3B(g)$	-240.4 ± 0.4		
$(C_3H_7O)_3B(liq)$	-268.9 ± 1.2		
$(C_3H_7O)_3B(g)$	-257.1 ± 1.5		

IV. DISCUSSION

A. *Preamble*

Bond energies are among the most common functions readily derivable from thermochemical data. Such terms, as opposed to bond dissociation energies (see below), have little theoretical foundation and hence their interpretation is not always unambiguous. However, they persist in the literature and are the most widely accepted index of bond strength. Such terms must not be identified with the ease of heterolytic bond breakage under normal reaction conditions. Bond energies are conventionally quoted in terms of enthalpies at 298.15°K.

To accent the distinction between bond energy (E) and bond dissociation energy (D), consider the compound $(Me_2N)_3B$. The following reactions indicate the difference between the above terms and also illustrate how another function, the mean bond dissociation energy ($\bar{D}$), arises:

$$(Me_2N)_3B(g) = 6\ C(g) + 18\ H(g) + 3\ N(g) + B(g)$$

$$\Delta H_1 = 18\ E(C\!-\!H) + 6\ E(C\!-\!N) + 3\ E(B\!-\!N) \qquad (52)$$

$$(Me_2N)_3B(g) = (Me_2N)_2B(g) + NMe_2(g)$$

$$\Delta H_2 = D([Me_2N]_2B\!-\!NMe_2) \qquad (53)$$

$$(Me_2N)_3B(g) = 3\ Me_2N(g) + B(g)$$

$$\Delta H_3 = 3\ \bar{D}(Me_2N\!-\!B) \qquad (54)$$

Bond energies are calculated from the following thermochemical cycle:

$$(NMe_2)_3B(g) \xrightarrow{\ \Sigma E\ } 6\ C(g) + 18\ H(g) + 3\ N(g) + B(g)$$

$$\left\uparrow \Delta H(liq \to g) \right. \qquad\qquad \left\uparrow \Sigma \Delta H_f^\circ(\text{elements, g}) \right. \qquad (55)$$

$$(NMe_2)_3B(liq) \xleftarrow[\ \Delta H_f^\circ[(NMe_2)_3B,\ liq]\ ]{} 6\ C(graphite) + 9\ H_2(g) + \tfrac{3}{2}\ N_2(g) + B(c)$$

Hence:

$$\Sigma E = -\Delta H_f^\circ[(NMe_2)_3B,\ gas] + \Sigma \Delta H_f^\circ[\text{elements, g}] \qquad (56$$

The composite term constituting the right-hand side of the equation is conventionally called the enthalpy of atomization of the compound. This must not be confused with the enthalpies of atomization of the elements, e.g. $\Delta H_f^\circ[B, g]$, accurate values of which are also required for bond energy calculations.

Estimates of the standard enthalpies of formation of radicals (or radical ions) are often available from spectroscopic, kinetic, electron impact or equilibrium studies. Using such data, $\bar{D}$ is determined from:

$$3\ \bar{D}(B\!-\!NMe_2) = 3\ \Delta H_f^\circ[NMe_2, g] + \Delta H_f^\circ[B, g] - \Delta H_f^\circ[(NMe_2)_3B, gl. \qquad (57)$$

Data for standard enthalpies of formation of selected radicals and elements are collected in Table 13.

TABLE 13. ENTHALPIES OF FORMATION OF SOME RADICALS AND ATOMS
IN THE GAS PHASE

Species[†]	ΔH_f°	Species[†]	ΔH_f°
Me	34±1	OH	9.3±1
Et	25.7±1	OCH_3	2±2
n-Pr	21±2	OC_2H_5	−6.7±2
		nC_3H_7O	−11±2
$n-C_4H_9$	17±2	nC_4H_9O	−17±2
		Me_2N[‡]	27±3
C_6H_5	72±2		
$c-C_6H_{11}$	12±3		
B	132.8±4	Br	26.74±0.07
O	59.56±0.03	I	25.54±0.01
C	+170.89±0.45	S	66.7±0.5
N	113±1	H	52.100±0.001
F	18.86±0.2		
Cl	28.92+0.03		

† all radical data from ref. 131 and all atom data from refs. 24 and 67 except where otherwise indicated.
‡ $\Delta H_f^\circ[NMe_2H]g^{(47)} = -6.6$ kcalmole^{-1}.
$D(Me_2N-H)^{(131)} = 86\pm3$.

With the growth of thermochemical data, it soon became clear that the concept of constant, transferable and additive bond energies was unacceptable. More sophisticated schemes, involving writing the atomization energy of a molecule in terms of bond energies and interaction energies between adjacent bonds, were soon developed. Such bond energy schemes have been the subject of a review by Skinner.[132]

B. *Mean Bond Dissociation Energies*

For molecules of the type BO_3 where Q = halogen, alkoxy, dimethylamino, hydroxyl, alkyl and aryl we have derived $\bar{D}$ values from the data in Tables 12 and 13 and the following equation:

$$3\,\bar{D}(B-Q) = 3\,\Delta H_f^\circ[Q, g] + \Delta H_f^\circ[B, g] - \Delta H_f^\circ[BQ_3, g] \qquad (58)$$

TABLE 14. SOME MEAN BOND DISSOCIATION
ENERGIES IN KCAL

Q	$\bar{D}$	Q	$\bar{D}$
F	153.7 ± 1.4	ø	106.0 ± 2.5
Cl	105.3 ± 1.2		
Br	87.4 ± 1.2	OMe	118.0 ± 2.4
I	68.4 ± 1.4	OEt	117.7 ± 2.4
OH	132.8 ± 1.7		
CH_3	88.1 ± 2.5	OPr	119.0 ± 2.4
C_2H_5	82.6 ± 2.1	OBu	118.9 ± 2.4
n-Pr	83.5 ± 2.6	Me_2N	91 ± 3.3
n-Bu	83.9 ± 2.5		

If errors in $\Delta H_f^\circ[B, g]$, $\Delta H_f^\circ[Q, g]$ and $\Delta H_f^\circ[BQ_3, g]$ are δ_B, δ_Q and δ_Δ then the error in $\bar{D}(B-Q)$ is $\frac{1}{3}(\delta_B^2 + 9\delta_Q^2 + \delta_\Delta^2)^{\frac{1}{2}}$.

The error in these $\bar{D}$ values is rather high: this is principally due to the high error in $\Delta H_f^\circ[B, g]$. It will be appreciated that differences between $\bar{D}$ values exclude this term and the error decreases accordingly.

C. Bond Energies (E)

Such terms are necessarily calculated by assuming the transferability of other bond energies. Such a procedure will only be justified in special cases, e.g. when considering E values in a series of compounds and drawing conclusions from their differences. Such a case is the value of $E(B—B)$ in B_2F_4, B_2Cl_4, $B_2(OH)_4$ assuming $E(B—X)$ in B_2X_4 is the same as in BX_3. We may write

$$E(B—B) + 4 E(B—X) = 2 \Delta H_f^\circ[B, g] + 4 \Delta H_f^\circ[X, g] - \Delta H_f^\circ(B_2X_4, g) \quad (59)$$

and

$$3 E(B—X) = \Delta H_f^\circ[B, g] + 3 \Delta H_f^\circ[X, g] - \Delta H_f^\circ[BX_3, g] \quad (60)$$

hence

$$E(B—B) = \tfrac{2}{3}\Delta H_f^\circ[B, g] - \Delta H_f^\circ[B_2X_4, g] + \tfrac{4}{3}\Delta H_f^\circ[BX_3, g] \quad (61)$$

X in B_2X_4	$E(B—B)$ in kcal
F	69.7 ± 2.8
Cl	76.5 ± 2.9
OH[†]	72.6 ± 3.7

It would appear from these limited data that as the electronegativity of X increases, $E(B—B)$ decreases.

[†] Calculated assuming $\Delta H(c \to g, B_2(OH)_4)$ was the same as $(HO)_3B$ times the ratio of their molecular weights (35.2 ± 1.5 kcalmole^{-1}).

It is interesting to compare values of $E(B—Ø)$ in $ØBCl_2$, $Ø_2BCl$, $ØBBr_2$, and $Ø_2BBr$ assuming $E(B—X)$ in $Ø_{3-n} BX_n$ $(n = 1, 2)$ is the same as $E(B—X)$ in BX_3.

Following the same procedure as above we derive

for $ØBX_2$: $E(B—Ø) = \Delta H_f^\circ[Ø, g] + \frac{1}{3}\Delta H_f^\circ[B, g] - \Delta H_f^\circ[ØBX_2, g] + \frac{2}{3}\Delta H_f^\circ[BX_3, g]$

$$(62)$$

for $Ø_2BX$: $E(B—Ø) = \Delta H_f^\circ[Ø, g] + \frac{1}{3}\Delta H_f^\circ[B, g] - \frac{1}{2}\Delta H_f^\circ[Ø_2BX, g] + \frac{1}{6}\Delta H_f^\circ[BX_3, g]$

$$(63)$$

	$ØBCl_2$	$ØBBr_2$	$Ø_2BCl$	$Ø_2BBr$
$E(B—ø)$	115.8 ± 2.5	114.5 ± 2.5	111.2 ± 2.4	109.0 ± 2.4

cf. $\bar{D}(B—Ø) = 106.0 \pm 2.5$

These results have been discussed[108] in terms of the π-contribution to the B—Ø bond and the molecular geometry of the arylboron halides.

D. *Cyclic Strain and Resonance*

When considering cyclic systems, the bond energy equation contains an additional term (P) which, if positive, may be interpreted as the strain energy of the cyclic system and, if negative, may be identified with the resonance energy:

$$\Sigma E—P = -\Delta H_f^\circ[\text{cyclic system, g}] + \Sigma \Delta H_f^\circ[\text{elements, g}] \quad (64)$$

The boroxine ring has been considered in this context.[111] Similar calculations for the borazole ring indicate extensive resonance stabilization. Assuming bond energy transferability and using $\bar{D}(N—B) \sim 91$ (Table 14), $\bar{D}(B—Cl) = 105.3$ (Table 14), $\bar{D}(B—H) \sim 88$ (from $\Delta H_f^\circ[BH_3, g]^{[67]}$ $= 25.5 \pm 10$ kcalmole^{-1}), $\bar{D}(N—Me) = 75.3$ kcalmole^{-1} (from $\Delta H_f^\circ[NMe_3, g]^{47} = -11.0$ kcalmole^{-1}) and $\bar{D}(N—H) = 93.4$ kcalmole^{-1} (from $\Delta H_f^\circ[NH_3, g]^{[38]} = -11.0$ kcalmole^{-1}) we derive P values for N-methylborazole, borazole, and B-trichloroborazole of -177, -158 (or -83 using combustion calorimetric value for $\Delta H_f^\circ[N_3B_3H_6, liq]$) and -78 kcal respectively.

E. *Bond Dissociation Energies*

Very little work is available for bond dissociation energies of stable boron compounds. Steele *et al.*[133] have investigated methyl, ethyl, isopropyl and vinylboron difluoride by a mass spectrometric technique. Combination of the bond dissociation energies (derived from the appearance potentials) with $\Delta H_f^\circ[BF_2, g] = -135$ kcalmole^{-1} (assumed) estimates of $\Delta H_f^\circ[RBF_2, g]$ were made.

TABLE 15. SOME APPROXIMATE DATA FOR SUBSTITUTED
BORON DIFLUORIDES

Compound	$\Delta H_f^\circ(g)$ kcalmole^{-1}	$D(R-BF_2)$
CH_3BF_2	-199 ± 3	95–100
$CH_2{=}CHBF_2$	-171 ± 8	111 ± 3
$EtBF_2$	-209 ± 8	101 ± 3
$i-PrBF_2$	-212 ± 8	96 ± 3

Using a similar technique Fehlner and Koski[134] studied the dissociation:

$$BH_3CO(g) \longrightarrow BH_3(g) + CO(g) \tag{65}$$

and derive $D(BH_3 - CO) = 23.1 \pm 2$ kcalmole^{-1}.

There are extensive data for dissociation energies of transient species and the reader is referred to reference 24 and Gaydon.[135]

V. REFERENCES

1. C. T. MORTIMER and H. D. SPRINGALL, *Endeavour*, **23**, 22 (1964).
2. *Experimental Thermochemistry*, Vol. I (ed. F. D. ROSSINI), 1956, and Vol. II (ed. H. A. SKINNER), Interscience, 1962.
3. H. STEINBERG, "Organoboron Chemistry", Vol. I *(Boron–Oxygen and Boron–Sulphur Compounds)* and Vol. II *(Boron–Nitrogen and Boron–Phosphorus Compounds)*, Wiley, 1964, 1966.
4. R. M. ADAMS, *Boron, Metallo–Boron Compounds and Boranes*, Interscience, 1964.
5. W. GERRARD, *The Organic Chemistry of Boron*, Academic Press, 1961.
6. *Boron–Nitrogen Chemistry*, Advances in Chemistry Series, *American Chemical Society*, **42**, 1964.
7. See, for example, E. J. PROSEN, W. H. JOHNSON and F. Y. PERGIEL, *J. Res. Natl. Bur. Stand.*, **61**, 247 (1958).
8. S. R. GUNN, *J. Phys. Chem.*, **69**, 1010 (1965).
9. H. E. ROBSON and P. W. GILLIES, *J. Phys. Chem.*, **68**, 983 (1964).
10. D. L. HILDENBRAND and W. F. HALL, *J. Phys. Chem.*, **68**, 989 (1964).
11. D. SMITH, A. S. DWORKIN and E. R. VAN ARTSDALEN, *J. Am. Chem. Soc.*, **77**, 2654 (1955).
12. A. W. SEARCY and C. E. MYERS, *J. Phys. Chem.*, **61**, 957 (1957).
13. P. O. SCHISSEL and W. S. WILLIAMS, *Bull. Am. Phys. Soc.*, **4**, 139 (1959).
14. P. A. AKISHIN, D. O. HIKITIN and L. N. GOROKHOV, *Proc. Acad. Sci. U.S.S.R.*, **129**, 1075 (1959).
15. YU. A. PRISELKOV, Y. A. SAPOZHNIKOV and A. V. TSEPLYAEVA, *Izv. Akad. Nauk, S.S.S.R. Otd. Tekhn. Nauk. Met. i Toplivo*, **1**, 134, (1960).

16. W. A. CHUPKA, quoted in *N.B.S. Report* No. 7093, U.S. Govt. Printing Office, Washington, D.C., 1961.
17. G. VERHAEGEN and J. DROWART, *J. Chem. Phys.*, **37**, 1367 (1962).
18. C. ALCOCK and P. GRIEVESON, *Thermodynamics of Nuclear Materials*, International Atomic Energy Agency, Vienna, Austria, p. 571, 1962.
19. R. G. PAULE and J. L. MARGRAVE, *J. Phys. Chem.*, **67**, 1368 (1963).
20. P. O. SCHISSEL and O. C. TRULSON, *J. Phys. Chem.*, **66**, 1492 (1962).
21. M. J. LINEVSKY, G. M. KIBLER, T. F. LYON and V. J. DE SANTIS, *Refractory Materials Research*, WADD-TR-60-646, Part IV, General Electric Co., Ohio, August 1964.
22. D. L. HILDENBRAND, *Thermodynamic Properties of Propellant Combustion Products*, QLR-64-10, Philco. Corporation, California.
23. R. P. BURNS, A. J. JASON and M. G. INGRAM, *J. Chem. Phys.*, **46**, 394 (1967).
24. *JANAF Thermochemical Tables*, the Dow Chemical Company, Midland, Michigan, 1964, now distributed by Clearinghouse for Federal, Scientific and Technical Information, Springfield, Virginia.
25. M. V. KILDAY and E. J. PROSEN, *J. Am. Chem. Soc.*, **82**, 5508 (1960).
26. M. V. KILDAY and E. J. PROSEN, *J. Res. Natl. Bur. Stand.*, **68A**, 127 (1964).
27. N. D. SOKOLOVA, S. M. SKURATOV, A. M. SHEMONAEVA and V. M. YULDASHEVA, *Russ. J. Inorg. Chem.*, **6**, 395 (1961).
28. D. J. MESCHI, W. A. CHUPKA and J. BERKOWITZ, *J. Chem. Phys.*, **33**, 530 (1960).
29. J. A. BLAUER and M. FARBER, *J. Phys. Chem.*, **68**, 2357 (1964).
30. W. D. GOOD and M. MÅNSSON, *J. Phys. Chem.*, **70**, 97 (1966).
31. M. V. STACKELBERG, F. QUATRAM and J. DRESSEL, *Z. Electrochem.*, **43**, 14 (1937).
32. W. H. EVANS, D. D. WAGMAN and E. J. PROSEN, *National Bureau of Standards Report*, 4943, 1956.
33. L. G. FASOLINO, *J. Chem. Eng. Data*, **10**, 373 (1965).
34. J. SMISKO and L. S. MASON, *J. Am. Chem. Soc.*, **72**, 3679 (1950).
35. A. FINCH, P. J. GARDNER and I. J. HYAMS, *Trans. Farad. Soc.*, **61**, 649 (1965).
36. A. L. MCCLOSKEY, R. J. BROTHERTON and J. L. BOONE, *J. Am. Chem. Soc.*, **83**, 4750 (1961).
37. F. D. ROSSINI et al., *Selected Values of Chemical Thermodynamic Properties*, National Bureau of Standards Circular 500, 1952.
38. D. D. WAGMAN et al., *Selected Values of Chemical Thermodynamic Properties*, National Bureau of Standards, Technical Notes 270–1 (1965) and 270–2 (1966).
39. S. S. WISE, J. L. MARGRAVE, H. M. FEDER and W. N. HUBBARD, *J. Phys. Chem.*, **65**, 2157 (1961).
40. P. GROSS, C. HAYMAN, D. L. KEVI and M. C. STEWART, *Fulmer Research Institute Report* R146/4/23, 1960.
41. W. N. HUBBARD, H. M. FEDER, E. GREENBERG, J. L. MARGRAVE, E. RUDZITIS and S. S. WISE, *Symposium on Thermodynamics and Thermochemistry*, Lund, Sweden, 1963.
42. G. K. JOHNSON, H. M. FEDER and W. N. HUBBARD, *J. Phys. Chem.*, **70**, 1 (1966).
43. D. A. SCARPIELLO and W. J. COOPER, *J. Chem. Eng. Data*, **9**, 364 (1964).
44. H. A. SKINNER and N. B. SMITH, *Trans. Farad. Soc.*, **49**, 601 (1953).
45. S. R. GUNN and L. G. GREEN, *J. Phys. Chem.*, **64**, 61 (1960).
46. L. G. FASOLINO, *J. Chem. Eng. Data*, **10**, 371 (1965).
47. LANDOLT–BÖRNSTEIN, 6th edition, Vol. 2, Part 4 (ed. K. SCHÄFER and E. LAX), Springer-Verlag (1961).
48. W. H. JOHNSON, R. G. MILLER and E. J. PROSEN, *J. Res. Natl. Bur. Stand.*, **62**, 213, (1959).
49. H. A. SKINNER and N. B. SMITH, *Trans. Farad. Soc.*, **51**, 19 (1955).
50. S. SUNNER and S. THORÉN, *Acta Chem. Scand.*, **18**, 1528 (1964).

51. A. STOCK and E. KUSS, *Ber.*, **47**, 3113 (1914).
52. H. A. SKINNER, *Rec. Trav. Chim.*, **73**, 991 (1954).
53. H. A. SKINNER and P. B. HOWARD, *J. Chem. Soc.* (A), 1536 (1966).
54. V. TIENSUU, Thesis (Cornell Univ.), 1962.
55. T. H. S. HIGGINS, E. C. LEISEGANG, C. J. G. RAW and A. J. ROSSOUW, *J. Chem. Phys.*, **23**, 1544 (1955).
56. S. R. GUNN and R. H. SANBORN, *J. Chem. Phys.*, **33**, 955 (1960).
57. A. K. HOLLIDAY and A. G. MASSEY, *Chem. Rev.*, **62**, 303 (1962).
58. S. R. GUNN, L. G. GREEN and A. I. VON EGIDY, *J. Phys. Chem.*, **63**, 1787 (1959).
59. G. URRY, T. WARTIK, R. E. MOORE and H. I. SCHLESINGER, *J. Am. Chem. Soc.*, **76**, 5293 (1954).
60. S. R. GUNN and L. G. GREEN, *J. Phys. Chem.*, **65**, 178 (1961).
61. J. BLAUER and M. FARBER, *Trans. Farad. Soc.*, **60**, 301 (1964) and *T. Chem. Phys.*, **39**, 158 (1963).
62. R. F. PORTER, D. R. BIDINOSTI and K. F. WATTERSON, *J. Chem. Phys.*, **36**, 2104 (1962).
63. E. M. MAGEE, *J. Inorg. Nucl. Chem.*, **22**, 155 (1961).
64. W. D. DAVIS, L. S. MASON and G. STEGEMAN, *J. Am. Chem. Soc.*, **71**, 2775 (1949).
65. V. B. PARKER, *Thermal Properties of Aqueous Uni-valent Electrolytes*, National Standard Reference Data Series, National Bureau of Standards-2, 1965.
66. W. H. JOHNSON, R. J. SCHUMM, I. H. WILSON and E. J. PROSEN, *J. Res. Nat. Bur. Stand.*, **65A**, 97 (1961).
67. *JANAF Thermochemical Tables*, Addendum PB 168 370–1, distributed by Clearinghouse for Federal, Scientific and Technical Information, 1966.
68. J. L. BILLS and F. A. COTTON, *J. Phys. Chem.*, **64**, 1477 (1960).
69. J. D. COX and D. HARROP, *Trans. Farad. Soc.*, **61**, 1328 (1965).
70. J. D. COX, National Physical Laboratory, U.K., private communication.
71. L. SHARTSIS and W. CAPPS, *J. Am. Ceram. Soc.*, **37**, 27 (1954).
72. L. SHARTSIS and E. NEWMAN, *J. Am. Ceram. Soc.*, **31**, 213 (1948).
73. N. P. NIES and G. W. CAMPBELL in ref. 4.
74. See BLiO$_2$ in ref. 24.
75. R. M. RULON and L. S. MASON, *J. Am. Chem. Soc.*, **73**, 5491 (1951).
76. B. POST in ref. 4.
77. M. WRIGHT and P. N. WALSH, *The Vaporization of* MgB$_4$(c), Technical Research Report OMCC–HEF–55, Ohio State University Research Foundation, 1958.
78. G. K. JOHNSON, E. GREENBERG and J. L. MARGRAVE, *J. Chem. Eng. Data*, **12**, 137 (1967).
79. G. V. SAMSONOV and L. Y. MARKOVSKII, Chemistry of borides, *Usp. Khim.*, **25**, 190 (1956).
80. L. BREWER and H. HARALDSEN, *J. Electrochem. Soc.*, **102**, 399 (1955).
81. J. M. LEITNAKER, M. G. BOUMAN and P. W. GILLIES, *J. Electrochem. Soc.*, **109**, 441 (1962).
82. E. J. HUBER, *J. Chem. Eng. Data*, **11**, 430 (1966).
83. H. I. SCHLESINGER, R. T. SANDERSON, and A. B. BURG, *J. Am. Chem. Soc.*, **62**, 3421 (1940).
84. P. GROSS, C. HAYMAN and M. C. STUART, *Proc. Brit. Ceram. Soc.*, 39 (1967).
85. C. J. THOMPSON and G. C. SINKE, Dow Chemical Co. Report, ARPA 1–164, January 1961.
86. G. L. GAL'CHENKO, A. N. KORNILOV and S. M. SKURATOV, *Zh. Neorg. Khim.*, **5**, 1282 (1960).
87. A. S. DWORKIN, D. J. SASMOR and E. R. VAN ARTSDALEN, *J. Chem. Phys.*, **22**, 837 (1954).

88. D. L. HILDENBRAND and W. F. HALL, *J. Phys. Chem.*, **67**, 888 (1963).

89. H. DREGER, V. V. DADAPE and J. L. MARGRAVE, *J. Phys. Chem.*, **66**, 1556 (1962).

90. S. S. WISE, J. L. MARGRAVE, H. M. FEDER and W. N. HUBBARD, *J. Phys. Chem.*, **70**, 7 (1966).

91. G. L. GAL'CHENKO, B. I. TIMOFEEV, D. A. GEDAKYAN, YA. K. GRINBERG and Z. S. MEDVEDEVA, *Izv. Akad. Nauk SSSR Neorg. Materially*, **2**, 1410 (1966).

92. A. FINCH, P. J. GARDNER and K. K. SEN GUPTA, unpublished results.

93. B. C. SMITH and L. THAKUR, *Nature*, **208**, 74 (1965).

94. M. V. KILDAY, W. H. JOHNSON and E. J. PROSEN, *J. Res. Natl. Bur. Stand.*, **65A**, 101 (1961).

95. E. R. VAN ARTSDALEN and A. S. DWORKIN, *J. Am. Chem. Soc.*, **74**, 3401 (1952).

96. W. V. HOUGH, G. W. SCHAEFFER, M. DZURUS and A. C. STEWART, *J. Am. Chem. Soc.*, **77**, 864 (1955).

97. H. I. SCHLESINGER, D. M. RITTER and A. B. BURG, *J. Am. Chem. Soc.*, **60**, 1296 (1938).

98. A. W. LAUBENGAYER and C. W. J. SCAIFE, *J. Chem. Eng. Data*, **11**, 172 (1966).

99. H. A. SKINNER and N. B. SMITH, *J. Chem. Soc.*, 4025 (1953); *ibid.*, 2324 (1954).

100. G. L. GAL'CHENKO, M. M. AMMAR, S. M. SKURATOV, Y. N. BUBNOV and B. MIKHAILOV, *Vest. Mosk. Univ. Ser. II Khim.*, **20**, 10 (1965); *C. A.* 14145h (1965).

101. W. H. JOHNSON, M. V. KILDAY and E. J. PROSEN, *J. Res. Natl. Bur. Stand.*, **65A**, 215 (1961).

102. L. H. LONG and R. G. W. NORRISH, *Phil. Trans. Roy. Soc. London*, A **241**, 587 (1949).

103. A. E. POPE and H. A. SKINNER, *J. Chem. Soc.*, 3704 (1963).

104. G. L. GAL'CHENKO and R. M. VARUSHCHENKO, *Zhur. Fiz. Khim.*, **37**, 2513 (1963); *C.A.* 4878 f. (1964).

105. H. A. SKINNER and T. F. S. TEES, *J. Chem. Soc.*, 3378 (1953).

106. E. A. HASELEY, A. B. GARRETT and H. H. SISLER, *J. Phys. Chem.*, **60**, 1136 (1956).

107. J. E. BENNETT and H. A. SKINNER, *J. Chem. Soc.*, 2472 (1961).

108. A. FINCH, P. J. GARDNER, E. J. PEARN and G. B. WATTS, *Trans. Farad. Soc.*, **63**, 1880, (1967).

109. H. A. SKINNER in *Adv. Organometallic Chem.* (ed. F. G. A. STONE and R. WEST), **2**, 49 (1964).

110. G. N. CHREMOS, H. WEIDMANN, and H. K. ZIMMERMAN, *T. Org. Chem.* **26**, 1683 (1961).

111. A. FINCH and P. J. GARDNER, *Trans. Farad. Soc.*, **62**, 3314 (1966).

112. A. FINCH, P. J. GARDNER and G. B. WATTS, *Chem. Comm.*, 1054, (1967)

113. A. S. DWORKIN and E. R. VAN ARTSDALEN, *J. Am. Chem. Soc.*, **76**, 4316 (1954).

114. H. A. SKINNER and N. B. SMITH, *J. Chem. Soc.*, 3930 (1954).

115. M. V. KILDAY, W. H. JOHNSON and E. J. PROSEN, *J. Res. Natl. Bur. Stand.*, **65A**, 435 (1961).

116. H. C. BROWN and B. C. SUBBA RAO, *J. Org. Chem.*, **22**, 1136 (1957); *J. Amer. Chem. Soc.*, **81**, 6428 (1959).

117. V. E. WIBERG and W. SÜTTERLIN, *Z. Anorg. Allgem. Chem.*, **202**, 1 (1931); *ibid.*, **202**, 22 (1931).

118. H. A. SKINNER and J. E. BENNETT, *J. Chem. Soc.*, 2150 (1962).

119. W. J. COOPER and J. F. MASI, *J. Phys. Chem.*, **64**, 682 (1960).

120. A. B. BURG and H. I. SCHLESINGER, *J. Am. Chem. Soc.*, **55**, 4023 (1933).

121. A. FINCH and P. J. GARDNER, *J. Chem. Soc.*, 2985 (1964).

122. P. J. GARDNER, unpublished results.

123. A. FINCH, P. J. GARDNER and E. J. PEARN, *Trans. Farad. Soc.*, **62**, 1072 (1966).

124. T. CHARNLEY, H. A. SKINNER and N. B. SMITH, *J. Chem. Soc.*, 2288 (1952).

125. J. H. S. GREEN, *Quart. Rev.*, **15**, 125 (1961).

126. A. SNELSON and H. A. SKINNER, *Trans. Farad. Soc.*, **57**, 2125 (1961).
127. H. A. SKINNER and A. SNELSON, *Trans. Farad. Soc.*, **56**, 1776 (1960).
128. A. SOMMER, P. N. WALSH and D. WHITE, *J. Chem. Phys.*, **33**, 297 (1960).
129. A. FINCH, P. J. GARDNER and G. B. WATTS, *Trans. Farad. Soc.*, **63**, 1603 (1967).
130. G. L. GAL'CHENKO, B. I. TIMOFEEV and S. M. SKURATOV, *Russ. J. of Inorg. Chem.*, **5**, 1279 (1960).
131. J. A. KERR, *Chem. Rev.*, **66**, 465 (1966).
132. H. A. SKINNER and G. PILCHER, *Quart. Rev.*, **17**, 264 (1963).
133. W. C. STEELE, L. D. NICHOLS and F. G. A. STONE, *J. Am. Chem. Soc.*, **84**, 1154 (1962).
134. T. P. FEHLNER and W. S. KOSKI, *J. Am. Chem. Soc.*, **87**, 409 (1965).
135. A. G. GAYDON, *Dissociation Energies and Spectra of Diatomic Molecules*, Chapman & Hall, 1953.

4

SOME RECENT DEVELOPMENTS IN BORON-NITROGEN CHEMISTRY

by Heinrich Nöth

Institut für Anorganische Chemie der Universität, Marburg/Lahn, West Germany

CONTENTS

I. INTRODUCTION

The enormous interest which boron–nitrogen chemistry has enjoyed during the last two decades has led to the synthesis of an astonishing number of new compounds, many of which have been completely new in type. A result of these advances has been a deeper understanding of the fundamental principles involved in this field.

The first attempt to systematize our knowledge of BN-containing compounds is due to E. Wiberg,[320] who in 1948 emphasized the relations that exist between isoelectronic BN- and CC-groups. Compounds comparable in this sense are not only isoelectronic but are often also isosteric. The sometimes remarkable similarities between such a pair of isoelectronic compounds implies a similar charge distribution and therefore also a comparable bonding situation. This is a result of the fact that B, C and N obey the octet rule, their coordination numbers not exceeding four, except in a few electron deficient compounds. However, since B, C and N use sp-, sp^2- and sp^3-hybrid orbitals in forming covalent bonds, coordination numbers of two and three are also possible and the latter is quite commonly encountered in B–N compounds.

Since Wiberg's outline of the chemistry of boron–nitrogen compounds,[320] where a simple nomenclature was able to embrace all known types of compounds, the number of these types has increased enormously, as has the problems of nomenclature. Meanwhile many of Wiberg's postulates have been verified, although some of the ideas put forward twenty years ago have required modification. Nevertheless the principles of the late nineteen forties continue to be of great value for further advances in the synthetic branch of the field, and also as a rough guide for the correlation of new data.

Within the last few years a number of reviews have appeared dealing with various aspects of compounds containing boron–nitrogen bonds. Thus the books by K. Niedenzu and J. W. Dawson[204] and also by H. Steinberg and R. J. Brotherton[295] deal solely with boron–nitrogen chemistry. *The Chemistry of Boron and its Compounds*, edited by E. L. Muetterties,[197] also contains two useful chapters on this topic (one by K. Niedenzu and J. W. Dawson and the other by M. F. Lappert). Other sources of material on

Abbreviations used: Me = CH_3, Et = C_2H_5, Pr = C_3H_7, i-Pr = iso-C_3H_7, Bu = C_4H_9, t-Bu = tert-C_4H_9, Ph = C_6H_5, py = NC_5H_5, dipy = o, o'-dipyridyl, quin = quinoline, iquin = isoquinoline, mes = mesityl, pz = 1-pyrazolyl; R = H or (and) organic group, X, Y any atom or group.

B–N chemistry include Volume I[296] in this series,[18, 67, 129] and a collection of papers presented at a symposium on boron–nitrogen chemistry in 1963.[97] The *Annual Surveys of Organometallic Chemistry*[277], also contain a great deal of summarized information on the progress made in this area. A short account of some developments in aminoborane chemistry is given by K. Niedenzu[202] and polymeric BN systems have been included in an article on polymeric boron compounds.[324] Most of the results cited or discussed in these reviews cover the literature up to 1964, and only in some instances are more recent data considered. The chemistry of the azidoboranes is excellently dealt with in a 1967 review by P. I. Paetzold[242]. M. F. Lappert and B. Prokai[297] have recently given a detailed description of insertion reactions of unsaturated substrates into BN bonds as one aspect of the wide applicability of this type of reaction. Therefore, these two topics are not included here except when necessary.

The 1966 review of Rüff, which appeared in Colburn's *Developments in Inorganic Nitrogen Chemistry* is already outdated,[51]

The subsequent discussion of boron–nitrogen compounds is restricted to the developments in this field since 1964, and most of the literature up to July 1967 has been consulted. However, this review does not attempt to cover the whole area exhaustively—thus the chemistry of borazines, as well as of the other BN-heteroaromatics[67] is excluded. Heterocyclic systems containing B–N linkages are mentioned at the appropriate places. Only those aspects which the author believes are significant for the future development of the field are considered in detail.

Also, it is hoped that the treatment given keeps closely enough to the definition of boron–nitrogen chemistry, namely the formation and breaking of B–N bonds and the structural and physicochemical properties of this bond. However, the stabilization of other bonds due to the presence of B–N linkages in the same molecule is also included in this definition.

II. THE COORDINATE BORON–NITROGEN BOND

A. *Introduction*

The tendency of borane and borane derivatives $BH_{3-n}X_n$ to act as Lewis acids is due to their electron sextet. That they coordinate with bases such as amines, ethers, etc., is well known, and the influence of inductive, mesomeric and steric effects on the strength of the donor–acceptor bond[296] has been studied extensively. One of the problems related to these adducts is the so-called "acid- or base-strength reversal". For instance, BH_3 is a stronger acid towards PMe_3 than towards NMe_3, while the reverse is true for BF_3. It is often very difficult experimentally to obtain reliable data on dissociation equilibria of type (1) in the gas phase, for example as a result of decomposition reactions.

$$D + BX_3 \rightleftharpoons D \cdot BX_3 \tag{1}$$

(D = donor molecule, such as an amine)

Hence a number of other methods have been used to study these equilibria and to derive qualitative or even quantitative values relating to orders of Lewis acid and base strength. A further complication arises if reactions (1) are carried out in solvents, in so far as the stoichiometry of the reaction of a base D with a borane BX_3 is then not necessarily $1:1$ (see Section III).

In this section some new data are collected which relate to the formation or the breaking of coordinate boron–nitrogen bonds, but overlap with some of the results discussed in Section III cannot be avoided completely.

B. *Amine-Boranes* $(D \cdot BH_3)$ *and Related Compounds*

1. *Formation*

The products of the reaction of a nitrogen base D with diborane in the absence of a polar solvent cannot be predicted unequivocally. Nevertheless it is safe to say, that tertiary amines or bulky nitrogen bases yield simple borane adducts $D \cdot BH_3$. Other amines can induce unsymmetrical cleavage of diborane, which will be discussed in Section III.

Use of polar solvents, for instance diethylether or particularly tetrahydrofurane, THF (with which diborane forms $THF \cdot BH_3$), leads to amine-boranes by the simple base displacement scheme (2). In this way, fairly large quantities of $H_3N \cdot BH_3$ have been obtained.[282, 292, 293] This com-

$$THF \cdot BH_3 + D \longrightarrow D \cdot BH_3 + THF \tag{2}$$

pound may also be prepared simply by passing diborane through an ammonia solution in ether, dioxane, alcohol or even water.[77] The hydrolytically quite stable compound crystallizes with a face-centered orthorhombic unit cell with $a = 7.22$, $b = 7.38$ and $c = 9.23$ Å.

On the other hand, a simple BH_3 adduct is said to be obtained by passing diborane directly into isopropylamine at $-30°$ to $0°$, and the product is claimed to be $iPrNH_2 \cdot BH_3$.[77] However, at present there is too little physical evidence for full confirmation of this structure.[218] Since

$MeNH_2$ and B_2H_6 yield $[H_2B(NMeH_2)_2]BH_4$ the formation of $iPrNH_2 \cdot BH_3$ would then be due to steric factors affecting the mode of cleavage of diborane.

Amine-boranes are also formed by reducing organic acid amides with diborane. Thus diborane and $Me_2NC(O)H$ form a primary product $HC(O)NMe_2 \cdot BH_3$ at $-30°$ which explodes on warming to room temperature; it is reduced by excess diborane to $Me_3N \cdot BH_3$. Similarly $MeC(O)NMe_2$ and diborane yield liquid ethyldimethylamine-borane, $EtMe_2N \cdot BH_3$, of m.p. $12.5–13.5°$ and b.p. $96–99°$.[78]

Carbonyl-borane undergoes a displacement reaction with trimethyl-amine to give trimethylamine-borane according to (3), but with NH_3,

$$H_3B \cdot CO + NMe_3 \longrightarrow H_3B \cdot NMe_3 + CO \tag{3}$$

NH_2Me and $NHMe_2$ in ether solutions boranocarbamates

$$Me_3N \cdots \overset{\overset{\displaystyle H \diagdown \diagup H}{|}}{\underset{\displaystyle H}{B}} \cdots CO$$

(I)

$[Me_{2-n}H_nNC(O)BH_3]^-$ are produced.[45] The latter precipitate as the corresponding ammonium or methylammonium salts. This displacement reaction, which in the gas phase is of second order, can be regarded as a typical S_N2 reaction, the activation energy being 8.65 ± 0.3 kcalmole for the temperature range $207–273°K$. An intermediate (I) with a penta-coordinated boron atom is postulated for reaction (3).[104] At temperatures higher than $273°K$ dissociation of borane-carbonyl becomes predominant.[104]

There is evidence for the existence of a compound $H_3B \cdot CO \cdot NMe_3$, stable at $-100°$ in dimethylether solution, but it need not necessarily have the structure (I).[45]

Usually, if diborane comes into contact with an appropriate amine then cleavage of diborane results. To date this cleavage has always been exclusively symmetrical when a tertiary amine is employed. If ditertiary amines are used, mono- or bis(borane)-adducts may form. In the case of tetramethylethylenediamine and under normal conditions the only product obtained is the $1:2$ adduct according to (4) even in the presence of excess amine.[188]

$$Me_2N—CH_2—CH_2—NMe_2 + B_2H_6 \longrightarrow H_3B \cdot Me_2N—CH_2—CH_2—NMe_2 \cdot BH_3 \tag{4}$$

$$H_3B \cdot Me_2N—CH_2—CH_2—NMe_2 \cdot BH_3 + Me_2N—CH_2—CH_2—NMe_2$$
$$\rightleftharpoons 2\, H_3B \cdot Me_2N—CH_2—CH_2—NMe_2 \tag{5}$$

This 1 : 2 adduct dissolves in tetramethylethylenediamine at 60° according to (5) and from the supercooled solution the monoadduct separates at −23°. This in turn disproportionates at room temperature to give the 1 : 2 adduct and the free amine again by the reverse reaction.[85] Similar behavior is observed in the system N,N'-dimethyldiethylenediamine/diborane. While $MeN(CH_2CH_2)_2 NMe \cdot 2 BH_3$ can be sublimed unchanged, the liquid monoadduct (m.p. 14–16°) decomposes readily to the free base and the bis(borane) adduct.[86]

The only stable monoadduct isolated up to now, namely $N(CH_2CH_2)_3N \cdot BH_3$, resulted from the base exchange reaction (6), and its resistance to disproportionation was attributed to lattice stabilization

$$H_3B \cdot N(CH_2CH_2)_3N \cdot BH_3 + Me_2NCH_2CH_2NMe_2$$

$$\xrightarrow{70°} N(CH_2CH_2)_3N \cdot BH_3 + Me_2NCH_2CH_2NMe_2 \cdot BH_3 \tag{6}$$

effects, since it decomposes into $N(CH_2CH_2)_3N \cdot 2 BH_3$ and $N(CH_2CH_2)_3N$ only on heating above its melting point.[86]

At the present time there is no obvious reason to explain the predominate formation of the bis-adducts, except for the assumption that this leads to a better packing in the crystal lattice in comparison with the monoadducts. Certainly there should be no drastic differences in the free energies of BN bond formation in these cases, and if there are any they should become smaller as the carbon chain between the nitrogen atoms increases. From this point of view it would be worth investigating the behaviour of $CH_2(NMe_2)_2$, $CH(NMe_2)_3$ and $C(NMe_2)_4$ as bases towards BH_3, and indeed $CH_2(NMe_2)_2 \cdot 2 BH_3$ was found to be stable and almost involatile.[8]

Ethylenediamine and diborane give the adduct (II)[127] whose struc-

$$H_3B \overset{\displaystyle CH_2-CH_2}{\underset{\displaystyle N}{\overset{\displaystyle N}{\Big|}}} BH_3$$

(II)

ture has been confirmed by means of ^{11}B-n.m.r. Similarly when hexamethyltriethylenetetramine is treated with diborane one mole of diborane is consumed per mole of the amine in the absence of a solvent. However, in the presence of ether four BH_3 groups add to the tetramine. Density measurements indicate that the packing of the molecules is better in the latter compound, which is the more stable one.[312] Analogous adducts with

N : B ratios of 1:1 were also prepared using tetramethylethylenediamine, pentamethyldiethylenetriamine and hexamethyltripropylenetetramine.[312]

If there are no carbon atoms between the two nitrogen atoms of the diamines, than two series of borane adducts can be prepared readily. These are hydrazine-(mono)boranes (III) and hydrazine-bis(boranes) (IV). There seems to be no preference for diadduct formation, and this can be interpreted in terms of electronic repulsion and increased steric hindrance.

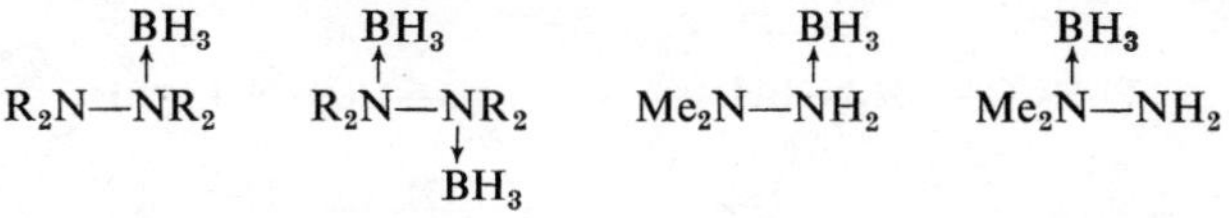

$$
\begin{array}{cccc}
\mathrm{BH_3} & \mathrm{BH_3} & \mathrm{BH_3} & \mathrm{BH_3} \\
\uparrow & \uparrow & \uparrow & \uparrow \\
\mathrm{R_2N{-}NR_2} & \mathrm{R_2N{-}NR_2} & \mathrm{Me_2N{-}NH_2} & \mathrm{Me_2N{-}NH_2} \\
& \downarrow & & \\
& \mathrm{BH_3} & & \\
\text{(III)} & \text{(IV)} & \text{(V)} & \text{(VI)}
\end{array}
$$

(R = H, Me, Ph, etc.)

If one mole of BH_3 is added to $MeHNNH_2$ or Me_2NNH_2, then the two isomers (V) and (VI) could form. ^{11}B n.m.r. indicates that only the isomer (VI) is present. This is expected for reasons of basicity although addition to the NH_2 group should be favored sterically.[216] This conclusion is also in agreement with i.r. studies on these adducts.[26, 214] On the other hand i.r. evidence favors the $H_3B \cdot NH_2$ group in $PhHNNH_2 \cdot BH_3$ and $Ph_2NNH_2 \cdot BH_3$.[17] The fact that here the H_2N group in the corresponding hydrazine is the preferred site for BH_3 attachment is to be expected on the basis of mesomeric and steric affects.

The question of at which site BH_3 will add to the cage compound (VII), 2,6,7-trimethyl-4-methyl-2,6,7-triaza-1-phosphabicyclo[2,2,2]-octane, is, in

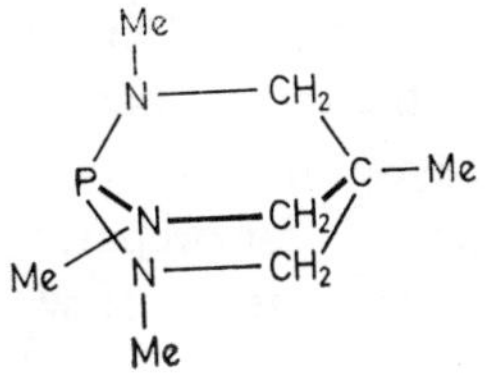

(VII)

some respects, similar to the case of the hydrazine-boranes. It has been found that both a 1:1 and a 1:2 BH_3 adduct form, while only 1:1 products result when the (P-)oxide or (P-)sulphide of (VII) is treated with diborane.

From these and n.m.r. data it was concluded that in the 1:1 adduct the BH_3 group is bonded to the P atom, while in the 1:2 case BN bonding also occurs.[149] The P attack of BH_3 is consistent with the "reverse" base strength order $R_3P| > |R_3N$ for BH_3. On the other hand, the oxide of (VII) is not reduced by diborane as is the case for $OP(NMe_2)_3$[149] or Ph_3PO.[135]

The coordination site problem stressed above also plays a part in the newly discovered synthesis of urea-boranes from organic isocyanates and the amine-boranes $MeNH_2 \cdot BH_3$, $EtNH_2 \cdot BH_3$, $t\text{-}BuNH_2 \cdot BH_3$ and $Et_2NH \cdot BH_3$ as demonstrated by (7). The question as to whether these possess structure (VIII) or (IX) is not yet solved.[53]

$$PhNCO + RH_2N \cdot BH_3 \longrightarrow PhHN—CO—NHR \cdot BH_3 \tag{7}$$

$$
\begin{array}{cc}
\underset{\underset{H_3B}{\downarrow}}{\overset{\overset{H}{|}}{Ph—N}}—C—\underset{}{\overset{\overset{H}{|}}{N}}—R & \underset{\overset{}{}}{\overset{\overset{H}{|}}{Ph—N}}—C—\underset{\underset{BH_3}{\downarrow}}{\overset{\overset{H}{|}}{N}}—R \\
H_3B \qquad O & O \qquad BH_3 \\
\text{(VIII)} & \text{(IX)}
\end{array}
$$

An equilibrium between the two isomers is considered as a possibility to explain the absence of NH band splittings in the i.r. spectra, but certainly B–O coordination is not present.[53]

Silylamines are in general weaker bases than alkylamines. Nevertheless they quickly absorb diborane and the following compounds have been obtained recently: $H_3SiNMe_2 \cdot BH_3$ (at $-80°$), $H_2Si(NMe_2)_2 \cdot 2BH_3$ (at $-84°$, decomposing irreversibly at $-46°$), $HSi(NMe_2)_3 \cdot 2BH_3$ (at $-135°$, decomposing at $-46°$), $HSi(NMe_2)_3 \cdot BH_3$ (volatile at $20°$) and $Si(NMe_2)_4 \cdot BH_3$ (at $-135°$) slight dissociation observed). Although $Si(NMe_2)_4 \cdot 2 BH_3$ seems stereochemically possible, no indications of its formation were found. From dissociation pressures at $-64°$ it was concluded that $HSi(NMe_2)_3$ is a stronger base than $H_2Si(NMe_2)_2$ in spite of the greater steric hindrance to BH_3 adduct formation by the former base. Thus π-bonding is an essential factor in explaining the base strengths of silylamines.[8]

So far, one of the best methods for preparing amine-boranes consists of allowing an amine hydrochloride to react with a metal borohydride as demonstrated by equation (8).

$$MBH_4 + R_{3-n}H_nN \cdot HCl \longrightarrow H_2 + MCl + R_{3-n}H_nN \cdot BH_3 \tag{8}$$

There are a number of techniques available today for carrying out this reaction. G. W. Schaeffer and E. R. Anderson[269] introduced this reaction using $LiBH_4$, $Me_3N \cdot HCl$ and ether. It was later extended to dialkylamine- and alkylamine-boranes by H. Nöth and H. Beyer[218] who also employed $NaBH_4$ or $Ca(BH_4)_2$ and others such as THF, monoglyme or dioxane.

If the protons attached to the nitrogen atom are relatively acidic, as is usually the case in aromatic amines as compared with aliphatic ones, then the borane adduct is unstable, losing hydrogen rather readily. Therefore neither $PhNH_2 \cdot BH_3$, o-phenylenediamine-borane nor 1,8-diaminonaphthaline-borane[95] were obtained, but the compounds (X) and (XI) were isolated in the latter two cases. The existence of $PhNH_2.BH_3$ is at least open to discussion (see page 221).

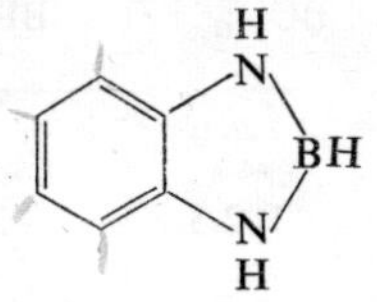

(X) 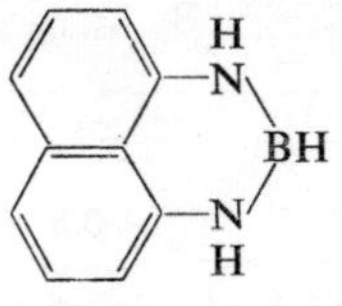   (XI)

No $H_2NCH_2CH_2NH_2 \cdot BH_3$ was formed from $H_2NCH_2CH_2NH_2 \cdot HCl$ and $NaBH_4$. Instead only polymeric materials resulted[95] which were similar to those obtained from the decomposition of ethylenediamine-borane.[77]

Often the decomposition of a suitable metal borohydrides with aqueous H_2S in the presence of tertiary amines according to (9) is useful for synthesizing amine-boranes.[274]

$$[Zn(NH_3)_4] (BH_4)_2 + 2 \, py + H_2S \longrightarrow 2 \, py \cdot BH_3 + 2 \, H_2 + 4 \, NH_3 + ZnS \qquad (9)$$

The reaction between $NaBH_4$ and hexamethylenetetramine at 0° in water yields insoluble $(CH_2)_6N_4 \cdot BH_3$ although this amine has four coordination sites.[177]

Consecutive reactions are observed, when the amine-borane prepared by reaction (8) has sites for hydroboration, and this method can be used to prepare BN heterocyclic compounds. An example worthy of mention is the formation of 14,16,18-tribora-13,15,17-triazatriphenylene (XIII), which results by heating the product (XII) formed from $LiBH_4$ and $CH_2{=}CH{-}CH_2NH_2 \cdot HCl$ to 300° followed by dehydrogenation with Pd on charcoal.[57]

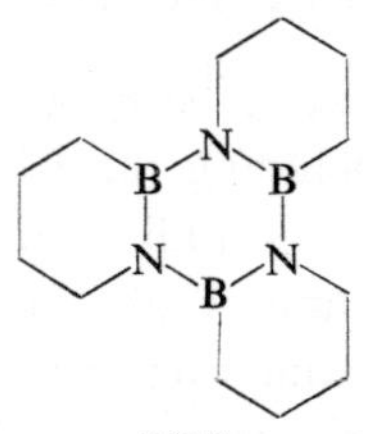

(XII) 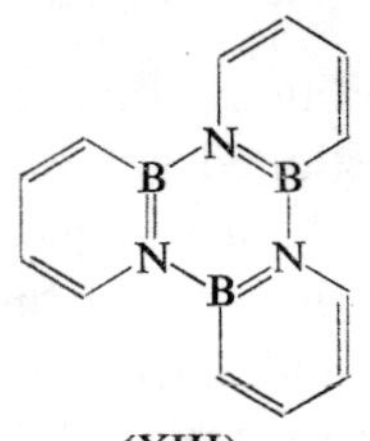   (XIII)

2. Physical Properties

The heat of formation of $Me_3N \cdot BH_3$ has been redetermined, and the most accurate value is -31.44 ± 0.06 kcal/mole. From this value and the heat of reaction of BF_3 with $HF \cdot 3.75$ H_2O the following ΔH_f° values are now recommended:[105]

Compound	$B_2H_{6\,(g)}$	$B_{(s.,\,amorph.)}$	$BCl_{3\,(l)}$	$BF_{3\,(g)}$	$B(OH)_{3\,(s.,cryst}$
ΔH_f° [kcal/mole]	$+8.6$	$+0.8$	-102.3	-269.7	-261.7

The dipole moments of the series $Me_{3-n}H_nN \cdot BH_3$ have been redetermined in various solvents[313] and found to be 5.05 D for $H_3N \cdot BH_3$, 5.15 (5.19) D for $MeNH_2 \cdot BH_3$, 4.99 (4.87) D for $Me_2NH \cdot BH_3$ and 4.69 (4.45) D for $Me_3N \cdot BH_3$ (older values in parentheses[219]). A careful analysis of these data revealed[314] that the assumption of constant bond moments[219] cannot be maintained as a consequence of the lone pair contribution to the dipole moment in the free base. Therefore the prediction of a greater dipole moment for $H_3N \cdot BH_3$ than for $MeH_2N \cdot BH_3$ was erroneous.

By ^nB n.m.r. it was shown[259] that Me_3N in $Me_3N \cdot B_3H_7$ is fixed to one B-atom, while this is not so for THF in $THF \cdot B_3H_7$ due to the weaker coordinate bond. Other n.m.r. data are discussed in Section II, G.

3. Chemical Properties

Hydrolysis of amine-boranes in neutral aqueous solution is rather slow but acids and also alcohols increase the rate. Although the reactions (10) or (11) seem to involve mainly the B—H bonds, kinetic measurements emphasize the great influence of the strength of the B—N bond upon the rate of hydrolysis.

$$R_3N \cdot BH_3 + 3\,H_2O \longrightarrow R_3N + 3\,H_2 + B(OH)_3 \qquad (10)$$

$$R_3N \cdot BH_3 + 3\,ROH \longrightarrow R_3N + 3\,H_2 + B(OR)_3 \qquad (11)$$

Solvolysis reactions using excess propanol have been performed for a series of substituted pyridine-boranes[263] including the BH_3 adducts of pyridine, α-, β-, and γ-picoline, 2,4- and 2,6-lutidine and γ-collidine (2,4,6-trimethylpyridine). In these pseudo first-order reactions it was found that 2- and 2,6-substitution in the pyridine ring increased the reaction rate due to strain relief in the transition state (1.1–3.5 kcal/mole).

Thus the rates of solvolysis for pyridine $\cdot BH_3$, α-picoline $\cdot BH_3$ and 2,6-lutidine $\cdot BH_3$ were in the ratio $1.7 : 4.27 : 7.29$, respectively. Apart from these cases, however, there is a direct relationship between K_b for the base and the rate of solvolysis of its BH_3 adduct. From these data it may be concluded that the first step in the reaction is the dissociation (12), followed by the fast solvolysis (13).

$$\text{py} \cdot BH_3 \xrightarrow{\text{slow}} \text{py} + BH_3 \tag{12}$$

$$BH_3 + \text{n-PrOH} \xrightarrow{\text{fast}} \text{products} \tag{13}$$

While the breaking of the B—N bond in the pyridine-boranes determines the rate of the reaction, a more complex behavior was observed in the hydrolysis of arylamine-boranes.[126] In this latter case the rate law is expressed by (14), where $[D \cdot BH_3]$ represents the concentration of the amine-borane.

$$\frac{-d[D \cdot BH_3]}{dt} = [D \cdot BH_3] \, (k_1 + k_2[H^+]) \tag{14}$$

Thus, there is an acid-dependent and an acid-independent step in the solvolysis. The activation energy for the acid-independent solvolysis is $E_A = 21.8$ kcal/mole, and for the acid-dependent reaction $E_A = 21.7$ kcal/mole, but their activation entropies have opposite signs, $S_A = -0.8$ and $+0.6$ cal/mole^{-1} deg^{-1} respectively. Again the acid-independent hydrolysis step is described by equation (12) (but with py replaced by an arylamine), and there is the expected relationship between solvolysis rate and Hammett's σ values for substitutes phenylamine-boranes. However, in the acid-dependent step of the hydrolysis there is concurrent breaking of the B—N bond while a proton is transferred to the nitrogen atom. Thus the formation of the activated complex (XIV) is rate determining.

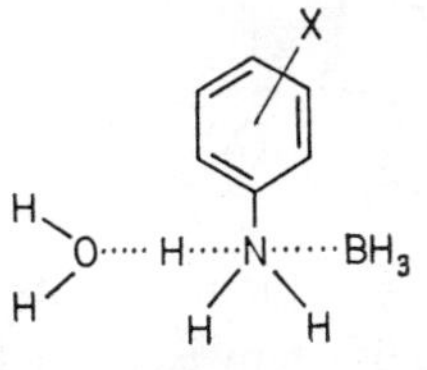

(XIV)

It is therefore clear from these kinetic data that the BN-bond of amine-boranes is easily broken and the heats of formation of these types of complexes are very similar, which is in accord with equation (12).

From this point of view the redistribution reactions (15), which were used to prepare the series of triethylamine-haloboranes may be explained in terms of BN-bond breaking before H/X exchange occurs.[76]

$$2 \, Et_3N \cdot BH_3 + Et_3N \cdot BX_3 \longrightarrow 3 \, Et_3N \cdot BH_2X \qquad (15)$$
$$X = Cl, \, Br$$

Certainly the mechanism of halogenating amine-boranes by *N*-halogeno-succinamide (16) differs from that described by (15). It is a non-radical process, which leads in the case of pyridine-borane, quinoline-borane and aromatic amine-boranes to a mixture of products,[70] and is straightforward only for trialkylamine-boranes.

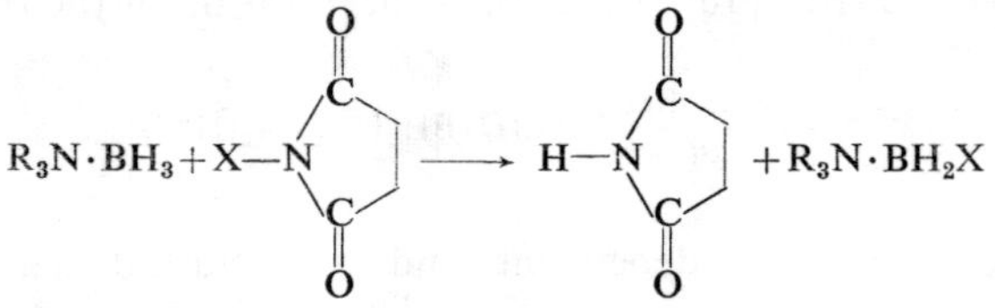

$$X = Cl, \, Br$$

Since it is known that "positive halogen" attacks amine-boranes with elimination of HX, an intermediate in reaction (16) may be a trialkyl-amine-succinamidylborane (XV), whose BN-bond is broken by the HX formed thereby giving the products.

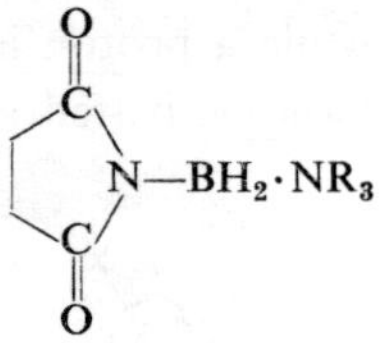

(XV)

Amine-boranes are versatile reducing agents and the reduction of Ph_3PO by these compounds proceeds stepwise, the first and quite complicated stage (17) involves a displacement reaction. This is followed by reaction (18).

$$3 \, Ph_3PO + 5 \, H_3B \cdot NR_3 \xrightarrow{120°} 3 \, Ph_3P \cdot BH_3 + 3 \, H_2 + B_2O_3 + 5 \, NR_3 \qquad (17)$$
$$3 \, Ph_3PO + 2 \, Ph_3P \cdot BH_3 \longrightarrow 5 \, Ph_3P + 3 \, H_2 + B_2O_3 \qquad (18)$$

C. Amine-Boranes of Type $D \cdot BH_2X$ and $D \cdot BHX_2$

The observation that alkoxyboranes $(RO)_{3-n}BH_n$ are weaker Lewis acids than mercaptoboranes $(RS)_{3-n}BH_n$ can be rationalized in terms of stronger B—O than B—S back-bonding. As an example, 1,3,2-dioxaborolidine-trimethylamine is less stable than 1,3,2-dithiaborolidine-trimethylamine,[2, 322] and the 1,3,2-dioxaborolane is a slightly weaker acid than the 1,3,2-dioxaborolidine. While it is well known that PMe_3 is a stronger base towards BH_3 than is NMe_3, the same is also true for 1,3,2-dithiaborolidine. The equilibrium (19) at 25° in THF is displaced 66 per cent to the right in favor of the PMe_3 adduct, while in the system

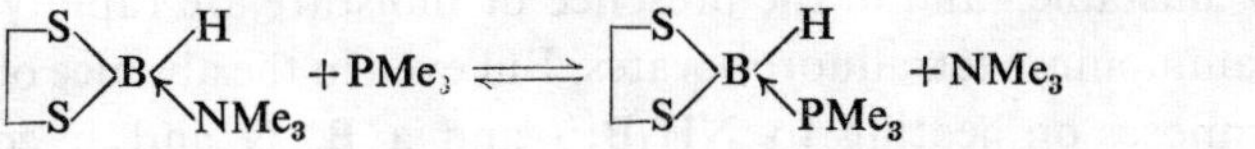

$$\begin{array}{c} S \\ S \end{array}\!\!\!\!>\!\!B\!<\!\!\begin{array}{c} H \\ NMe_3 \end{array} + PMe_3 \rightleftharpoons \begin{array}{c} S \\ S \end{array}\!\!\!\!>\!\!B\!<\!\!\begin{array}{c} H \\ PMe_3 \end{array} + NMe_3 \qquad (19)$$

$H_3B/PMe_3/NMe_3$ in THF at 25° the analogous equilibrium lies 80 per cent to the right.[322] These observations suggest that BH_3 is a stronger acid than the dithiaborolidine. This base strength reversal is not observed with 1, 3, 2-dioxaborolidine;[2, 322] here $Me_3N > Me_3P$ still holds. BH_3 and dithiaborolidine are therefore "soft acids" in the sense of the Schwarzenbach–Pearson definition. Further factors contributing to the acid strength of boranes are discussed in Section II, G.

If a substituted borane possesses both an acceptor and a donor site, then internal association can be expected. A new example of this type is the four-membered ring compound 1,2-azaboretidine obtained from ethyleneimine and $NaBH_4$ in acetic acid[3] according to (20).

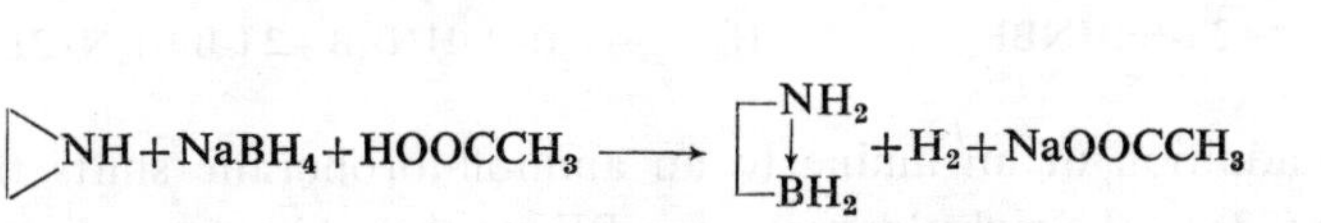

$$\triangleright NH + NaBH_4 + HOOCCH_3 \longrightarrow \begin{array}{c} NH_2 \\ \downarrow \\ BH_2 \end{array} + H_2 + NaOOCCH_3 \qquad (20)$$

This is a rather toxic compound and is related to 1,1-dimethyl-1,2-azaborolidine (XVI), prepared from N,N-dimethylallylamine and $H_3B \cdot NMe_3$.[166]

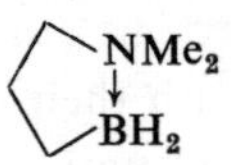

$$\begin{array}{c} NMe_2 \\ \downarrow \\ BH_2 \end{array}$$

(XVI)

D. *Amine-Boron Trihalides*

1. *Amine-Boron Trifluorides*

For a long time there have been conflicting reports on various amine-trifluoroboranes. Recent work has shown how these compounds can be prepared, and also that alkylammonium tetrafluorborates are readily formed as their decomposition products. Previously, these latter products had often been referred to as $R_{3-n}NH_n \cdot BF_3$ compounds. Under strictly anhydrous conditions $BF_3 \cdot OEt_2$ reacts with one equivalent of NH_3 or an alkylamine $R_{3-n}NH_n$ ($R = Me$, Et; $n = 3, 2, 1$) giving the crystalline 1:1 adducts.[99, 266, 291] The adducts with NH_3 and the amines are comparatively unstable, and in the presence of moisture are rapidly converted to alkylammonium tetrafluoroborates. But even in the absence of air $NH_3 \cdot BF_3$ decomposes on heating to NH_4BF_4 and a B, N and F containing polymer; at temperatures up to $250°$ no boron nitride is formed.[266] The ethylamine–borontrifluorides are also rather unstable as shown by I. G. Ryss and O. Donskaya.[265] The liquid adduct $Et_3N \cdot BF_3$, which is best obtained from its two components in the presence of petroleum ether, transforms to solid $[Et_3NH]BF_4$ under the influence of traces of water. Some of the more stable amine borontrifluorides, such as $MeNH_2 \cdot BF_3$ or $t\text{-}BuNH_2 \cdot BF_3$ are formed during the aminolysis of diethylaminodifluoroborane.[99] Thus reactions of types (21) and (22) occur, in contrast to the behavior of other aminohaloboranes (e.g. chlorides or bromides), because of the relatively high thermodynamic stability of the BF_3 adducts.

$$3\,Et_2NBF_2 + 5\,MeNH_2 \longrightarrow (MeHN)_3B + 2\,MeNH_2 \cdot BF_3 + 3\,Et_2NH \quad (21)$$

$$3\,t\text{-}BuHNBF_2 + 2\,t\text{-}BuNH_2 \longrightarrow (t\text{-}BuHN)_3B + 2\,t\text{-}BuH_2N \cdot BF_3 \quad (22)$$

The addition of an amine to an aminofluoroborane shifts the equilibrium (23) to the right by removing BF_3 as the adduct.

$$3\,R_2NBF_2 \rightleftharpoons (R_2N)_3B + 2\,BF_3 \quad (23)$$

Of course this is an extremely simplified description of a rather complicated situation; to mention only one other factor, the aminofluoroboranes are themselves in equilibrium with their dimers.

If in reaction (24) the base strength of the donor D is low, then only an equilibrium is established, and the adduct dissociates reversibly.

$$D + BF_3 \rightleftharpoons D \cdot BF_3 \quad (24)$$

In the case of amines D this dissociation is observed with very weak bases. Thus aniline, dimethylaniline and methyldiphenylamine[252] all add BF_3. Methyldiphenylamine first forms an emerald green solid, m.p. 87°, of composition $MeNPh_2 \cdot 0.58 BF_3$, and only after long contact with BF_3 is white $MeNPh_2 \cdot BF_3$, dec. > 100°, formed.[252] The appearance of colored products is not unusual, and may be attributed to pseudo anilinium ions or to charge transfer.[268] The work by R. S. Satchell and D. N. Satchell on $PhNH_2 \cdot BF_3$ clearly demonstrates the adduct nature of this compound[268] and rules out the rather exotic "π-complex" formulation proposed by W. Gerrard and E. F. Mooney[89] in 1960. They also found that an equilibrium (25) is established between substituted anilines and ether with BF_3. The pk values for these base displacements roughly parallel the k_b values for protonation of the bases in aqueous solution.[268]

$$D \cdot BF_3 + Et_2O \rightleftharpoons Et_2O \cdot BF_3 + D \qquad (25)$$

TABLE 1. EQUILIBRIUM CONSTANTS FOR REACTION (25) AT 25° IN DIETHYLETHER

D	p$_k$
3-Nitroaniline	−2.65
2-Methyl-5-nitroaniline	−2.03
4-Chloro-3-nitroaniline	−2.12
2,5-Dichloroaniline	−1.16
3-Methyl-4-nitroaniline	−1.81
4-Nitroaniline	−1.46
NN-Dimethyl-4-nitroaniline	−0.13
N-Phenylaniline	−0.11
3,5-Dinitroaniline	−0.45
2-Nitroaniline	−0.02
4-Nitro-1-naphthylamine	−0.55

Little is known about the behavior of boron trifluoride towards bases that possess several coordinating sites due to different atoms. Since the base strength towards BF_3 is $R_2O > R_2S$ and $R_3N > R_3P$, one might expect BF_3 to coordinate to the nitrogen atoms in the compounds $(Me_2N)_2S$, $(Me_2N)_2SO$, $(Me_2N)_2SO_2$, $(Me_2N)_3P$, $(Me_2N)_3PO$, etc. Unfortunately cleavage of the element–nitrogen bond often hinders the study of the mode of addition. Although the products formed are consistent with BF_3 attacking at the nitrogen atoms, this does not prove that this is necessarily the first reaction step. The compound $(Me_2N)_2S$ reacts with

BF_3 in petroleum ether yielding a crystalline 1 : 1 adduct, which decomposes exothermically at 70–80°. The ^1H-n.m.r. spectrum shows a single peak for the adduct, and this is considered to be evidence for S rather than N-coordination.[228] It was also shown in this study that the product reported earlier[38] as $(Me_2N)_2S \cdot 2BF_3$, whose composition was taken to demonstrate B–N coordination, was in fact a mixture of products.[228]

On the other hand, S_4N_4 in CH_2Cl_2 solution gives an almost quantitative yield of $S_4N_4 \cdot BF_3$ when treated with BF_3.[321] Since S_4N_4 is a weak base, BF_3 can be removed *in vacuo*, and the values $\Delta H° = 15.0$ kcalmole and $\Delta S° = 31$ cal/mole^{-1} deg^{-1} were determined for reaction (26).

$$S_4N_4 \cdot BF_{3 \text{ solid}} \rightleftharpoons S_4N_{4 \text{ solid}} + BF_{3 \text{ gas}} \qquad (26)$$

In ureas and thioureas the O, S or N atoms are possible coordination sites for the boron halides. With N-di- and tri-substituted ureas at $-78°$ two moles of BF_3 are consumed, but from CH_2Cl_2 solutions stable 1 : 1 adducts separate. These adducts may also be obtained by reaction (27), if R' is an aromatic nucleus. Since the BF_3 adducts still retain the ability to form complexes with transition metal ions (through O- or S-coordination), and on the basis of somewhat inconclusive i.r., u.v. and n.m.r data, it has been postulated that there is N-coordination in these BF_3 adducts.[101]

$$R'NH_2 \cdot BF_3 + RNCO \longrightarrow RNH\text{—}CO\text{—}NHR' \cdot BF_3 \qquad (27)$$

It is of course possible to replace F atoms in $F_3B \cdot NR_3$ by other substituents, and the question arises as to whether this occurs directly, or *via* free BF_3 due to the equilibrium (24). For the conversion of $F_3B \cdot NEt_3$ into $Cl_3B \cdot NEt_3$ with BCl_3 it has been found that for the temperature range $-80°$ to 60° there predominantly is a direct replacement according to (28a) (64.8 per cent at 0°). However, at temperatures higher than 60° BN-bond cleavage (70.3 per cent at 60°) becomes the principal reaction as shown in (28b).[52]

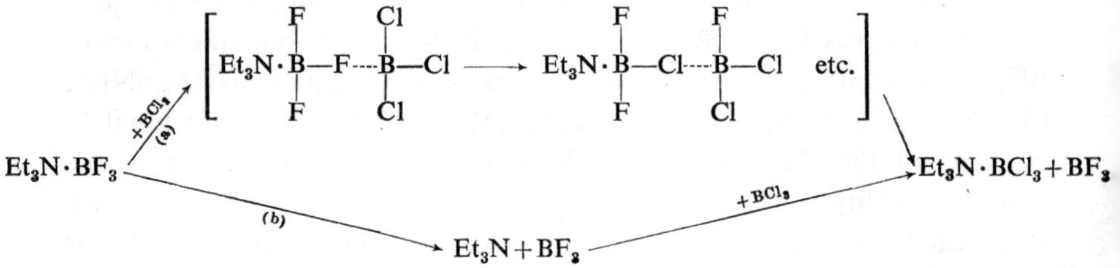

$$(28)$$

Furthermore, in the hydrogenation of $Me_3N \cdot BF_3$ with $LiBH_4$, which proceeds according to (29), a BN cleavage was noted by means of an isotopic labelling technique.[112]

$$Me_3N \cdot BF_3 + 3\, LiBH_4 \longrightarrow Me_3N \cdot BH_3 + 3/2\, B_2H_6 + 3\, LiF \qquad (29)$$

2. *Amine-Boron Trichlorides*

The simplest and most widely adopted method for preparing amine adducts of the boron halides is the reaction between equimolar amounts of amine and BX_3 under appropriate conditions in the presence or absence of an inert solvent. Most of the reactions are strongly exothermic, and if not strictly controlled, there can be deviations from the stoichiometry as shown in scheme (30) for a secondary amine and BCl_3.

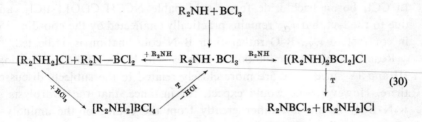

In studying the thermal decomposition of alkylammoniumtetrahalogenoborates[40, 41, 43], tetraphenylborates[58] and phenyltrihalogenoborates[58] by differential thermal analysis, I. M. Butcher *et al.* always noted the formation of an amine-boron halide as an intermediate.

The equilibrium (26) suggests that stronger Lewis acids should displace BF_3 from $S_4N_4 \cdot BF_3$ even more readily than in (28). This was indeed demonstrated, using BCl_3 or $SbCl_5$ (equation (31)).

$$S_4N_4 \cdot BF_3 + ECl_n \longrightarrow S_4N_4 \cdot ECl_n + BF_3 \qquad (31)$$

For these adducts N-coordination is postulated by analogy with the known structure of $S_4N_4 \cdot SbCl_5$. This assumption was substantiated in so far as $S_4N_4 \cdot BCl_3$ reacts with $SbCl_5$ to form $S_4N_4 \cdot BCl_3 \cdot SbCl_5$, whose i.r. spectrum can be interpreted in terms of structure (XVII).[321]

$$\left[\begin{array}{ccc} S & N & S \\ N & Cl-B-Cl & N \\ S & N & S \end{array} \right] SbCl_6$$

(XVII)

The question of the coordination site also arises in the adducts of organometallic azides and ureas with boron halides. From variations in $\nu_{as}(N_3)$ and $\nu_{sym}(N_3)$ in the free organometallic azides and in the compounds $Ph_3SiN_3 \cdot BBr_3$, $Ph_3GeN_3 \cdot BBr_3$, $Ph_3SnN_3 \cdot BBr_3$, $MePh_2SiN_3 \cdot BBr_3$, $Me_3SiN_3 \cdot BBr_3$ and $Me_3SnN_3 \cdot BBr_3$, it was concluded that BBr_3 adds to the α–N atom, and not the γ–N atom.[298] This is consistent with the structures of $[X_2BN_3]_3$ ($X = Cl$, Br), which also involve coordination to the α–N atoms.[247]

BCl_3 and BBr_3 are much more reactive towards substituted ureas than BF_3 (see Section II, D 1). A number of CH_2Cl_2-soluble products are formed, and these seem to be similar in structure to the BF_3 adducts. The insoluble ones may contain ionic structures.[101]

Furthermore, cyanoacetic esters offer boron halides two sites of attack. In CCl_4 boron trichloride forms sublimable $NCCH_2COOEt \cdot BCl_3$, and due to the fact that ν_{CN} remains practically unaffected by the coordination in contrast to ν_{CO}, B–O rather than B–N coordination is indicated.[120]

Reactions of the amine-boron halides will be mentioned in Sections III and IV since these are more closely related to the subjects discussed there. However, one would expect, for instance, that the hydrolysis of $R_3N \cdot BCl_3$ should not differ greatly from the course of the aminolysis of these compounds. Now, for reaction (32) a pseudo first-order rate constant was determined in aqueous–alcoholic solution.

$$D \cdot BCl_3 + 3\,H_2O \longrightarrow D \cdot HCl + 2\,HCl + B(OH)_3 \qquad (32)$$

Data relating to this type of hydrolysis are found in Table 2 for ethanol–water solutions.

TABLE 2. DATA RELATING TO THE HYDROLYSIS OF SOME AMINE-BORON
TRICHLORIDES IN 25% (v/v) ETHANOL–WATER

	Rel. rate	E_A kcal/mole	ΔH_f kcal/mole	pK_b
$py \cdot BCl_3$	1.00	12	41.0	5.15
α–pic $\cdot BCl_3$	0.46	13	4.9	5.97
γ–pic $\cdot BCl_3$	3.54	10	44.9	6.02
$Me_3N \cdot BCl_3$	0.08	—	52.9	9.72

The $+I$ effect enhances the stability of the adduct in γ-pic $\cdot BCl_3$ although the compound hydrolyses more rapidly than $py \cdot BCl_3$. Since the hydrolysis rate increases for $Me_3N \cdot BCl_3$ in solvents of high dielectric constants, an

S_N1 reaction mechanism is likely, and for the pyridine adducts mechanism (33) is preferred over any other.[111]

$$D \cdot BCl_3 \underset{\text{slow}}{\rightleftharpoons} [D \cdot BCl_2]Cl \xrightarrow[\text{fast}]{3 H_2O} D \cdot HCl + 2 HCl + B(OH)_3 \qquad (33)$$

E. *Amine-Organoboranes*

The basicities of many amines are expressed not only in terms of pK values, but are also very often related to BMe_3 as the reference acid.[296] From gas phase dissociation studies and from displacement reactions, B. J. Aylett and L. K. Peterson[8] concluded that bis(dimethylamino) methane is a weaker base than trimethylamine, the equilibrium of reaction (34) lying 88 per cent to the right at room temperature.

$$CH_2(NMe_2)_2 \cdot BMe_3 + Me_3N \rightleftharpoons CH_2(NMe_2)_2 + Me_3N \cdot BMe_3 \qquad (34)$$

Steric strain in the adduct with $CH_2(NMe_2)_2$ was considered to be responsible for this. On the other hand there is a 100 per cent conversion according to (35).

$$H_2Si(NMe_2)_2 \cdot BMe_3 + Me_3N \longrightarrow H_2Si(NMe_2)_2 + Me_3N \cdot BMe_3 \qquad (35)$$

Generally, aminosilanes are weaker bases than amines, but the series of base strength $Me_3N > H_3SiNMe_2 > H_3SiNEt_2 > NEt_3$[9] can hardly be rationalized in terms of steric effects alone. $HSi(NMe_2)_3$ and $Si(NMe_2)_4$ form no BMe_3 adducts at all, even at $-78°$, due to steric crowding.[9] The behavior of these aminosilanes towards BH_3[8] as an acid has been described in Section II, B, 1.

A most interesting base displacement was observed by treating $K[Me_3B{-}PMe_2{-}BMe_3]$ with ammonia.[299] At room temperature reaction (36) was complete within 66 hours. It is most likely that NH_3 first removes one of the two Me_3B groups as $Me_3B \cdot NH_3$, which then transfers a proton to the Me_2P group liberating Me_2PH, as shown in (37).

$$K[Me_3B{-}PMe_2{-}BMe_3] + NH_3 \longrightarrow K[Me_3B{-}NH_2{-}BMe_3] + Me_2PH \qquad (36)$$

$$[Me_3B{-}PMe_2{-}BMe_3]^- + NH_3 \longrightarrow Me_3B{-}PMe_2^- + H_3N \cdot BMe_3$$

$$\longrightarrow Me_3B \cdot HPMe_2 + H_2NBMe_3 \longrightarrow Me_2PH + [Me_3B{-}NH_2{-}BMe_3]^- \qquad (37)$$

While the action of simple amines on BX_3 compounds has been studied extensively, this is not so with diamines. W. Brüser and K. H. Thiele[34] have reported recently that $Me_2NCH_2CH_2NMe_2 \cdot 2BMe_3$ and $Et_2NCH_2CH_2NEt_2 \cdot 2BMe_3$ are stable only at $< -40°$, the latter dissociates completely into its components at room temperature, while the former dissociates slightly in benzene solution and splits off BMe_3 at its melting point. On the other hand 2,2'-dipyridine gives no adduct with BMe_3 even at $-78°$, and this is explained by steric hindrance as well as by reluctance of the boron atom to achieve a coordination number 5. It is worth mentioning that $Me_3B \cdot Me_2NCH_2CH_2NMe_2 \cdot BMe_3$ is an isomer of $Me_3N \cdot Me_2BCH_2CH_2BMe_2 \cdot NMe_3$.[287]

Tetracoordination of the boron atom is one way in which to stabilize certain bonds, for instance, by suppressing group-exchange and disproportionation reactions. Striking cases are the stabilization of Bu_2BCF_3 as its NH_3 adduct,[253] the stabilization of alkynylboranes by pyridine as $(RC\equiv C)_3B \cdot py$,[290] and the increase in resistance to oxidation and thermal decomposition of dimethylboron azide when converted to $Me_2BN_3 \cdot NMe_3$.[248] On the other hand, tetracoordination completely alters the structure and therefore also the properties of tricoordinate boron compounds. While 3-phenyl-3-benzoborazine (XVIII) was shown by u.v.- and n.m.r. spectroscopy to be a carbon–boron heterocycle with aromatic character, this property is completely lost in its weak Me_3N adduct because the vacant boron-p_z-orbital can no longer be part of the delocalized π-system.[152]

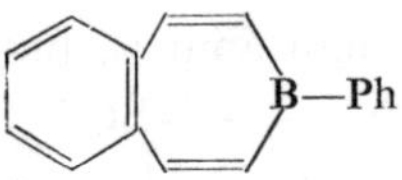

(XVIII)

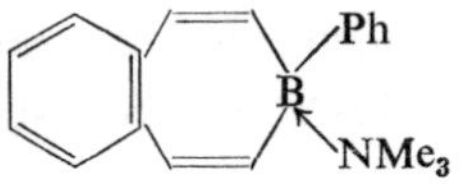

F. Internal B–N Coordination in Ring Systems

B–N coordinate bond formation quite commonly leads to the production of cyclic borane derivatives if the electron-donating and electron-accepting sites in the molecule are suitably placed so as to give four-, five- or six-membered ring systems. A well-known example is Letsinger's 2-aminoethyl diphenylborinate.[329] The formation of a three-membered ring system seems likely in $Me_2BCH_2NH_2$ (XIX) prepared elegantly by R. Schaeffer

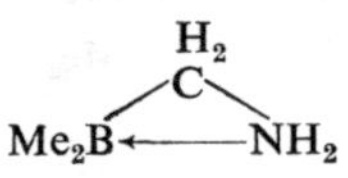

(XIX)

and L. J. Todd[270] through ammonolysis of Me_2BCH_2Cl. Reactions with NH_3 and BMe_3 reveal that dimethyl(aminomethyl) borane is a weaker acid than BMe_3 and a weaker base than NH_3. ^1H n.m.r. studies suggest that the three-membered ring system exists as a metastable form, and ^{11}B n.m.r. shows that the amount of monomeric $Me_2BCH_2NH_2$ decreases at lower temperatures. In contrast to these results the compound $H_2BCH_2NMe_2$ exists only as the dimeric six-membered heterocycle as a consequence of the higher acceptor and donor qualities of the H_2B- and Me_2N groups.

Five-membered heterocycles (XX) containing coordinate B–N links are numerous and have been extensively studied by B. M. Michailov and W. A. Dorochov.[166, 195] These workers obtained a large number of this

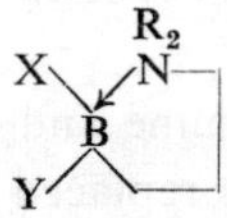

$$X = R, H, Cl, Br$$
$$Y = R, H, OR, NR_2, RS$$

(XX)

type by the hydroboration of allylamines, by the reaction of trialkyl-boranes with allylamines,[166, 195] by the addition of HX to 1,2-azaboro-les[196] and by heating a mixture of R_3B, $H_3B \cdot NR_3$ and an allylamine.[166] If $X = Y = R$, then exchange reactions using BX_3, $B(SR)_3$, ROH, etc., lead to new derivatives having other X and Y groups attached to boron.

B–N coordination is also observed in the dimeric O-substituted boryl-hydroxylamines (XXI), and O-diphenylboryl-, phenylthienylboryl- and

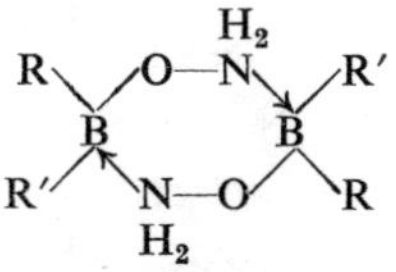

(XXI)

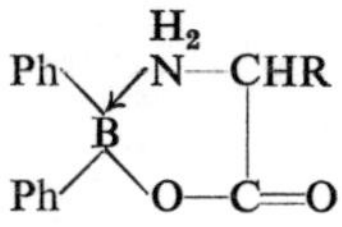

$(R = H, Me, CH_2—CHMe_2)$

(XXII)

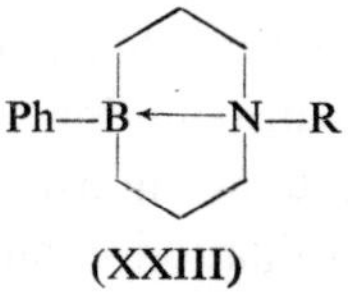

(XXIII)

dithienylborylhydroxylamine dimers have been obtained simply by mixing the borinic acid with H_2NOH in alcoholic solution.[260] Compounds of type (XXII), which also exhibit internal BN coordination, have been obtained by heating butyl diphenylborinate (Ph_2BOBu) with α-amino acids.[251] The compounds can be recrystallized from aqueous alcohol, thus showing their great hydrolytic stability. On the other hand, they react rapidly with ethanolamine to give aminoethyldiphenylborinate and liberate the α-amino acid.[288]

Another class of compounds in which internal BN coordination is favored sterically consists of the 1-aza-5-borabicyclo[3,3,0]octanes (XXIII). These are prepared by a hydroboration reaction between $PhBH_2 \cdot NEt_3$ and $(CH_2{=}CH{-}CH_2)_2$ NR.[44]

In the case of the hydroxylamine and α-amino acid derivatives (XXI) and (XXII), the R_2B group has replaced a hydrogen atom which is normally involved in strong bridge bonding. This situation is comparable to the pyrazole–pyrazabole case (see Sections IV, C and V, A).

From this point of view and also from a consideration of many reactions of BX_3 compounds of comparatively high Lewis acidity, the group X_2B can be regarded a "pseudohydrogen atom". This is a useful standpoint from which to consider many of the heterocycles having coordinate BN-bonds which have been synthesized by F. Umland and co-workers.[308-310] Most of these compounds were prepared from the corresponding boron acid or boron halide[267] with ring-formation favoring the reactions.[251] Typical examples (XXIV) to (XXVIII) are given below.

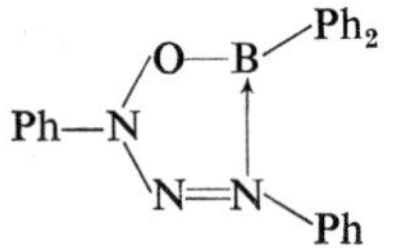

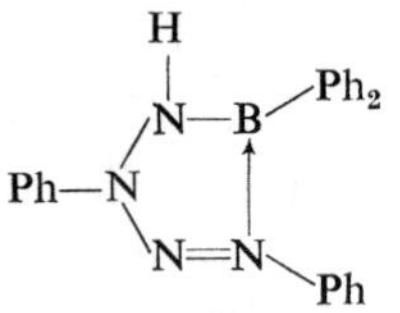

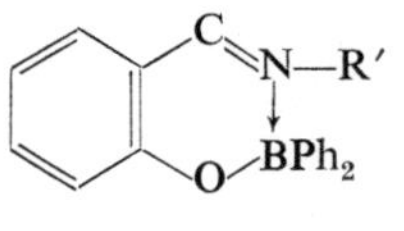

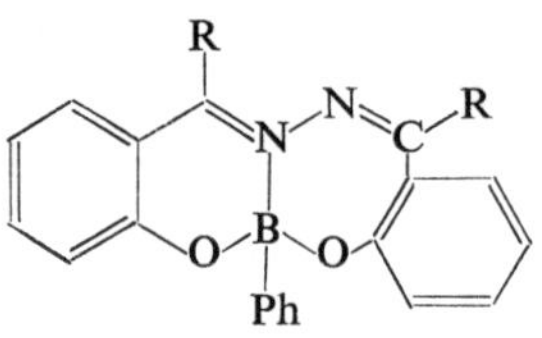

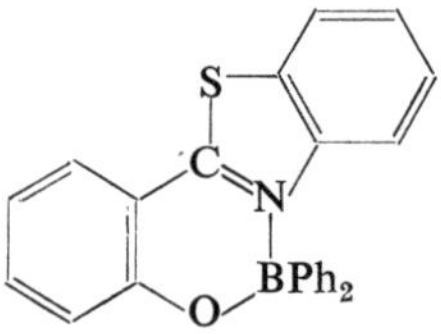

Again these molecules are exceptionally stable to hydrolysis, which points to a rather favorable charge distribution in these systems. Thus there can be no doubt that many interesting aspects are still to be revealed in the further study of BN coordination compounds.

G. *The Acidity Scale of the Boron Halides*

A comparison of the free energies of formation of $Me_3N \cdot BX_3$ compounds has shown that the order of increasing strength of the BN—bond is consistent with the order $BF_3 < BCl_3 < BBr_3$ for the acceptor power of the boron halides.[296] Meanwhile a great number of experiments have confirmed these results, and it is generally accepted that the heavier the halogen atom in the boron halide the stronger is its character as a Lewis acid. Thermochemical measurements are supplemented by more qualitative data obtained from shifts in the CO- or CN-stretching in the infrared spectra of BX_3-adducts with carbonyl or nitrile groups as well as from ^1H- and ^{11}B-n.m.r. data. Base displacement reactions provide another tool for obtaining information concerning acid and base strengths.

Some more recent thermochemical data that further substantiate earlier conclusions are:

$$BF_3 + py \longrightarrow py \cdot BF_{3\ solid} \quad -42.3\ \text{kcal}^{[147]} \tag{38}$$

$$BF_3 + CH_3CN \longrightarrow CH_3CN \cdot BF_{3\ solid} \quad -26.5\ \text{kcal}^{[184]} \tag{40}$$

$$BCl_3 + CH_3CN \longrightarrow CH_3CN \cdot BCl_{3\ solid} \quad -33.8\ \text{kcal}^{[184]} \tag{41}$$

$$BBr_3 + CH_3CN \longrightarrow CH_3CN \cdot BBr_{3\ solid} \quad -39.4\ \text{kcal}^{[184]} \tag{42}$$

Recently ^{11}B chemical shifts have often been used to study donor–acceptor interactions.[83] Since there is usually a great difference in the chemical shift for tri- and tetra-coordinate boron atoms, it is tempting to assume that the difference in chemical shift ($\Delta\delta$) between the free acceptor molecule and the donor–acceptor complex is a direct measure of the strength of the newly formed bond.[62, 84, 113, 185] This certainly would be so if the shielding of the boron nucleus were only a simple function of the differences in hybridization, but inductive and mesomeric effects also play a role as do anisotropy effects. Therefore conclusions drawn from n.m.r. evidence alone, even if they correlate more or less closely to other physical evidence, must be judged critically. As an example, the $\Delta\delta$ values and the heats of formation in the series $Me_3N \cdot BF_{3-n}(Me)_n$ are opposite in their trends:

	$Me_3N \cdot BF_3$	$Me_3N \cdot BF_2Me$	$Me_3N \cdot BFMe_2$	$Me_3N \cdot BMe_3$
ΔH kcal/mole[296]	20	20.25	18.3	17.6
$\Delta\delta$ [ppm][311]	11	22	49	86

This is of course due to resonance stabilization in the boron fluorides, which is greatest in BF_3, therefore leading to a better shielding of the boron nucleus in BF_3 than in $MeBF_2$ or Me_2BF. Using $\Delta\delta\,^1H$ as a measure of acid strength in the series $Me_3N \cdot BX_3$ (X = H (65.2), F (11.0), Cl (36.3), Br (42.5), I (47.3)), BH_3 should be the strongest acid. Since neighbor anisotropy effects, which increase from F to I, normally produce a low field shift of protons, as would also be expected from inductive effects alone, the difference in $\Delta\delta\,^1H$ between free and coordinated base cannot be considered due solely to the Lewis acid strength of the boron halides. Therefore a low field shift of the proton resonances in $Me_3N \cdot BX_3$[183] (X = H, F, Cl, Br) is not unequivocal. Hence reliable data follow only from thermochemical information obtained from gas phase or solution equilibria studies.

In this respect some results of S. G. Shore et al.[322] are noteworthy. They found that the equilibrium (43) in $CHCl_3$ or $C_6H_5NO_2$ solution lies fully on the side of $Me_3N \cdot BX_3$ for X = F, while the reverse is true for X = Cl.

$$Me_3N \cdot BX_3 + PMe_3 \rightleftharpoons Me_3N + Me_3P \cdot BX_3 \qquad (43)$$

These observations suggest that the stronger the Lewis acid, the greater is its tendency to reverse the normal order of the base strengths.[322] This is a hypothesis that strongly deserves further study.

Although there is no doubt as to the increase in Lewis acid strength in passing from BF_3 to BI_3, the base strength orders have still to be critically evaluated.[62]

Future studies of amine-boranes could profitably be directed towards obtaining more structural and thermodynamic data on $N \rightarrow B$ dative bonds, and the chemistry of these compounds also needs further investigation. That the latter area has certainly not been neglected will become evident in the two following sections.

III. AMINE–BORON CATIONS

A. *Introduction*

The literature prior to 1955 describes a number of BX_3 addition compounds whose composition could not be rationalized with a coordination number of four for boron. These compounds received little attention until R. W. Parry et al.[285] proved that the structure of the diammoniate of diborane is $[H_2B(NH_3)_2]BH_4$, resulting from unsymmetrical cleavage of diborane. Since then a great number of cationic boron species have been prepared and the types $[X_2B(amine)_2]^+$, $[XB(amine)_3]^{++}$ and $[B(amine)_4]^{+++}$ are now known. In retrospect it seems rather peculiar that these boron–nitrogen compounds went unrecognized for such a long time and that their chemistry remained virtually unknown until this last decade.

B. *Dihydridobis(amine)boron Cations*

1. *Formation*

The reaction of diborane with ammonia at low temperature can be made to proceed primarily according to equation (43)[285] and a large scale synthesis for $[H_2B(NH_3)_2]BH_4$ has been developed.[282]

$$B_2H_6 + 2\,NH_3 \longrightarrow [(H_2B(NH_3)_2]BH_4 \qquad (43)$$

The course of the reaction is explicable on the basis of an unsymmetrical cleavage of diborane by ammonia, in constrast to symmetrical cleavage as observed with trimethylamine, which leads to $Me_3N\cdot BH_3$. One of the obvious reasons for these differences in behavior of diborane must be the different steric requirements of the two nitrogen bases.

On the basis of ^{11}B n.m.r. evidence S. G. Shore and C. L. Hall[283] showed that the intermediates in the reactions of diborane with NH_3, $MeNH_2$, Me_2NH and Me_3N at $-70°$ in CH_2Cl_2 all have the composition $D\cdot B_2H_6$ (D = amine) with a single hydrogen bridge. Thus unsymmetrical or symmetrical cleavage proceeds through the same type of intermediate (XXIX).

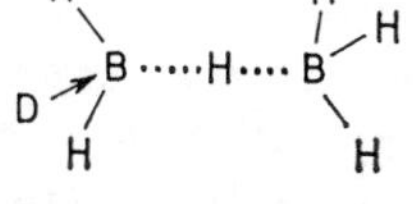

(XXIX) 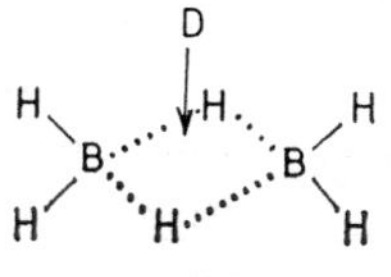   (XXX)

J. E. Eastham[74] suggested structure (XXX) for $D \cdot B_2H_6$, i.e. the donor molecule interacting with a vacant non-bonding orbital of the BH_2B four-center four-electron bond.

Considering the charge distribution in diborane, one would expect that a base D would attack one boron atom of diborane nucleophilically through one face of the two edge-combined tetrahedra of diborane (XXXI).

(XXXI)   (XXXII)

The charge distribution in the $D \cdot B_2H_6$ adduct would then be best if the two tetrahedra share a common corner, corresponding to the single bridge model (XXXII). An attack of D on diborane as suggested by Eastham would be subject to two objections: (i) attack of D on the BH_2B-system is sterically more hindered than the "face" attack, especially for NMe_3, (ii) such attack must result in a bending of the two tetrahedra along the common edge which would cause a mutual repulsion between the two terminal H-atoms (a) in (XXXI).

The single bridge compounds $D \cdot B_2H_6$ may also be prepared from diborane and $D \cdot BH_3$. By using $^{10}B_2H_6$ the products have the composition $D \cdot {}^{11}BH_3 \cdot {}^{10}BH_3$. If Eastham's structure is correct, then $D \cdot BH_3$ formed from its decomposition should contain equal amounts of $D \cdot {}^{11}BH_3$ and $D \cdot {}^{10}BH_3$; however, no ^{10}B enrichment in $D \cdot BH_3$ has been observed.[286] There is therefore no "opening" of the B—N bond.

Recently, O. T. Beachley[19] and S. G. Shore, C. W. Hickam, Jr. and D. Cowles[284] have shown independently that asymmetrical cleavage of diborane is also effected by $MeNH_2$ and Me_2NH as shown in (44).

$$2 R_{3-n}NH_n + B_2H_6 \longrightarrow [H_2B(NH_nR_{3-n})_2]BH_4 \qquad (44)$$

Since $D \cdot B_2H_6$ is an intermediate in the reaction, the attack of a second molecule of amine must occur at the DBH_2 part of the molecule with the formation of BH_4^- encouraging the production of the cation $[D_2BH_2]^+$.

O. T. Beachley[19] found that $[H_2B(NH_2Me)_2]BH_4$ rearranges slowly to give $MeNH_2 \cdot BH_3$. From this observation it was postulated that diborane is always cleaved asymmetrically, not only by $MeNH_2$ but also by other bases, and that the product isolated depends only upon its relative thermodynamic and kinetic stability. In the case of NMe_3 the rearrangement (45) proceeds rapidly and only $Me_3N \cdot BH_3$ is obtained, while for the NH_3 derivative the compound $[H_2B(NH_3)_2]BH_4$ is much more thermodynamically stable than $H_3N \cdot BH_3$, and therefore the salt remains unaltered.

$$[H_2BD_2]BH_4 \longrightarrow 2\,D \cdot BH_3 \qquad (45)$$

However S. G. Shore et al.[283] disagree with this hypothesis and further experiments are necessary to resolve the arguments. The fact that $[H_2B(t\text{-}BuNH_2)_2]Cl$ on reaction with $LiBH_4$ gives $t\text{-}BuNH_2 \cdot BH_3$[220] is consistent with Beachley's view, and also with the assumption that H/D rearrangement should be favored if D is a bulky amine, thus allowing a predissociation according to (46). Indeed $[H_2B(NMe_3)_2]Cl$ decomposes on heating into NMe_3 and $Me_3N \cdot BH_2Cl$.[188]

$$[H_2BD_2]^+ \rightleftharpoons D + [H_2BD]^+ \qquad (46)$$

Besides these studies related to the cleavage of diborane by amines, a number of methods are available today for synthesizing dihydridobis-(amine)boron ions. They are summarized in equations (47–57):

$$D \cdot BH_3 + D \cdot HX \longrightarrow [H_2BD_2]X + H_2^{(188,\ 220)} \qquad (47)$$

$$[H_2BD_2]^+ + 2\,D' \longrightarrow [H_2BD'_2]^+ + 2\,D^{(188)} \qquad (48)$$

$$4\,D \cdot BH_3 + 5\,B_2H_6 \longrightarrow [H_2BD_2]_2B_{12}H_{12} + 13\,H_2^{(188)} \qquad (49)$$

$$2\,D + R_2O \cdot BH_2X \longrightarrow [H_2BD_2]X + R_2O^{(220)} \qquad (50)$$

$$D \cdot BH_2X + D \longrightarrow [H_2BD_2]X^{(220)} \qquad (51)$$

$$D \cdot BH_3 + 2\,D + I_2 \longrightarrow [H_2BD_2]I + D \cdot HI^{(72,\ 262)} \qquad (52)$$

$$2\,D \cdot BH_3 + 2\,D + HgCl_2 \longrightarrow 2\,[H_2BD_2]Cl + Hg + H_2^{(262)} \qquad (53)$$

$$6\,D + 3\,I_2 + 2\,NaBH_4 \longrightarrow 2\,[H_2BD_2]I + 2\,NaI + 2\,D \cdot HI + H_2^{(262)} \qquad (54)$$

$$D \cdot BH_2SR + D \cdot BH_2X \longrightarrow [H_2BD_2]X + RSBH_2^{(179)} \qquad (55)$$

$$2\,D \cdot BH_2SR + R'X \longrightarrow [H_2BD_2]X + R'SR + RSBH_2^{(179)} \qquad (56)$$

$$[RSBH_2]_3 + 3\,D + 3\,D \cdot HX \longrightarrow 3\,[H_2BD_2]X + 3\,RSH^{(179)} \qquad (57)$$

(D or D′ = amines such as NH_3, RNH_2, R_2NH, Me_3N, py; R′ = CCl_3 $CHBr_2$, C_2H_5; X = Cl, Br.)

Reaction (47) is one of the most versatile and is carried out at 100–180°[188] or in boiling $CHCl_3$ solution.[220] By using different amines D, cations of the type $[H_2B(D)D']^+$ may be obtained readily by (47) or (51). The size of the amine molecules determines largely whether the cations are formed or not. Thus in reaction (47) amines such as Et_3N or N,N-dimethylcyclohexylamine do not react. Therefore, if $[H_2BD_2]^+$ contains a bulky amine, and D′ represents an amine of low steric requirements, the bulky amine is expelled as described by (48) and the cation with the less bulky amine is formed. That basicity plays a less important role than steric factors is demonstrated by reaction (58).

$$Me_3N \cdot BH_2Cl + 2\,py \longrightarrow [H_2Bpy_2]Cl + NMe_3 \qquad (58)$$

In displacements according to (48) diamines such as N, N'-dimethylpipera-zine and N, N, N', N'-tetramethylethylenediamine displace non-chelating amines.[188] Reactions of type (50) and (51) proceed rapidly, if D is a primary or secondary amine or ammonia.

The mechanism of the formation of dihydridobis(amine) boron ions for most of the reactions cited is not known exactly. However, it has been conclusively demonstrated that $D \cdot BH_2X$ is an intermediate in the forma-tion of the cation, a process for which both an S_N1 or S_N2 type reaction is conceivable.

The kinetics of reaction (52) involving $Me_3N \cdot BH_3$, py and I_2 has been studied.[262] It was found to be first order for I_2 and for $Me_3N \cdot BH_3$ and a large isotope effect $k_H/k_D = 2.03$ was observed. Due to this effect and to the fact that the rate is determined by the iodine concentration it was concluded that the rate-determining step is a hydride abstraction, followed by rapid conversion of the intermediate to $[H_2Bpy_2]^+$. The intermediate species may also be $py \cdot BH_2I$ or even $[H_2B(py)NMe_3]I$.[262]

$$Me_3N \cdot BH_3 + I_2 \xrightarrow[\text{slow}]{py/-HI} Me_3N \cdot BH_2I \xrightarrow[\text{rapid}]{py} [H_2Bpy_2]I \qquad (59)$$

It has been shown that the first step in reaction (56) and (57) is the for-mation of $D \cdot BH_2X$, and in (56) the nucleophilic attack of SR^- on the halocarbon $R'X$ may be rate determining.[179]

2. Structure

The dihydridobis(amine)boron cations contain boron atoms of coordin-ation number four as demonstrated by their ^{11}B n.m.r. spectra which show a $1 : 2 : 1$-triplet and a chemical shift in the range of 12–18 ppm upfield of $B(OCH_3)_3$.[188] In the i.r. spectra a doublet is observed for the BH_2 group in the 2350 to $2500/cm^{-1}$ region.[188, 220] At present no information seems available for BN bond distances and bond angles.

3. Reactions

One remarkable property of all dihydridobis(amine)boron cations is their hydrolytic stability. The extent of this stability depends on the shielding of the boron atom by the amine, the (bistertiary amine)boron cations being the most stable. Thus $[H_2B(NMe_3)_2]^+$ can be recovered from concentrated H_2SO_4, HCl, HNO_3 or 10 per cent aqueous NaOH even after heating to 100°.[188] Cations of the type $[H_2B(NH_2R)_2]^+$ are rather stable in acidic media, but are readily attacked by bases.[220] A comprehensive study of the hydrolytic stability of various salts of this type has been carried out by

T. A. Shchegoleva, V. D. Sheludyakov and B. M. Michailov.[279] They find that the reaction in 50 per cent aqueous alcohol is first order in the cation concentration and that the rate constant increases for the series H_2O < 50 per cent alcohol < absolute alcohol.

The data obtained indicate that the stability towards solvolysis increases for salts $[H_2B(amine)_2]Cl$ as follows:

amine = $t\text{-BuNH}_2$ < $C_6H_{13}NH_2$ < $C_5H_{11}NH_2$ < $i\text{-PrNH}_2$ < $BuNH_2$ <

$i\text{-BuNH}_2$ < Me_2NH < $PrNH_2$ < py < $EtNH_2$ < $PhCH_2NH_2$ < $MeNH_2$ < NH_3.

Thus for straight chain amines the tendency to hydrolyze increases with increasing chain length; branching at the α C atom facilitates hydrolysis and the most stable compound in the series investigated is $[H_2B(NH_3)_2]Cl$. As one would expect, the influence of the anion is of negligible effect on the rate constant; this has been demonstrated for $[H_2B(NHMe_2)_2]X$ (X = Cl, Br, SEt). These salts are more rapidly solvolyzed in alcohols than in water, but the rate decreases in the order CH_3OH > C_2H_5OH > C_3H_7OH > $i\text{-}C_3H_7OH$ > $t\text{-}C_4H_9OH$ for alcohols as the solvent. The isotope effect is normal. On the basis of all experimental evidence the first stage in the solvolysis process was postulated as an amine displacement reaction (60).

$$[H_2B(amine)_2]^+ + ROH \longrightarrow [H_2B(amine)ROH]^+ + amine \qquad (60)$$

The experimentally determined activation energies[279] in ethanol for $[H_2B(NHMe_2)_2]Cl$ (12.9 kcal/mole), $[H_2B(NHMe_2)_2]Br$ (13.9) and $[H_2B(NHMe_2)_2]SEt$ (12.2) lead to $\Delta G° = 21.7$, 22.7 and 21.5 kcal/mole respectively. The activation entropies are all highly negative (-29.4 to -31.2), thus implying a rather highly ordered transition state, the most likely one being (XXXIII).

$$\begin{bmatrix} R & H\;H \\ \diagdown & \diagdown\!\diagup \\ & O \cdots B \cdots D \\ H & \uparrow \\ & D \end{bmatrix}$$

(XXXIII)

A great number of salts of the $[H_2B(amine)_2]^+$ ion can be prepared by metathetic reactions or through ion exchange. Large anions such as PF_6^-, Br_3^-, $AuCl_4^-$, $B_{12}H_{12}^{2-}$, BPh_4^- yield salts sparingly soluble in water.[187] $FeCl_3$ and $AlCl_3$ add to $[H_2B(amine)_2]Cl$ complexing the chloride ion.[220] The salts $[H_2B(amine)_2]X$ in turn can be used to obtain simple monosubstituted amine-boranes in good yield by (61).[187] Temperatures of 120–200°

$$[H_2B(NMe_3)_2]X \xrightarrow{\;T\;} Me_3N \cdot BH_2X + NMe_3 \qquad (61)$$

are required. The compounds $Me_3N \cdot BH_2X$ with X = Cl, Br, N_3 have been obtained in this way. The decomposition of the carbonate $[H_2B(NMe_3)_2]_2 CO_3$ leads to traces of a liquid which may have the structure $Me_3N \cdot BH_2$—O—CO—$OBH_2 \cdot NMe_3$.[187]

The reducing power of the hydrogen atoms in $[H_2B(amine)]^+$ is rather low; Ag^+ or Au^{3+} is not converted to the metal.[188] This is certainly the result of the positive charge on the cation which should be responsible for a decrease in the H—B polarity. Halogens (except I_2), aqua regia, SF_5Cl, NCl_3 and ICl partly or fully replace the hydrogens by halogen atoms. Reactions (62—65) are illustrative.[188]

$$[H_2B(NMe_3)_2]Cl + 2\ ICl \xrightarrow{80°} [Cl_2B(NMe_3)_2]Cl + 2\ HI \tag{62}$$

$$[H_2B(NMe_3)_2]^+ + 2\ F_2 \xrightarrow[H_2O]{0°} [F_2B(NMe_2)_2]^+ + 2\ HF \tag{63}$$

$$[H_2B(NMe_3)_2]^+ + Br_2 \xrightarrow[H_2O]{} [HBrB(NMe_3)_2]^+ + HBr \tag{64}$$

$$[H_2B(NMe_3)_2]^+ \xrightarrow{S_2O_6F_2} [(FSO_3)_2B(NMe_3)_2]^+ \tag{65}$$

Optically active cationic boron compounds are produced by monohalogenating (XXXIV) to (XXXV),[188] but their resolution has not yet been

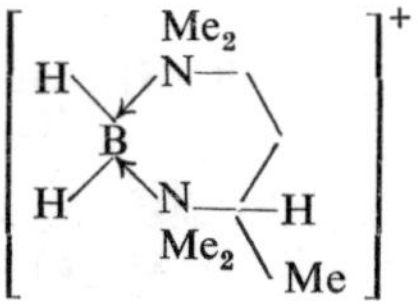

(XXXIV) 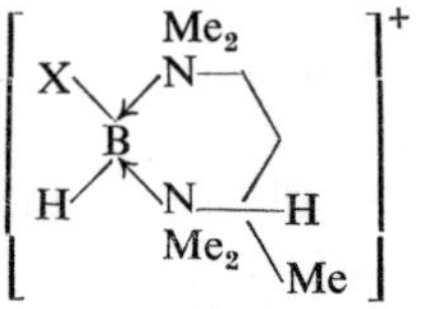   (XXXV)

described. G. E. Ryschkewitz and J. M. Garrett[264] using the d-antimonyltartrates, were able to resolve the cation $[ClHB(NMe_3)(\gamma\text{-pic})]^+$ (XXXVI), where γ-pic = γ-picoline, which was prepared according to the scheme (66). Specific rotations measured on selected crystals were $-5.1°$ and $+4.1°$.

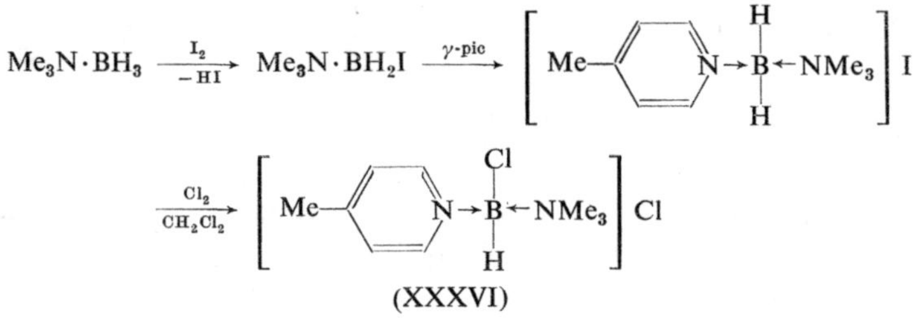

(XXXVI)

$[H_2B(NMe_3)_2]Cl$ is a most interesting compound since it can be used to prepare novel types of BN-containing heterocycles. Treatment with NaH in refluxing dimethoxyethane,[189] or with $NaBH_4$[190] or $LiBu$[190] gives 1,1,4,4-tetramethyl-1,4-diaza-2,5-diboracyclohexane (XXXVII) (yields ranging from 5–30 per cent). Thus NaH, $NaBH_4$ or LiBu act as a proton abstracting agents, and the intermediate formed loses trimethylamine after rearrangement and leads to dimerization according to (67).

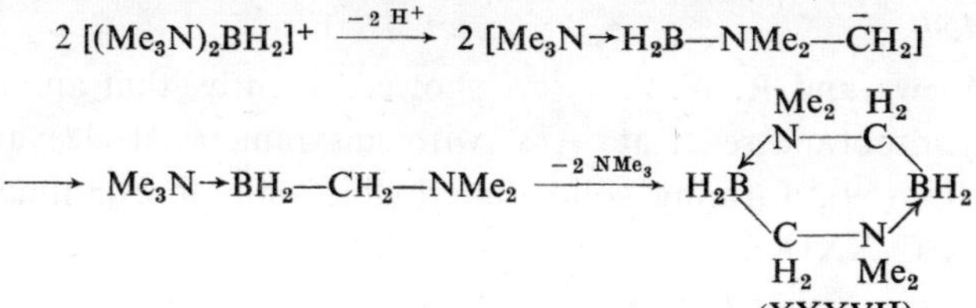

$$2 \ [(Me_3N)_2BH_2]^+ \xrightarrow{-2\ H^+} 2 \ [Me_3N{\rightarrow}H_2B{-}NMe_2{-}\overset{-}{C}H_2]$$

$$\longrightarrow \ Me_3N{\rightarrow}BH_2{-}CH_2{-}NMe_2 \xrightarrow{-2\ NMe_3} \qquad\qquad\qquad (67)$$

This cyclohexane analog does not dissociate below 170°.[190] The reaction to form the heterocycle (XXXVII) from $[H_2B(NMe_3)_2]Cl$ and LiBu is quite sensitive to the stoichiometry of the reactants employed;[186] the best yields being obtained using a 1 : 1.5 mole ratio. The intermediate compound $Me_3N \cdot BH_2CH_2NMe_2$, which is a liquid decomposing slowly at room temperature according to (67), can be isolated. The compound absorbs diborane to give solid $Me_3N \cdot BH_2CH_2NMe_2 \cdot BH_3$ and reacts with HCl to form $[Me_3N \cdot BH_2CH_2NHMe_2]Cl$. On the other hand, if a 1:2 mole ratio of $[H_2B(NMe_3)_2]Cl : LiBu$ is used then the novel compound (XXXVIII) results.[186]

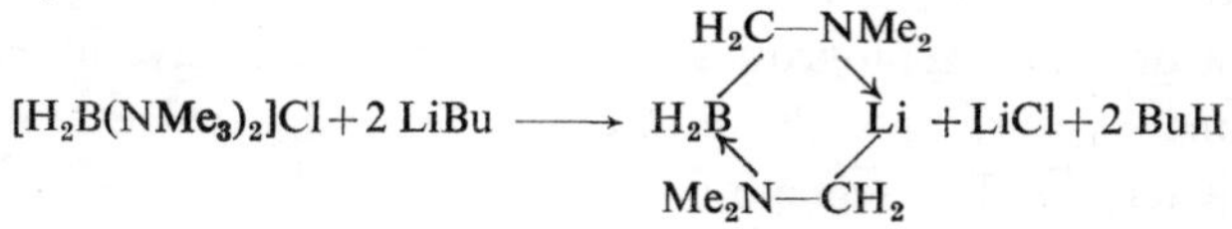

$$[H_2B(NMe_3)_2]Cl + 2 \ LiBu \longrightarrow \qquad\qquad + LiCl + 2 \ BuH \qquad (68)$$

This observation is consistent with a further deprotonation of the intermediate $Me_3N \cdot BH_2CH_2NMe_2$. If this type of reaction can be extended to other organometallics a new series of organometallic heterocycles and/or polymers could possibly be prepared.

C. Diorganobis(amine)boron cations

1. Introduction

The hydrogen bridge of diborane may be cleaved by amines either symmetrically or unsymmetrically. Because a similar bonding situation exists in alkyl- and aryldiboranes, one might expect analogous behavior of these species towards amines. However, due to the larger size of the organic group compared with the hydrogen atom, the course of the reaction may be different and perhaps more sterically controlled than in the case of diborane.

2. Formation

P. C. Moews and R. W. Parry[191] showed recently, that ammonia and tetramethyl-diborane react at $-78°$ with unsymmetrical cleavage of the BH_2B bridge (69), forming solid dimethylbis(amine)borondimethyldihydridoborate (XXXIX).

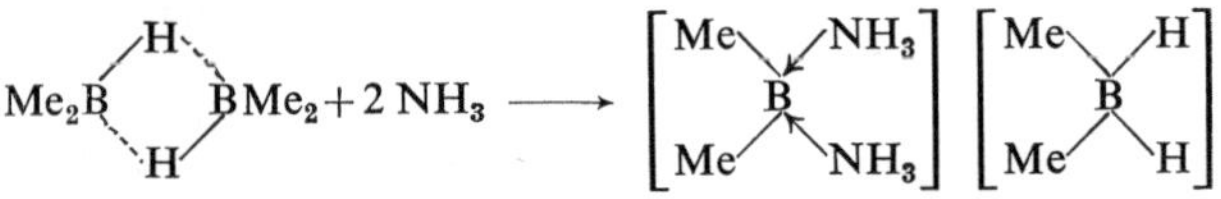

$$Me_2B \underset{H}{\overset{H}{<}} BMe_2 + 2\,NH_3 \longrightarrow \begin{bmatrix} Me & NH_3 \\ & B & \\ Me & NH_3 \end{bmatrix} \begin{bmatrix} Me & H \\ & B & \\ Me & H \end{bmatrix} \qquad (69)$$

$$(XXXIX)$$

In contrast to $[H_2B(NH_3)_2]BH_4$ this salt melts with decomposition at room temperature and hydrogen and Me_2BNH_2 are formed. On the other hand, if reaction (69) is carried out in Me_2O at $-78°$, the only product is $Me_2BH \cdot NH_3$; this is due to initial formation of $Me_2O \cdot BHMe_2$ followed by a base displacement.

The salt (XXXIX) is readily converted with HCl or HBr in diethylether at $-78°$ to $[Me_2B(NH_3)_2]X$ (where X = Cl, Br).[191] Actually, salts containing one or two organogroups attached to boron have already been described many years ago, the first correctly recognized example being $[Ph_2Bdipy]ClO_4$.[61] The scheme (70) summarizes the synthetic routes to these types of boron cations, D representing an amine (donor) molecule.

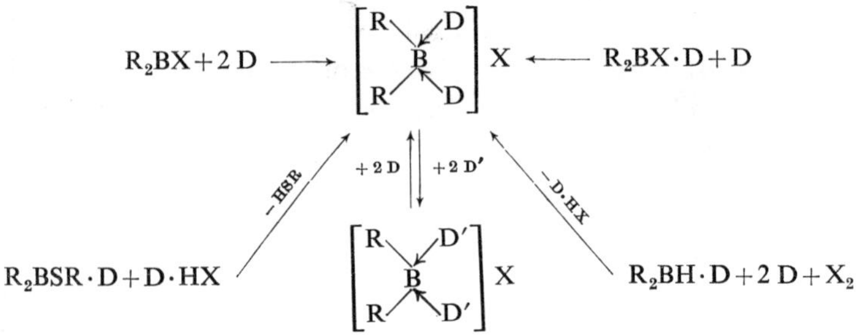

$$R_2BX + 2\,D \longrightarrow \begin{bmatrix} R & D \\ & B & \\ R & D \end{bmatrix} X \longleftarrow R_2BX \cdot D + D$$

$$R_2BSR \cdot D + D \cdot HX \quad \overset{-HSR}{\nearrow} \quad \underset{+2\,D \,\|\, +2\,D'}{\begin{bmatrix} R & D' \\ & B & \\ R & D' \end{bmatrix} X} \quad \overset{-D \cdot HX}{\searrow} \quad R_2BH \cdot D + 2\,D + X_2 \qquad (70)$$

The most simple and straightforward route to dialkyl- and diarylbis-(amine)boron salts is the direct union of two moles of the amine with the diorganoboron halide. So far only chlorides, bromides or iodides have been used.[13, 156, 170–71, 173, 175, 180, 214, 278] The success of the reaction is strongly dependent on steric factors. Thus using Ph_2BCl, no cations are formed with trialkylamines, but with pyridine[214] or dipyridyl[13, 61] the corresponding cations are formed. From Me_2BCl and NH_3, $MeNH_2$ and Me_2NH the salts $[Me_2BD_2]Cl$[214] can be prepared. The reaction with NH_3 had indirectly been observed by H. J. Becher in 1952.[23] If dibutylboron chloride is treated with NH_3[156, 178] or an amine such as $MeNH_2$[278] or in petroleum ether at low temperatures, boron cations are formed. However, under similar conditions the reaction of dibutylboron chloride with Me_2NH, Et_2NH, $n\text{-}BuNH_2$ or $t\text{-}BuNH_2$[156] is predominantly a simple aminolysis. Studies of the reaction of Ph_2BCl with Me_2NH at various temperatures indicated that the yield of $[Ph_2B(NHMe_2)_2]Cl$ (together with $Me_2NH \cdot HCl$) decreased as the reaction temperature increased.[214] It is evident that if there is steric crowding due to R, the organic group attached to boron, then the steric requirements of D must be kept low to permit formation of the diamine-boron ions, and vice versa. If pyridine is used as the base, then rather bulky R groups may still remain at the boron atom, for instance the cyclohexyl- or hexyl-(1,1,2-trimethylpropyl) groups.[71] Under these conditions it may be necessary to further stabilize the salts with an appropriate anion, the perchlorate being the most suitable one.[61, 71]

Chelating donors such as dipyridine or ethylenediamine enhance the stability of the cations. This is clearly demonstrated by the easy replacement of NH_3 from $[Me_2B(NH_3)_2]X$ by $H_2NCH_2CH_2NH_2$, the reaction being practically quantitative at room temperature within 1 hour.[191] Also the rather high hydrolytic stability of $[Bu_2Bdipy]I$ is attributable to this factor. Since this yellow solid salt yields a colorless aqueous solution, the color is considered to be due to a charge transfer from the iodide to the π-system of the complexed dipyridine.[13]

3. Reactions

The chemistry of the diorganobis(amine)boron salts has received little attention thus far, though it is known that anions are readily exchanged for ClO_4^-, BPh_4^-, N_3^-, NCS^-, $Me_2PS_2^-$[13] and that the compound $[Ph_2Bpy_2]Cl$ loses one molecule of pyridine in a vacuum at 80–100°.[214] More important for preparative purposes is the fact that thermal decomposition of bis(amine)boron salts containing NH_3, primary or secondary amines yield the same products as the aminolysis of the diorganoboron halides (71).[156, 214]

$$[R_2B(NH_2R)_2]X \longrightarrow R_2BNHR + RNH_2 \cdot HX \qquad (71)$$

D. *Other Amine-Boron Cations*

It has already been mentioned that the cations $[H_2B(amine)_2]^+$ can be converted via $[XHB(amine)_2]^+$ to $[X_2B(amine)_2]^+$ by strong halogenating reagents,[188] and the series of compounds $[H_{2-n}X_nB(NMe_3)_2]^+$ are hydrolytically and thermally the most stable type of boron cations. Also, if there is a decrease of steric hindrance in $[R_2B(amine)_2]^+$, for instance by substituting R by H, then much more stable cations result. Indeed, the cations $[RHB(amine)_2]^+$ and $[XHB(amine)_2]^+$ do form readily. Further routes to these compounds are summarized by equations (72) and (73), reaction (74) competing with (73).

$$RBH_2 \cdot D + 2\,D + I_2 \longrightarrow [RHBD_2]I + D \cdot HI \quad [72] \tag{72}$$
$$(R = Ph,\ C_6H_{11};\ D = py,\ quinoline)$$
$$(R_2N)_2BH + 2\,HX \longrightarrow [XHB(NHR_2)_2]X \quad [221] \tag{73}$$
$$(R_2N)_2BH + 2\,HX \longrightarrow (R_2N)BHX + R_2NH \cdot HX \quad [221] \tag{74}$$

Only when R = Me is the cationic boron species obtained by the scheme (73); when R = Et, Pr or Bu reaction (74) predominates at room temperature.[221]

The addition of HX to bis(amino)boranes $(R_2N)_2BY$ according to (75) is a very versatile route to various bis(amine)boron cations. However, if Y is a bulky group, for instance NMe_2, $SiMe_3$, $SnMe_3$, Ph, Pr or Bu, then HCl cleaves one BN bond by a reaction analogous to (75).

$$(R_2N)_2BY + 2\,HX \longrightarrow [XYB(NHR_2)_2]X \tag{75}$$
$$(R = Me,\ X = Cl,\ Y = Cl^{[226]}\ Me,\ Et,^{[222]}\ Ph;^{[227]}$$
$$R = Me,\ X = Br,\ Y = Me^{[222]})$$

From the experimental data relating to reactions of type (73) and (75) it appears they are primarily kinetically rather than thermodynamically controlled. Thermodynamic factors seem to be more involved in the reactions of the boron halides of various kinds with amines. The chelate effect greatly enhances cation formation, even if bulky groups like Me_2N are attached to the boron atom.[13, 275] Equation (76) demonstrates this clearly.

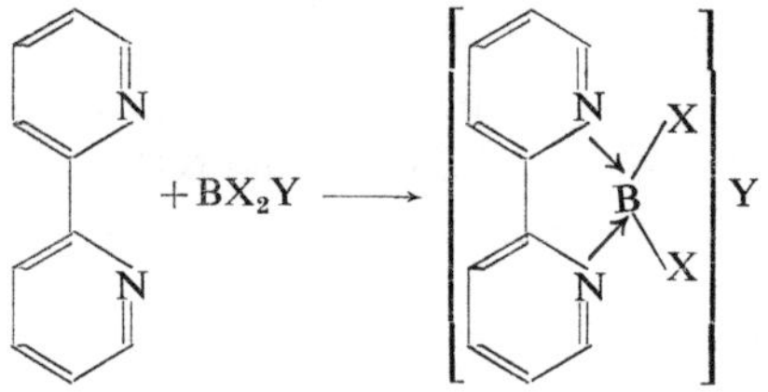

$$\tag{76}$$

$$X = Ph,\ Cl,\ F,\ NMe_2;\quad Y = Cl,\ Br,\ I,\quad X_2 =$$

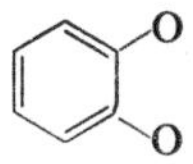

The weaker the B—Y bond, the more readily the reaction proceeds and dipy and BI_3 also react in a 2 : 1 molar ratio, indicating the formation of a spiro complex (XL)[214]. Most of the complexes with dipy as a ligand are of a yellow to orange color,[13] and the u.v. spectra indicate intraionic charge transfer.

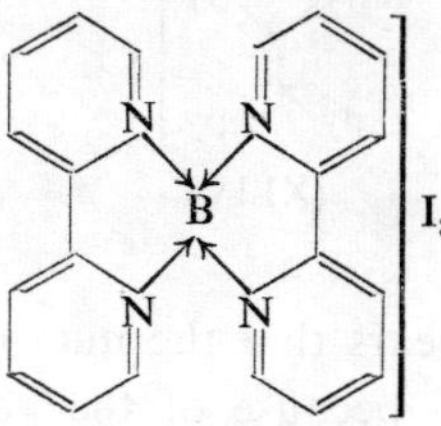

(XL)

Also, if a boron atom forms part of a ring system, cations are readily obtained. The interesting reaction between 2-isocyanato-1,3,2-benzodioxaborole and dibutylamine leads to the 2-dibutylamine derivative and also to the compound (XLl) as shown in (77).[144]

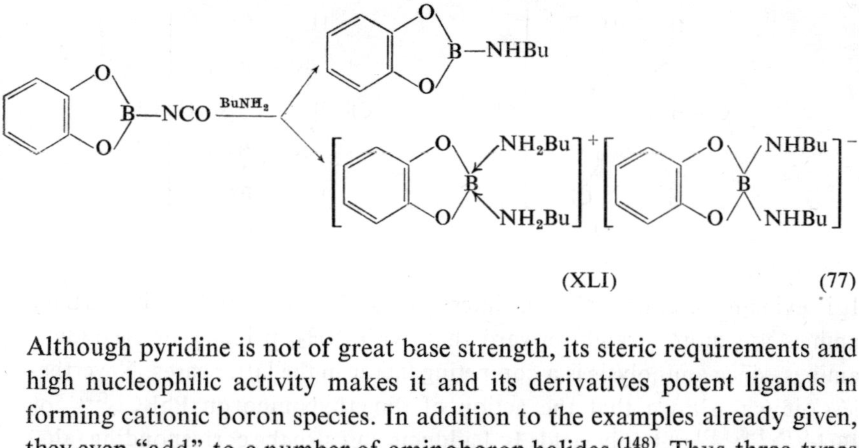

(XLI)   (77)

Although pyridine is not of great base strength, its steric requirements and high nucleophilic activity makes it and its derivatives potent ligands in forming cationic boron species. In addition to the examples already given, they even "add" to a number of aminoboron halides.[148] Thus three types of complexes (XLII, XLIII and XLIV) are observed. Me_2NBCl_2 gives type (XLIII), while the crowding on the boron atom in $i\text{-}Pr_2NBCl_2$ hinders the formation of (XLIII) and the reaction therefore stops at stage (XLII). The same is true if γ-picoline is used as the base, but complex formation is completely suppressed if α-picoline or 2,6-lutidine are used. This demonstrates again the influence of steric factors. It is therefore the more surprising that pyridine and Et_2NBCl_2 yield $[Et_2NBpy_3]Cl_2$, a compound of type (XLIV).

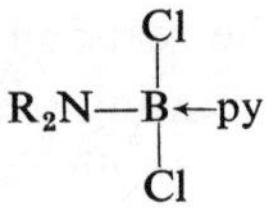

(XLII)

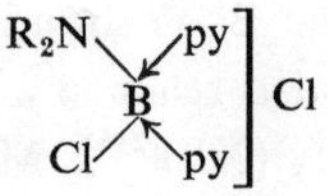

(XLIII)

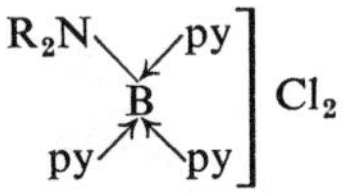

(XLIV)

In all these reactions it appears that the nucleophilic displacement of Cl from the boron is achieved because of the gain in lattice energy. This assumption seems reasonable since the stoichiometry of these reactions depends not only on steric effects but also on solubilities and the latter can be controlled by using various solvents or by altering the mode of addition.

Thus *B*-trichloroborazine reacts with excess pyridine in benzene to give $(ClBNH)_3 \cdot 4\,py$, while addition in the reverse manner yields $(ClBNH)_3 \cdot 2\,py$. Structures (XLV) and (XLVI) were proposed for these two products.[148]

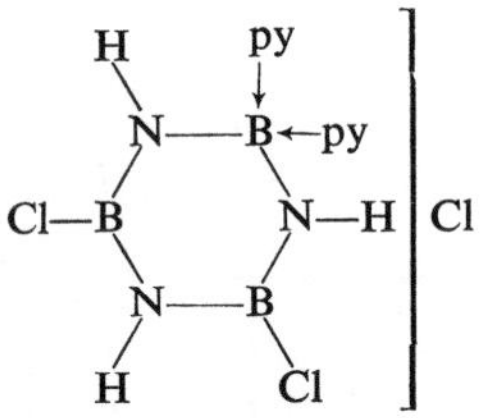

(XLV)

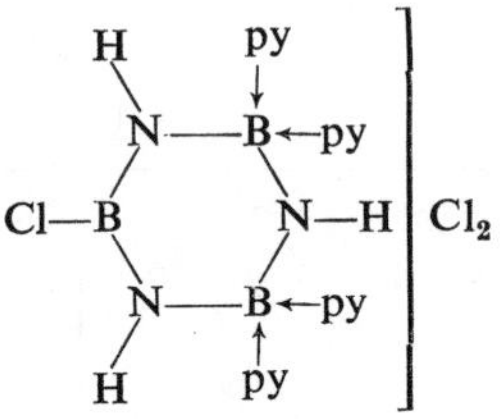

(XLVI)

It is evident, of course, that the interactions of boron halides with tertiary amines are more straightforward than with NH_3, primary or secondary amines, since aminolysis is a competing factor in the latter cases. Nevertheless, it was shown that the action of dimethylamine on $PhBCl_2$[227] or BCl_3[178, 180, 237] not only yields 1 : 1 adducts, but also the corresponding salts, as shown by schemes (78) and (79), which also demonstrate some reactions of these compounds.

$$PhBCl_2 + 2\,HNMe_2 \longrightarrow [PhClB(NHMe_2)_2]Cl \xrightarrow{T} PhBCl(NMe_2) + Me_2NH \cdot HCl$$

$$[PhClB(NHMe_2)_2]Cl \xrightarrow{NEt_3} PhB(NMe_2)_2 \qquad (78)$$

$$[PhClB(NHMe_2)_2]Cl \xrightarrow{FeCl_3} [PhClB(NHMe_2)_2]FeCl_4$$

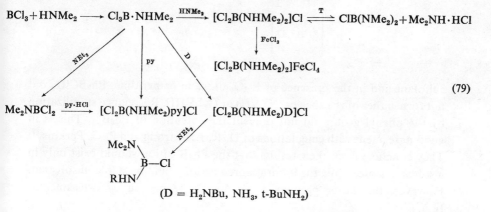

(79)

$(D = H_2NBu, NH_3, t\text{-}BuNH_2)$

There are indications that BBr_3 behaves like BCl_3 towards $HNMe_2$.[178] From these results a better understanding of the factors influencing the aminolysis of the boron halides has been achieved. S. Prasad and N. P. Singh[256-7] found that the reactions of BCl_3 with a large number of substituted primary aromatic amines, aminobenzoic acids and aminophenols were often in a 1 : 2 or 1 : 3 ratio. These results may well indicate formation of compounds analogous to types (XLIII) and (XLIV) At the present time, we know extremely little of the various steps involved in the reactions of primary amines with the boron halides. However, there is little doubt that cationic boron species are involved whose composition and properties are virtually unknown. From this we may deduce that an even more complex behavior is to be expected for reactions of ammonia with the boron halides. It is obvious therefore that further detailed studies of the chemistry of various types of $[XYBD_2]Z$ salts should be helpful in this respect.

E. *The Structures of Boronium Ions*

The compounds so far dealt with can be regarded as ligand-stabilized boronium ions X_2B^+. In fact, not only amines but also phosphines, arsines, ethers and dialkylsulfides have been useful for this purpose. That the types $[X_2BD_2]^+$ and $[XBD_3]^{++}$ predominate is to be attributed to the great stability of tetrahedrally coordinated boron atoms. As R. B. Moodie, B. Ellul and T. M. Connor[192] have pointed out, one should also expect boronium ions of the type (XLVII) and (XLVIII).

Although they discussed only the formation of diphenylboronium ions, this is a question of wider relevance. In a solvent of low basicity such as

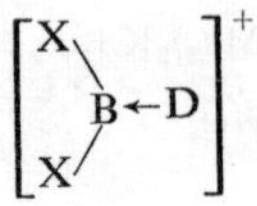

$$\left[\begin{array}{c} X \\ \diagdown \\ B \leftarrow D \\ \diagup \\ X \end{array}\right]^{+} \qquad [X-B-X]^{+}$$

(XLVII)   (XLVIII)

sulfolane and in the presence of Et_2O, dioxane or pyridine, Ph_2BClO_4 did not form either of the species (XLVII) or (XLVIII). Resonance stabilization by the phenyl group did not prevent the solvation of Ph_2B^+. This is in good agreement with calculations of D. R. Armstrong and P. G. Perkins.[7] They conclude from their results that the Ph_2B cation should exist only in a solvated state, while the 9-borafluoren cation could be stable unsolvated. However, this latter prediction remains to be proven by experimentation.

On the other hand, existence of (XLVIII) in the form of $[B(NR_2)_2]^+$ seems possible. It is well known that the BN linkage can be of a high bond order and strong π-bonding and steric shielding of the boron atom can hinder a nucleophilic attack and could contribute to the stability of such a cation in this manner. None the less, attempts in these laboratories to prepare salts of this type of cation, e.g. $[(R_2N)_2B]SbF_6$, have been unsuccessful to date.

IV. THE CHEMISTRY OF SIMPLE BORON–NITROGEN COMPOUNDS

A. *Introduction*

The combination of boron with nitrogen leads to a number of types of compounds depending on the ratio B : N. Among those which already have been thoroughly investigated are the trisaminoboranes $B(NR_{,2})_3$ bisaminoboranes $XB(NR_2)_2$ and monoaminoboranes X_2BNR_2. The substituents X and R may vary widely, but most of the results described deal with a small range of R groups (alkyl, aryl or hydrogen), while X can represent a rather wide variety of atoms or groups. Much less attention has been devoted so far to two other types of simple boron nitrogen compounds, the diborylamines $X_2B-NR-BX_2$ and the triborylamines $(X_2B)_3N$.

Although boron-substituted hydrazines and BN-containing heterocycles are part of the above classification, their properties differ to a greater or lesser extent from those of the simple aminoboranes. They are, however, included in the next section, except for the di-, tri-, tetra- and oligomeric boronimides, to which only occasional reference is made.

B. *Aminoboranes*

1. *Introduction*

$B(NR_2)_3$, $XB(NR_2)_2$ and $X_2B(NR_2)$ show many analogies in their preparations and properties. Nevertheless the chemical properties alter in this series due to the changing B—N bond strength, which is greatest in the X_2BNR_2 compounds.[94]

2. *Preparation*

(a) *Introduction.* Within the last few years no really new methods for the preparation of aminoboranes have been discovered. So far, the amino-lysis of an appropriate boron halide is still the route most widely used. In addition, B—H and B—C bond cleavage of borane or a trialkylborane by an amine has received new attention due to *Trofimenko's* preparation of pyrazaboles.[300-4] The transamination technique is now universally recognized as being very versatile.[137, 205, 209, 211-12, 307, 315] Since the princi-ples underlying the method have been amply described elsewhere,[202 204, 295,] only some new results relating to the formation of cyclic BN com-pounds will be quoted here. The transformation of B—S into B—N bonds by reacting mercaptoboranes or sulfidoboranes with amines is of increas-ing importance.[123, 174, 280] Group exchange reactions using metal ami-des[1, 88, 90-92, 316] are among the more recently studied routes for prepar-ing B–N-containing compounds.

(b) *Aminolysis of boron halides.* The formation of bis(amine)boron cations in the course of the aminolysis of boron halides is now well recog-nized and is often responsible for low yields of aminoboranes. It was always assumed that the first step in the reaction of a boron halide with an amine is formation of the 1 : 1 complex from which dehydrohalogena-tion or boron cation formation is induced by the addition of more amine. These ideas have been confirmed by many experiments including the isolation of the intermediates and studies of their reactions.[22, 275] How-ever, if the Lewis acidity of the boron halide and the Lewis basicity of the amine decrease, the 1 : 1 complex may not form at all, or only in an extremely small concentration. A kinetic study by J. C. Lockhart[154] reveals this. It was found that with excess of boron halides the rate of aminoborane formation was equal to the rate of amine consumption irrespective of the solvent used. The reaction is first order for both amine and boron halide, and from the results obtained it was concluded that for reaction (80) $k_2/k_1 > 100$.

$$RNH_2 + PhBCl_2 \xrightleftharpoons[k_{-1}]{k_1} RNH_2 \cdot PhBCl_2 \xrightarrow{k_2} RNH-BPhCl + HCl \qquad (80)$$

A synchronous bond fission and bond formation takes place and the "adduct" is not so much an intermediate but rather the transition state. In the presence of complexing solvents such as CH_3CN, substitution occurs directly or with the complexed boron halide; in this specific case CH_3CN displacement from $CH_3CN \cdot PhBCl_2$ is indicated. Table 3 records some relevant rate constants.

TABLE 3. KINETIC DATA RELATING TO THE AMINOLYSIS OF $PhBCl_2^{(154)}$

Boron halide	Amine	$k(\text{min}^{-1})$	Solvent
$PhBCl_2$	2,4-Dinitronaphthylamine	0.0335	CH_3CN
$PhBBr_2$	2,4-Dinitronaphthylamine	0.290	CH_3CH
$PhB(OBu)Cl$	2,4-Dinitronaphthylamine	0.0228	CH_3CN
$PhBCl_2$	o-Nitroaniline	2	Et_2O
$PhBCl_2$	2,4-Dinitroaniline	0.0677	Et_2O
$PhBCl_2$	2,4-Dinitronaphthylamine	0.0027	Et_2O
$PhBCl_2$	2,4-Dinitronaphthylamine	0.269	Benzene

A first-order rate constant for the dehydrochlorination of o-$MeC_6H_4NH_2 \cdot BCl_3$ to o-toluidinoboron dichloride was also determined by J. S. Turner et al.[305] Although one would expect further HCl elimination from o-$MeC_6H_4NHBCl_2$ to give the borazine $(o$-$MeC_6H_4NBCl)_3$, triethylamine when used as the deprotonating agent acts on the aminoborane according to (81). However, one of the "aromatic protons" is also eliminated, and the first product is most probably the intermediate (XLIX) given in equation (81).

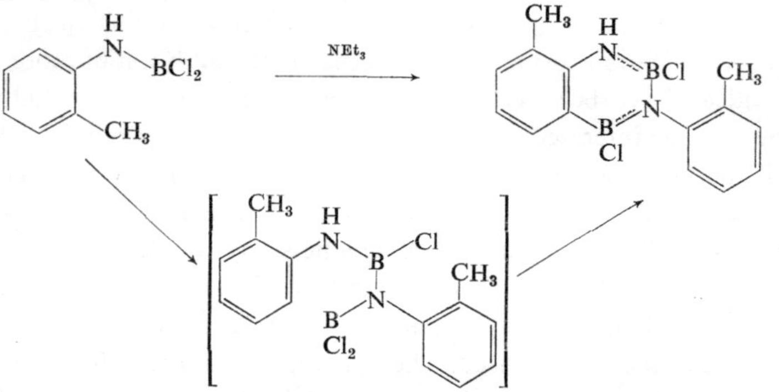

(81)

(XLIX)

Such condensation reactions are not uncommon in the dehydrohalogenation of arylaminoboron halides, and M. J. S. Dewar and his school
have made extensive use of this fact.[67] Steric factors, of course, play an
important role, influencing the extent of the reaction, the nature of the
products and the rate of their formation. A typical example of a more
complex aminolysis is found in the action of boron trichloride on
2-aminopyridine.[199] If this reaction is carried out in a 1 : 3 mole ratio then
tris(boryl-2-pyridylamino)borane (L) is formed. The halogen atoms of
this compound can be displaced by hydrogen or alkyl groups in the usual
manner.

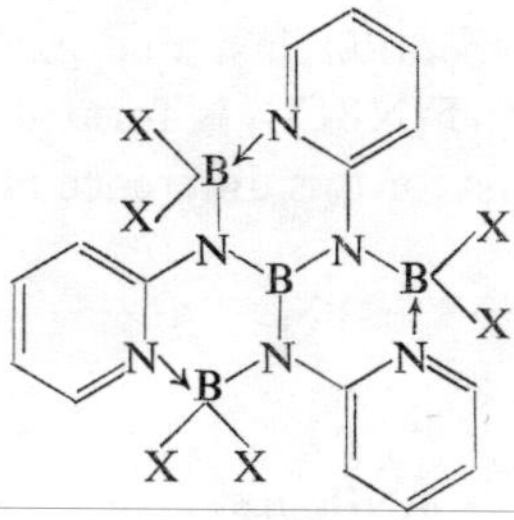

X = Cl, H, R

(L)

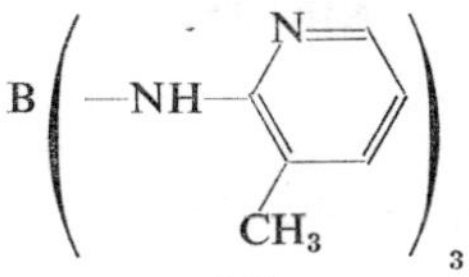

(LI)

If 3- or 4-aminopyridine reacts with BCl_3 under the same conditions
followed by reaction with a Grignard reagent, only mixtures of products
result because the steric requirements for forming low molecular weight
compounds containing B–N coordinate bonds are lacking. On the other
hand, if 2-amino-3-methylpyridine is used as a base, tris(3-methyl-2-
pyridylamino)borane (LI) forms due to the (sterically induced) instability
of a tris(N-dialkylboryl-3-methyl-2-pyridylamino)borane.[199]

The influence of various steric factors exerted by amines on the course
of the aminolysis and its products has been made by R. K. Bartlett et al.[16]
Similar studies starting with alkylammonium tetrahaloborates have been
carried out by W. Gerrard's group.[60] One of the important results
obtained is that the use of t-butylamine not only allows the isolation of
t-BuNHBCl$_2$ and (t-BuNH)$_2$BCl (the first volatile monoalkylaminoboron
halides to be characterized), but also of [t-BuNBCl]$_4$.[305a] Nevertheless,

17*

N-t-butylborazine itself can be prepared by the thermal decomposition of t-BuNH$_2$·BH$_3$[162] at 360°, the resulting borazine being exceptionally stable towards hydrolysis. It would be most interesting to learn if [t-BuNBH]$_3$ can be chlorinated to [t-BuNBCl]$_3$, and if so whether its properties can be compared with those of the borazocine system.[305a]

Today we have a good knowledge of the dehydrohalogenation of amine-boron halides, as described in scheme (82). Nevertheless, much remains to be done that leads to an understanding of more subtle differences, to explain low yields, the catenation and/or cyclization of the amino- and iminoboranes formed, and to discover how to suppress or encourage competing reactions. To mention just one example: pentafluoroaniline reacts with BCl$_3$ but no [C$_6$F$_5$NBCl]$_3$ is formed in boiling toluene unless Et$_3$N is present.[161] Reasons for this difference in behavior are speculative.

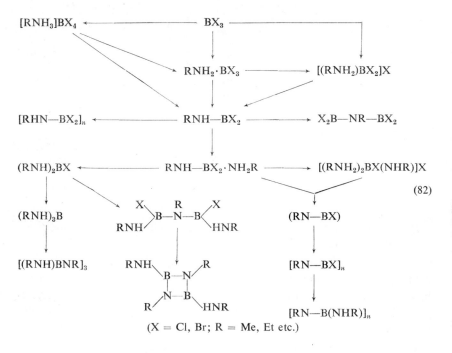

(82)

(X = Cl, Br; R = Me, Et etc.)

The aminolyses of BCl$_3$, BBr$_3$ and BI$_3$ by the weak base (CF$_3$)$_2$NH are accompanied by side reactions, and only (CF$_3$)$_2$NBCl$_2$ and (CF$_3$)$_2$NBBr$_2$, both probably dimeric, are isolated. Reaction with BI$_3$ immediately leads to CF$_3$N=CF$_2$, BF$_3$ and I$_2$.[98] The bis(trifluoromethyl)-aminoboron dihalides decompose above room temperatures with evolution of BF$_3$.

The syntheses of aminovinylboranes such as $CH_2{=}CHB(NMe_2)_2$, $CH_2{=}CHBBr(NMe_2)$,[80] $CH_2{=}CHB(Me)NMe_2$ and $CH_2{=}CHBPh$ $(NRR')_2$[81] from the appropriate vinylboron halide and amine are more conventional. The compounds are analogous to the butadienes, although this is a formalism not as yet sufficiently substantiated by physical data.[134]

The interesting new inorganic heterocyclic system, the boraphospho-nitriles (LII), has been obtained by treating organoboron halides with tetraphenyldiaminodiphosphorus nitride chloride. The new compounds are quite stable to hydrolysis and are not assumed to possess aromaticity.[281]

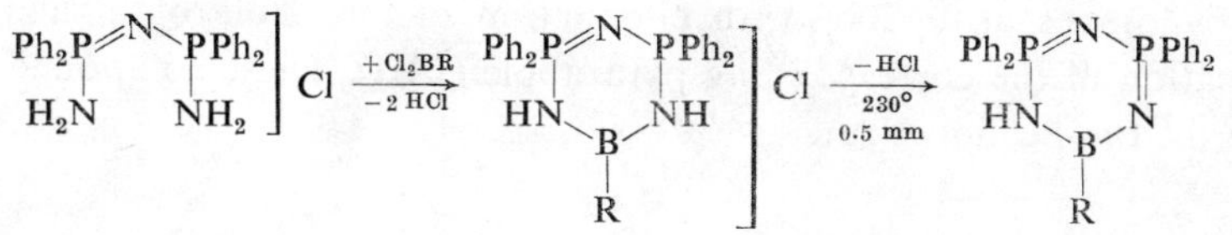

(LII)

Certain aminoboranes are unstable und stabilize through ligand ex-change. Examples are the production of $B(NMe_2)_3$ in the decomposition of $(Me_2N)_2BF$ (83)[325] or from reaction (84).[48]

$$4\,(Me_2N)_2BF \longrightarrow 2\,B(NMe_2)_3 + [Me_2NBF_2]_2 \qquad (83)$$

$$Cl_2B{-}CH{=}CH{-}BCl_2 + 8\,HNMe_2$$

$$\longrightarrow B(NMe_2)_3 + \frac{1}{n}\,[{-}B(NMe_2){-}CH{=}CH{-}]_n + 4\,Me_2NH\cdot HCl \qquad (84)$$

The high strength of the B—F bond usually inhibits its aminolysis, the reaction normally stopping after formation of the adduct $R_{3-n}H_nN\cdot BF_3$. However, it has been shown recently that under more severe conditions the BF_3 adducts of secondary amines decompose at 250–300° according to (85).

$$2\,Et_2NH\cdot BF_3 \longrightarrow [Et_2NH_2]BF_4 + Et_2NBF_2 \qquad (85)$$

In this way diethylaminodifluoroborane was formed. Powdered metals (Al, Mg, Zn) increase the yield, perhaps by reacting with the ammonium salt formed[265]. Alternatively it may be that the reaction in the presence of a metal proceeds via the metal amide or its BF_3 complex, $M[Et_2NBF_3]_n$, and that the latter anion loses F^- to give the aminoboron difluoride. Somewhat related to reaction (85) is the ready decomposition (86) of RHN—BF_2 and the reaction (87) of methylamine with Et_2NBF_2. The driving force of reaction (87) must be the rather high energy of formation of

$MeNH_2 \cdot BF_3$, which together with the resonance stabilization of $B(NHMe)_3$ accounts for the ready aminolysis of the B—F bond.[99]

$$3\ [RHNBF_2]_2 \longrightarrow 2\ [RNH_3]BF_4 + \frac{4}{n}\ [RNBF]_n \tag{86}$$

$$3\ Et_2NBF_2 + 5\ MeNH_2 \longrightarrow (MeHN)_3B + 2\ MeNH_2 \cdot BF_3 + 3\ Et_2NH \tag{87}$$

Some other aspects of the aminolysis of boron fluorides are discussed in Section II, C.

(c) *Aminolysis of B—C and B—H bonds.* Pyrazole reacts readily with trialkyboranes at $> 200°$ with elimination of one mole of alkane and formation of the corresponding pyrazabole (LIII). These compounds are described in Section V, A.

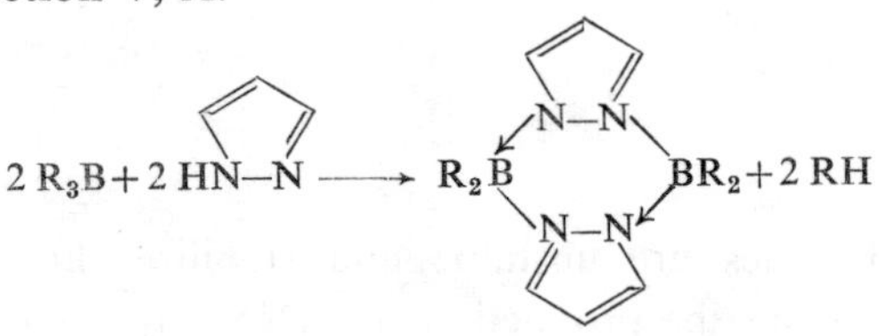

$$2\ R_3B + 2\ HN{\overset{\frown}{\longrightarrow}}N \longrightarrow R_2B \underset{N-N}{\overset{N-N}{\big\langle}} BR_2 + 2\ RH \tag{88}$$

(LIII)

Triazole[300] and imidazole[300] react in an analogous fashion. Azomethine derivatives also interact with trimethylborane. Initially the unstable adduct $Ph_2C{=}NH \cdot BMe_3$ is formed; its heat of dissociation is only ~ 5 kcal/mole since Ph_2CNH is intermediate in basicity between R_3N and RCN. Heating the adduct at 160–200° for 2 weeks yields methane and $Ph_2C{=}N{-}BMe_2$. This compound is monomeric in the gas phase. No such derivative was obtained using BEt_3 or BPh_3.[254] On the basis of the breakdown pattern in the mass spectrum a rather weak B—N bond is indicated in $Ph_2C{=}N{-}BMe_2$ which is isoelectronic with the allene system.

Reactions according to equation (89) proceed more readily than those analogous to (88).

$$\rangle BH + HN\langle \longrightarrow \rangle B{-}N\langle + H_2 \tag{89}$$

This well-known reaction type still has great merits as is shown in the synthesis of certain diborylamines (see Section IV, D, 1). Although (89) appears to be a clear-cut reaction, unexpected products may result. If o-aminodiphenyl is heated with triethylamine borane to 60°, N-o-diphenylborazine (LIV) forms, which on further heating to 400–410° loses more hydrogen to yield 1,2 : 3,4 : 5,6-tris(2,2′biphenylene)borazine (LV).[132]

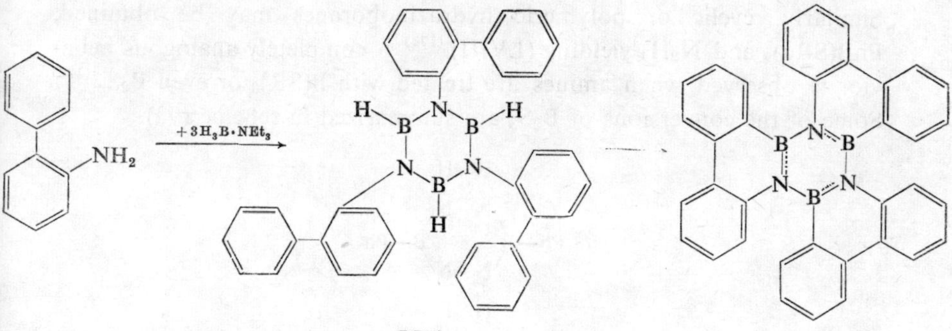

(LIV)   (LV)

(90)

This highly condensed ring system is a very interesting one, but from the point of view of reactivity the products of reaction (91) are more surprising.[134]

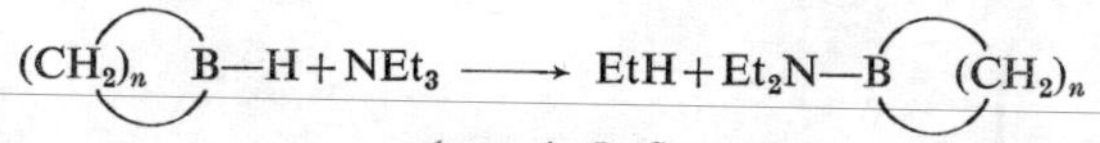

$$(CH_2)_n \quad B-H + NEt_3 \longrightarrow EtH + Et_2N-B \quad (CH_2)_n \qquad (91)$$

(n = 4, 5, 6)

Somewhat similar is the following N—C bond cleavage by trialkylboranes. Refluxing BMe₃, BBu₃ or B(CH₂Ph)₃ with bis(1, 3-diphenyl-2-imidazolidinylidene) in toluene gives derivatives of 1,3,2-diazaborolane as shown in (92).[117]

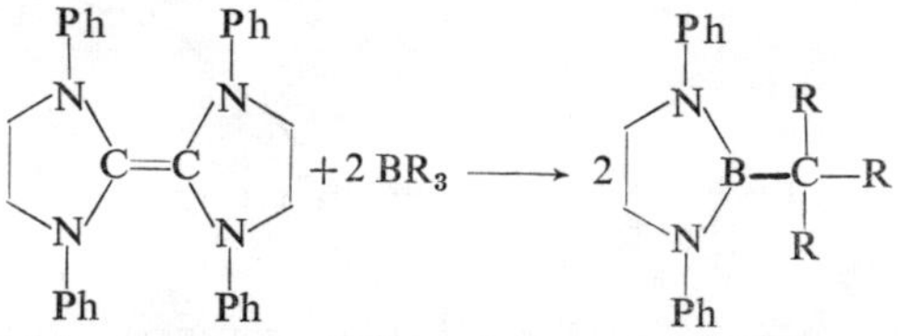

(92)

(d) *Aminolysis of B—S bonds.* The use of boron–sulfur compounds as starting materials for the preparation of various aminoboranes is largely due to B. M. Mikhailov *et al.*[168, 172] This method has now become of wider interest because it gives ready access to a number of aminoborane-type compounds. Mikhailov and Kosminskaya[174] showed that 1,6-hexylenediamine and i-C₅H₁₁B(SBu)₂ yield a resinous polymer (LVI), or according to the stoichiometry the thirty-six-membered ring compound (LVII).

$$H_2N-(CH_2)_6-NH[-B(i-C_5H_{11})-NH-(CH_2)_6-NH]_4-H$$
(LVI)

$$[-B(i-C_5H_{11})-NH-(CH_2)_6-NH]_4$$
(LVII)

Similarly, cyclic or polymeric hydrazinoboranes may be obtained, $PhB(SBu)_2$ and N_2H_4 yielding (LVIII).[174] A completely analogous behavior is observed when amines are treated with $B(SR)_3$ or even B_2S_3.[123] Some of the conversions of B_2S_3 are summarized in scheme (93).

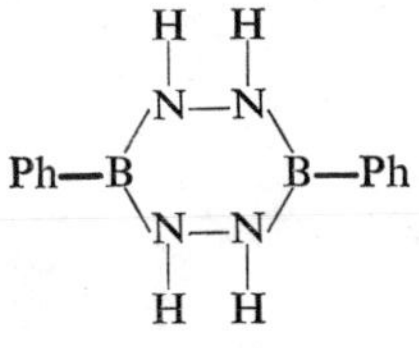
H H
N—N
Ph—B B—Ph
N—N
H H

(LVIII)

Ph_2NH and $O_2N\text{-}C_6H_4NH_2$ failed to react.[123] B—N bonds are also formed if R_2BSR' adds to acetonitrile. This is followed by a rearrangement with SR′ group migration (94a), and is an example of an insertion reaction.[169]

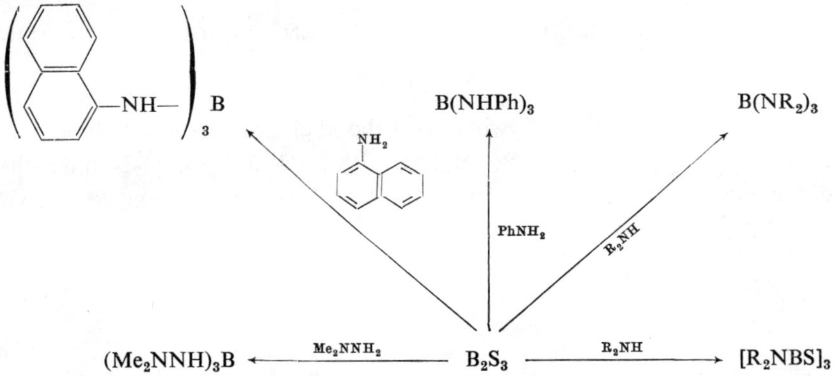
—NH— B
3
B(NHPh)3
B(NR2)3
NH2
PhNH2
R2NH
(Me2NNH)3B
Me2NNH2
B2S3
R2NH
[R2NBS]3

(93)

Another very interesting reaction is that of $B(SR)_3$ with azines, which opens a route to enaminoboranes according to (94b).[69]

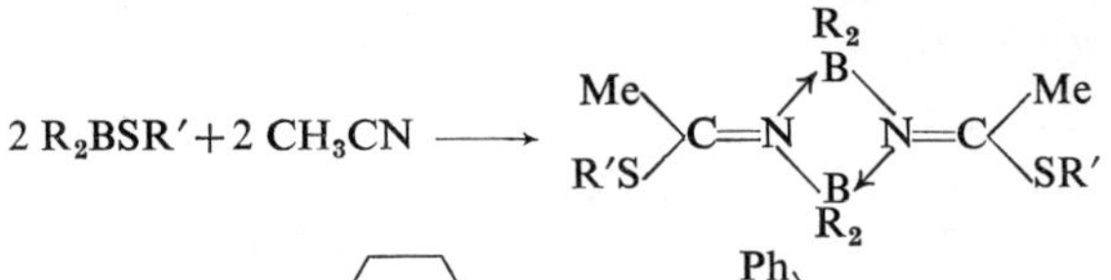
2 R2BSR′+2 CH3CN
R2
B
Me
Me
C=N
N=C
R′S
SR′
B
R2

(94a)

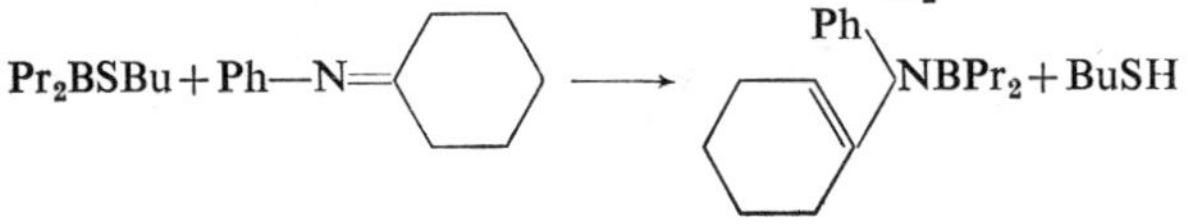
Pr2BSBu+Ph—N=
Ph
NBPr2+BuSH

(94b)

While some of these reactions are typical, most of them offer no advantage over the aminolysis of boron halides or aminoboranes, although they can be carried out in a homogeneous medium. Due to the greater strength of the B—N bond compared with the B—S bond and also as a result of the high volatility of H_2S or RSH (R = Me, Et) the reactions do give high yields. However, experimentally there will always be the problem of objectionable odor to be dealt with.

On the other hand, aminolysis reactions of this type are very useful in preparing diborylamines, and it is in this area that they are of great value.

(e) *Aminolysis of B—O bonds.* It has already been mentioned that compounds containing coordinate BN bonds can be prepared from certain N-containing compounds and either borinic or boronic acids or their esters. On the other hand, it is often not easy to convert $\rangle$B—O-

into $\rangle$B—N$\langle$ bonds, except in group exchange reactions. But if heterocycles are formed in these reactions, this factor helps to drive the equilibrium (95) to the right.[68, 114, 308]

$$\rangle B—OH + H_2N—R \rightleftharpoons \rangle B—NHR + HOH \qquad (95)$$

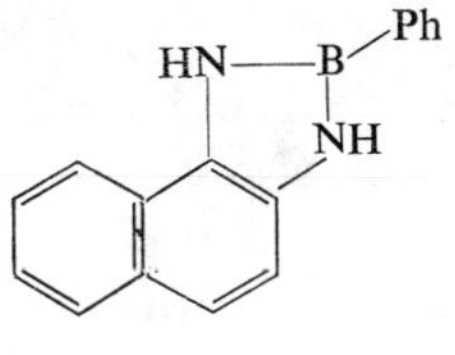

(LIX)[114] 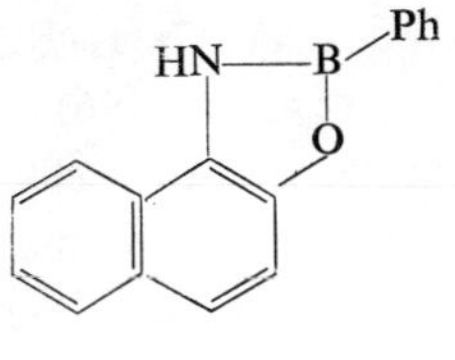   (LX)[114] 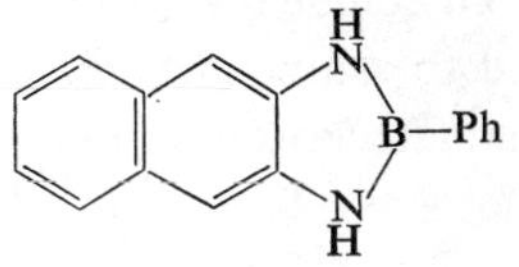   (LXI)[114]

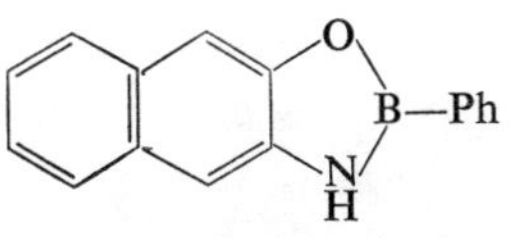

(LXII)[114] 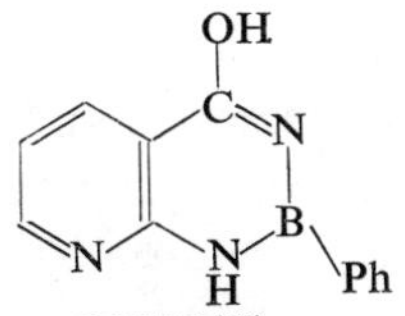   (LXIII)[68] 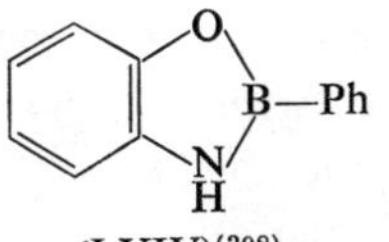   (LXIV)[308]

Thus the compounds (LIX)–(LXIV) were readily accessible from PhB(OH)$_2$.

(f) *Group exchange and cleavage reactions.* A widely used reaction for preparing "mixed" aminoboranes is the treatment of an aminoborane B(NR$_2$)$_3$ or XB(NR$_2$)$_2$ with another boron compound BX$_3$ to induce an exchange of the substituents. Very often the desired compounds are formed quantitatively, as demonstrated by equations (96) and (97). On the other hand, the mixed aminoboranes are sometimes formed only in equilibrium with all the other possible members of the exchange series.[155]

$$BX_3 + 2 B(NR_2)_3 \longrightarrow 3 XB(NR_2)_2 \qquad (96)$$

$$2 BX_3 + B(NR_2)_3 \longrightarrow 3 X_2BNR_2 \qquad (97)$$

The following new types of aminoboranes have been obtained by means of such exchange reactions: $(R_2N)_2BF$,[238] $(RHN)BF_2$,[99] $(R_2N)_2BI$, R_2NBI_2,[15] $RB(NMe_2)F$,[311] $RB(NMe_2)Cl$,[311] $RB(NMe_2)H$,[311] $R_2NB(SR')Cl$,[55] $R_2NB(SR')Br$,[55] $RB(NMe_2)RS'$,[311] $(RO)B(NR_2)Cl$,[214] $R_2NBClBr$.[102] Usually alkyl exchange does not occur if amino groups are present in the molecule, while the exchange of the R_2N group from $B(NR_2)_2$ or $XB(NR_2)_2$ proceeds quickly.

The stability towards disproportionation of these newly synthesized "mixed" aminoboranes depends mainly on two factors: (a) increase of B—N bond strength and (b) steric effects due to the group R. Thus $(Me_2N)_2BF$ is unstable, decomposing readily according to (83), while $(Et_2N)_2BF$ and $(Bu_2N)_2BF$ can be stored unchanged indefinitely.[238] Also the bulky i-Pr or t-Bu groups are responsible for the stability of dimeric $[RNHBF_2]_2$. Only when $(MeNH)_3B$, $(EtNH)_3B$, $(PrNH)_3B$ or $(i\text{-}BuNH)_3B$ are treated with BF_3 are the adducts $RNH_2 \cdot BF_3$ formed.[99]

Of course the group exchange reactions (96) and (97) need not be restricted to aminoboranes, but amides of other elements can also be used to aminate boron compounds. Since aminosilanes can be readily prepared they have been quite closely examined in their behavior towards boron halides. The Si—N cleavage reactions, first observed in 1952 by Burg and Kuljan,[39] are still recommended procedures for making silylamino-boranes, aminoboranes, boretidines and borazines, diborylamines and a number of other types of compounds. Although the main results were known prior to 1964, some very interesting observations have been made since then.

E. Abel and co-workers[1] found by using Me_3SiNR_2 that a maximum of three R_2N groups can be transferred to BCl_3 as shown in (98). In a similar manner Et_2NBBr_2 and Bu_2NBI_2 were prepared.[1]

$$BCl_3 + n\, R_2NSiMe_3 \longrightarrow BCl_{3-n}(NR_2)_n + n\, Me_3SiCl \qquad (98)$$
$$(R = Et, Bu)$$

Phenylboron dichloride cleaves both SiN bonds of heptamethyldisilazane as well as those in $[Me_2SiNMe]_3$, the products being Me_3SiCl or Me_2SiCl_2 and the borazine $[PhBNMe]_3$. Even the B—Cl bonds of trichloroborazine cleave the Si—N bond of Me_3SiNEt_2 to give $[Et_2NBNH]_3$ and Me_3SiCl.[1]

Because the reactivity of the M—N bond increases in the series R_3Si—$NR_2' < R_3Ge$—$NR_2' < R_3Sn$—NR_2' it is not unexpected that Me_3Sn—NMe_2 possess a higher amination potential than Me_3SiNMe_2. This is clearly demonstrated by the 92 per cent conversion of $BF_3 \cdot OEt_2$ into $B(NMe_2)_3$ by means of Me_3SnNMe_2,[88] while Me_3SiNMe_2 only forms a labile adduct. Furthermore, treatment with Me_3SnNMe_2 converts Bu_3B to Bu_2BNMe_2 and Ph_3B to Ph_2BNMe_2 in good yields.[88] Group exchange reactions between the aminostannane and Ph_2BOMe or $B(OMe)_3$ give Ph_2BNMe_2 or $B(NMe_2)_3$ almost quantitatively. These conversions cannot be attributed solely to volatility factors, but must be due to great differences in the free energies of the systems involved.[88]

Silazane cleavage by boron halides becomes more complicated when compounds of the type R_3SiNHR, $(R_2SiNH)_3$, R_3Si—NH—SiR_3 or R_3SiNH_2 are employed. Not only does SiN cleavage occur according to (99) and (100)[215] but also dehydrohalogenations take place due to $>B$—Cl $+ HN<$ condensation reactions. The formation of diborylamines by (100)

$$R_2BCl + Me_3SiNHSiMe_3 \longrightarrow R_2B\text{—}NH\text{—}SiMe_3 + Me_3SiCl \qquad (99)$$

$$2\,R_2BCl + Me_3SiNHSiMe_3 \longrightarrow R_2B\text{—}NH\text{—}BR_2 + 2\,Me_3SiCl \qquad (100)$$

will be dealt with in more detail in Section IV, D. Thus, although Abel et al.[1] obtained the expected yield of Et_3SiCl from the 1 : 1 reaction between BCl_3 and $EtHNSiMe_3$, the other product did not analyse for $EtHN$—BCl_2.

Despite the fact that one may obtain $Cl_2BNHSiMe_3$ or $(HNBCl)_3$ from BCl_3 and $(Me_3Si)_2NH$, the unsymmetrical triaminoborane (LXV) can be isolated if an excess of $(Me_3Si)_2NH$ is employed, NH_4Cl being another major product. The compound (LXV) is an isomer of $B(NHSiMe_3)_3$, the

$$[Me_3Si]_2N\text{—}B\begin{array}{c} \diagup NHSiMe_3 \\ \diagdown NH_2 \end{array}$$

(LXV)

$$[Me_3Si]_2N\text{—}B\begin{array}{c} \diagup NHSiMe_3 \\ \diagdown Cl \end{array}$$

(LXVI)

expected product of SiN cleavage of $(Me_3Si)_2NH$ by BCl_3, and its structure has been fully confirmed by IR and ^1H n.m.r. spectroscopic data.[316] The compound (LXVI), a product similar to (LXV), results when $BCl_3 \cdot NEt_3$ is used as the boron halide source,[316] the presence of NEt_3 favoring dehydrohalogenation. The same compound (LXVI) is also formed from

[Me$_3$Si]$_2$NBCl$_2$ and (Me$_3$Si)$_2$NH under certain conditions. [Me$_3$Si]$_2$NBCl$_2$ results from the action of BCl$_3 \cdot$NEt$_3$ on (Me$_3$Si)$_2$NH.[318]

Dehydrohalogenation competes with Si—N bond cleavage in these reactions, especially in the presence of NR$_3$ as an HCl acceptor, and the NH$_2$ group in (LXV) most likely forms by the reaction $>$B—NH—SiMe$_3$ + HCl → $>$B—NH$_2$ + Me$_3$SiCl. Low yields of silylaminoboranes are indications of these and other side reactions, which can be silylamino- or silylhalide eliminations. Therefore an almost quantitative yield of (PhBNH)$_3$ from PhBCl$_2$ and (Me$_3$Si)$_2$NH is explicable, because any type of elimination from intermediates like PhBCl(NHSiMe$_3$), PhB(NH$_2$)NHSiMe$_3$ or PhB(NHSiMe$_3$)$_2$ will lead to the borazine. On the other hand, the 21 per cent yield of PhB(NMe$_2$)NHSiMe$_3$ obtained from the 1:1 reaction of PhB(NMe$_2$)Cl with hexamethyldisilazane points to the formation of a high proportion of other materials.[122]

Dehydrohalogenation in reactions of the above type can be avoided if either N(SiMe$_3$)$_3$ or MN(SiMe$_3$)$_2$ (M = Li, Na) are used for preparing the silylaminoboranes since the steric effects connected with the bulky (Me$_3$Si)$_2$N group determine in part the reaction products. Typical preparations are represented by (101), but to date B[N(SiMe$_3$)$_2$]$_3$ has not been prepared.

$$BX_3 + n\,MN(SiMe_2)_2 \longrightarrow X_{3-n}B[N(SiMe_2)_2]_n + n\,MX \qquad (101)$$
$$X = F,^{[92,\,261]}\ Cl;^{[92]}\quad n = 1, 2;\quad M = Li,^{[261]}\ Na^{[90,\,92]}$$

When this synthesis was attempted in boiling xylene, reaction (102) occurred, leading to another example of a boretidine.

$$2[(Me_3Si)_2N]_2BCl + 2\,NaN(SiMe_3)_2 \longrightarrow 2\,NaCl + 2\,N(SiMe_3)_3$$
$$+ [(Me_3Si)_2N—B{=}NSiMe_3]_2 \qquad (102)$$

It is worth noting that [(Me$_3$Si)$_2$N]$_2$BF is formed in preference to (Me$_3$Si)$_2$NBF$_2$,[90, 261] the former decomposing at 200° with elimination of Me$_3$SiF to give [(Me$_3$Si)$_2$NB=NSiMe$_3$]$_2$. In contrast, in the aminoboron fluoride series the compounds R$_2$NBF$_2$ are much more readily formed than (R$_2$N)$_2$BF. The formation of (Me$_3$Si)$_2$NBF$_2$ from BF$_3$ and [(Me$_3$Si)$_2$N]$_2$BF requires 2–4 hours heating to 95°,[261] while in the aminoborane case the reaction is rapid at room temperature.[238] This reversed stability order can be attributed to the extra energy gained through formation of R$_2$NBF$_2$ dimers.

All the above-mentioned silylaminoboron compounds as well as the type R$_2$NBCl[N(SiMe$_3$)$_2$][91] are liable to lose Me$_3$SiX upon heating, although rather drastic conditions are often necessary (> 240°). Thus

$Cl_2BN(SiMe_3)_2$ decomposes to $[ClBNSiMe_3]_3$,[90] and $PhB(OEt)N(SiMe_3)_2$ to $[PhBNSiMe_3]_3$, but in the case of $XB[N(SiMe_3)_2]_2$ the bulky $(Me_3Si)_2N$ groups prevent the triborazine formation and favor the boretidines.[90] Rather unexpected is the result of reaction (103), which can be rationalized in terms of a methyl group migration from silicon to boron.[91]

$$2\ Et_2NBCl[N(SiMe_3)_2] + 2\ NaN(SiMe_3)_2 \longrightarrow 2\ NaCl + [Me_2Si-NSiMe_3]_2$$
$$+ 2\ MeBNEt_2[N(SiMe_3)_2] \qquad (103)$$

All the reactions leading to silylaminoboranes, N-silylborazines and other compounds containing the Si–N–B group are of great interest since studies can be made of various effects not exhibited by the alkylaminoboranes. Further reactions of the silylaminoboranes will be discussed in section IV, D, 2.

3. Oligomerization of Aminoboranes

Many of the aminoboranes exist not only as monomeric species but tend to form dimers, trimers, etc., and even polymers. Work prior to 1964 on R_2NBF_2, R_2NBCl_2, R_2NBBr_2, $R_2NBR'Cl$, R_2NBH_2, R_2NBHR', H_2BNR_2 and R_2BNHR' showed[†] that the tendency to form dimers is determined mainly by three factors: (i) a strong electrophilic site at the boron atom, (ii) a strong basic site at the nitrogen atom, (iii) steric requirements (the less bulky the group on B or N the greater is the tendency to polymer formation). From this it becomes apparent that H_2B-NH_2, the monomeric form of which has been detected to date only by mass spectrometry, shows the greatest tendency to be polymeric and the highest degree of association. Böddeker, Shore and Bunting[29] obtained $[H_2NBH_2]_n$ with $n = 2,3,4,5$ together with higher polymers from the action of $NaNH_2$ upon $[H_2B(NH_3)_2]BH_4$, i.e. by deprotonation of the cation. The pentameric $[H_2NBH_2]_5$ is the product most soluble in liquid ammonia, and $[H_2NBH_2]_4$ is the least characterized member of the series. While $[H_2NBH_2]_5$ is quite stable, hydrolytically more so than thermally, the dimer transforms on heating to the trimer, which therefore is thermodynamically the more stable species. The pentamer $[H_2NBH_2]_5$ decomposes at 125—145° to give polymeric $[H_2NBH_2]_n$ and the volatile compounds B_2NH_7 and H_3NBH_3. The variety of degrees of association of $[H_2NBH_2]$ is unusual for the aminoboranes and is only approached in this respect by $[MeHNBH_2]_n$.

Although it had been suggested earlier that aminoborane dimers or trimers, etc., are formed via the monomers, recent work has shown that

† For literature references see ref. 142.

the degree of association of the product depends very much on the reaction conditions and on the types of intermediates formed during the specific reaction. While the reaction between $NaBH_4$ and $Me_2NH \cdot HCl$ in ether leads to $Me_2NH \cdot BH_3$ it gives the linear $Me_2NH \cdot BH_2NMe_2 \cdot BH_3$ in diglyme[109] which on heating converts to $[Me_2NBH_2]_2$ with H evolution. Analogous intermediates are formed using $MeNH_2 \cdot HCl$ and NH_4Cl instead of $Me_2NH \cdot HCl$.[109] However, such an intermediate compound is not produced in the partial pyrolysis of $Me_2NH \cdot BH_3$, and none was detected in the reactions of $Me_2NH \cdot HCl$ with $[Me_2NBH_2]_2$ and of $[H_2B(NH_2Me)_2]Cl$ with $H_3B \cdot NMe_2H$.[21] It was concluded, therefore that due to steric effects no ionic species are involved in the formation of $[Me_2NBH_2]_2$, although the degree of association of the compound seems (at least in part) to be determined by the precursors. For instance, it is known that not only a dimer $[Me_2NBH_2]_2$ but also a trimer $[Me_2NBH_2]_3$ exists. The trimer is obtained by treatment of the dimer with B_5H_9.[37] The first step in this rather unexpected transformation of $[Me_2NBH_2]_2$ into $[Me_2NBH_2]_3$ is the removal of a BH_3 group from B_5H_9, the products formed are $Me_2NB_2H_5$ and $Me_2NBH_2 \cdot B_4H_6$. It is the latter compound that supports the growth with Me_2NBH_2 of an aminoborane chain (polyborazane), which then cyclizes to $[Me_2NBH_2]_3$ reforming $Me_2NBH_2 \cdot B_4H_6$.[37] This compound decomposes irreversibly in two ways: (i) with loss of BH_3 as $Me_2NB_2H_5$ to yield $[Me_2NBH_2 \cdot B_3H_3]_x$, (ii) with loss of H_2 to give involatile $[Me_2NBH_2 \cdot B_4H_4]_n$.

A rather more complicated picture must be drawn to describe the formation of methylaminoborane polymers. Schaeffer and Gaines[326] were able to show the existence of the two expected conformational isomers of $[MeHNBH_2]_3$, and O. T. Beachley Jr.[21] in a number of intriguing experiments revealed a great deal of the route to these compounds. The reaction scheme (104) summarizes the results.

That the intermediates do indeed determine the size of the oligomer formed is seen from the results of M. P. Brown and R. W. Heseltine[33] who investigated the gas phase reaction of Me_2NBH_2 with $MeNH_2$. In this case the conditions were quite different from those prevailing in *Beachley*'s experiments, and the transamination at $> 60°$ led to a transparent film of polymeric methylaminoboranes, non-volatile in a vacuum up to $160°$. It is assumed that the blocking of coordination sites by the Me_2NH formed, i.e. $Me_2NH \cdot BH_2 - NHMe \ldots$, is the reason for the polymerization, although in the presence of traces of Me_2NH at $50–75°$ and after 12 hours about 50 per cent of the polymer converts to the trimer. Therefore the trimer is thermodynamically more stable than the polymeric material.[33] In solu-

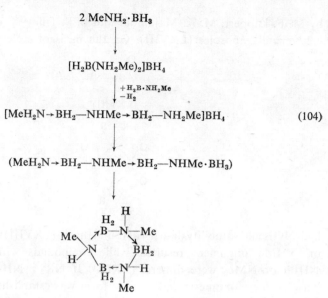

$$2\ MeNH_2 \cdot BH_3$$

$$\downarrow$$

$$[H_2B(NH_2Me)_2]BH_4$$

$$\downarrow \begin{array}{l} +H_3B \cdot NH_2Me \\ -H_2 \end{array}$$

$$[MeH_2N \rightarrow BH_2 - NHMe \rightarrow BH_2 - NH_2Me]BH_4 \qquad (104)$$

$$\downarrow$$

$$(MeH_2N \rightarrow BH_2 - NHMe \rightarrow BH_2 - NHMe \cdot BH_3)$$

$$\downarrow$$

tion[32, 33] disproportionation and aminolysis, as well as hydrogen transfer of the type (105),

$$RNH_2 \cdot BH_3 + Me_2NBH_2 \longrightarrow \tfrac{1}{3}[RNHBH_2]_3 + Me_2NH \cdot BH_3 \qquad (105)$$

determine the course of the reactions and (LXVII) is a likely intermediate; other products isolated are $Me_2N(MeHN)BH$, unstable, volatile $(MeHN)_2BH$[33] and $RNH_2 \cdot BH_2NMe_2 \cdot BH_3$.[32] The hydrogen transfer (105) proceeds most likely via monomeric Me_2NBH_2 in a concerted manner.

(R = Pr)
(LXVII)

The new results described above shed light on the polymerization mechanism of the aminoboranes $R_{2-n}H_nNBH_2$. However, the situation with respect to all the other aminoboranes is still far from being completely understood despite the general trends indicated on page 261. In most cases kinetic and thermodynamic data are lacking, and so it is necessary to rely on qualitative observations only.

Further studies on the polymerization of aminoboranes were carried out

by M. F. Lappert, M. K. Majumdar and B. P. Tilley[142] using 2-amino-1,
3,2-benzodiazaboroles (LXVIII). On the basis of molecular weights and

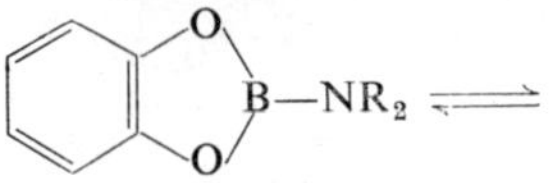

(LXVIII)

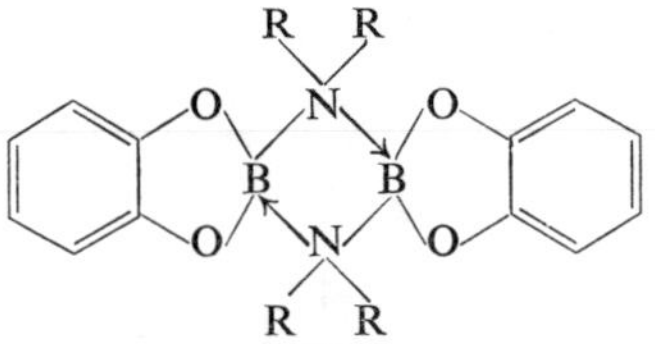

(LXIX)

hydrolysis and solubility data it was found that in (LXVIII) when $NR_2 = NH_2$
or NHMe oligomers resulted while compounds with $NR_2 = NHEt$,
NHPh or NMe_2 were dimeric (LXIX). If $NR_2 = NH\text{-i-Pr}$ equilibrium
between the monomeric and dimeric form was established, and the compound having $NR_2 = N(\text{i-Pr})_2$ was monomeric only. These results demonstrate clearly the importance of steric effects, since in all the compounds
studied the acceptor site was kept unchanged. Evidently the steric factors
are more significant than basicity effects. No association was observed
in boron substituted pyrimidines.[153]

On the other hand, if the donor site is kept unchanged and the acceptor
site altered, the tendency to oligomerization diminishes as the acceptor
properties become weaker. This is apparent in recent studies on boron
azides: $(Cl_2BN_3)_3$ and $(Br_2BN_3)_3$ are known only as the trimers,[247]
Me_2BN_3 in the liquid state exhibits a temperature-dependent degree of
association,[248] and the bis(dialkylamino)boron azides, $(R_2N)_2BN_3$, are
monomeric.[242, 249]

Some quantitative data relating to the equilibrium (106) in the liquid
state have been obtained from ^{11}B n.m.r. studies.[241] Table 4 lists equilibrium constants at 100° and reaction enthalpies for the dissociation of
various aminoboranes according to (106).

$$2\,R_2NBX_2 \rightleftharpoons [R_2NBX_2]_2 \tag{106}$$

The high thermal stability of the dimers in the liquid state, and also some
experimental difficulties involved did not allow an exact study of equilibrium (106) for Me_2NBH_2, $\langle\ \rangle NBH_2$ and Me_2NBF_2. In general, the
dimers are more stable than the monomers by 10–20 kcal/mole, with,
predictably, the least stabilization for compounds having weak acceptor

Table 4. Data Related to the Monomer–Dimer Equilibrium of Aminoboranes

	$K_{100°}$	$-\Delta H°$	$-\Delta S°$
Me_2NBH_2	> 100	—	—
Et_2NBH_2	63	13.2	27.0
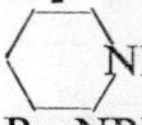 ⬡NBH_2	> 100	—	—
Bu_2NBH_2	5.8	15.4	37.8
Et_2NBF_2	3.6	18.9	48.3
Bu_2NBF_2	0.18	15.9	46.1
Me_2NBCl_2	0.085	10.8	33.9
Me_2NBBr_2	0.01	—	—
Me_2NBHMe	0.78	11.0	30.0
Me_2NBFMe	0.0055	8.8	34.0
$Me_2NBClMe$	0.0035	11.9	43.1
$Me_2NBBrMe$	0.0035	18.3	60.2
H_2NBMe	0.006	10.7	39.2
H_2NBEt_2	0.021	9.4	33.2
$MeHNBMe_2$	0.002	10.7	40.8

sites such as R_2B. In spite of the rather narrow energy range, the equilibrium constants differ by a factor of 10^5. The greatest differences in the $K_{100°}$ values are found when Me_2N- and Et_2N-boranes are compared, and this demonstrates the importance of steric factors. In fact, among the series of dialkylaminoboron difluorides, the only truly monomeric species thus far obtained is the volatile liquid i-Pr_2NBF_2.[216] $(Me_3Si)_2NBF_2$[261] and $(Me_3Si)_2NBCl_2$[92] also are monomers due to the bulky Me_3Si groups and probably also due to SiN π-bonding which decreases the basicity of the nitrogen atom.

The rate of dimerization of aminoboranes, at least qualitatively, conforms approximately with the stability of the dimers. Aminoboranes and aminodifluoroboranes dimerize rapidly, while the dimerization of Me_2NBCl_2 and of Me_2NBBr_2 is a rather slow process.[142]

A different situation holds for the equilibrium (106) in the gas phase. In this case monomeric species predominate and Me_2NBCl_2, Me_2NBBr_2 and Me_2NBI_2[15] are present only as monomers. There is still a monomer–dimer equilibrium even in the gas phase for Me_2NBF_2. The strength of the BN bond in Me_2NBF_2, to which a rather high bond order has to be attributed, is responsible for this. The following infrared data for the BN stretching mode, which is mixed with an NC_2 mode, illustrates this. If the wave numbers are taken as a direct measure of the BN bond strength, the fluoride would have the strongest BN bond. On the other hand, the

	Me_2NBF_2 [cm^{-1}]	Me_2NBCl_2 [cm^{-1}]	Me_2NBBr_2 [cm^{-1}]	Me_2NBI_2 [cm^{-1}]
$\nu^{10}BN$	1595	1548	1543	1538
$\nu^{11}BN$	1562	1527	1517	1508

acceptor properties of Me_2NBF_2 must then be rather high, and this accounts for the case of dimerization.

Structural data for dimeric aminoboranes indicate that the molecules all have a planar B_2N_2 ring system of C_{2h} symmetry.

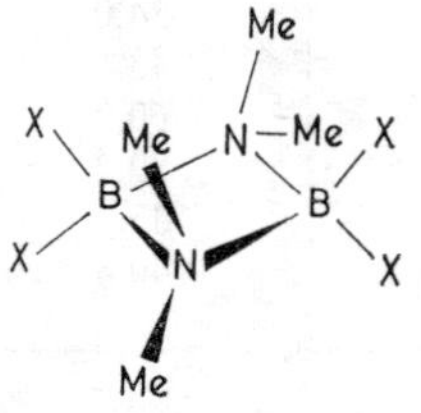

	$B-N^{x)}$	$B-X^{x)}$	$N-C^{x)}$	⊰BNB	⊰NBN	⊰BX$_2$
$X = H^{xx)}$	1.61	1.1	1.47	–	–	109.5°
$X = F^{(110)}$	1.601	1.364	1.469	83.3°	91.8°	113.2°
		1.345				
$X = Cl^{(116)}$	1.591	1.83	1.505	–	–	–

x) in Å-units.
xx) as cited in (295), p. 48.

(LXX)

Similarly Et_2NBF_2 is dimeric in the solid state,[103] the crystals being tetragonal. The dimethylaminoboron halides show long-range proton–boron coupling, and B—H and B—F coupling is also observed[14, 110, 183] The rather short B—B distances in the B_2N_2 rings led Hess[116] to speculate on the possibility of three-center bonding in these compounds. This seems rather unlikely, and another explanation may lie in a reduced electronic repulsion by the boron atoms.

4. Addition Reactions

(a) *Introduction.* The ready oligomerization of many monomeric aminoboranes is attributed to the presence of both electron acceptor and electron donor sites in these molecules. When the acceptor property becomes stronger the aminoboranes not only react with themselves but also add to bases such as amines. Examples relating to R_2NBCl_2 and borazines have been described in Section III, D. Compounds of the type $B(NR_2)_3$ and $(R_2N)_2BCl$ do not add amines because internal coordination (back bonding) and steric effects lower the Lewis acidity considerably. The fact that $B(NHMe)_3$ readily adds $NHMe^-$ to form $B(NHMe)_4^-$ demonstrates again the importance of steric factors.[239]

While the acceptor properties of boron–nitrogen systems have recieved considerable attention, it is only within the last few years that they have been recognized as potential ligands. In principle, two types of donor interaction with an acceptor molecule, seem possible (i) simply through the nitrogen lone pair and (ii) by π-donation of a π-B—N electron pair. At present it seems that both types of interaction do, in fact, occur, and so it is necessary to include results obtained using borazoles and other BN heterocycles.

(b) *σ-Donor–acceptor interaction of aminoboranes.* On the basis of resonance, steric and inductive effects, a decreasing Lewis basicity would be expected in the aminoborane series $(R_2N)_3B$, $(R_2N)_2BX$, R_2NBX_2. Hence $(Me_2N)_3B$ should be the most favorable aminoborane with which to form adducts with various acids MX_n. The adducts will, of course, only be stable if ligand exchange is prevented, and this should be the case if the M—N bond is weaker than the B—N bond.

Recently a number of workers have found that aminoboranes do indeed act as σ-donors.[87, 136, 146, 231, 271-2] The reactions of $B(NMe_2)_3$ with $AlCl_3$[272] $AlBr_3$,[272] $GaCl_3$,[136] $GaBr_3$,[136] $SnCl_4$,[87, 146, 271] $TiCl_4$, $ZrCl_4$, $VOCl_3$,[146, 331] $ZnCl_2$,[271] $CdCl_2$,[271] and $HgCl_2$[146, 271, 231] yielded only 1:1 adducts. Only with $FeCl_3$ was a 1:2 adduct, $B(NMe_2)_3 \cdot 2\,FeCl_3$,[271] obtained. Treatment of $B_2(NMe_2)_4$ with $AlCl_3$ in ether or with $SnCl_4$ in petroleum ether precipitated $B_2(NMe_2)_4 \cdot 2\,AlCl_3$ and $B_2(NMe_2)_4 \cdot SnCl_4$.[272]

The various above-mentioned metal halides also gave 1:1 adducts with $(Me_2N)_2BCl$,[271] $(Me_2N)_2BPh$,[146, 271] $(i\text{-}Pr_2N)_2BPh$,[146] $(Me_2N)_2BMe$[136] and Me_2NBMe_2.[136] No adduct was isolated from Et_2NBPh_2[271] or Me_2NBCl_2[146] and $SnCl_4$, thus demonstrating the decreased basicity of the monoaminoboranes. No reaction was observed between $B(NMe_2)_3$ or $(Me_2N)_2BCl$ and $Cr(CO)_6$, $Mo(CO)_6$ or $Ni(CO)_4$.[146]

Certain borazines also can act as σ-donors; the first reported examples being $(MeBNMe)_3 \cdot TiCl_4$, $(MeBNMe)_3 \cdot 3\,TiCl_4$ and $(MeBNMe)_3 \cdot AlBr_3$.[231] While $(MeBNMe)_3 \cdot 3\,TiCl_4$ dissociates completely in benzene solution, as shown by ^1H n.m.r. spectra and cryoscopic work, the $AlBr_3$ complex is stable in solution as also is $(MeBNMe)_3 \cdot TiCl_4$. $SnCl_4$ reacted with $(MeBNMe)_3$ to give the 1:1 adduct[231] but did not react with $(ClBNH)_3$.[146] $(HBNMe)_3$ and $SnCl_4$ gave no stable adduct, but reacted, however, to give $Me_3N_3B_3H_2Cl$ (20 per cent yield), $Me_3N_3B_3HCl_2$ (60 per cent) and $Me_3N_3B_3Cl_3$ (20 per cent).[6] By-products from the reaction were HCl and $SnCl_2$, and part of the HCl produced added to the $(HBNMe)_3$. At a temperature of 100°, $SnBr_4$ reacted with $(HBNMe)_3$ to yield $Me_3N_3B_3H_2Br$ (10 per cent), $Me_3N_3B_3HBr_2$ (80 per cent) and $Me_3N_3B_3Br_3$ (10 per cent). SnI_4 proved to be unreactive towards the borazine even at

$155°.$[6] These results indicate the decrease in Lewis acidity from $SnCl_4$ to SnI_4 and the products can be explained in terms of an addition-elimination process. No complex formation was observed between $(HBNMe)_3$ and $HgCl_2$.[6]

The structures of most of these aminoborane complexes are not yet fully established. In many cases a $30–100$ cm^{-1} increase in ν_{BN} stretch in the infrared occurs on going from free to complexed borazine. This is attributed to a strengthening of one BN bond due to the coordination of one or two of the other Me_2N groups. While Nöth et al.[231] postulated a six-coordinate titanium atom for $B(NMe_2)_3 \cdot TiCl_4$, Lappert and Srivastava[146] concluded on the basis of i.r.-evidence five-coordinate tin for $(Me_2N)_2BCl \cdot SnCl_4$. Products like $Me_2NBMe_2 \cdot MX_3$ may have either structure (LXXI), (LXXII) or (LXXIII) and the evidence available at present favors an equilibrium between (LXXI) and (LXXIII).[136]

$$Me_2B—NMe_2 \rightarrow MX_3$$
$$\text{(LXXI)}$$

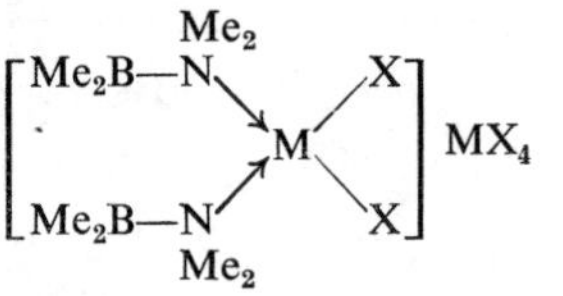

$$\begin{bmatrix} Me_2B—N \overset{Me_2}{\diagdown} & & X \\ & M & \\ Me_2B—N \diagup & & X \end{bmatrix} MX_4$$
$$Me_2$$
$$\text{(LXXII)}$$

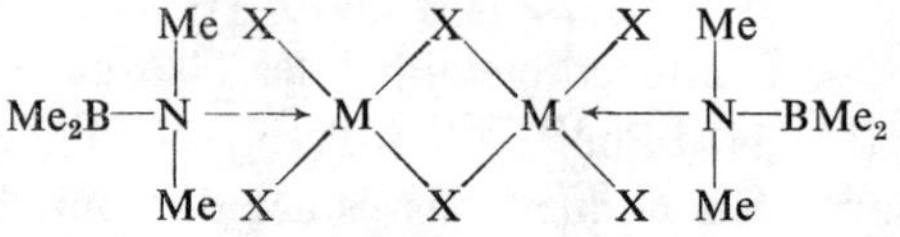

$$\text{(LXXIII)}$$

Thermal decomposition of the adducts proceeds at temperatures $> 150°$, and the course of the reactions are sometimes curious. In many cases no simple dissociation or ligand exchange occurs but rather an extensive decomposition takes place. As an example, $PhB(NEt_2)_2 \cdot AlBr_3$ on pyrolysis yields a rather large amount of $PhBH(NEt_2)$ besides some $PhBBr(NEt_2)$.[87]

(c) *π-Donor–acceptor interaction of aminoboranes.* That aminoboranes or the borazines donate electrons by their π-system is demonstrated by a number of charge transfer complexes recently described. It had first been shown by K. A. Muskat and B. Kirson[198] that phenylborazines give $1:1$ charge transfer complexes with tetracyanoethylene, but it appears that in these cases the charge transfer is not due solely to the borazine system.

Since the charge transfer behavior of olefins with iodine is well known, a comparison with some isoelectronic aminoboranes should differentiate between an olefin-like or amine-like behavior of the aminoboranes.

A study of the reactions of $C_5H_{10}NBMe_2$, Me_2NBPr_2 and Me_2NBMe_2 with I_2 in CCl_4[75] by the method of continuous variations demonstrated not only the formation of $1:1$ complexes but also the amine-like character of the aminoborane in these adducts.

At present there is no evidence that aminoboranes can be involved in π-complex formation, although there is convincing evidence that borazines do act in this way. P. Prinz and H. Werner[258] successfully prepared the first transition metal–borazine π-complex by treating $(CH_3CN)_3Cr(CO)_3$ with $(MeBNMe)_3$. If the acetonitrile formed according to (107) is continually removed in a vacuum, the sublimable, orange-yellow hexamethylborazine chromium-tricarbonyl is formed. The properties of this com-

$$(MeBNMe)_3 + (CH_3CN)_3Cr(CO)_3 \longrightarrow (MeBNMe)_3Cr(CO)_3 + 3\ CH_3CN \quad (107)$$

pound resemble those of hexamethylbenzenechromiumcarbonyl quite closely, while they differ markedly from those of tris(anilino)chromium-tricarbonyl.[258] Although the planarity of the $(MeBNMe)_3$ ring system in this complex has still to be established, there is little doubt that at least a quasi π-bonded borazine system exists in this compound. Actually, D. A. Brown[31] on the basis of MO calculations has predicted that this compound is stable and is the most stable in the series $(MeNBX)_3 \cdot Cr(CO)_3$ (X = H, Cl, Me).

BN analogs of cyclopentadienyl-metal complexes are also accessible. The anion (LXXIV) is isoelectronic with $C_5H_5^-$. The lithium salt reacts with $FeCl_2$ in a $2:1$ molar ratio and brown crystals of $Fe(B_2N_3Me_2Ph_2)_2$

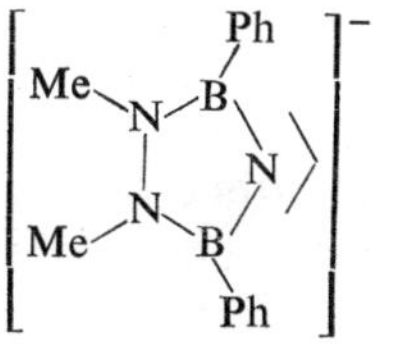

(LXXIV)

can be isolated which are soluble in benzene, rather resistant to hydrolysis and diamagnetic when freshly prepared.[229] These, and some other properties also indicate a BN π-complex analog of ferrocene, although the suggested structure needs confirmation by X-ray work.

These new results indicate some "aromatic" character both in borazines and in the cyclo-1,3,4-triaza-2,5-diborines, and these new compounds may have opened the door to the area of complex transition metal chemistry involving inorganic heterocycles.

5. Substitution Reactions

(a) *Introduction.* The B—N bond in aminoboranes is fairly reactive. M. F. Lappert and M. K. Majumdar[141] have summarized earlier results and added new material which demonstrates again a reactivity order $BCl > BOR > BNR_2$ towards carbonium ion substitution.

Three types of reactions are especially interesting in preparative boron chemistry (i) addition reactions of the aminoboranes (see the preceding paragraph), (ii) insertion reactions, (iii) cleavage reactions. The latter may be divided into two categories, namely those which involve proton transfer (whatever differences there may be in the reaction mechanisms) and exchange of groups (scrambling reactions). The latter have already in part been dealt with in the discussion of the preparation of various aminoboranes.

(b) *Insertion reactions.* Insertion reactions have recently been summarized authoritatively[327] and only a few minor additions are appropriate. The general scheme for insertions (108) also holds for CO_2 and CS_2.[56]

$$B{-}NR_2 + X{=}E{=}Y \longrightarrow B{-}X{-}E(Y){-}NR_2 \qquad (108)$$

CO_2 reacts with a wide range of aminoboranes yielding carbamatolboranes, the only exceptions being the aminodialkylboranes. (Heating Bu_2BNH_2 or Bu_2BNMe_2 with CO_2 to 80–100° gave only low yields of $(Bu_2B)_2O$.) Normally, as many CO_2 groups are inserted as there are B—N bonds present.

The insertion of CS_2 is much more restricted. While $B(NMe_2)_3$ and CO_2 readily give $B(OCONMe_2)_3$, CS_2 does not react at all. The reason for this is attributed to great steric interaction in the transition state which favors CS_2 elimination rather than addition. Indeed, in reaction (109) only tris(dimethylamino)borane is obtained.

$$(Me_2N)_2BCl + NaSC(S)NMe_2 \longrightarrow B(NMe_2)_3 + NaCl + CS_2 \qquad (109)$$

On the other hand, $B(SC(S)NMe_2)_3$ can be prepared from BCl_3 and $NaSC(S)NMe_2$; thus there is no steric problem concerning the stability of the tris(dimethyldithiocarbamoyl)borane.[275] Bisaminoboranes add CS_2 according to (108),[56] while monoaminoboranes do not. This is consistent with the observation that only aminoboranes R_2BNMe_2 are obtained when R_2BCl reacts with $NaSC(S)NMe_2$.[275]

Novel types of B—N compounds, borenylbisamidines, borylamidines and borylguanidines result when carbodiimides are treated with aminoboranes:

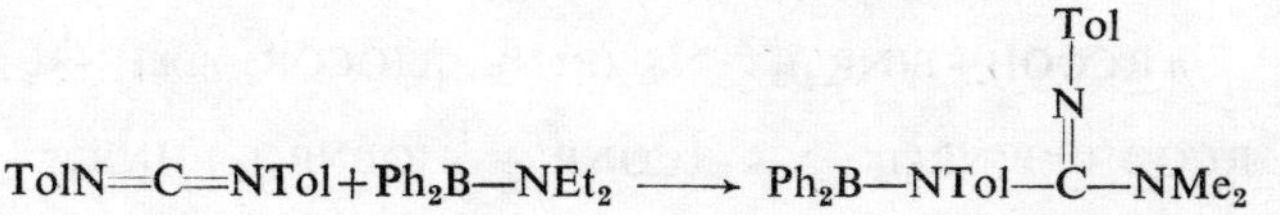

$$\text{TolN}{=}\text{C}{=}\text{NTol} + \text{Ph}_2\text{B}{-}\text{NEt}_2 \longrightarrow \text{Ph}_2\text{B}{-}\text{NTol}{-}\underset{\underset{\displaystyle\text{N}}{|}}{\overset{\text{Tol}}{\underset{|}{\text{C}}}}{-}\text{NMe}_2 \qquad (110)$$

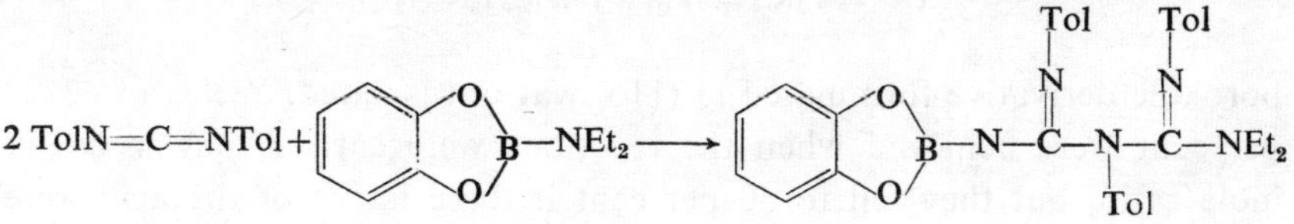

(111)

$$\text{TolN}{=}\text{C}{=}\text{NTol} + (\text{i-Pr}_2\text{N})_2\text{BNHBu} \longrightarrow \text{i-Pr}_2\text{NH} \quad + (\text{LXXV}) \quad (112)$$

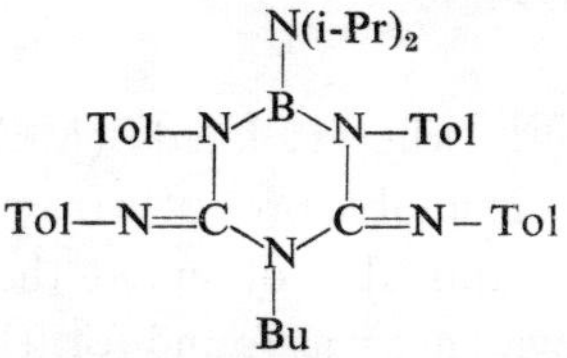

(LXXV)

If amine elimination is possible, derivatives of the new cyclo-1,3,5,2-tri-azaborine (LXXV) are obtained. Again in reactions (110)–(112) steric factors are rate-determining, $B(NEt_2)_3$ and $PhB(NEt_2)_2$ do not react at all, and the B–NHR group is more reactive than the B–NR$_2$ group.[121] Cyclic boron substituted β-ketoenolates are formed in the reaction of aminoboranes with diketene.[119]

(c) *Protolytic reactions.* The BN bond in aminoboranes in the widest sense is quite labile with respect to protolytic attack, the mechanisms of the reactions probably varying considerably. The reaction scheme (113) permits discussion of a wide range of reactions under one general clas-sification.

$$\rangle\text{B}{-}\text{NR}_2 + \text{HY} \longrightarrow \rangle\text{B}{-}\text{Y} + \text{R}_2\text{NH} \qquad (113)$$

Although a great deal is known about the behavior of hydrogen halides towards the aminoboranes (see Section III, D) the action of carboxylic acids on aminoboranes has only recently been investigated closely. P. Nelson and A. Pelter[201] found that an exothermic reaction took place and the acid amide was produced slowly at room temperature.

Thus, instead of (114) the reaction (115) was observed. The

$$n\,RCOOH + B(NR_2')_3 \longrightarrow (R_2'N)_{3-n}B(OCOR)_n + n\,HNR_2' \qquad (114)$$

$$RCOOH + B(NR_2')_3 \longrightarrow RCONR_2' + \frac{1}{3}[OBNR_2']_3 + HNR_2' \qquad (115)$$

$$(R' = PhCH_2,\ Ph,\ t\text{-}Bu,\ CH_3(CH_2)_4)$$

boroxole derivative formulated in (115) was not isolated. Yields of 62–87 per cent were achieved when the reactions were carried out in a 1:1 mole ratio, but they fell to 30 per cent if three moles of the acid were used. The postulated mechanism (116)[201] finds partial support in the

$$RCOOH + B(NR_2')_3 \longrightarrow [(R_2'N)_2BNHR_2']^+RCOO^- \longrightarrow R_2'NH + RCOOB(NR_2')_2$$

$$\xrightarrow{\ H^+\ } RCONHR_2'^+ + R_2'NBO + R_2'NH \longrightarrow R_2'NH_2^+ + RCONR_2' + R_2'NBO$$

$$(116)$$

results of Schweizer,[275] who isolated $(MeC(O)O)_3B \cdot NHMe_2$ and $(PhC(O)O)_3B \cdot NHMe_2$ as insoluble products from the reaction of RCOOH with $B(NMe_2)_3$ in ether, and also observed the formation of $HNMe_2$. On the other hand, no evidence was found for $[(R_2'N)_2BNHR_2']^+RCOO^-$, since $(RC(O)O)_3B \cdot NHMe_2$ was always obtained irrespective of the proportions of starting materials used.

Alcoholysis of aminoboranes provides an excellent route to boric acid esters, although these compounds can usually be prepared in a less expensive way. The example (117) shows that the i-Pr$_2$N—B bond is more readily attacked than the B—NCS bond.[144] This is probably due to the bulky nature of the leaving group, as compared with reaction (118).

$$(i\text{-}Pr_2N)_2B{-}NCS + ROH \longrightarrow i\text{-}Pr_2N(RO)B{-}NCS + i\text{-}Pr_2NH \qquad (117)$$

$$(Et_2N)_2BNCX + 2\,BuOH \longrightarrow Et_2NB(OBu)_2 + Et_2NH \cdot HNCX \qquad (118)$$

$$(X = O,\ S)$$

Reaction (119), which affords ready access to NN-dialkylhydroxylamino-boranes, has some preparative value.[216] The compounds $(Me_2N)_2BONEt_2$ and $Me_2NB(ONEt_2)_2$ cannot be isolated and the equilibrium (120) lies far to the left.

$$R_{3-n}B(NMe_2)_n + n\,HONEt_2 \longrightarrow R_{3-n}B(ONEt_2)_n + n\,Me_2NH \qquad (119)$$

$$(Me_2N)_3B + B(ONEt_2)_3 \rightleftharpoons (Me_2N)_2BONEt_2 + Me_2NB(ONEt_2)_2 \qquad (120)$$

This situation is reminiscent of the $B(OMe)_3/B(NMe)_3$ system, where neither $Me_2NB(OMe)_2$ nor $(Me_2N)_2BOMe$ can be isolated although their presence is readily detected by ^{11}B n.m.r. spectroscopy.[239] It should be noted in this connection that in many displacement reactions a rather rapid exchange of groups occurs and the position of equilibrium is shifted to the side of the most volatile compound under the condition of distillation. Clearly the results of such experiments cannot give values for equilibrium constants.

As has already been mentioned in Section IV, B, 2 (d), mercaptoboranes may be used as starting materials for preparing BN compounds. This is due to the fact that the B—S bond energy is less than the B—N bond energy. Another example of such a synthesis is given by (121).[176]

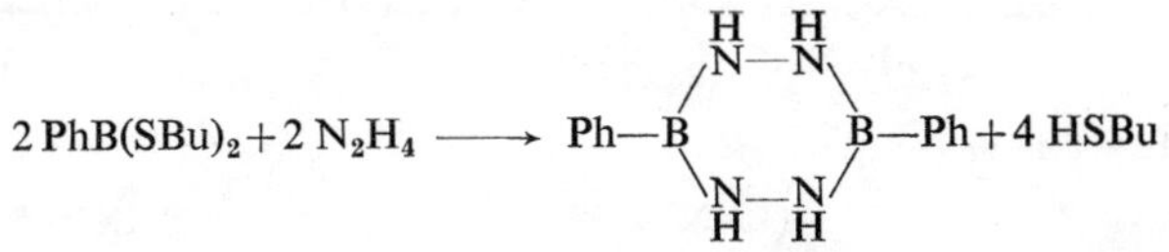

$$
2\,PhB(SBu)_2 + 2\,N_2H_4 \longrightarrow Ph-B \underset{\underset{H\ \ H}{N-N}}{\overset{\overset{H\ \ H}{N-N}}{\Big\langle \Big\rangle}} B-Ph + 4\,HSBu \qquad (121)
$$

Reactions of this type also can be reversed, and the aminoboranes serve as a means by which to establish boron–sulfur bonds by "protolytic" processes. Reactions (122)–(124)[208] represent some illustrative examples, ring formation favoring the reactions.

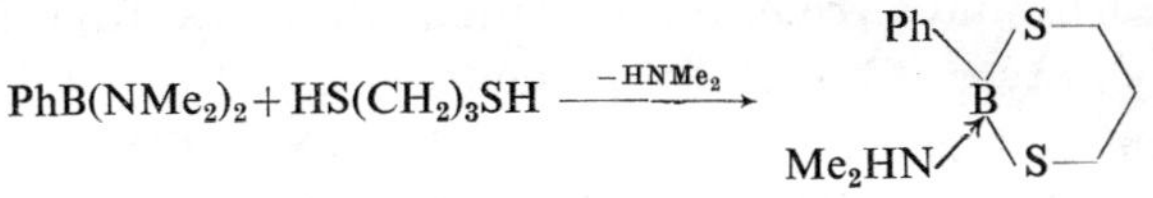

$$
PhB(NMe_2)_2 + HS(CH_2)_3SH \xrightarrow{\ -HNMe_2\ } \qquad (122)
$$

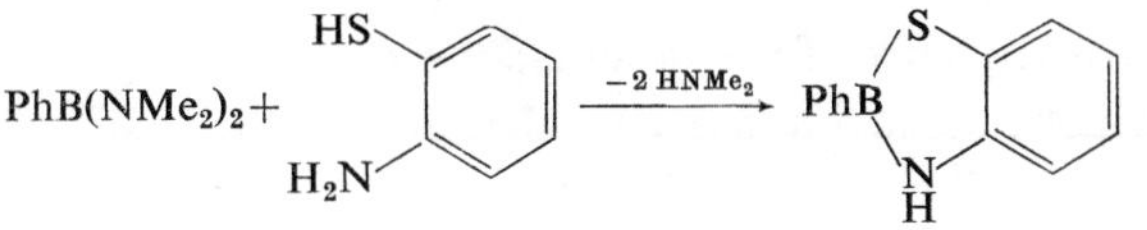

$$
PhB(NMe_2)_2 + \ \underset{H_2N}{\overset{HS}{\bigcirc}} \xrightarrow{\ -2\,HNMe_2\ } \qquad (123)
$$

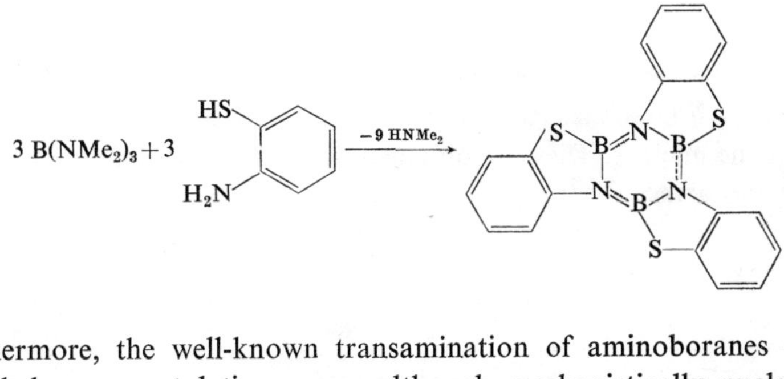

$$
3\,B(NMe_2)_3 + 3\ \underset{H_2N}{\overset{HS}{\bigcirc}} \xrightarrow{\ -9\,HNMe_2\ } \qquad (124)
$$

Furthermore, the well-known transamination of aminoboranes can be regarded as a protolytic process although mechanistically nucleophilic attack by the incoming amine is rate determining and not the proton transfer. The heterocycles (LXXVI–LXXXII) are among those new compounds recently prepared via transaminations.

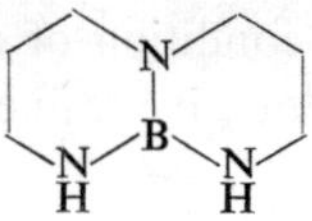

(LXXVI)[80] 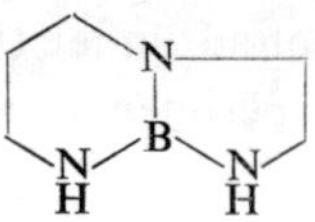   (LXXVII)[212] 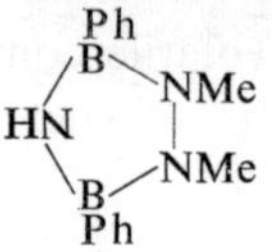   (LXXVIII)[211] 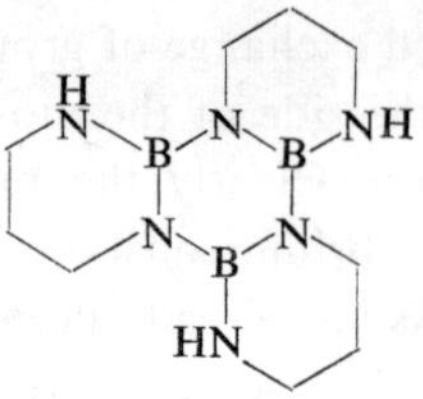   (LXXIX)[209]

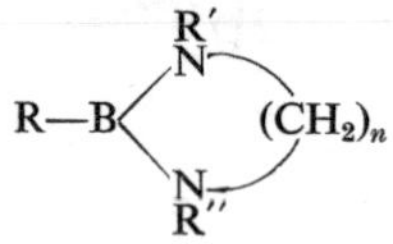

n = 2,3,4   R = Me,H

(LXXX)[209]   (LXXXI)[209] 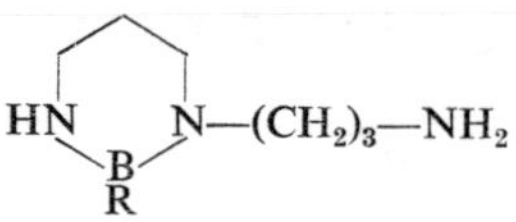   (LXXXII)[209]

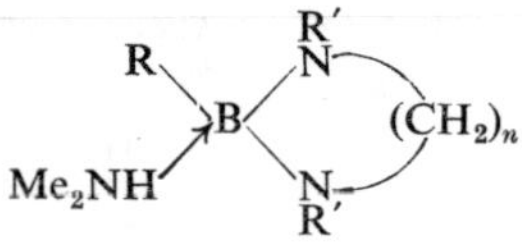

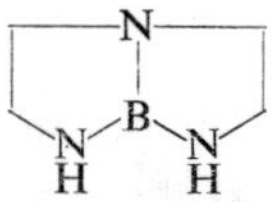

(LXXXIII)[212]

The intermediates involved in the formation of (LXXX) from $RB(NMe_2)_2$ are probably of type (LXXXI) (or its tautomer), and if $n > 4$, the formation of polymers dominates. The compound (LXXXIII)[212] was not obtained from $B(NMe_2)_3$ and $HN(CH_2CH_2NH_2)_2$; only polymers resulted. This is certainly a consequence of the sp^2 character of the nitrogen atom in aminoboranes, which would impose considerable strain in (LXXXIII). The ring size has little effect on the BN bond strength in the diazaborolidines and diazaborolanes[315] which have been extensively studied and reviewed.[203] If unsaturated diamines are used to transaminate $RB(NMe_2)_2$, then rather low yields are obtained even in the most favorable case (LXXXIV)[315] Diamines can also be used to prepare polymers containing the borazine nucleus. They are obtained by treating diamines with B-dialkylaminoborazines.[306-7]

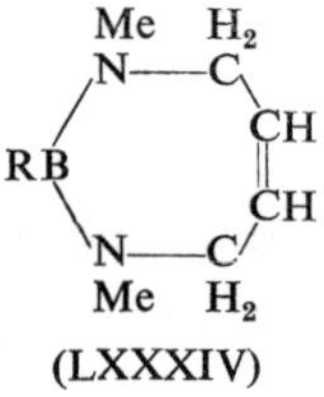

(LXXXIV)

Some very interesting reactions of (LXXVI) have been reported such as ring expansion by means of PhNCO or PhNCS (these are similar to the

(LXXXV)

preparation of boron substituted ureas and thioureas from aminoboran-es [28, 327]) and substitution reactions on the NH bonds[82, 217] affording, for instance, the system (LXXXV).[217] Borazines also may be transamina-ted, although as shown in (125), only adduct formation and proton transfer occur and no amine elimination resulted in this specific case.[137]

$$(HBNH)_3 + 3\ PhNH_2 \longrightarrow (HBNH)_3 \cdot 3\ PhNH_2 \longrightarrow 3\ HB\begin{smallmatrix}NH_2\\\\NHPh\end{smallmatrix} \quad (125)$$

Borazines, on the other hand, are formed by heating bis(primary amino)boranes and although it was reported that $PhB(NH\text{-}i\text{-}Pr)_2$ gave rise to polymeric materials it has now been shown conclusively that the only product is indeed the borazine $(PhBN\text{-}i\text{-}Pr)_3$.[276]

It is clear that instead of organic amines and amides, inorganic amides can be used in transamination reactions. No reaction was observed be-tween S_7NH or $S_4N_4H_4$ and aminoboranes, and action of $SO_2(NH_2)_2$ or $HN(SO_2NH_2)_2$ led to unidentified BO-containing materials.[208] On the other hand, the interaction of Et_2BNEt_2 with Et_3SiNH_2 under reflux yield-ed $Et_2BNHSiEt_3$ and $(Et_3SiNH)_2BEt$.[205] Similarly Ph_3SiNH_2 and $PhB(NMe_2)_2$ gave $PhB(NHSiPh_3)_2$.[122] An amino group exchange is also to be expected and this should be observed under milder reaction condi-tions than those employed to bring about the transamination.

The conversion of a B—N bond of an aminoborane to a B—P or B—As bond by a suitable phosphine or arsine has not yet been achieved; whether ring formation would be helpful in this respect remains to be investigated.

In conclusion, it is to be expected that many more new types of boron compounds will be accessible via "protolytic reactions". A final example in this field is the formation of $PhC\!\equiv\!CB(NMe_2)_2$ from $B(NMe_2)_3$ and phenyl-acetylene.[275]

(d) *Cleavage reactions.* Treatment of trisaminoboranes and bisamino-boranes with other borane derivatives yields "mixed products" $B(NR_2)XY$ by group exchange. These reactions, of course, involve B—N bond break-ing and making which proceed most likely through a transition state involving a tetracoordinate boron atom. A number of reactions of this type have already been discussed and need not be repeated here (see Section IV, B, 2(f)).

H. K. Hofmeister and J. R. van Wazer[118] investigated the system $BPh_3/B(NMe_2)_3$ at 200° and found Ph_2BNMe_2 to be about 100 times more stable towards disproportionation than $PhB(NMe_2)_2$, the former being more stable by 4.3 kcal/mole than the latter. A similar exchange of groups is observed in reaction (126), which is determined mainly by volatility factors, so that nothing is known about the actual position of the equilibrium,[181] the same being true for a OR/SR' exchange.

$$2 Pr_2BNHPh + C_6H_{13}B(OMe)_2 \longrightarrow 2 Pr_2BOMe + C_6H_{13}B(NHPh)_2 \qquad (126)$$

An important process for the preparation of organoboron halides uses aminoboranes as the starting material. Equations (127) and (128) represent two methods which were applied for synthesizing methyl- and ethylboron fluorides, but other organoboron halides can also be prepared[240] by this route.

$$R_2B—NR'_2 + BF_3 \longrightarrow R_2BF + R'_2NBF_2 \qquad (127)$$

$$RB(NR'_2)_2 + 2 BF_3 \longrightarrow RBF_2 + 2 R'_2NBF_2 \qquad (128)$$

In the case of the fluorides the driving force in these reactions is almost certainly the dimerization of the aminoboron difluoride. These do react at $> 100°$ with $B(NR_2)_3$ to give $(R_2N)_2BF$,[238] and the high reaction temperature suggests that it is the monomeric species that is involved in this process.

Contrary to earlier statements it has now been found that reaction (129) proceeds irreversibly, while (130) is reversible at 83–127° with $\Delta G° = 4.11$ kcalmole.[36]

$$B(NMe_2)_3 + Me_2NBH_2 \longrightarrow 2 (Me_2N)_2BH \qquad (129)$$

$$(Me_2N)_2BH + Me_3N \cdot BH_3 \longrightarrow 2 Me_2NBH_2 + NMe_3 \qquad (130)$$

The irreversibility of (129) was attributed to the higher base strength of $B(NMe_2)_3$ compared with $(Me_2N)_2BH$. This does not seem a particularly convincing argument, and perhaps a more likely explanation is that there is better resonance stabilization and less steric hindrance in $(Me_2N)_2BH$ as compared with $(Me_2N)_3B$.

Aminoboranes are useful reagents in enamine syntheses.[201] They react readily with ketones, 1,3-diketones, and β-ketoesters:

$$R-CO-CH_2R' + B(NR_2)_3 \longrightarrow R-\underset{\underset{NR_2}{|}}{C}=CH-R' + HNR_2 + OBNR_2 \quad (131)$$

$$CH_3-CO-CH_2-CO-CH_3 + B(NR_2)_3 \longrightarrow CH_3-CO-CH=\underset{\underset{NR_2}{|}}{C}-CH_3 \quad (132)$$

$$+ HNR_2 + OBNR_2$$

$$CH_3-CO-CH_2COOR + B(NR_2)_3 \longrightarrow CH_3-\underset{\underset{NR_2}{|}}{C}=CH-COOR + HNR_2 + OBNR_2$$

$$(133)$$

6. Stabilization by the B—N Bond

The low electron density at the boron atom of aminoboranes is largely compensated by back bonding from the nitrogen lone-pair. Furthermore the boron atom in a dialkylaminoborane is subject to steric shielding by the alkyl groups, and this effect becomes more important the greater the number of R_2N groups bonded to the B atom.

The combination of these two factors permits the rationalization of a number of effects, which may be termed stabilization by the B—N bond. This stabilizing influence has three consequences (i) suppression of poly-merization (ii) suppression of exchange processes and (iii) stabilization of "unusual" bonds.

Factor (i), the suppression of polymerization of aminoboranes to dimers, trimers or polymers, is largely dependent upon the size of the alkyl groups attached to the nitrogen. However, electronic factors are also important, even though these effects are often outweighed by the oligomerization energy in many R_2NBX_2 compounds. These factors have been discussed in Section IV, B, 3. Thus the non-dimerization of $(RHN)_2BY$ is due to elec-tronic effects. The compounds $(R_2N)_2BH$, $(RHN)_2BH$, and $(R_2N)_2BF^{[238]}$ are only monomeric. The azides $R_2NB(N_3)_2$ and $(R_2N)_2BN_3$ (R = Me, Et)$^{[249]}$ are also monomeric, and, furthermore, they are subject to the stabi-lization factors (ii) and (iii) in that they are quite thermally stable (up to 200°) and do not rearrange to $B(NR_2)_3$ and $B(N_3)_3$.

The stabilizing influence (ii) is exhibited by the compound $Me_2NBBrCl$,$^{[102]}$ which is prepared from Me_2NBCl_2 and Me_2NBBr_2 at > 100° (134).

$$Me_2NBBr_2 + Me_2NBCl_2 \longrightarrow 2 Me_2NBBrCl \quad (134)$$

Only after the mixed halide has remained for 1 week at room temperature do the first crystals of $[Me_2NBCl_2]_2$ appear. Although mixtures of various boron halides equilibrate rapidly, the redistribution of $Me_2NBBrCl$ is a slow process. The slow rate of this reaction can be readily explained if it is assumed that the transition state (LXXXVI) is the preferred one. The

(LXXXVI)

rather rapid equilibration of Et_2NBCl_2 with Et_2NBBr_2,[102] which precludes the isolation of pure $Et_2NBBrCl$, can also be understood in terms of the nature of the transition state. If it is assumed that the bulky Et_2N group leads to a transition state similar to (LXXXVI) but with one R_2N group replaced by X, i.e. with a BNBX rather than a BNBN ring system, then ready halogen exchange is to be expected.

The suppression of polymerization due to the presence of a B—N bond is also applicable to cyanoboranes.[27] Reaction of $(Me_2N)_2BCl$ with AgCN yields the feebly associated $(Me_2N)_2BCN$ which shows no acceptor properties towards NMe_3. While $Et_2NB(CN)_2$ is monomeric, Me_2N—$B(CN)_2$ is dimeric. The latter compound is best obtained from Me_3SiCN and Me_2NBCl_2 by decomposition of the primary product $Me_3SiCN \cdot B(NMe_2)(CN)_2$ at 100° *in vacuo*. The stability towards association of these cyanoboranes is therefore due to low Lewis acidity. Other stable B—CN compounds are the *B*-cyanoborazines, which have been prepared from a number of *B*-chloroborazines and AgCN in acetonitrile[106] or benzene[107], and compounds containing tetracoordinated boron atoms, for example the above mentioned adducts of $Me_2NB(CN)_2$ with Me_3SiCN.[27]

For a long time only trimeric, tetrameric or polymeric phosphinoboranes were known,[295] and Ph_2BPPh_2, which had been claimed to be monomeric,[328] proved to be a dimer.[216] The first phosphinoboranes shown to be monomeric by molecular weight measurements[235-6] and by ^{11}B n.m.r.[239] data include the following: $(Me_2N)_2BPEt_2$, $Et_2NB(PEt_2)_2$, $Et_2NB(Cl)PEt_2$, $(Et_2P)Et_2NB$—$BNEt_2(PEt_2)$, $Et_2NB(Bu)PEt_2$, $Me_2NB(Bu)PEt_2$, $PhP[BNEt_2(PEt_2)]_2$ and $PhP[B(NMe_2)_2]_2$.[235-36]

It can be seen that the presence of only one B—N bond is sufficient to completely inhibit the polymerization of these phosphinoboranes. However, in the case of $R_2NB(R')PR_2$ and especially for $R_2NB(H)PR_2$, there is a tendency to disproportionate to $(R_2N)_2BR'$ and polymeric $(R_2P)_2BR'$ or $(R_2P)_2BH$.[236] In a sense this disproportionation is a violation of factor (ii), the suppression of exchange due to the preference for a transition state of type (LXXXVI). However, it is unreasonable to suppose that this preference for a BNBN ring in the transition state will always be shown. Clearly it depends upon the size and nature of X and Y as to how effectively these groups compete with the R_2N group for a ring position in the reaction intermediate. If X or Y is small, for instance H, then there is a fairly high probability of the R_2P group replacing one R_2N group in the ring system in (LXXXVI), and this can then lead to the observed disproportionation. Thus steric factors can favor the exchange even though the presence of the B—N linkage tends to stabilize the molecule. All other known phosphinoboranes, except perhaps the poorly characterized Me_2BPH_2, are polymeric and this emphasizes the importance of effects (i) and (ii).

The stabilizing influence (iii) of the B—N bond can also be used to prepare compounds containing boron directly bonded to elements of group IV. Among the organoboranes, vinylboranes are much more reactive than alkylboranes, and alkynylboranes are even less stable. For instance, $B(C\equiv CH)_3$ decomposes above $-30°$[46] and compounds of the type $RC\equiv CBBu_2$ decompose spontaneously at room temperature.[46] In marked contrast the alkynylbis(amino)boranes, $RC\equiv CB(NR_2')_2$ $\left(R=Ph,\right.$ $CH_2=\overset{|}{C}-CH_3,$ ⬡—, Me and $R'=Me, Et\left.\right)$ are distillable liquids[289] readily prepared according to (135).

$$RC\equiv C-Na + ClB(NR_2')_2 \xrightarrow{-60°} NaCl + RC\equiv CB(NR_2')_2 \qquad (135)$$

The compound $RC\equiv CB(OR)_2$ also can be synthesized, the strong B—O bond showing a stabilizing effect similar to that due to the R_2N group.

The instability of free alkynylboranes seems to be related to a Lewis acid induced polymerization of the alkynyl part of the molecule and this might possibly lead to organocarboranes.

The instability of trifluoromethyl- and perfluoroalkylboranes, which rather quickly evolve BF_3, is of another origin; it is due to F-abstraction by the highly electrophilic boron atom in these types of compounds. In contrast to this, perfluorophenylboranes can be prepared rather readily and a variety of types are known.[49] The B—C bond is rapidly cleaved by the action of base, as would be expected for a strongly polarized bond. The only compound containing the CF_3 group which has been known for some time is $Me_3Sn[CF_3BF_3]$, where stabilization is achieved by tetracoordination of the boron atom. More recently, CF_3BF_2 was found to be produced in low yield, according to (136).

$$\text{Bu}_2\text{BK}\cdot\text{NEt}_3{}^\dagger+\text{CF}_3\text{I} \xrightarrow{-\text{KI}} \text{Bu}_2\text{BCF}_3\cdot\text{NEt}_3 \xrightarrow[-\text{NEt}_3,\cdot\text{HCl}]{\text{HCl}} \text{Bu}_2\text{BCF}_3$$

$$\xrightarrow{\text{BF}_3} \text{CF}_3\text{BF}_2+\text{Bu}_2\text{BF}+\text{BuBF}_2 \qquad (136)$$

† The exact nature of Bu_2BK is still unresolved.

The compound decomposes rapidly into BF_3 (F-abstraction) and non-volatile materials.[253] The preparation of trifluoromethylboranes from CF_3HgI and BBr_3 was unsuccessful.[253] Bu_2BCF_3 itself is also unstable, but coordination with amines prevents decomposition.

The compound $\text{B(CF}{=}\text{CF}_2)_3$ is also of rather low stability decomposing at 100° to give $\text{CF}_2{=}\text{CFBF}_2$ and BF_3.[294] However, if the $\text{CF}_2{=}\text{CF}$ group is attached to a borazine nucleus thermal stability is greatly enhanced. Thus, the compounds (LXXXVII) and $(\text{MeNBCF}{=}\text{CF}_2)_3$, which are

Me   Me
| |
/B——N\
Me—N   B—CF=CF₂
\B——N/
| |
Me   Me

(LXXXVII)

prepared from $\text{LiCF}{=}\text{CF}_2$ and the appropriate borazine,[128] are stable at 300°. In these two compounds F-abstraction is suppressed due to the low acceptor power of the boron atoms.

There is a formal resemblance between borazines and bis(amino)boranes in that both classes of compounds have a B atom bonded to two nitrogen atoms. This analogy is relevant to the stabilization of bonds between boron and a fluoroalkyl group. For instance, the compound $(\text{Me}_2\text{N})_2\text{BC}_3\text{F}_7$ can be isolated as a stable liquid by treating $(\text{Me}_2\text{N})_2\text{BCl}$ with LiC_3F_7 at $-50°$ or from the action of sodium amalgam upon a mixture of $\text{C}_3\text{F}_7\text{I}$ and $(\text{Me}_2\text{N})_2\text{BBr}$.[50] The $\text{B}-\text{C}_3\text{F}_7$ bond can also be stabilized by means of the benzodioxoborole ring system.[50]

In 1962[224] some simple molecules containing a B—Si bond were described. Since they all contained at least one B—N bond, their stability was attributed to the presence of this bond. Further work in this area[225] has substantiated these findings. Scheme (137) summarizes the preparations and chemical properties of these compounds.

While $\text{Me}_2\text{NB(SiPh}_3)\text{Bu}$ and $\text{Et}_2\text{NB(SiPh}_3)_2$ can be prepared, all attempts to produce $(\text{RO})_2\text{BSiR}_3$, $\text{B(SiR}_3)_3$ or $\text{R}'_2\text{BSiR}_3$ were unsuccessful.[225] The stabilities of silylboranes follow from those of the organosilylboranes. Again either a tetracovalent boron atom or at least one amino group

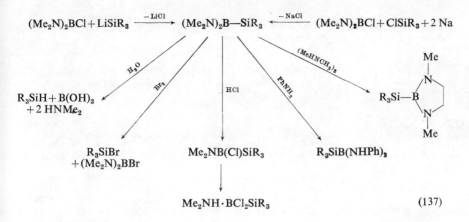

$$(137)$$

attached to the boron atom is necessary to ensure stability. By using H_3SiK and the appropriate chloroborane, the following compounds were obtained: $H_3SiBH_2 \cdot NEt_3$, $H_3SiB(NMe_2)_2$, $H_3SiB(NEt_2)_2$, $H_3SiB(Bu)NMe_2$.[5] Hydride abstraction occurs immediately if the acceptor properties of the boron atom increase, and therefore Bu_2BSiH_3 is unstable.[5]

The thermal stabilities of the amino(organosilyl)boranes are fairly high; many of the compounds can be heated to 150° without decomposition. However, when the B—Si bond is replaced by a B—Sn bond there is a marked reduction in thermal stability implying decreasing bond strength in the series B—C, B—Si, B—Sn, although no quantitative data are available to support this conclusion. Organostannylboranes have only been mentioned briefly in the literature.[223] Their preparation is indicated in (138).

$$(Me_2N)_{3-n}BCl_n + n\,LiSnR_3 \longrightarrow n\,LiCl + (Me_2N)_{3-n}B(SnR_3)_n \qquad (138)$$

Their isolation in a pure state is difficult.[115] Chemically, organostannylboranes behave similarly to the organosilylboranes, except that $(Me_2N)_2BSnR_3$ decomposes readily at temperatures $> 100°$ into $B_2(NMe_2)_4$ and $R_3Sn—SnR_3$.

The presence of B—N bonds in a molecule makes possible the synthesis of organogermylboranes.[115] The co-dehalogenation of $(Me_2N)_2BCl$ and Me_3GeCl with Na/K alloy gave $(Me_2N)_2B—GeMe_3$, b.p. 49–51°/1 mm.

It is widely known that the most readily accessible diboron compound is $B_2(NMe_2)_4$. This is a very versatile compound which can be used to prepare many new diborane(4) derivatives. These compounds are stable if each boron atom carries one R_2N group or if the boron atoms are tetrahedrally coordinated. New reactions carried out with $B_2(NMe_2)_4$ are reviewed in the schematic representation.

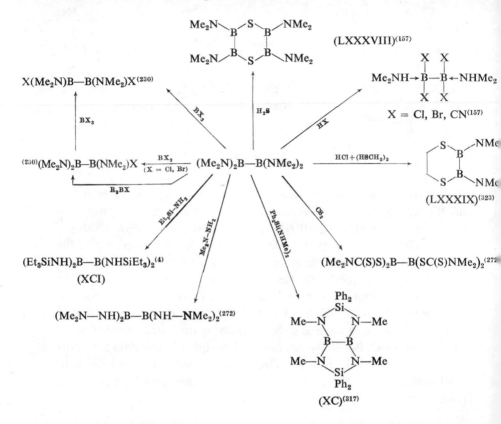

It has been reported that (LXXXIX), a new ring system, is both hydrolytically and thermally very stable.[323] This is surprising if one compares it with (LXXXVIII) or $B_2(NMe_2)_2(SR)_2$ which do not have good hydrolytic stability. High thermal stability up to 280° is also reported for (XCI), which forms in only 20 per cent yield. That no further condensation occurs is most likely due to its rather rigid structure.[4] The compound $Ph_2Si(NHMe)_2$ does not react with $PhB(NMe_2)_2$ to give a four-membered BN_2Si ring system and this was taken as evidence for the five-membered ring structure in (XC). This compound is formed as indicated in 12 per cent yield after 4 days at 120–170°.

One of the most fascinating reactions of $B_2(NMe_2)_4$ was reported by E. P. Schramm.[273] The alkylation of the diboron compound proceeds according to equation (139).

$$3 \text{ B}_2(\text{NMe}_2)_4 + 4 \text{ (AlMe}_3)_2 \longrightarrow \text{Al}_4\text{B}(\text{NMe}_2)_3\text{Me}_6 + 5 \text{ Me}_2\text{BNMe}_2 + 2 \text{ [Me}_2\text{NAlMe}_2]_2$$

$$(139)$$

yielding a yellow, fibrous product of the composition $\text{Al}_4\text{B}(\text{NMe}_2)_3\text{Me}_6$, for which structure (XCII) was suggested on the basis of physical data. This novel structure needs further confirmation, since there seems to be an error in the reported ^1H n.m.r. data which served as partial evidence for structure (XCII).

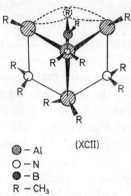

(XCII)

$\bigotimes$ – Al
$\bigcirc$ – N
$\bigodot$ – B
R – CH_3

It is logical to assume that if the great stability of $\text{B}_2(\text{NMe}_2)_4$ is due to favorable electronic and steric shielding of the boron atoms, then a series of dialkylaminopolyboranes should exist containing simple B—B bonds. Thus, indeed, the dehalogenation of $(\text{R}_2\text{N})_2\text{B}—\text{B}(\text{NR}_2)\text{Cl}$ alone, or in the presence of $(\text{R}_2\text{N})_2\text{BCl}$ led to a number of dialkylaminopolyboranes. According to (140) yields up to 40 per cent of $\text{B}_4(\text{NMe}_2)_6$ were achieved.[115]

$$(\text{Me}_2\text{N})_2\text{B}—\text{B}(\text{NMe}_2)\text{Cl} + 2 \text{ Na} + \text{Cl}(\text{Me}_2\text{N})\text{B}—\text{B}(\text{NMe}_2)_2 \longrightarrow 2 \text{ NaCl}$$

$$+ (\text{Me}_2\text{N})_2\text{B}—\text{B}(\text{NMe}_2)—\text{B}(\text{NMe}_2)—\text{B}(\text{NMe}_2)_2 \qquad (140)$$

The members of the series $\text{B}_n(\text{NR}_2)_{n+2}$ are colorless but have a yellow color from $\text{B}_6(\text{NMe}_2)_8$ upwards. Table 5 records physical data for these types of compounds. The ^{11}B n.m.r. signals for these polyboranes are rather broad and centered at -36 and -54 ppm (external $\text{BF}_3 \cdot \text{OEt}_2$).

TABLE 5. BOILING RANGE OF SOME DIALKYLAMINOPOLYBORANES

	b.p./1 mm	n_{20}		b.p./1 mm	n_{20}
$\text{B}_3(\text{NMe}_2)_5$	80–83°	1.4788	$\text{B}_3(\text{NEt}_2)_5$	106°	1.4728
$\text{B}_4(\text{NMe}_2)_6$	98–102°	1.4912	$\text{B}_4(\text{NEt}_2)_6$	125°	1.4780
$\text{B}_5(\text{NMe}_2)_7$	120–125°	1.5062	$\text{B}_3(\text{NMe}_2)_4\text{NEt}_2$	85–87°	—
$\text{B}_6(\text{NMe}_2)_8$	129–135°	1.5121			

19*

The thermal stability of these polyboranes decreases as the B chain length increases. Their chemical behavior resembles $B_2(NMe_2)_4$.[115] It should be pointed out that a dehalogenation procedure applied to B-chloroborazines leads to B, B'-bisborazyls and B,B'-polyborazinyls.[108] If the B—N bond is responsible for stabilization of the B—B, B—Si, and B—Sn bonds, it should be possible to form boron–metal bonds stabilized in this way. This hypothesis led to the synthesis of the first boron–manganese compound $(Me_2N)_2B$—$Mn(CO)_5$,[232, 234] obtained from $(Me_2N)_2BCl$ and $NaMn(CO)_5$. Subsequently various dialkylaminoboron transition metal compounds have been prepared, and it appears that the above mentioned hypothesis needs considerable refinement. A recent summary on boron-transition metal coordination compounds[233] discusses this in more detail than can be given here, especially since this topic is beyond the scope of this article. However it can be noted that if the metal has only few d-electrons available B—N bonds play a role by stabilizing the metal–boron bond. This bond may be quite weak, as is indicated by the observation that the yellow solution obtained by reducing $(Me_2N)_2BBr$ with sodium amalgam in cyclohexane decomposes to $B_2(NMe_2)_4$, $B(NMe_2)_3$ and Hg, indicating the possible existence of $[(Me_2N)_2B]_2Hg$.[50]

C. Unusual Oxidation States of Boron in B–N Compounds

The aminopolyboranes are examples of boron compounds of "unusual" oxidation state. Another type of B–N compound, namely certain boron radicals, also have a formal oxidation state lower than three and stabilization is achieved by electron delocalization.

The first stable boron free radicals were reported by R. Köster et al.[131] These workers isolated and investigated a number of radicals of the type R_2B-py and R_2B-iquin (R = Et, Ph; iquin = isoquinoline) prepared by reducing R_2BCl·py in tetrahydrofuran or diethylether with lithium. The compounds are all deeply colored (see Table 6).

TABLE 6 COLOR OF SOME ORGANOBORON RADICALS

Et_2B·py	Et_2B·quin	Et_2B·iquin	Ph_2B·py
dark green	yellow	red	deep blue

A reported dimesitylboron-pyridine radical[151] was shown to be the radical anion $(mes)_3B^-$, which is stable for a long period in liquid ammonia.[319] S. I. Weissman and H. van Willigen[319] assume that this anionic radical is formed through disproportionation of $(mes)_2B$.

The pyridine stabilized radicals show e.s.r. hyperfine structure in contrast to the compounds $R_2B \cdot N$⟨ ⟩—R′ which give rise to a 10-line spectra, indicating coupling of the electron spin with three boron atoms. The fact that hyperfine splitting shows the interaction of the unpaired electron with three boron atoms suggests that the B atoms are incorporated into the π-system of the heterocycles. These compounds are all associated in ether or benzene, and association decreases from $n = 3$ to $n = 1.5$ in benzene solution in the series R = PhCH$_2$ > Me > Et > Pr > i-Pr > t-Bu. No association was observed for (XCIII). The

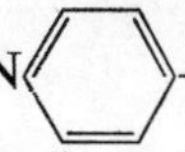

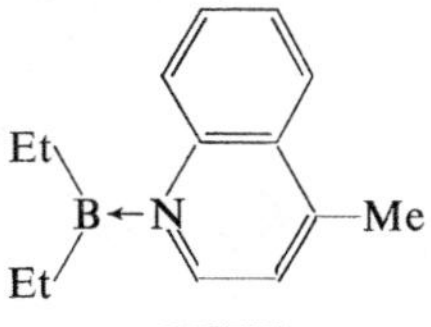

(XCIII)

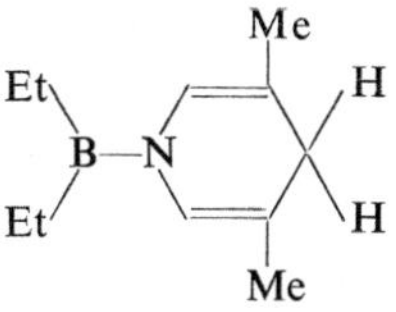

(XCIV)

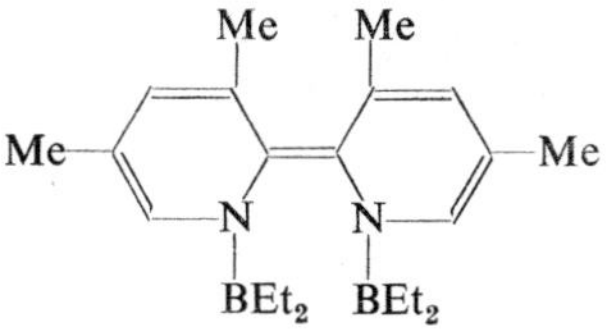

(XCV)

R_2B radicals containing m-substituted dialkylpyridines as ligand are rather unstable and they decompose to (XCIV) (diamagnetic liquids) and to yellow crystalline $N'N$-bis (diethylboryl)-3,5,-3′,5′-tetramethyl-2,2′-di(1,2-dihydropyridine) (XCV).[130] Products analogous to (XCIV) and (XCV) result if Et$_2$BCl·py is dehalogenated in boiling tetrahydrofuran.[130]

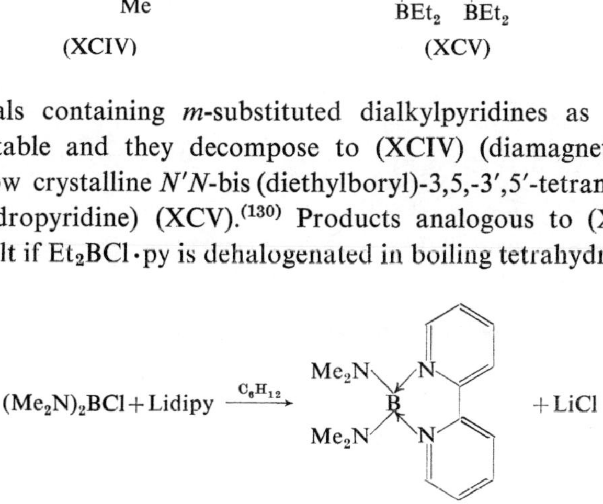

$$(Me_2N)_2BCl + Lidipy \xrightarrow{C_6H_{12}} \quad + LiCl \qquad (141)$$

(XCVI)

Although $Ph_2B \cdot py$ is a stable crystalline compound, the stability of these types of radicals may be further increased if chelating groups are introduced. The reactions (141) and (142) lead to the air-sensitive compounds (XCVI) and (XCVII), the latter being diamagnetic, as expected for a B(I) species.[138]

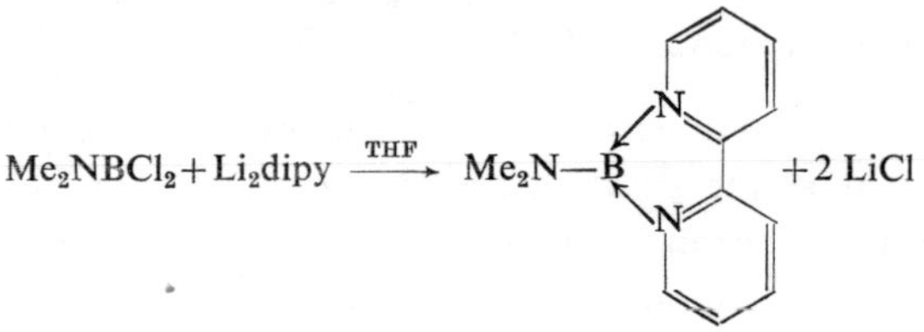

$$Me_2NBCl_2 + Li_2dipy \xrightarrow{THF} Me_2N\text{—}B \qquad +2\,LiCl \qquad (142)$$

(XCVII)

While (XCVII) can be prepared in THF as a solvent, the synthesis of the paramagnetic species (XCVI), (XCVIII) and (XCIX) require a hete-

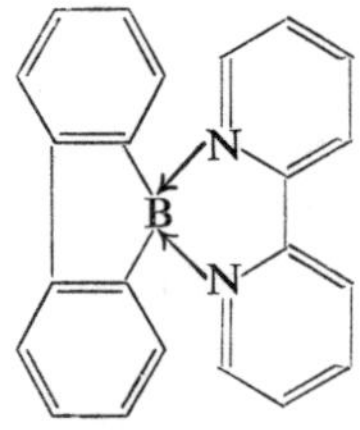

(XCVIII)

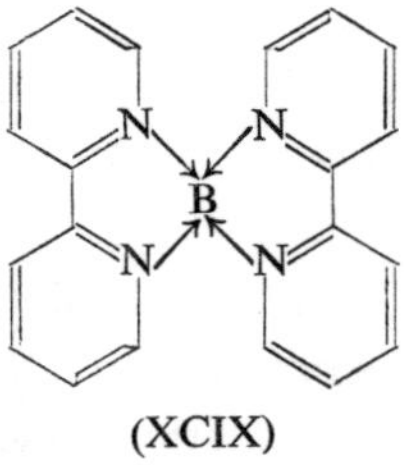

(XCIX)

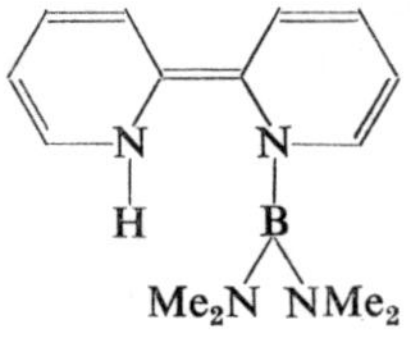

$Me_2N \quad NMe_2$

(C)

rogenous reaction in a hydrocarbon solvent such as cyclohexane. E.s.r. indicates that the unpaired electron in (XCVI) and (XCVIII) couples with the nitrogens and the boron atom, while in (XCIX) four nitrogens and one boron atom are involved. Three coordinated (XCVII) can add one molecule of Me_2NH and most likely THF also; these adducts are colored but diamagnetic and there is evidence that their structure is similar to (C).[138] Therefore, the situation resembles closely that of the pyridine stabilized radicals,[131] since bis-dihydropyridine derivatives are formed. It is, of course, difficult to assign a definite oxidation state to the boron atoms in these compounds; the formal oxidation state is $+II$, $+I$ and 0 in (XCVI), (XCII) and (XCIX) respectively.

D. *Diborylamines*

1. *Introduction*

The term diborylamine is used for compounds containing a B–N–B linkage, although from this point of view, borazines would also belong to this class. However, the name diborylamine should be restricted to compounds with the general formula X_2B—NR—BX_2. It is to be expected that their stability will depend on the nature of X. In resonance formalism diborylamines can be represented by (143), their boron atoms being less shielded electronically by the nitrogen lone-pair than in X_2B—NR_2.

$$X_2B \leftharpoondown NR—BX_2 \longleftrightarrow X_2B—NR \rightleftharpoons BX_2 \tag{143}$$

Therefore the boron atoms should show stronger electron acceptor properties than in comparable aminoboranes.

2. *Preparation*

There are a number of methods for synthesizing diborylamines:

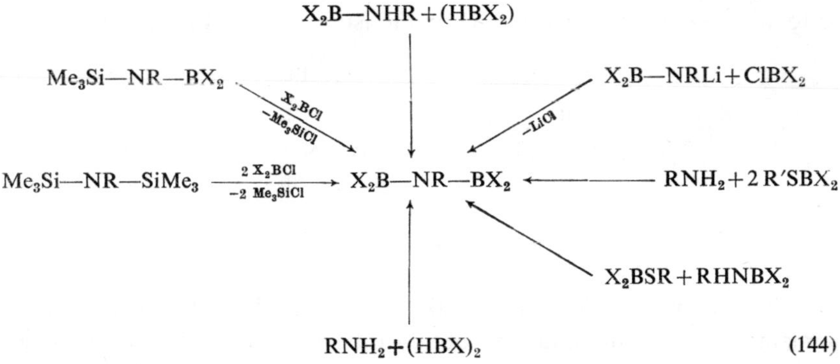

$$X_2B—NHR + (HBX_2)$$

$$Me_3Si—NR—BX_2 \qquad\qquad X_2B—NRLi + ClBX_2$$

$$\text{(} \tfrac{X_2BCl}{-Me_3SiCl} \text{)} \qquad \text{(} -LiCl \text{)}$$

$$Me_3Si—NR—SiMe_3 \xrightarrow[-2\ Me_3SiCl]{2\ X_2BCl} X_2B—NR—BX_2 \longleftarrow RNH_2 + 2\ R'SBX_2$$

$$X_2BSR + RHNBX_2$$

$$RNH_2 + (HBX)_2 \tag{144}$$

The tetraalkyldiborylamines R_2B—NH—BR_2 were prepared via the silazane cleavage reaction,[215] and in the synthesis of Me_2B—NH—BMe_2 the system $(EtMe_2Si)_2NH$ Me_2BBr was used to achieve a good separation of the products.[311] However, this route is not always applicable, for example for the preparation of R_2B—NMe—BR_2, because the reaction between Et_2BCl and Me_3Si—NMe—$SiMe_3$ stops at the Et_2B—NMe—$SiMe_3$ stage.[311]

The SiN cleavage reaction should offer a pathway not only to symmetrical but also to unsymmetrical diborylamines. However, in the following two examples only borazines, instead of the expected diborylamines[311], were obtained.

$$Me_2NBMeNHSiMe_3 + BrBMe_2 \begin{cases} Me_3SiBr + Me_2NBMeNHBMe_2 \\ Me_3SiBr + Me_2NBMe_2 + \frac{1}{3}(MeBNH)_3 \end{cases}$$ (145a)

$$Me_2NBMeCl + Me_3SiNHBMe_2 \begin{cases} Me_3SiCl + Me_2NBMeNHBMe_2 \\ Me_3SiCl + Me_2NBMe_2 + \frac{1}{3}(MeBNH)_3 \end{cases}$$ (145b)

Use of $Et_2BNMeLi$ and $ClBEt_2$ gave a low yield of $Et_2BNMeBEt_2$.[311] R. Köster and K. Iwasaki[133] devised a versatile route to diborylamines, but it seems to be restricted to the reaction of borolanes with either a primary amine or a borolanylamine. If alkylboranes are used instead of the borolanes (146), heterocycles are formed which contain the B–N–B grouping.[133]

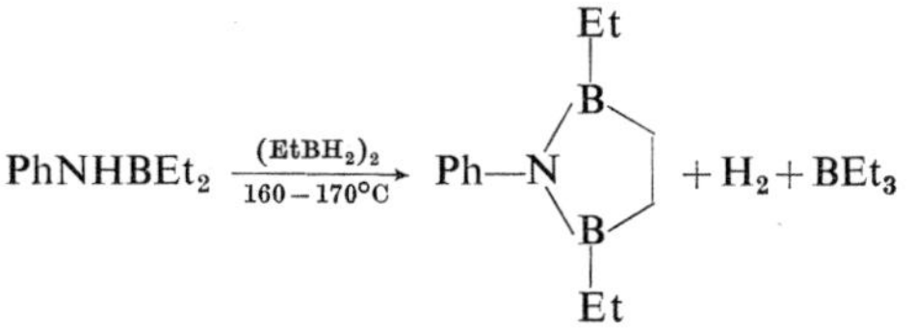

$$PhNHBEt_2 \xrightarrow[160-170°C]{(EtBH_2)_2} Ph—N\overset{\overset{\displaystyle Et}{|}{\underset{\underset{\displaystyle Et}{|}{B}}{\overset{B}{\diagdown}}} + H_2 + BEt_3$$ (146)

When propyl boranes are used as starting materials it is uncertain as to whether the product is (CI) or the isomer (CII). In the presence of

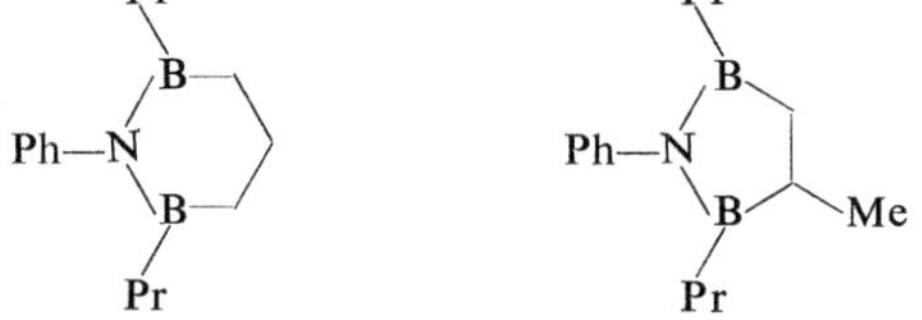

(CI)   (CII)

a tertiary amine, the Köster–Iwasaki reaction leads to the formation of borazines (147). This failure to obtain diborylamines (148) is attributed to a catalytic effect of the tertiary amine which induces the migration of a boron substituent (149).

$$\tfrac{1}{2}(R_2BH)_2 + R'NHBR_2 \xrightarrow[\text{120-130°}]{\text{NR}_3'} \tfrac{1}{3}[RB—NR']_3 + BR_3 + H_2 \qquad (147)$$

$$R_2BCl + R'NHBR_2 \xrightarrow[-R_3'N\cdot HCl]{\text{NR}_3''} R_2B—NR'—BR_2 \qquad (148)$$

$$3\,R_2BCl + 3\,R'NHBR_2 + 3\,NR_3'' \longrightarrow (RBNR')_3 + 3\,BR_3 + 3\,[HNR_3'']Cl \qquad (149)$$

On the other hand, (t-BuNH)$_2$BCl and t-BuNH$_2$ do react to give the diborylamine (t-BuNH)$_2$BN(t-Bu)B(BH-t-Bu)$_2$ or, starting from BCl$_3$, (t-BuHN)$_2$BN(t-Bu)BCl(NH-t-Bu).[140] The presence of the bulky t-butyl group is responsible for this exceptional aminolysis route. However, if the proper reaction conditions can be found, diborylamines might be prepared by way of scheme (82).

A very simple route to tetraalkyldiborylamines is the controlled aminolysis of mercaptoboranes using either ammonia or an appropriate aminoborane.[311] This is a modification of the procedure of B. M. Mikhailov et al.[182] leading to a 1,2,7-azadiboracycloheptane derivative. The diborylamines which have been described so far are shown in Table 7.

TABLE 7. SOME PROPERTIES OF DIBORYLAMINES

Me$_2$BNHBMe$_2$	71–72°/720[311]	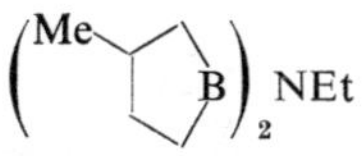
Et$_2$BNHBEt$_2$	56–58°/9[311]	121–27°/11.5[133]
Pr$_2$BNHBPr$_2$	80–84°/3[215]	
Bu$_2$BNHBBu$_2$	75–78°/0.01[215]	
Me$_2$BNMeBMe$_2$	104–07°/720[311]	137–38°/0.4[133]
Et$_2$BNMeBEt$_2$	65–68°/8[311]	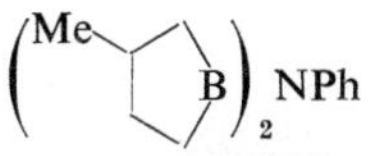
Me(Me$_2$N)BNHB(NMe$_2$)Me	77–78°/13[311]	
Ph(Me$_2$N)BNHB(NMe$_2$)Ph	158–60°/3[122]	98.5–100°/0.3[133]
(t-BuHN)ClBN-t-BuBCl(NH-t-Bu)	82–86°/0.02[140]	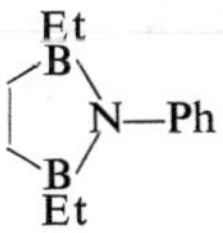 75°/0.3[133]
(t-BuHN)$_2$BN-t-BuB (NH-t-Bu)$_2$	98–100°/0.01[140]	
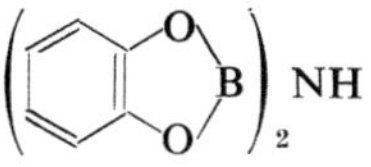	200°/2[145] subl.	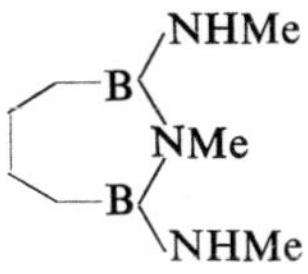 (182)

Pure diborylamines are rather stable compounds. Compounds of the type $[(R_2N)RB]_2NH$ decompose at about $100°$ with formation of borazine while $(R_2B)_2NH$ resists decomposition even on heating to $> 100°$.[311] The most stable diborylamine is bis(benzodioxaborolanyl)amine which sublimes unchanged at $200°$.[145] These observations lead to the conclusion that the diborylamines X_2BNHBX_2 increase in stability when the tendency of X to migrate is reduced. The unstable compound $Me_2NBMeNHBMe_2$ decomposes to Me_2NBMe_2 and $(MeBNH)_3$, perhaps by an intramolecular process involving migration of an Me_2N group.[311]

The decomposition of $(Et_2B)_2NH$ into BEt_3 and $(EtBNH)_3$ is not catalysed by NEt_3, PPh_3 or $SnCl_4$, while Et_2BCl induces the decomposition even at room temperature.[311]

3. Reactions

The diborylamines represent an extremely reactive system which is useful for many preparative purposes.[145, 182, 211, 311] Some of the reactions of tetraalkyldiborylamines are shown in (150).

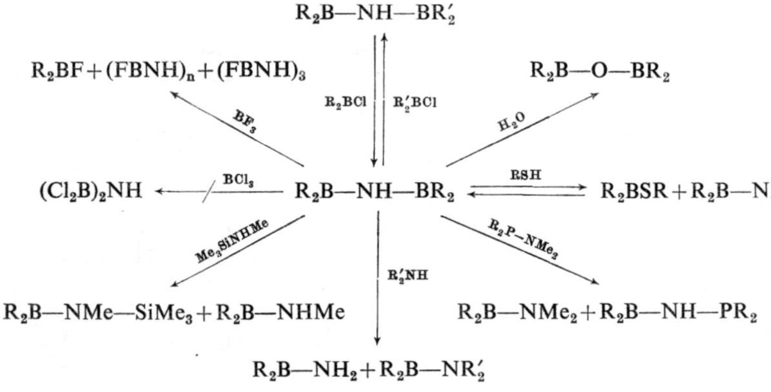

$$\tag{150}$$

If the diborylamines carry a functional group, such as Me_2N, ring formation is easily achieved by transaminations.[122]

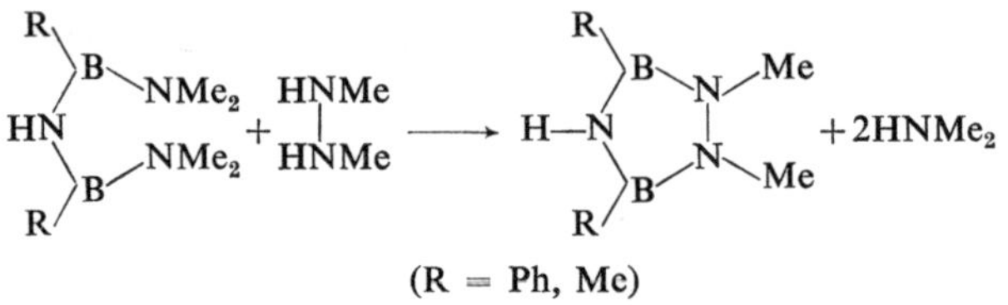

$$\tag{151}$$

$$(R = Ph, Me)$$

4. Structure

Structural investigations on diborylamine point to a C_{2v} symmetry for the molecules $Me_2BNHBMe_2$ and $Me_2BNMeBMe_2$.[311] The introduction of a second Me_2B group leads to a deshielding of the boron nucleus as compared with Me_2BNH_2, Me_2BNHMe or Me_2BNMe_2, the difference in $\delta^{11}B$ between the two classes of compounds is ~ 11 ppm. This is in accord with the proposed resonance formulation, which implies sp^2 hybridization for both the boron and nitrogen atoms. If the C_{2v}

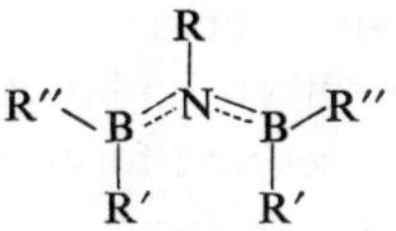

(CIII)

symmetry is correct, then the two alkyl groups on the boron atom are magnetically non-equivalent if rotation around the BN bonds is restricted as shown in (CIII). At room temperature only a single signal was found for the B–CH$_3$ groups of $(Me_2B)_2NH$, but at $-55°$ the expected two signals were observed. While this does not, of course, prove the C_{2v} symmetry it is at least consistent with it, and i.r. and Raman spectral data can only be interpreted on the basis of this symmetry.[311]

E. Triborylamines

Members of the family of triborylamines are exceedingly rare. Lappert and Majumdar[145] obtained the first example according to (152), This compound has a planar BN$_3$ framework.[139]

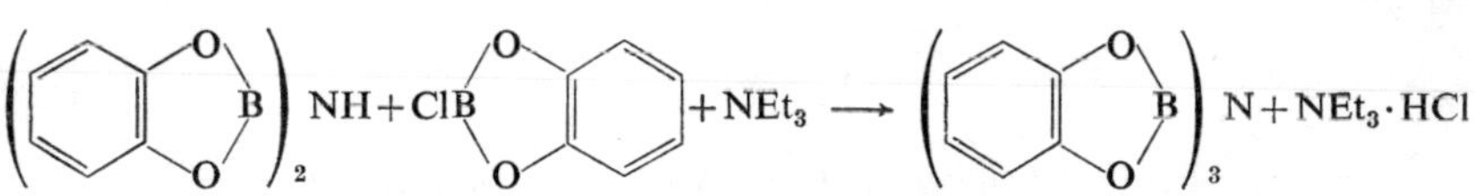

(152)

The crystalline compound is sufficiently volatile for mass spectroscopic investigations, where it forms doubly and triply charged ions in abundance.[145] If one defines a triborylamine as a compound containing the NB$_3$ group, then the diborazines, for example (CIV), also belong to this class, at least formally. Diborazines can be prepared rather readily as described by (153).[160]

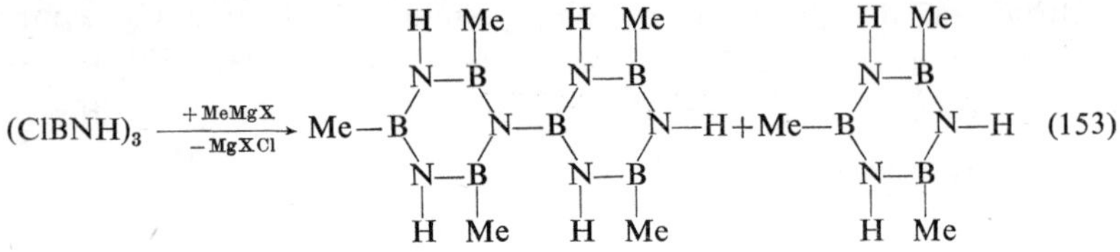

$$(ClBNH)_3 \xrightarrow[-MgXCl]{+MeMgX}$$

(153)

While B-chloro-N-methylborazine and MeMgX give yields of diborazines not exceeding 70 per cent due to side reactions,[79] the B-chloroborazines $(ClBNH)_3$ rather readily give condensation products[30, 160] containing the BN_3 moiety. This is most likely an effect due to the NH group which can react with the Grignard reagent to give a Mg—N bond, which can then react with the Cl–B group to form a new B—N bond.

Similarly, the metallation of (LXXVIII) by LiR, followed by treat-

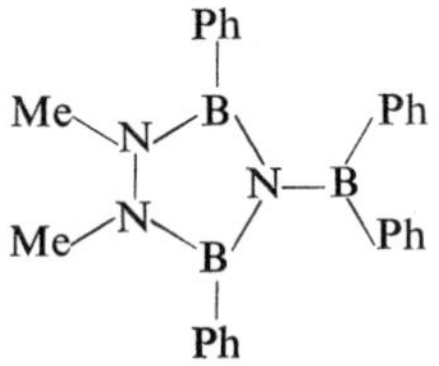

(CIV)

ment with Ph_2BCl gives the triborylamine (CIV).[229] However, the search for a hexaalkyltriborylamine $(R_2B)_3N$ has so far been unsuccessful as far as the isolation of a pure compound is concerned. This is not unexpected if the great reactivity and lability of the diborylamines are considered. Nevertheless it has been shown by ^1H and ^{11}B n.m.r. spectroscopy that $(Me_2B)_3N$ is formed in the reaction of $Me_2BN(SiMe_3)_2$ with Me_2BBr.[311]

F. Physicochemical and Structural Investigations

1. Introduction

The three topics which continue to be of importance to the understanding of BN bonds in general and aminoborane chemistry in particular are: (i) the strength of the BN bond, (ii) the geometry prevailing in the aminoborane and (iii) the charge distribution in these compounds.

2. Vibrational Spectra

The force constants of the BN stretching vibrations in aminoboranes are a measure of the BN bond strengths. These force constants indicate a fairly high BN bond order in the aminoboranes, and values in Table 8 demonstrate this.

TABLE 8. FORCE CONSTANTS, f, AND BOND ORDERS, b, IN BN COMPOUNDS[94]

	f[mdyn/Å]	$b_I{}^a$	$b_{II}{}^b$
$X_3B \cdot NR_3$	2.8	0.7	1.0
$[OBNR_2]_3$	4.0	1.0	1.3
$B(NR_2)_3$	6.0	1.5	1.7
$[HBNH]_3$	6.3	1.6	1.8
R_2BNCO	6.6	1.7	1.9
X_2BNY_2	7.0	1.8	2.0
BN_2^{3-}	7.0	1.8	2.0
$BN_{(monomer)}$	8.3	2.1	2.3

^{a} Bond order $b = 1$ for $f = 3.9$ mdyn/Å.
^{b} Bond order $b = 1$ for f in $X_3B \cdot NR_3$, bond order $= 2$ for BN_2^{3-}.

In most papers the authors discuss only the observed BN stretching frequencies, and these are only an approximate measure of the bond strength A. Meller[159] has summarized most of the data published and therefore only a few additions are necessary here.

H. J. Becher and H. T. Baechle[25] have shown, by investigating ^{10}B-enriched aminoboranes, that the A_1 BN modes couple with the N—CH_3 or N—CH_2 A_1 modes of aminoboranes such as Cl_2BNR_2, $ClB(NR_2)_2$ and $B(NR_2)_3$ (R = Me, Et). Full assignments are given for these compounds. In the case of Cl_2BNEt_2, the presence of rotational isomers is indicated in the spectra and a force constant of $f_{BN} = 7-8$ mdyn/Å was calculated for this molecule. In the dimethylamino(vinyl)bromoboranes the BN bonding is weakened. One reason for this is a π-bonding interaction of the boron atom with the vinyl group.[207]

Substitution of NMe for NPh in aminoboranes causes a characteristic lowering of the BN frequencies of X_2B—NRR', but a similar substitution at the boron atom has only little effect. Although one might expect the existence of cis–trans isomers in RR'B—NR''R''' these could not be observed in their i.r. spectra.[20]

A comparison of the BN stretching frequencies in bis(dimethylamino) boranes $(Me_2N)_2BX$[63, 96, 238] indicates an increase of the BN bond strength in the series X = Me < H < I < Br < Cl < F < CN. Infrared data also reveal a high B—N bond order in the 1,3,2-diazaborolidines.[63]

3. N.M.R.-spectroscopic Studies

In aminoboranes there is a barrier to rotation about the BN bond due to the high BN bond order. This barrier has been estimated from ^1H n.m.r. studies on CH_2=$CHBBr(NMe_2)$ to have a magnitude of 14.0 ± 0.4 kcal/mole.[207] Cis–trans isomerism has already been demonstrated for $Me(Cl)BNMePh$ and $Ph(PhCH_2)BN(CH_2Ph)Ph$. H. J. Becher and H. T. Baechle[20] found another example in $Me(EtO)BNMePh$, the ratio of the quantities of the two forms at room temperature is very high (10–20 : 1) and several hours are needed to reach equilibrium between the two forms.

The same authors[24] also investigated the ^1H n.m.r. spectra of phenyl substituted aminoboranes such as $Me_2BNMePh$, Me_2BNPh_2, $PhMeBNMe_2$, Ph_2BNMe_2, $Ph_2BNMePh$ and $PhMeBNPh_2$. From the chemical shift data it was concluded that the phenyl groups are tilted by 30–60° out of the C_2BN plane, and two different conformers of $MePhBNPh_2$ were observed. In connection with these studies on cis–trans isomerization, dipole moment studies were also carried out.[22] The addition of individual bond moments obtained from symmetrical aminoboranes gave excellent agreement between calculated and experimental values for unsymmetrically substituted aminoboranes. Furthermore, it was shown by these data that the trans-isomer of $MeClBNMePh$ is more stable at room temperature than the cis-isomer.

^{11}B n.m.r. spectra[47, 239] of aminoboranes revealed that the chemical shift observed is strongly dependent upon the BN π-bonding interaction, which is strongest in the monoaminoboranes. Furthermore, it has been shown[47, 239] that the observed shift is proportional to the calculated BN bond order (and force constant), and also to the electron density on the boron atom, as calculated by a LCAO–MO method.[47] The trends observed are consistent with the i.r. data, i.e. N substitution increases and B substitution decreases the shielding of the boron atoms, if anisotropy effects are neglected.[239]

4. Mass Spectral and other Studies

In the series $BCl_{3-n}(NMe_2)_n$ the ionization potential increases quite regularly from $B(NMe_2)_3$ (7.6 eV) to BCl_3 (11.7 eV).[12, 143] If the ionization potential is equal to the energy of the highest occupied electronic level in these compounds, than the data fit with the theories of BN and BCl π-bonding in which BN π-bonding is considered much more important than BCl π-bonding.[12] Furthermore, u.v. data indicate by the high intensities of the bands observed that n → π^* transitions are predominant.[143]

That the BN bond in the aminoboranes is quite strong is also apparent from the fragmentation patterns of aminoboranes.[136] M. J. S. Dewar et al.[65] have even observed the doubly charged cations $[B(NMe_2)_3]^{++}$, $[PhB(NMe_2)_2]^{++}$ and $[PhB(NMeCH_2)_2]^{++}$, which appear at 70 eV. The stability of these ions is attributed to a favorable charge distribution, and all these ions have isoelectronic organic counterparts.

The diamagnetic susceptibility of the BN bond for compounds with $B[NC_2]$, $B[NC(H)]$ and BNH_2 frameworks was found to be $\chi_{B-N} = -8.9 \pm 1.0 \times 10^{-6}$.[150]

LCAO–MO[47] and SCF–MO[194, 255] calculations were carried out for various aminoboranes, and electron densities, bond orders, electronic transitions, etc., were calculated. These calculations confirm, in general, results arrived at earlier by J. Kaufmann and R. Hoffmann. The following figures give the π-BN bond orders as calculated by the SCF–MO[255] treatment:

	$B(NMe_2)_3$	$B_2(NMe_2)_4$	H_2BNMe_2	Me_2BNHMe	Me_2BNMe_2
π-BN bond order	0.489	0.589	0.719	0.534	0.567

These results are in accord with the trends already discussed above, and it is worth noting the low π-bond order of the B—B bond in $B_2(NMe_2)_4$, which was calculated to be 0.079.[255]

Accurate structural data are still lacking for aminoboranes of the type X_2BNR_2 and $XB(NR_2)_2$, but the structure (CV) of a triaminoborane has been reported.[35]

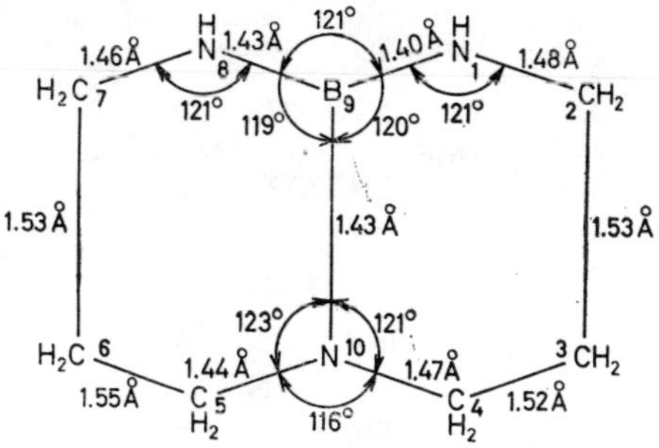

(CV)

Although 1,8,10,9-triazaboradecalin is certainly a very special triamino-borane, the structural results are important because they give figures for the BN bond distances in these compounds. The triazaboradecalin is almost planar, and thus the assumption of a planar system[206] on the basis of n.m.r. and i.r. spectroscopy is now substantiated. It is interesting to note the two different BN distances in this molecule and the completely planar BN_3 system. The atoms C_2 and C_7 lie 0.2 Å and atoms C_3 and C_6 0.5 Å out of this plane. The bond angles on the boron and nitrogen atoms are almost identical and deviate slightly from the expected 120° only for the N_{10} atom. No NH...N interaction was observed and the BN distances are nearly the same as in the borazines.

V. B–N–N– AND PYRAZABOLE CHEMISTRY

Although the chemistry of hydrazinoboranes is very similar to that of the aminoboranes, the two classes of compounds show some very interesting differences. Nevertheless, there have only been a few reports since 1964[97] on this topic[11, 213] and none of these could be regarded as a contribution indicating new trends. The only development that should be mentioned briefly is the ready formation of heterocyclic systems containing the B–N–N grouping.[166] Typical examples are (CVI)–(CXII).

(CVI)

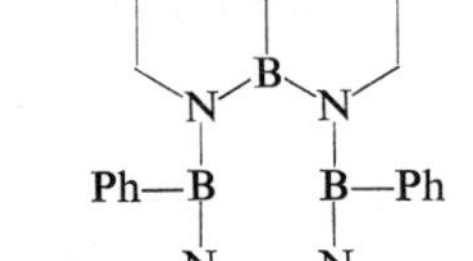

(CVII)

(CVIII)[174, 142]

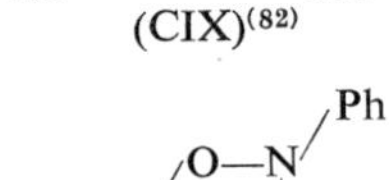

(CIX)[82]

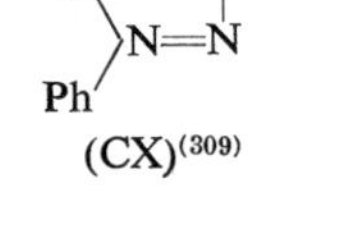

(CX)[309]

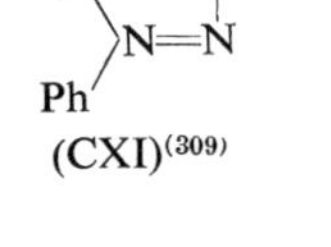

(CXI)[309]

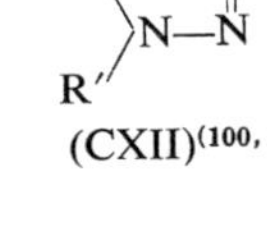

(CXII)[100, 193]

The derivatives of the cyclotetrazenoborane (CXII)[100, 194] have received a thorough theoretical treatment[193] and also have been studied spectroscopically.[73]

First described in 1966, compounds with the B–N–N structural unit such as the pyrazaboles and related compounds are exceptional with respect to their high thermal stability, their ease of formation and their versatility as chelate ligands.[301, 304, 124-5] Research in boron chemistry since 1924 has shown that hydrogen bridge protons in many compounds can be replaced by an X_2B group (X = F, OR, R, etc.). It is also known that rather strong H-bridges exist in pyrazole, favoring a dimeric structure. S. Trofimenko[302] showed that it is easy to replace these hydrogens by a variety of X_2B groups. The new compounds of type (CXIII) are called pyrazaboles.

Treatment of pyrazole with $Me_3N \cdot BH_3$ in boiling toluene afforded crystalline (CXIII) together with H_2 and Me_3N. In this way a large num-

(CXIII)

ber of substituted pyrazaboles (X = H, Me, Ph, CF_3, Br and Y = H, Cl, Br, Me, CN, NO_2) were prepared. Pyrazoles with strong electron withdrawing groups react slowly due to the decreased nucleophilic potential of the second nitrogen atom, and in these cases it is advantageous to use diborane in tetrahydrofuran instead of $Me_3N \cdot BH_3$. When trialkyl- or triarylboranes were used to replace the N-hydrogen atom of pyrazole, temperatures of 120–150° were required to force the reaction (154) to completion.[302]

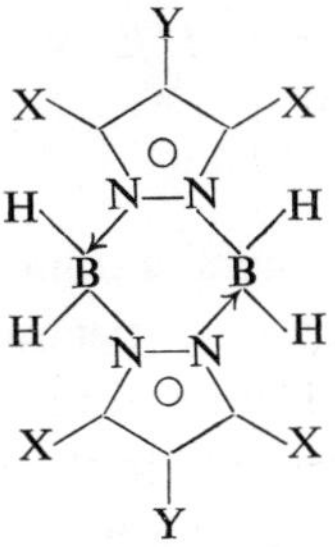

$$2 R_3B + 2 HN - N \longrightarrow [R_2B - N - N]_2 + 2 RH \qquad (154)$$

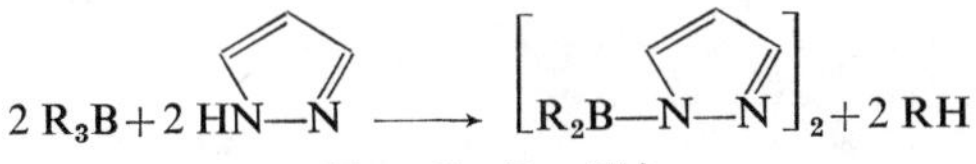

(R = Et, Bu, Ph)

The tetraalkylpyrazaboles are stable to air oxidation. This is attributed to a very good shielding of the boron and nitrogen atoms by the R groups and also to an extremely good charge distribution. The presence of tetracoordinated B atoms in the pyrazoles was confirmed by ^{11}B n.m.r. spectroscopy. The ^{11}B signal was found in the -3 to -1 ppm region for the B-alkylated species, and at $+10$ to $+13$ ppm (BF$_3$·OEt$_2$ standard) for the parent compounds (CXIII). These are the same regions as those in which the signals for (Et$_2$BNHMe)$_2$ (-2.1 ppm) and (H$_2$BNR$_2$)$_2$ ($+8$ ppm) are found.[311]

The pyrazaboles are in fact dimeric dihydro- or diorganoborylpyrazoles, and therefore the following series of pyrazolylboranes and borates are accessible: X$_2$Bpz, XB(pz)$_2$, Bpz$_3$, B(pz)$_4^-$, XB(pz)$_3^-$, X$_2$Bpz$_2^-$, X$_3$Bpz$^-$ (pz = 1-pyrazolyl).

The anionic species are readily obtained by treating molten pyrazole at 90–120° with KBH$_4$ (155).[303]

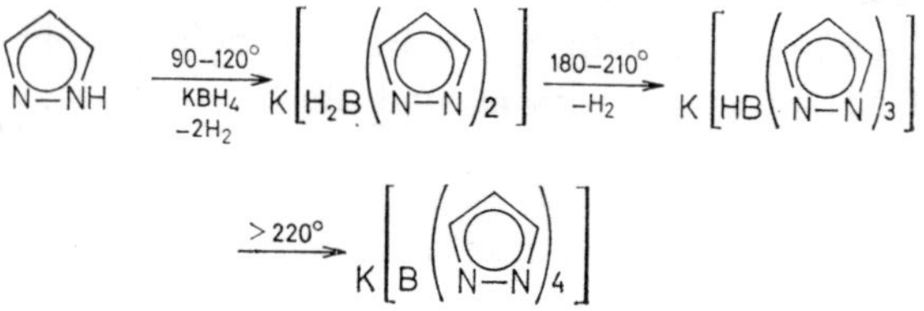

From the various potassium salts, a number of other salts (Li, Na, Cs, NMe$_4$) were prepared. The ^{1}H n.m.r. spectra of these salts are quite sensitive both to a change in the solvent and in the cation (chelate formation). That "internal" coordination occurs in these compounds is demonstrated by the fact that the lithium salts can be sublimed. The lithium salts are prepared in the same manner as the potassium analogs but with LiBH$_4$ as the reducing agent. The compound Na[HB(pz)$_3$] is also sublimable.[303]

Both hydrolytic and oxidative stability increase with an increase in the number of pz groups in the anions. Even the free acids, such as H[H$_2$B (pz)$_2$]·4 H$_2$O or HBpz$_4$, can be isolated.

The 1-pyrazolylborates are extremely good complexing agents and readily form metal chelates in aqueous or alcoholic solutions. These chelates, which are of various types depending on the nature of the metal ion and the type of 1-pyrazolylborate, are shown below.

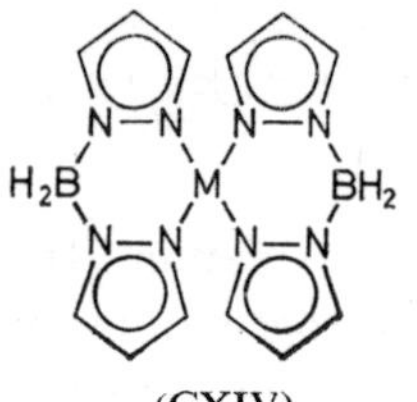

(CXIV)    (CXV)

Most of these neutral complexes are sublimable or soluble in non-polar solvents.[303] They are very interesting compounds and their properties have been examined quite closely;[124, 125] for instance, Fe^{2+} complexes of either high spin or low spin state or of an intermediate spin can be prepared by altering the substituents on the pyrazolyl group.[124] In compounds of the type $M[B(pz)_4]_2$ ($M = Co^{2+}$, Cd^{2+}, Pd^{2+}, Hg^{2+}) one pz group per boron atom does not coordinate.[303]

Other types of complexes involve pyrazolylborate-metal carbonyls,[304] or pyrazolylborate-π-allyl-metal carbonyls, which are extremely stable.[304] However, the coordination chemistry of this new ligand system is beyond the scope of this chapter, as also is a discussion of the substitution reactions carried out by replacing the hydrogen atoms in $[H_2B(pz)_2]^-$ or $[HB(pz)_3]$ by other groups.[301]

The favorable structures of the pyrazolylborate group and of the pyrazaboles are responsible for their rather unusual properties. That the position of the second nitrogen is of prime importance is shown by comparison with imidazolylboranes and imidazolylborates.[300] Heating BEt_3 with

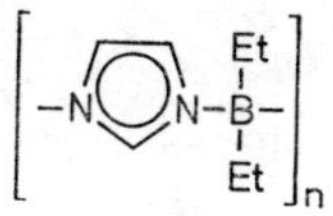

$$\left[-N \overset{\displaystyle\frown}{\underset{\displaystyle\smile}{}} N - \overset{\textstyle Et}{\underset{\textstyle Et}{\overset{|}{\underset{|}{B}}}} - \right]_n$$

(CXVI)

midiazole gave an elastomer of 40–80 repeating (CXVI) units. Although these polymers are stable in air, they are much more readily attacked by acid or base than the pyrazaboles. The reaction between BEt_3 and 2-methylimidazole yielded a strain-free tetramer, which was unaffected by hot alkali, acid or dichromate.[300]

VI. MONOMERIC BORONIMIDES

A. *Introduction*

In 1948 E. Wiberg[320] postulated that the formation of borazines (borazoles) from the reaction of diborane, boron halides or alkylboranes with ammonia or an alkylamine should proceed stepwise according to (155).

$$H_3N \cdot BX_3 \xrightarrow{-HX} H_2N-BX_2 \xrightarrow{-HX} HN=BX \longrightarrow \frac{1}{3}[HNBX]_3 \qquad (155)$$

$$(X=H, Cl, Br, R)$$

20*

It was assumed that dehydrogenation, dehydrohalogenation or dehydro-alkylation led to monomeric boronimides as reaction intermediates, which rapidly trimerized to the thermodynamically favored borazines $[HNBX]_3$. However, this scheme cannot readily explain why the polymerization of a monomeric boronimides should not give rise to cyclic dimers, tetramers or other oligomers or polymers. Nevertheless, it is known that under the conditions of the high-temperature pyrolyses, cyclic dimers and tetramers of boronimides are converted to the triborazines although the conversion rate is rather slow. At present we know nothing definite about the mechanisms of these ring expansion and contraction reactions, but it is likely that monomeric boronimides play a role in them.

Comparison of the monomeric boronimides with the isoelectronic acetylene system is valuable. In simple terms the bonding situation in a boronimides XBNY can be considered to involve both the B and N atoms in a state of sp hybridization. Two π-bonds are formed using the p_y- and p_z-orbitals of both atoms, and a BN σ-bond results from overlap of the two sp hybrid orbitals as sketched in (CXVII).

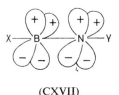

(CXVII)

This scheme would involve strong back bonding from the nitrogen lone electron pair to the empty boron p-orbitals, and it is to be expected that despite the polarity $\overset{+\delta}{B}\text{—}\overset{-\delta}{N}$ of the σ-bond, this back donation could lead to a considerable increase of negative charge on the boron atom. In this simple picture it is assumed that there is no π-bonding interaction in the B—X and the N—Y bonds. However, if there were such interaction the bonding situation might change, for instance by forcing the nitrogen to assume sp^2 hybridization in order to decrease electronic repulsions.

One might expect to be able to stabilize monomeric boronimides, XBNY, as was done by P. I. Paetzold et al.,[243, 250] by making X a strongly electron-withdrawing and Y an electron donating group, and also by suitable application of steric factors acting so as to suppress oligomerization.

B. *Monomeric Boronimides as Reaction Intermediates*

In 1963 P. I. Paetzold[246] was able to demonstrate that a monomeric boronimide was a reaction intermediate in the pyrolysis of the diphenyl-boronazide–pyridine adduct. The thermal decomposition of the compound at 200° proceeds according to (156):[245]

$$3 \, Ph_2BN_3 \cdot py \longrightarrow (PhNBPh)_3 + 3 \, N_2 + 3 \, py \tag{156}$$

However, if the decomposition of the azide is carried out in the presence of 2,5-diphenyltetrazole, which at the temperature of the reaction eliminates N_2 to form diphenylnitrilimine, the latter compound acts as a 1,3-dipolar scavenger for any monomeric boronimide intermediate. The novel ring system, 1,2,4,5-tetraphenyl-1-bora-2,3,5-triazacyclopent-3-ene, was formed in this way in 12 per cent yield according to (157).

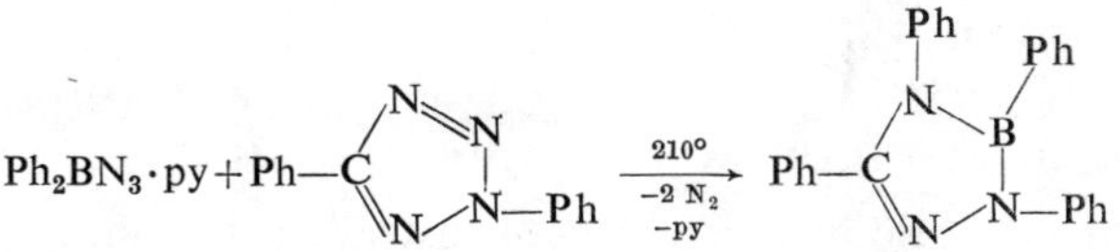

$$Ph_2BN_3 \cdot py + Ph{-}C \underset{\substack{N \\ }}{\overset{\substack{N \\ N}}{}} N{-}Ph \xrightarrow[\substack{-2\,N_2 \\ -py}]{210°} Ph{-}C \underset{N}{\overset{N}{}} B \tag{157}$$

The formation of the heterocycle is consistent with the existence of an intermediate monomeric PhBNPh, but it can not be taken to prove that this species is indeed the reactive intermediate.

Some time later Greenwood and Morris[100] synthesized derivatives of the boratetrazole heterocyclic system by the action of phenyl azide on aminoboranes or decaborane. The former reaction yields the product as described by (158), and since aniline-borane readily loses hydrogen it is reasonable to assume that monomeric PhNBH adds to the 1,3-dipolar phenylazide.

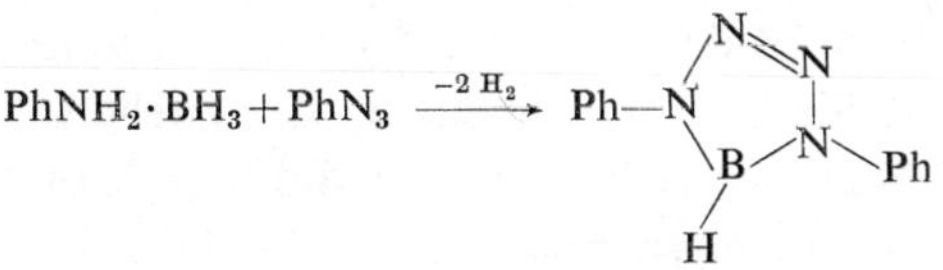

$$PhNH_2 \cdot BH_3 + PhN_3 \xrightarrow{-2\,H_2} Ph{-}N \underset{B}{\overset{N}{}} N{-}Ph \tag{158}$$

Paetzold[242] suggested that 1,3-dipolar cycloaddition reactions may be a very useful and perhaps general method for trapping monomeric boronimides. If this is so then a large number of new boron-containing heterocyclic systems should become readily accessible.

C. *Preparation of Monomeric Boronimides*

Application of the previously mentioned hypothesis for stabilizing monomeric boronimides by placing an electron-withdrawing group on the boron atom and an electron-donating group on the nitrogen atom led to the preparation of the first authentic monomeric sample by Paetzold and Simson[250] in 1966. These workers employed the method of Kolbezen, i.e. refluxing a 1:1 molar mixture of pentafluorophenylborondichloride and either mesidine or anisidine for two days in benzene, toluene or xylene. One mole of HCl is quickly eliminated from the mixture, indicating the rapid formation of $C_6F_5BCl(NHR)$. A second mole of HCl is evolved quite slowly according to (159). The monomeric boronimides thus obtained in 70–85 per cent yield are volatile in high vacuum at 120°. Other products isolated in low yield include cyclodiborazines(diazaboretidines), and pentafluorophenylbis(arylamino)boranes as shown in (160).

$$C_6F_5BCl_2 + H_2NR \underset{\overset{\displaystyle -2\,HCl}{\nearrow}}{\overset{}{}}\begin{cases} C_6F_5B{=}NR & (159) \\[2mm] [C_6F_5B{-}NR]_2 + C_6F_5B(NHR)_2 & (160) \end{cases}$$

$$(R = C_6H_3(CH_3)_3{}^{(250)}, C_6H_4OCH_3{}^{(250)}, C_6H_4SCH_3{}^{(244)}, C_6F_5{}^{(244)})$$

If aniline or aliphatic amines are used instead of the amines mentioned above, no monomeric boronimides but rather borazines and polymers are obtained.[244] The replacement of $C_6F_5BCl_2$ by $C_6F_5BBr_2$ results in a more rapid rate of HX evolution[244] due to the weaker B—Br bond.

As yet unexplained is the fact that $PhBCl_2$ and $H_2NC_6F_5$ yield monomeric $PhBNC_6F_5$.[243] On the other hand, a mixture of BBr_3 and $H_2NC_6F_5$ readily evolves HBr to give monomeric $C_6F_5NHBBr_2$ and crystals of m.p. 118–120° which were assumed to be $[C_6F_5NBBr]_3$.[200] (A better characterized sample of these crystals melted at 250°).[93] Certainly no monomeric boronimides was produced in the reaction. Neither could it be prepared by the reaction between $C_6F_5NH_2$ and BCl_3[93] or $Et_3N \cdot BCl_3$[107, 161] in boiling nonane, the only definite product isolated was $C_6F_5NHBCl_2$. Similarly treatment of $C_6F_5NH_2$ with $Et_3N \cdot BBr_3$ in the presence of NEt_3 as HBr acceptor did not lead to a monomeric boronimide.[165] No depolymerization of the borazine ring system is observed when $(ClBNC_6F_5)_3$ reacts with Grignard reagents such as $C_6F_5CH_2MgBr$[164] or C_6F_5MgI[93, 165] or C_6F_5MgX.[165] In addition, the reactions of $(ClBNH)_3$ or $(ClBNR)_3$ with LiC_6F_5 or C_6F_5MgX led not to the monomeric C_6F_5BNH but to the borazines $(C_6F_5BNH)_3$ and $(C_6F_5BNR)_3$ respectively.[158, 163, 165]

D. *Reactions of Monomeric Boronimides*

Not surprisingly the monomeric boronimides are highly reactive compounds. Their Lewis acidity is not very pronounced since the crystalline 1:1 adducts of C_6F_5BNR with pyridine or trimethylamine readily lose the base. On the other hand, they are rapidly attacked by HCl in benzene resulting in cleavage of the BN bond:[250]

$$C_6F_5BNR + 3\ HCl \longrightarrow RNH_2 \cdot HCl + C_6F_5BCl_2 \qquad (161)$$

While C_6F_5BNmes (mes = mesityl) does not react with 1,3-dipolar compounds due to steric reasons, in general 1,3-dipolar addition provides a convenient route to new heterocyclic systems as indicated below.

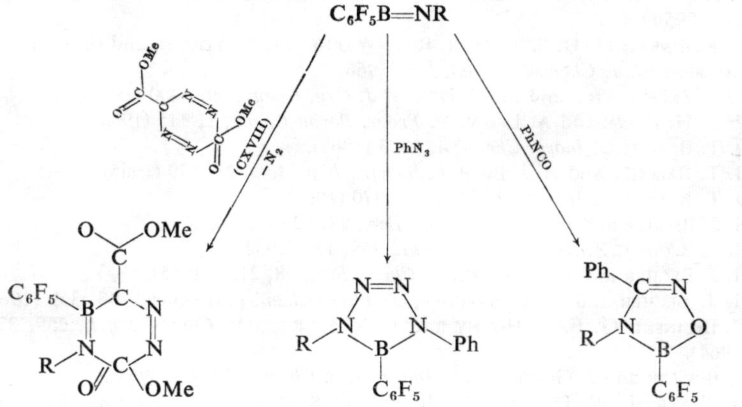

$$(162)$$

The structure of the monomeric boronimides has not yet been determined. As well as being monomeric in organic solvents[250] the compounds have a strong Raman line at 1703 to 1710 cm^{-1} indicating a high BN bond order. Comparison of the Raman and infrared spectra seems to indicate that these compounds may not be linear as are the isoelectronic acetylenes.[244] However, further study is needed in order to obtain a definite picture of the structures of these compounds containing two-coordinate boron atoms. The only other boron species with coordination number two is the ion $[BN_2]^{3-}$.

Without doubt this intriguing area of boron–nitrogen chemistry is of great interest and merits further investigation.

VII. REFERENCES

1. E. W. ABEL, D. A. ARMITAGE, R. P. BUSH and G. R. WILLEY, *J. Chem. Soc.*, 62 (1965).
2. G. E. MCACHRAN and S. G. SHORE, *Inorg. Chem.*, **5**, 2044 (1966).
3. S. AKERFELD and M. HELLSTRÖM, *Acta Chem. Scand.*, **20**, 1418 (1966).
4. A. L. ALSOBROOK, A. L. COLLINS and R. L. WELLS, *Inorg. Chem.*, **4**, 253 (1965).
5. E. AMBERGER and R. RÖMER, Z. *Anorg. Allg. Chem.*, **345**, 1 (1966).
6. G. A. ANDERSON and J. J. LAGOWSKI, *Chem. Comm.*, 649 (1966).
7. D. R. ARMSTRONG and P. G. PERKINS, *J. Chem. Soc.*, 1026 (1966).
8. B. J. AYLETT and L. K. PETERSON, *J. Chem. Soc.*, 4043 (1965).
9. B. J. AYLETT and J. EMSLEY, *J. Chem. Soc.*, 652 (1967).
10. M. AZZARO, *Bull. Soc. Chim. France*, 2201 (1964).
11. M. S. BAINS, *Canad. J. Chem.*, **44**, 534 (1966).
12. J. C. BALDWIN, M. F. LAPPERT, J. B. PEDLEY, P. N. K. RILEY and R. D. SEDGWICK, *Inorg. Nucl. Chem. Letters*, **1**, 57 (1965).
13. L. BANFORD and G. E. COATES, *J. Chem. Soc.*, 3564 (1964).
14. A. J. BANISTER and N. N. GREENWOOD, *J. Chem. Soc.*, 1534 (1965).
15. A. J. BANISTER, N. N. GREENWOOD, B. P. STRAUGHAN and J. WALKER, *J. Chem. Soc.*, 995 (1964).
16. R. K. BARTLETT, H. S. TURNER, R. J. WARNE, M. A. YOUNG, and (in part) I. J. LAWRENSON, *J. Chem. Soc.* (A), 479 (1966).
17. R. J. BAUMGARTEN and M. C. HENRY, *J. Org. Chem.*, **29**, 3400 (1964).
18. C. E. H. BAWN and A. LEDWITH, *Progr. Boron Chem.*, **1**, 345 (1964).
19. O. T. BEACHLEY, *Inorg. Chem.*, **4**, 1823 (1965).
20. H. T. BAECHLE and H. J. BECHER, *Spectrochim. Acta*, **21**, 579 (1965).
21. O. T. BEACHLEY, Jr., *Inorg. Chem.*, **6**, 870 (1967).
22. H. J. BECHER and H. DIEHL, *Chem. Ber.*, **98**, 526 (1965).
23. H. J. BECHER, Z. *Anorg. Allg. Chem.*, **268**, 133 (1952).
24. H. J. BECHER and H. T. BAECHLE, *Chem. Ber.*, **98**, 2159 (1965).
25. H. J. BECHER and H. T. BAECHLE, Z. *Phys. Chem. (Frankfurt)*, **48**, 359 (1966).
26. C. BELINSKI, G. F. C. HORNY and F. X. L. KERALY, *Compt. Rend.*, **259**, 3737 (1964).
27. E. BESSLER and J. GOUBEAU, Z. *Anorg. Allg. Chem.*, **352**, 67 (1967).
28. H. BEYER, J. W. DAWSON, H. JENNE and K. NIEDENZU, *J. Chem. Soc.*, 2115 (1964).
29. K. W. BÖDDEKER, S. G. SHORE and R. K. BUNTING, *J. Amer. Chem. Soc.*, **88**, 4396 (1966).
30. J. L. BOONE and G. W. WILLCOCKSON, *Inorg. Chem.*, **5**, 311 (1966).
31. D. A. BROWN and C. G. MCCORMACK, *Chem. Comm.*, 383 (1967).
32. M. P. BROWN, R. W. HESELTINE and D. W. JOHNSON, *J. Chem. Soc.* (A), 597 (1967).
33. M. P. BROWN and R. W. HESELTINE, *J. Inorg. Nucl. Chem.*, **29**, 1197 (1967).
34. W. BRÜSER and K. H. THIELE, Z. *Anorg. Allg. Chem.*, **349**, 310 (1967).
35. G. J. BULLEN and N. H. CLARK, *Chem. Comm.*, 670 (1967).
36. A. B. BURG and J. S. SANDHU, *Inorg. Chem.*, **4**, 1467 (1965).
37. A. B. BURG and J. S. SANDHU, *J. Amer. Chem. Soc.*, **89**, 1626 (1967).
38. A. B. BURG and H. W. WOODROW, *J. Amer. Chem. Soc.*, **76**, 219 (1954).
39. A. B. BURG and R. J. KULIAN, *J. Amer. Chem. Soc.*, **72**, 3103 (1950).
40. I. M. BUTCHER, B. R. CURRELL, W. GERRARD and G. K. SHARMA, *J. Inorg. Nucl. Chem.*, **27**, 817 (1965).
41. I. M. BUTCHER and W. GERRARD, *J. Inorg. Nucl. Chem.*, **27**, 823 (1965).

42. I. M. BUTCHER and W. GERRARD, *J. Inorg. Nucl. Chem.*, **27,** 2114 (1965).
43. I. M. BUTCHER, B. R. CURRELL, and G. K. SHARMA, *J. Inorg. Nucl. Chem.*, **28,** 2137 (1966).
44. G. B. BUTLER, G. L. STATTON and W. S. BREY, Jr., *J. Org. Chem.*, **30,** 4194(1965).
45. J. C. CARTER and R. W. PARRY, *J. Amer. Chem. Soc.*, **87,** 2354 (1965).
46. E. C. ASHBY and W. E. FOSTER, *J. Org. Chem.*, **29,** 3225 (1964).
47. M. R. CHAKRABARTY, C. C. THOMPSON, Jr. and W. S. BREY, Jr., *Inorg. Chem.*, **6,** 519 (1967).
48. C. CHAMBERS and A. K. HOLLIDAY, *J. Chem. Soc.*, 3459 (1965).
49. R. D. CHAMBERS and T. CHIVERS, *J. Chem. Soc.*, 3933 (1965).
50. T. CHIVERS, *Chem. Comm.*, 157 (1967).
51. CH. E. COLBURN, (Editor), *Developments in Inorganic Nitrogen Chemistry*, Elsevier Publishing Co., Amsterdam, 1966. Vol. 1; J. K. RUFF, p. 470.
52. T. D. COYLE, *Proc. Chem. Soc. (London)*, 172 (1963).
53. R. H. CRAGG and N. N. GREENWOOD, *J. Chem. Soc.* (A), 961 (1967).
54. R. H. CRAGG, M. F. LAPPERT and B. P. TILLEY, *J. Chem. Soc.*, 2108 (1964).
55. R. H. CRAGG, M. F. LAPPERT and B. P. TILLEY, *J. Chem. Soc.* (A), 947 (1967).
56. R. H. CRAGG, M. F. LAPPERT, H. NÖTH, P. SCHWEIZER and B. P. TILLEY, *Chem. Ber.*, **100,** 2377 (1967).
57. G. C. CULLING, M. J. S. DEWAR and P. A. MARR, *J. Amer. Chem. Soc.*, **82,** 1125 (1964).
58. B. R. CURRELL and M. KHODABOCUS, *J. Inorg. Nucl. Chem.*, **28,** 371 (1966).
59. B. R. CURRELL, W. GERRARD and M. KHODABOCUS, *Chem. Comm.*, 77 (1966).
60. B. R. CURRELL, W. GERRARD and M. KHODABOCUS, *J. Organometallic Chem.*, **8,** 411 (1967).
61. J. M. DAVIDSON and M. FRENCH, *Chem. and Ind.*, 750 (1959).
62. P. G. DAVIES and E. F. MOONEY, *Spectrochim. Acta*, **22,** 953 (1966).
63. J. W. DAWSON, P. FRITZ and K. NIEDENZU, *J. Organometallic Chem.*, **5,** 211 (1966).
64. J. W. DAWSON, P. FRITZ and K. NIEDENZU, *J. Organometallic Chem.*, **5,** 13 (1966).
65. M. J. S. DEWAR and P. RONA, *J. Amer. Chem. Soc.*, **87,** 5510 (1965).
66. M. J. S. DEWAR and J. L. VON ROSENBERG, Jr., *J. Amer. Chem. Soc.*, **88,** 358 (1966).
67. M. J. S. DEWAR, *Progr. Boron Chem.*, **1,** 235 (1964).
68. A. DORNOW and D. WILLE, *Chem. Ber.*, **98,** 1505 (1965).
69. V. A. DOROKHOV and B. M. MIKHAILOV, *Izvest. Akad. Nauk SSSR, Ser. Khim,.* 364 (1966).
70. J. E. DOUGLASS, *J. Org. Chem.*, **31,** 962 (1966).
71. J. E. DOUGLASS, G. R. ROEHRIG and O. H. MA, *J. Organometallic Chem.*, **8,** 421 (1967).
72. J. E. DOUGLASS, *J. Amer. Chem. Soc.*, **86,** 5431 (1964).
73. A. J. DOWNS and J. II. MORRIS, *Spectrochim. Acta*, **22,** 957 (1966).
74. J. E. EASTHAM, *J. Amer. Chem. Soc.*, **89,** 2236 (1967).
75. I. D. EUBANKS and J. J. LAGOWSKI, *J. Amer. Chem. Soc.*, **88,** 2425 (1966).
76. J. N. G. FAULKS, N. N. GREENWOOD and J. H. MORRIS, *J. Inorg. Nucl. Chem.* **29,** 329 (1967).
77. E. M. FEDNEVA, V. I. ALPATOVA and V. I. MIKHEEVA, *Russ. J. Inorg. Chem.*, **9,** 56 (1964).
78. E. M. FEDNEVA, V. N. KONOPLEV and V. D. KRASNOPEROVA, *Russ. J. Inorg. Chem.*, **11,** 1094 (1966).
79. E. M. FEDNEVA, I. V. KRYOKOVA and V. I. ALPATOVA, *Russ. J. Inorg. Chem.*, **11,** 1101 (1966).

80. P. Fritz, K. Niedenzu and J. W. Dawson, *Inorg. Chem.*, **3**, 626 (1964).
81. P. Fritz, K. Niedenzu and J. W. Dawson, *Inorg. Chem.*, **3**, 778 (1964).
82. P. Fritz, K. Niedenzu and J. W. Dawson, *Inorg. Chem.*, **4**, 886 (1965).
83. D. F. Gaines and R. Schaeffer, *J. Amer. Chem. Soc.*, **86**, 1505 (1964).
84. P. N. Gates, E. J. McLaughlan and E. F. Mooney, *Spectrochim. Acta*, **21**, 1445 (1965).
85. A. R. Gatti and T. Wartik, *Inorg. Chem.*, **5**, 329 (1966).
86. A. R. Gatti and T. Wartik, *Inorg. Chem.*, **5**, 2075 (1966).
87. I. Geisler, Diplomarbeit, Univ. Marburg, 1967.
88. T. A. George and M. F. Lappert, *Chem. Comm.*, 463 (1966).
89. W. Gerrard and E. F. Mooney, *J. Chem. Soc.*, 4028 (1960).
90. P. Geymayer and E. G. Rochow, *Monatsh.*, **97**, 429 (1966).
91. P. Geymayer and E. G. Rochow, *Monatsh.*, **97**, 437 (1966).
92. P. Geymayer, E. G. Rochow and U. Wannagat, *Angew. Chem.*, **76**, 499 (1964).
93. O. Glemser and G. Elter, *Z. Naturforsch.*, **21b**, 1132 (1966).
94. J. Goubeau, *Angew. Chem.*, **78**, 565 (1966).
95. J. Goubeau and H. Schneider, *Annalen*, **675**, 1 (1964).
96. J. Goubeau, E. Bessler and D. Wolff, *Z. Anorg. Allg. Chem.*, **352**, 285 (1967).
97. R. F. Gould, (Editor), *Boron–Nitrogen Chemistry*, American Chemical Society, Washington, *Advanc. Chem. Ser.*, **42** (1964).
98. N. N. Greenwood and K. A. Hooton, *J. Chem. Soc.* (A), 751 (1966).
99. N. N. Greenwood, K. A. Hooton and J. Walker, *J. Chem. Soc.* (A), 21 (1966).
100. N. N. Greenwood and J. H. Morris, *J. Chem. Soc.*, 6205 (1965).
101. N. N. Greenwood and B. H. Robinson, *J. Chem. Soc.* (A), 511 (1967).
102. N. N. Greenwood and J. Walker, *Inorg. Nucl. Chem. Letters*, **1**, 65 (1965).
103. N. N. Greenwood and J. Walker, *J. Chem. Soc.* (A), 959 (1967).
104. J. Grotewold, E. A. Lissi and A. E. Villa, *J. Chem. Soc.* (A), 1034, 1038 (1966).
105. S. R. Gunn, *J. Phys. Chem.*, **69**, 1010 (1965).
106. V. Gutmann, A. Meller and E. Schaschel, *Monatsh.*, **95**, 1188 (1964).
107. V. Gutmann and A. Meller, *Österr. Chem. Ztg.*, **66**, 324 (1965).
108. V. Gutmann, A. Meller and R. Schlegel, *Monatsh.*, **95**, 314 (1964).
109. G. A. Hahn and R. Schaeffer, *J. Amer. Chem. Soc.*, **86**, 1503 (1964).
110. A. C. Hazell, *J. Chem. Soc.* (A), 1392 (1966).
111. G. S. Heaton and P. N. K. Riley, *J. Chem. Soc.* (A), 952 (1966).
112. C. W. Heitsch, *Inorg. Chem.*, **3**, 767 (1964).
113. C. W. Heitsch, *Inorg. Chem.*, **4**, 1019 (1965).
114. R. Hemming and D. G. Johnston, *J. Chem. Soc.*, 466 (1964).
115. K. H. Hermannsdörfer, Dissertation, Univ. München, 1966.
116. H. Hess, *Z. Kristallographie*, **118**, 366 (1963); *Acta Cryst.*, **16**, A 74 (1963).
117. G. Hesse and A. Haag, *Tetrahedron Letters*, 1123 (1965).
118. H. K. Hofmeister and J. R. van Wazer, *J. Inorg. Nucl. Chem.*, **26**, 1209 (1964).
119. J. R. Horder and M. F. Lappert, *Chem. Comm.*, 485 (1967).
120. S. C. Jain and R. Rivest, *Canad. J. Chem.*, **42**, 1079 (1964).
121. R. Jefferson, M. F. Lappert, B. Prokai and B. P. Tilley, *J. Chem. Soc.* (A), 1584 (1966).
122. H. Jenne and K. Niedenzu, *Inorg. Chem.*, **3**, 68 (1964).
123. S. Jerumanis and J. M. Lalancette, *J. Org. Chem.*, **31**, 1531 (1966).
124. J. P. Jesson, S. Trofimenko and D. R. Eaton, *J. Amer. Chem. Soc.*, **89**, 3158 (1967).
125. J. P. Jesson, S. Trofimenko and D. R. Eaton, *J. Amer. Chem. Soc.*, **89**, 3148 (1967).
126. H. C. Kelly, F. R. Marchelli and M. B. Giusto, *Inorg. Chem.*, **3**, 431 (1964).

127. H. C. KELLY and J. O. EDWARDS, *Inorg. Chem.*, **2**, 226 (1963).
128. A. J. KLANICA, J. P. FAUST and C. S. KING, *Inorg. Chem.*, **6**, 840 (1967).
129. R. KÖSTER, *Progr. Boron Chem.*, **1**, 289 (1964).
130. R. KÖSTER, H. BELLUT and E. ZIEGLER, *Angew. Chem.*, **79**, 241 (1967).
131. R. KÖSTER, G. BENEDIKT and H. W. SCHRÖTTER, *Angew. Chem.*, **76**, 649 (1964).
132. R. KÖSTER, S. HATTORI and Y. MORITA, *Angew. Chem.*, **77**, 719 (1965).
133. R. KÖSTER and K. IWASAKI, *Advanc. Chem. Ser.*, **42**, 148 (1964).
134. R. KÖSTER, W. LARBIG and G. W. RÖTERMUND, *Annalen*, **682**, 21 (1965).
135. R. KÖSTER and Y. MORITA, *Angew. Chem.*, **77**, 589 (1965).
136. P. KONRAD, Dissertation, Univ. Marburg, 1968.
137. A. KREUTZBERGER and F. C. FERRIS, *J. Org. Chem.*, **30**, 360 (1965).
138. M. A. KUCK and G. URRY, *J. Amer. Chem. Soc.*, **88**, 426 (1966).
139. M. F. LAPPERT, Private communication.
140. M. F. LAPPERT and M. K. MAJUMDAR, *Advanc. Chem. Ser.*, **42**, 209 (1964).
141. M. F. LAPPERT and M. K. MAJUMDAR, *J. Organometallic Chem.*, **6**, 316 (1966).
142. M. F. LAPPERT, M. K. MAJUMDAR and B. P. TILLEY, *J. Chem. Soc.* (A), 1590 (1966).
143. M. F. LAPPERT, J. B. PEDLEY, P. N. K. RILEY and A. TWEEDALE, *Chem. Comm.*, 788 (1967).
144. M. F. LAPPERT, H. PYSZORA and M. RIEBER, *J. Chem. Soc.*, 4256 (1965).
145. M. F. LAPPERT and G. SRIVASTAVA, *Proc. Chem. Soc.*, 120 (1964).
146. M. F. LAPPERT and G. SRIVASTAVA, *Inorg. Nucl. Chem. Letters*, **1**, 53 (1965).
147. M. F. LAPPERT and J. K. SMITH, *J. Chem. Soc.*, 7102 (1965).
148. M. F. LAPPERT and G. SRIVASTAVA, *J. Chem. Soc.* (A), 602 (1967).
149. B. L. LAUBE, R. D. BERTRAND, G. A. CASEDY, R. D. COMPTON and J. G. VERKADE, *Inorg. Chem.*, **6**, 173 (1967).
150. J. P. LAURENT and G. GROS, *C.A.*, **63**, 14360 (1965); *Compt. Rend.*, **261**, 724 (1965).
151. J. LEFFLER, E. DOLAND and T. TANIGAKI, *J. Amer. Chem. Soc.*, **87**, 928 (1965).
152. A. J. LEUSINK, W. DRENTH, J. G. NOLTES and G. J. M. VAN DER KERK, *Tetrahedron Letters*, 1263 (1967).
153. T. K. LIAO, E. G. PODREBARAC and C. C. CHENG, *J. Amer. Chem. Soc.*, **86**, 1869 (1964).
154. J. C. LOCKHART, *J. Chem. Soc.* (A), 809 (1966).
155. J. C. LOCKHART, *Chem. Rev.*, **65**, 131 (1965).
156. S. LUKAS, Dissertation, Univ. München, 1962.
157. S. C. MALHORTA, *Inorg. Chem.*, **3**, 862 (1964).
158. A. G. MASSEY and A. J. PARK, *J. Organometallic Chem.*, **2**, 461 (1964).
159. A. MELLER, *Organometallic Chem. Rev.*, **2**, 1 (1967).
160. A. MELLER and H. EGGER, *Monatsh.*, **97**, 790 (1966).
161. A. MELLER, V. GUTMANN and M. WECHSBERG, *Inorg. Nucl. Chem. Letters*, **1**, 79 (1965).
162. A. MELLER and E. SCHASCHEL, *Inorg. Nucl. Chem. Letters*, **2**, 41 (1966).
163. A. MELLER, M. WECHSBERG and V. GUTMANN, *Monatsh.*, **96**, 388 (1965).
164. A. MELLER, M. WECHSBERG and V. GUTMANN, *Monatsh.*, **97**, 1163 (1966).
165. A. MELLER, M. WECHSBERG and V. GUTMANN, *Monatsh.*, **97**, 619 (1966).
166. B. M. MIKHAILOV, V. A. DOROKHOV and N. V. MOSTOWOI, *Izvest. Akad. Nauk SSSR, Ser. Khim.*, 223 (1965).
167. B. M. MIKHAILOV, V. A. DOROKHOV and N. V. MOSTOWOI, *Zhw. Obshchei. Khim., Sonderdruck Moskau*, 228 (1965).
168. B. M. MIKHAILOV, V. A. DOROKHOV and T. A. SHCHEGOLEVA, *Izvest. Akad. Nauk SSSR, Ser. Khim.*, 446 (1963).
169. B. M. MIKHAILOV, V. A. DOROKHOV and I. P. YAKOLEV, *Izvest. Akad. Nauk, SSSR, Ser. Khim.*, 332 (1966).

170. B. M. MIKHAILOV and N. S. FEDOTOV, *Izvest. Akad. Nauk SSSR, Ser. Khim.*, 1590 (1960).
171. B. M. MIKHAILOV, N. S. FEDOTOV, T. A. SHCHEGOLEVA and V. D. SHELUDYAKOV, *Doklady. Akad. Nauk SSSR*, **145**, 340 (1962).
172. B. M. MIKHAILOV and A. F. GALKIN, *Izvest. Akad. Nauk SSSR, Ser. Khim.*, 575 (1963).
173. B. M. MIKHAILOV and T. K. KOZMINSKAYA, *Izvest. Akad. Nauk SSSR, Ser. Khim.* 1703 (1963).
174. B. M. MIKHAILOV and T. K. KOZMINSKAYA, *Izvest. Akad. Nauk SSSR, Ser. Khim.*, 439 (1965).
175. B. M. MIKHAILOV, T. K. KOZMINSKAYA and L. V. TARASOVA, *Doklady Akad. Nauk SSSR*, **160**, 615 (1965).
176. B. M. MIKHAILOV and T. K. KOZMINSKAYA, *Izvest. Akad. Nauk SSSR, Ser. Khim.*, 439 (1965).
177. V. I. MIKHEEVA and S. E. OSTROVITTYANOVA, *Russ. J. Inorg. Chem.* **11**, 2159 (1966).
178. B. M. MIKHAILOV, T. A. SHCHEGOLEVA and V. D. SHELUDYAKOV, *Izvest. Akad. Nauk SSSR, Ser. Khim.*, 2165 (1964).
179. B. M. MIKHAILOV, T. A. SHCHEGOLEVA and V. D. SHELUDYAKOV, *Doklady Akad. Nauk*, **158**, 738 (1963).
180. B. M. MIKHAILOV, V. D. SHELUDYAKOV and T. A. SHCHEGOLEVA, *Izvest. Akad. Nauk SSSR, Ser. Khim.* 1698 (1962).
181. B. M. MIKHAILOV and L. S. VASIL'EV, *Zhur. Obshchei Khim.*, **35**, 1073 (1965).
182. B. M. MIKHAILOV, L. S. VASIL'EV and A. YA. BEZMENOV, *Izvest. Akad. Nauk SSSR, Ser. Khim.*, 712 (1965).
183. J. M. MILLER and M. ONYSZCHUK, *Canad. J. Chem.*, **42**, 1518 (1964).
184. J. M. MILLER and M. ONYSZCHUK, *Canad. J. Chem.*, **43**, 1877 (1965).
185. J. M. MILLER and M. ONYSZCHUK, *Canad. J. Chem.*, **44**, 899 (1966).
186. N. E. MILLER, *J. Amer. Chem. Soc.*, **88**, 4284 (1966).
187. N. E. MILLER, B. L. CHAMBERLAND and E. L. MUETTERTIES, *Inorg. Chem.*, **3**, 1064 (1964).
188. N. E. MILLER and E. L. MUETTERTIES, *J. Amer. Chem. Soc.*, **86**, 1033 (1964).
189. N. E. MILLER and E. L. MUETTERTIES, *Inorg. Chem.*, **3**, 1196 (1964).
190. N. E. MILLER, M. D. MURPHY and D. L. REZNICKE, *Inorg. Chem.*, **5**, 1832 (1966).
191. P. C. MOEWS, Jr. and R. W. PARRY, *Inorg. Chem.*, **5**, 1552 (1966).
192. R. B. MOODIE, B. ELLUL and T. M. CONNOR, *Chem. and Ind.*, 767 (1966).
193. J. H. MORRIS and P. G. PERKINS, *J. Chem. Soc.* (A), 576 (1966).
194. J. H. MORRIS and P. G. PERKINS, *J. Chem. Soc.* (A), 580 (1966).
195. N. V. MOSTOVOI, V. A. DOROKHOV and B. M. MIKHAILOV, *Izvest. Akad. Nauk SSSR, Ser. Khim.*, 90 (1966).
196. N. V. MOSTOVOI, V. A. DOROKHOV and B. M. MIKHAILOV, *Izvest. Akad. Nauk SSSR, Ser. Khim.*, 70 (1966).
197. E. L. MUETTERTIES (Editor), *The Chemistry of Boron and its Compounds*, John Wiley & Sons, Inc., New York, 1967.
198. K. A. MUSZKAT and B. KIRSON, *Israel J. Chem.*, **2**, 63 (1964).
199. K. NAGASAWA, T. YOSHZAKI and H. WATANABE, *Inorg. Chem.*, **4**, 275 (1965).
200. V. S. V. NAYAR and R. D. PEACOCK, *Nature*, **207**, 603 (1965).
201. P. NELSON and A. PELTER, *J. Chem. Soc.*, 5142 (1965).
202. K. NIEDENZU, *Angew. Chem.*, **76**, 168 (1964).
203. K. NIEDENZU, *Allg. und Prakt. Chem.*, *(Wien)*, **17**, 596 (1966).
204. K. NIEDENZU and J. W. DAWSON, *Boron–Nitrogen Compounds*, Springer Verlag, Berlin, 1965.
205. K. NIEDENZU, J. W. DAWSON, P. FRITZ and H. JENNE, *Chem. Ber.*, **98**, 3050 (1965).
206. K. NIEDENZU, J. W. DAWSON and P. FRITZ, *Z. Anorg. Allg. Chem.*, **342**, 297 (1966).

207. K. NIEDENZU, J. W. DAWSON, G. A. NEECE, W. SAWODNY, D. R. SQUIRE and W. WEBER, *Inorg. Chem.*, **5**, 2161 (1966).
208. K. NIEDENZU, J. W. DAWSON, P. FRITZ and W. WEBER, *Chem. Ber.*, **100**, 1898 (1967).
209. K. NIEDENZU and P. FRITZ, *Z. Anorg. Allg. Chem.*, **340**, 329 (1965).
210. K. NIEDENZU and P. FRITZ, *Z. Anorg. Allg. Chem.*, **344**, 329 (1966).
211. K. NIEDENZU, P. FRITZ and H. JENNE, *Angew. Chem.*, **76**, 535 (1964).
212. K. NIEDENZU and W. WEBER, *Z. Naturforschg.*, **21b**, 811 (1966).
213. K. NIEDENZU, P. FRITZ and W. WEBER, *Z. Naturforschg.*, **22b**, 225 (1967).
214. H. NÖTH, *Habilitationsschrift*, Univ. München, 1961.
215. H. NÖTH, *Z. Naturforschg.*, **16b**, 618 (1961).
216. H. NÖTH, Unpublished results.
217. H. NÖTH and G. ABELER, *Angew. Chem.*, **77**, 506 (1965).
218. H. NÖTH and H. BEYER, *Chem. Ber.*, **93**, 928 (1960).
219. H. NÖTH and H. BEYER, *Chem. Ber.*, **93**, 939 (1960).
220. H. NÖTH, H. BEYER and H. J. VETTER, *Chem. Ber.*, **97**, 110 (1964).
221. H. NÖTH, V. A. DOROKHOV, P. FRITZ and F. PFAB, *Z. Anorg. Allg. Chem.*, **318**, 293 (1962).
222. H. NÖTH and P. FRITZ, *Z. Anorg. Allg. Chem.*, **322**, 297 (1963).
223. H. NÖTH and K. H. HERMANNSDÖRFER, *Angew. Chem.*, **76**, 377 (1964).
224. H. NÖTH and G. HÖLLERER, *Angew. Chem.*, **74**, 718 (1962).
225. H. NÖTH and G. HÖLLERER, *Chem. Ber.*, **99**, 2197 (1966).
226. H. NÖTH and S. LUKAS, *Chem. Ber.*, **95**, 1505 (1962).
227. H. NÖTH, S. LUKAS and P. SCHWEIZER, *Chem. Ber.*, **98**, 962 (1965).
228. H. NÖTH, G. MIKULASCHEK and W. RAMBECK, *Z. Anorg. Allg. Chem.*, **344**, 316 (1966).
229. H. NÖTH and W. REGNET, *Z. Anorg. Allg. Chem.*, **352**, 1 (1967).
230. H. NÖTH, H. SCHICK and W. MEISTER, *J. Organometallie. Chem.*, **1**, 401 (1964).
231. H. NÖTH, G. SCHMID and Y. CHUNG, *Proceedings 8, ICCC*, 180 (1964).
232. H. NÖTH and G. SCHMID, *J. Organometallic Chem.*, **5**, 109 (1966).
233. H. NÖTH and G. SCHMID, *Allg. und Prakt. Chem. (Wien)*, **17**, 610 (1966).
234. H. NÖTH and G. SCHMID, *Z. Anorg. Allg. Chem.*, **345**, 69 (1966).
235. H. NÖTH and W. SCHRÄGLE, *Chem. Ber.*, **97**, 2218 (1964).
236. H. NÖTH and W. SCHRÄGLE, *Chem. Ber.*, **97**, 2374 (1964).
237. H. NÖTH, P. SCHWEIZER and F. ZIEGELGÄNSBERGER, *Chem. Ber.*, **99**, 1089 (1966).
238. H. NÖTH and H. VAHRENKAMP, *Chem. Ber.*, **99**, 2757 (1966).
239. H. NÖTH and H. VAHRENKAMP, *Chem. Ber.*, **99**, 1049 (1966).
240. H. NÖTH and H. VAHRENKAMP, *J. Organometallic Chem.* **11**, 399 (1968).
241. H. NÖTH and H. VAHRENKAMP, *Chem. Ber.*, **100**, 3353 (1967).
242. P. I. PAETZOLD, *Fortschr. Chem. Forschg.*, **8**, 437 (1967).
243. P. I. PAETZOLD, *Angew. Chem.*, **79**, 583 (1967).
244. P. I. PAETZOLD, Private communication.
245. P. I. PAETZOLD, *Z. Anorg. Allg. Chem.*, **326**, 58 (1963).
246. P. I. PAETZOLD, *Z. Anorg. Allg. Chem.*, **326**, 64 (1963).
247. P. I. PAETZOLD, M. GAYOSO and K. DEHNICKE, *Chem. Ber.*, **98**, 1173 (1965).
248. P. I. PAETZOLD and H. J. HANSEN, *Z. Anorg. Allg. Chem.*, **345**, 79 (1966).
249. P. I. PAETZOLD and G. MAIER, *Angew. Chem.*, **76**, 343 (1964).
250. P. I. PAETZOLD and W. M. SIMSON, *Angew. Chem.*, **78**, 825 (1966).
251. M. PAILER and H. HUEMER, *Monatsh.*, **95**, 373 (1964).
252. A. A. PALKO, *J. Inorg. Nucl. Chem.*, **27**, 287 (1965).
253. TH. D. PARSONS, J. M. SELF and L. H. SCHAAD, *J. Amer. Chem. Soc.*, **89**, 3446 (1967).
254. I. PATTISON and K. WADE, *J. Chem. Soc.* (A), 1098 (1967).

255. P. G. Perkins and D. H. Wall, *J. Chem. Soc.* (A), 1207 (1966).
256. S. Prasad and N. Singh, *Z. Anorg. Allg. Chem.*, **346**, 217 (1966).
257. S. Prasad and N. Singh, *Z. Anorg. Allg. Chem.*, **350**, 332 (1967).
258. R. Prinz and H. Werner, *Angew. Chem.*, **79**, 63 (1967).
259. M. A. Ring, E. F. Witucki and R. C. Greenwough, *Inorg. Chem.*, **6**, 395 (1967).
260. H. J. Roth and B. Miller, *Arch. Pharm.*, **297**, 744 (1964).
261. C. R. Russ and A. G. MacDiarmid, *Angew. Chem.*, **76**, 500 (1964).
262. G. E. Ryschkewitz, *J. Amer. Chem. Soc.*, **89**, 3145 (1967).
263. G. E. Ryschkewitz and E. R. Birnbaum, *Inorg. Chem.*, **4**, 575 (1965).
264. G. E. Ryschkewitz and J. M. Garrett, *J. Amer. Chem. Soc.*, **89**, 4240 (1967).
265. I. G. Ryss and D. B. Donskaya, *Russ. J. Inorg. Chem.*, **12**, 2251 (1967).
266. I. G. Ryss and N. G. Parkhomenko, *Russ. J. Inorg. Chem.*, **11**, 55 (1966).
267. H. K. Saha, *J. Inorg. Nucl. Chem.*, **26**, 1617 (1964).
268. R. S. Satchell and D. P. N. Satchell, *J. Chem. Soc.* (B), 36 (1967).
269. G. W. Schaeffer and E. R. Anderson, *J. Amer. Chem. Soc.*, **71**, 2150 (1949).
270. R. Schaeffer and L. J. Todd, *J. Amer. Chem. Soc.*, **87**, 488 (1965).
271. R. Schell, Diplomarbeit, Univ. München, 1967.
272. H. Schick, Dissertation, Univ. München, 1966.
273. E. P. Schramm, *Inorg. Chem.*, **5**, 1291 (1966).
274. F. Schubert and K. Lang, U.S. Patent 2 994 698 (1961).
275. P. Schweizer, Dissertation, Univ. München, 1964.
276. J. A. Semlyen and P. J. Flory, *J. Chem. Soc.* (A), 191 (1966).
277. D. Seyferth and R. B. King, (Editors), *Annual Surveys of Organometallic Chemistry* Elsevier Publishing Co., Amsterdam.
278. T. A. Shchegoleva and B. M. Mikhailov, *Izvest. Akad. Nauk SSSR, Ser. Khim.*, 714 (1965).
279. T. A. Shchegoleva, V. D. Sheludyakov and B. M. Mikhailov, *Zhur. Obshchei. Khim.*, **35**, 1066 (1965).
280. V. D. Sheludyakov, T. A. Shchegoleva and B. M. Mikhailov, *Izvest. Akad. Nauk. SSSR, Ser. Khim.* 632 (1964).
281. F. G. Sherif and C. D. Schmulbach, *Inorg. Chem.*, **5**, 322 (1966).
282. S. G. Shore and K. W. Böddeker, *Inorg. Chem.*, **3**, 914 (1964).
283. S. G. Shore and C. L. Hall, *J. Amer. Chem. Soc.*, **88**, 5346 (1966).
284. S. G. Shore, C. W. Hickam, Jr. and D. Cowles, *J. Amer. Chem. Soc.*, **87**, 2755 (1965).
285. S. G. Shore and R. W. Parry, *J. Amer. Chem. Soc.*, **77**, 6084 (1955);
D. R. Schultz and R. W. Parry, *ibid.*, **80**, 4 (1958);
S. G. Shore and R. W. Parry, *ibid.*, **80**, 8, 12, 15 (1958).
286. S. G. Shore and C. L. Hall, *J. Amer. Chem. Soc.*, **89**, 3947 (1967).
287. D. F. Shriver and M. J. Biallas, *J. Amer. Chem. Soc.*, **89**, 1078 (1967).
288. I. H. Skoog, *J. Org. Chem.*, **29**, 492 (1964).
289. J. Soulié and P. Cadiot. *Bull. Soc. Chim. France*, 3846 (1966).
290. J. Soulié and P. Cadiot, *Bull. Soc. Chim. France*, 1981 (1966).
291. V. I. Spitsyn, I. D. Colli, R. A. Rodinov and T. G. Sevastyanova, *Doklad. Akad. Nauk SSSR*, **160**, 1101 (1965).
292. V. P. Sorokin, B. I. Vesnina and N. S. Klimova, *Russ. J. Inorg. Chem.*, **8**, 32 (1963).
293. V. P. Sorokin and B. I. Vesnina, *Russ. J. Inorg. Chem.*, **8**, 32 (1963).
294. S. L. Stafford and F. G. A. Stone, *J. Amer. Chem. Soc.*, **82**, 6283 (1960).
295. H. Steinberg and R. J. Brotherton, *Organoboron Chemistry*, Vol. 2, Interscience Publishers, New York, 1966.
296. H. Steinberg and A. L. McCloskey, (Editors), *Progress in Boron Chemistry*, Vol. 1, Pergamon Press, Oxford, 1964.

297. F. G. A. STONE and R. WEST, (Editors), *Advances of Organometallic Chemistry*, Vol. 5, Academic Press, New York, 1967, see chapter by M. F. Lappert and B. Prokai, p. 225.
298. J. S. THAYER and R. WEST, *Inorg. Chem.* **4**, 114 (1965).
299. N. R. THOMPSON, *J. Chem., Soc.* 6290 (1965).
300. S. TROFIMENKO, *J. Amer. Chem., Soc.*, **89**, 3903 (1967).
301. S. TROFIMENKO, *J. Amer. Chem. Soc.*, **88**, 1842 (1966).
302. S. TROFIMENKO, *J. Amer. Chem. Soc.*, **89**, 3165 (1967).
303. S. TROFIMENKO, *J. Amer. Chem. Soc.*, **89**, 3170 (1967).
304. S. TROFIMENKO, *J. Amer. Chem. Soc.*, **89**, 3904 (1967).
305. H. S. TURNER, R. J. WARNE and I. J. LAWRENSON, *Chem. Comm.*, 20 (1965).
(305a) H. S. TURNER and R. J. WARNE, *Advanc. Chem. Ser.*, **42**, 290, (1964).
306. J. M. TURNER, *J. Chem. Soc.* (A), 401 (1966).
307. J. M. TURNER, *J. Chem. Soc.* (A), 415 (1966).
308. F. UMLAND, *Angew. Chem.*, **79**, 583 (1967).
309. F. UMLAND and C. SCHLEYERBACH, *Angew. Chem.*, **77**, 169 (1965).
310. F. UMLAND and C. SCHLEYERBACH, *Angew. Chem.*, **77**, 426 (1965).
311. H. VAHRENKAMP, Dissertation, Univ. München, 1967.
312. F. E. WALKER and R. K. PEARSON, *J. Inorg. Nucl. Chem.*, **27**, 1981 (1965).
313. J. R. WEAVER and R. W. PARRY, *Inorg. Chem.*, **5**, 713 (1966).
314. J. R. WEAVER and R. W. PARRY, *Inorg. Chem.*, **5**, 718 (1966).
315. W. WEBER, J. W. DAWSON and K. NIEDENZU, *Inorg. Chem.*, **5**, 726 (1966).
316. R. L. WELLS and A. L. COLLINS, *Inorg. Nucl. Chem. Letters*, **2**, 201 (1966).
317. R. L. WELLS and R. W. NELSON, *Inorg. Nucl. Chem. Letters*, **1**, 149 (1965).
318. R. L. WELLS and A. L. COLLINS, *Inorg. Chem.* **5**, 1327 (1966).
319. S. I. WELSSMAN and H. VAN WILLIGEN, *J. Amer. Chem. Soc.* **87**, 2285 (1965).
320. E. WIBERG, *Naturwissenschaften*, **35**, 182, 212 (1948).
321. K. J. WYNNE and W. L. JOLLY, *Inorg. Chem.*, **6**, 107 (1967).
322. D. E. YOUNG, G. E. McACHRAN and S. G. SHORE, *J. Amer. Chem. Soc.*, **88**, 4390 (1966).
323. S. G. SHORE, *et al.*, *151st Mtg. Amer. Chem. Soc.*, Pittsburgh, Pa., March 22, 1966, Abstracts of Papers, p. H-15; *Nachr. Chem. Techn.*, **14**, 194 (1966).
324. V. A. ZAMYATINA and N. I. BEKASOVA, *Russ. Chem. Reviews*, 524 (1964).
325. A. B. BURG and J. BANUS, *J. Amer. Chem. Soc.*, **76**, 3903 (1954).
326. R. SCHAEFFER and D. F. GAINES, *J. Amer. Chem. Soc.*, **85**, 395 (1963).
327. M. F. LAPPERT and B. PROKAI, *Advanc. Organometallic Chem.*, **5**, 225 (1967).
328. G. E. COATES and J. A. LIVINGSTONE, *J. Chem. Soc.*, **1**, 1000 (1961).
329. R. L. LETSINGER and I. H. SKOOG, *J. Amer. Chem. Soc.*, **77**, 2491 (1955).

5

ORGANIC BORON–SULFUR COMPOUNDS

by B. M. Mikhailov

N. D. Zelinsky Institute for Organic Geochemistry, Moscow, U.S.S.R.

CONTENTS

I. INTRODUCTION

The first mention of a sulfur-containing organoboron compound was in 1878, when Councler[11] obtained a substance from triisobutylborate and phosphorus pentasulfide which he took to be triisobutylthioborate. In the half century beginning with 1908[8] only incidental investigations of organic boron–sulfur compounds were made; their systematic study dating from 1959. At present this branch of boron chemistry may be considered to the quite well explored. Simple and convenient methods are available for synthesizing the principal types of compounds with B—S bonds such as thioborates, esters of thioboronic and thioborinic acids, alkylthioboranes, compounds with mixed functions and various types of sulfur-containing heterocyclic boron compounds.

Quite extensive study has been made of the chemical properties of organic boron–sulfur compounds.

Their chemistry has been partly treated in a review article[27] and in Gerrard's monograph.[16] The book by Steinberg[116] published in 1964 discussed compounds with B—S, but not B—C bonds.

II. ORTHO- AND METATHIOBORATES

A. *Introduction*

By reacting triisobutyl borate with phosphorus pentasulfide a yellow substance was obtained which was assumed to be impure triisobutyl thioborate.[11] However, it is quite doubtful that this was a thioborate since it decomposed on distillation in vacuum whereas alkyl thioborates are stable under such conditions. The simplest representative of the class of thioborate esters, trimethyl thioborate, was synthesized in 1952 by treating boron tribromide with silver or lead thiocyanate in benzene solution.[18]

B. *Preparation of Thioborates*

1. *Thioborates from Boron Trihalides*

The cheapest and most convenient way of producing thioborates is by treatment of boron trichloride with mercaptans:

$$BCl_3 + 3\,RSH \longrightarrow B(SR)_3 + 3\,HCl \tag{1}$$

Despite the apparent simplicity of this reaction, the first attempts to prepare thioborates in this way were unsuccessful. Thus, reaction of boron trichloride and methyl mercaptan gave methylthioboron dichloride as a dimer $(CH_3SBCl_2)_2$ that did not react further with mercaptan.[18] The reaction with thiophenol also does not go to completion.[15] Later it was shown that substitution of all the chlorine atoms by alkylthio groups can be achieved only under more drastic conditions. On mixing boron trichloride with mercaptans in the cold the crystalline complex

$$Cl_3B \cdot S \diagdown_{H}^{R}$$

is formed which at 0–40° eliminates only one molecule of hydrogen chloride. In the case of the higher mercaptans with sufficiently high boiling points, alkyl thioborates could be obtained by subsequent heating of the reaction mixture at 80–150°.[40] In this way n-propyl thioborate (30 per cent) and n-butyl thioborate (67 per cent) were obtained. n-Amyl thioborate was prepared in 57 per cent yield by refluxing boron trichloride and amyl mercaptan in a solution of decane.[108]

The reaction between boron trichloride and mercaptans proceeds more smoothly in the presence of tertiary amines or pyridine. However, without the application of heat the yield, here too, is not very high; for example, it is only 34 per cent for trimethyl thioborate.[108] In contrast hours of reflux of an ether solution of a mixture of methyl mercaptan, boron trichloride and triethylamine increases the yield of trimethyl thioborate to 80 per cent.[40] The reaction proceeds as easily with other mercaptans (yields 70–90 per cent).[40]

$$BCl_3 + 3\ RSH + 3\ R_3'N \rightarrow B(SR)_3 + 3\ R_3NHCl \tag{2}$$

The reaction of boron tribromide with mercaptans was also investigated. In the case of methyl mercaptan a monosubstituted compound, the dimer of methylthioboron dibromide[18, 127] is formed; reaction with n-amyl mercaptan in the presence of pyridine gives n-amyl thioborate in 27 per cent yield; thiophenol gives triphenyl thioborate.[15, 128]

Thioborates also can be obtained by the action of metal mercaptides on boron trihalides. Trimethyl thioborate was synthesized by refluxing a benzene solution of boron trichloride and silver or lead (but not potassium) mercaptide[18] or by heating boron tribromide with sodium mercaptide in the absence of solvent.[7]

Boron thiocyanate was prepared from boron tribromide and silver thiocyanate.[8] Subsequently this compound was assigned the structure of an isothiocyanate.[26]

2. Thioborates from Diborane

Trimers of alkylthioboranes (1,3,5-cyclotriborothianes) formed by reaction between mercaptans and diborane react in turn with mercaptans at $100-120°$ to give alkyl thioborates.[84]

$$(RSBH_2)_3 + 6 \ RSH \xrightarrow[53-84\%]{100-120°} 3 \ B(SR)_3 + 6 \ H_2 \qquad (3)$$

Alkyl thioborates can also be obtained by heating dialkylthioboranes (4)[78, 79] or trialkylamine-boranes (5)[20, 25] with mercaptans.

$$(RS)_2BH + RSH \xrightarrow[70-78\%]{60-150°} B(SR)_3 + H_2 \qquad (4)$$

$$R_3N \cdot BH_3 + 3 \ RSH \xrightarrow[65-70\%]{60-160°} B(SR)_3 + R_3N + 3 \ H_2 \qquad (5)$$

3. Other Methods of Preparing Orthothioborates and their Derivatives and Metathioborates

A number of other reactions giving thioborates and their derivatives are known, most of which have no preparative significance. Treatment of methyl, hexyl or benzyl mercaptan and hydrogen chloride with tris(dialkylamino)borane yields the corresponding thioborates.[5, 108] Triethyl thioborate has been prepared from hydrogen sulfide and dibromo(ethylthio)borane.[118]

$$3 \ C_2H_5SBBr_2 + 3 \ H_2S \longrightarrow B(SC_2H_5)_3 + B_2S_3 + 6 \ HBr \qquad (6)$$

The polymer of methylthioborane undergoes disproportionation at $140°$ to trimethyl thioborate and diborane.[7] Trimethylamine-methylthioborane[7], methyl methylthioboronate and methyl diamethylthioborinate[6] also disproportionate to trimethyl thioborate on heating. Triphenyl thioborate is formed by reaction between potassium fluoroborate and aluminium thiophenolate.[23]

$$KBF_4 + Al(SC_6H_5)_3 \xrightarrow{260°} B(SC_6H_5)_3 + AlF_3 + KF \qquad (7)$$

Ethylene dithioglycol reacts with boron trichloride to give 2-chloro-1,3,2-dithiaborolane, which can be converted to the dimethylamino derivative.[9]

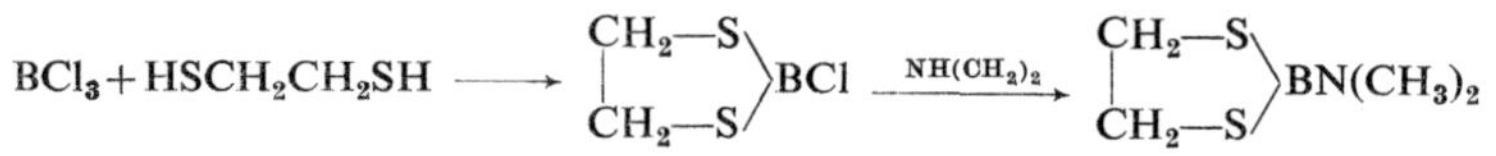

$$BCl_3 + HSCH_2CH_2SH \longrightarrow \begin{matrix} CH_2-S \\ | \\ CH_2-S \end{matrix}\Big\rangle BCl \xrightarrow{NH(CH_3)_2} \begin{matrix} CH_2-S \\ | \\ CH_2-S \end{matrix}\Big\rangle BN(CH_3)_2 \qquad (8)$$

2-Alkylthio-1,3,2-dioxaborolanes were prepared from reactions of the corresponding chlorine derivatives with mercaptans.[2, 3, 17]

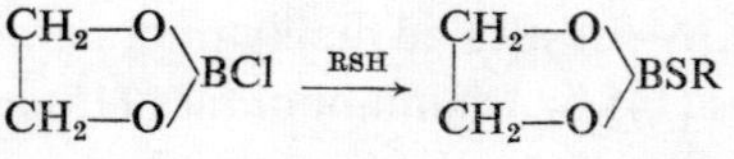

$$
\begin{array}{c}
\text{CH}_2\text{—O} \\
\text{CH}_2\text{—O}
\end{array}\!\!\!\text{BCl} \xrightarrow{\text{RSH}}
\begin{array}{c}
\text{CH}_2\text{—O} \\
\text{CH}_2\text{—O}
\end{array}\!\!\!\text{BSR} \qquad (9)
$$

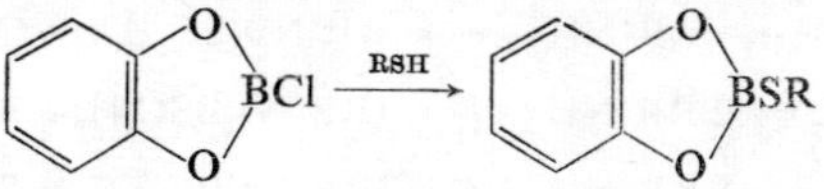

$$
\text{BCl} \xrightarrow{\text{RSH}} \text{BSR} \qquad (10)
$$

$$
R = n\text{—}C_4H_9, \quad n\text{—}C_8H_{17}
$$

Prolonged refluxing of a carbon disulfide solution of metaboric acid[117, 119] and triethyl borate produces the trimer of metathioboric acid,[126]

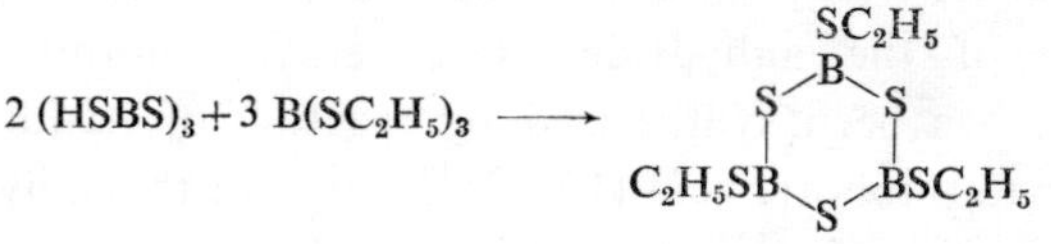

$$
2\,(HSBS)_3 + 3\,B(SC_2H_5)_3 \longrightarrow
\begin{array}{c}
SC_2H_5 \\
| \\
B \\
S \quad S \\
C_2H_5SB \quad BSC_2H_5 \\
S
\end{array} \qquad (11)
$$

which on distillation is converted to the dimer, $(C_2H_5SBS)_2$.

C. Properties of Thioborates

1. General

Trialkyl thioborates are liquids with a very unpleasant odor; triphenyl thioborate is a solid without a sharp melting point.

The B^{11} chemical shift in tri-n-butyl thioborate is -47.9 ppm with respect to that for trimethyl borate as zero reference.[20] The infrared spectra of the thioborate shows intense absorption bonds at 10–$11\,\mu$ and weaker ones at 13.25–$13.50\,\mu$.

2. Hydrolysis and Alcoholysis

Thioborates are readily hydrolyzed to boric acid and mercaptans[7, 20] and alcoholyzed to alkyl borates and mercaptans.[18]

$$
B(SR)_3 + 3\,R'OH \longrightarrow B(OR')_3 + 3\,RSH \qquad (12)
$$

3. Reactions with Amines

Trimethyl thioborate forms a $1:1$ complex with pyridine;[40] with triethyl thioborate the pyridine complex is unstable and on heating decomposes into the initial components.[111]

As shown by the example of triethyl thioborate, thioborates with primary or secondary amines undergo successive substitution of the alkylthio groups by the corresponding alkylamino groups. As a result, depending on the reactant ratio, N-substituted dialkylthio(amino)boranes (13), alkylthio(diamino)boranes (14) or triaminoboranes (15) are obtained.[111]

$$B(SC_2H_5)_3 + RR'NH \longrightarrow RR'NB(SC_2H_5)_2 + C_2H_5SH \qquad (13)$$

$$B(SC_2H_5)_3 + 2\ RR'NH \longrightarrow (RR'N)_2BSC_2H_5 + 2\ C_2H_5SH \qquad (14)$$

$$B(SC_2H_5)_3 + 3\ RR'NH \longrightarrow (RR'N)_3B + 3\ C_2H_5SH \qquad (15)$$

$$R = H,\ or\ alkyl$$

Diethylthio(phenylamino)borane, obtained by the action of one mole of aniline on triethyl thioborate, undergoes disproportionation at room temperature to tris(phenylamino)borane and triethyl thioborate.

The trimer of the anhydride of dimethylaminothioboronic acid, $[(CH_3)_2NBS]_3$, has been synthesized by reaction of tris(dimethylamino)borane with metathioboric acid (16)[120, 122] or with the anhydride of bromothioboronic acid (17);[122] it can also be obtained from metathioboric acid and $Al[N(CH_3)_3]_3$.[123]

$$(HSBS)_3 + B[N(CH_3)_2]_3 \longrightarrow [(CH_3)_2NBS]_3 + B(SH)_3 \qquad (16)$$

$$(BrBS)_3 + B[N(CH_3)_2]_3 \longrightarrow [(CH_3)_2NBS]_3 + BBr_3 \qquad (17)$$

$(CH_3NHBS)_3$ is produced by the action of methylamine on $(BrBS)_3$.[118]

$$(BrBS)_3 + 3\ NH_2CH_3 \longrightarrow (CH_3NHBS)_3 + 3\ HBr \qquad (18)$$

The reaction between triethyl thioborate and ethylene diamine afforded a polymer without ethylthio groups, for which the following structure was suggested:[111]

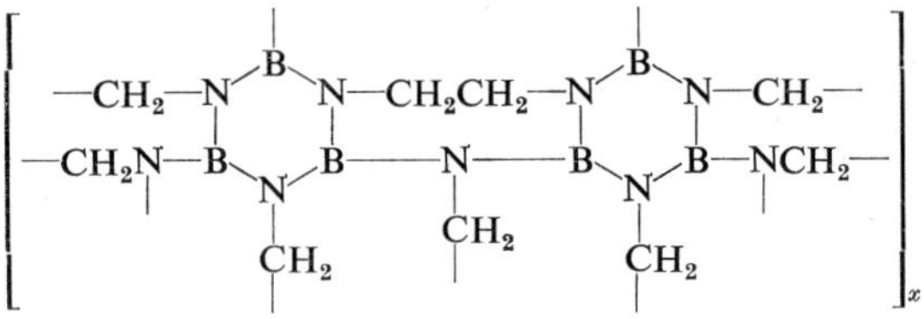

4. Double Decomposition Reactions with Boron Compounds

Trialkyl thioborates enter into double decomposition reactions with esters of organoboron acids[94, 95, 99] or trialkylboranes[6, 81, 94] to give esters of thioorganoboron acids.

These reactions will be discussed later (Section III, A, 1).

5. Reactions with Carbonyl Compounds and Vinyl Ethers

The high reactivity of thioborates can be seen from the fact that they undergo reaction with multifarious organic compounds, such as aldehydes, ketones, amides and vinyl ethers.

Thioborates react with aldehydes or ketones to give thioacetals or thioketals, respectively, and boron trioxide.[55] The reaction between triethyl thioborate and benzaldehyde in benzene solution proceeds spontaneously with evolution of heat, giving benzaldehyde diethylthioacetal in 91 per cent yield and boron trioxide.

$$3 \, C_6H_5CHO + 2 \, B(SC_2H_5)_3 \longrightarrow 3 \, C_6H_5CH(SC_2H_2)_2 + B_2O_3 \qquad (19)$$

On heating a benzene solution of acetone and tri-n-butyl thioborate on a water bath, acetone di-n-butylketal is formed in high yields:

$$3 \, CH_3COCH_3 + 2 \, B(SC_4H_9)_3 \longrightarrow 3 \, (CH_3)_2C(SC_2H_5)_2 + B_2O_3 \qquad (20$$

Under analogous conditions triethyl thioborate and acetophenone give smoothly acetophenone diethylthioketol. The reaction apparently proceeds by addition of the thioborate to the carbonyl group (21) and subsequent conversion of the product according to equation (22).

$$R_2CO + B(SR)_3 \longrightarrow R_2C{\overset{\textstyle OB(SR)_2}{\underset{\textstyle SR}{<}}} \qquad (21)$$

$$R_2C{\overset{\textstyle OB(SR)_2}{\underset{\textstyle SR}{<}}} + B(SR)_3 \longrightarrow R_2C(SR)_2 + (RS)_2BOB(SR)_2 \qquad (22)$$

$$\downarrow$$

$$B(SR)_3 + B_2O_3$$

Dimethyl formamide reacts with thioborate similarly to its reaction with aldehydes and ketones (23).

$$3 \, (CH_3)_2NCHO + 2B(SC_4H_9)_3 \longrightarrow (CH_3)_2NCH(SC_4H_9)_2 + B_2O_3 \qquad (23)$$

The relatively high yield (66 per cent) of bis(butylthio)dimethylaminomethane points to the possibility of using this reaction for preparation of amidomercaptals.

On heating thioborates with vinyl alkyl ether, vinyl alkyl sulfide and trialkyl borates are obtained.[13]

$$3 \, CH_2{=}CHOR + B(SR)_3 \xrightarrow[71\%]{100-150°} CH_2{=}CHSR + B(OR)_3 \qquad (24)$$

In this way vinyl butyl sulfide was prepared from tri-n-butyl thioborate and vinyl butyl ether. Similar reactions were carried out with n-butyl thioborate and vinyl ethyl ether.

It may be assumed that the thioborate adds to the double bond of the vinyl ether; the product then undergoes β-decomposition with elimination of vinyl alkyl sulfide.

$$CH_2=CHOR + B(SR)_3 \longrightarrow (RS)_2BCH_2CH{\diagup}^{OR}_{\diagdown SR} \longrightarrow$$
$$(RS)_2BOR + CH_2=CHSR \tag{25}$$
$$\downarrow$$
$$B(OR)_3 + B(SR)_3$$

The proposed mechanism is in good agreement with the tendency of β-alkoxy derivatives of boronic ester to eliminate ethylene.[32]

III. THIOBORONATES AND THIOBORINATES

A. *Introduction*

Like the thioborates, esters of thioboronic and thioborinic acids began to attract attention only in recent years. The first compounds of this class to be synthesized were dodecyl β-chlorovinylthioboronate and n-octyl chloro-β-chlorovinylthioborinate prepared in 1946.[109]

Later several more representatives of this class were described, but systematic study of the esters of thioboronic and thioborinic acids began only in 1959. The preparative methods for these compounds are based on the use of organoboron halides, trialkylboranes and esters of organoboron acids. Similarly to the thioborates, esters of organothioboron acids are very reactive and are successfully used for preparing a variety of types of organoboron compounds.

B. *Preparation of Thioboronates and Thioborinates*

1. *Synthesis of Thioboronates and Thioborinates from Organoboron Halides and Mercaptans*

One of the general methods for preparing esters of thioorganoboron acids consists in the reaction of mercaptans or mercaptides with organoboron halides. Both aliphatic and aromatic esters of the types $RB(SB)_2$,

R_2BSR and $RBCl(SR)$ can be obtained in this way. The method was first used to prepare n-octyl chlorovinylchlorothioborinate from n-octyl mercaptan and β-chlorovinylboron dichloride.[109] Subsequently it was found that whether only one or both halogen atoms of the alkylboron dihalides are replaced by alkylthio groups depends on the reactant ratio, temperature and duration of heating.

n-Butyl alkylthioboronates are smoothly formed on refluxing a mixture of alkylboron dichloride or alkylboron dibromide with excess n-butyl mercaptan for 10–15 hours:[61]

$$RBX_2 + 2\,n\text{-}C_4H_9SH \xrightarrow[80-90\%]{} RB(S\text{-}n\text{-}C_4H_9)_2 + 2\,HX \tag{26}$$

Similarly n-butyl phenylthioboronate was prepared using phenylboron dichloride.[61]

Heating a mixture of alkylboron dichloride and excess ethylmercaptan to boiling yields diethyl alkylthioboronate and products of the incomplete substitution of the chlorine atoms, ethyl alkylchlorothioborinates.[64]

$$RBCl_2 \xrightarrow{C_2H_5SH} RB(SC_2H_5)_2 + RBCl(SC_2H_5) \tag{27}$$

Esters of alkylchlorothioborinic acids on vacuum distillation undergo partial disproportionation to alkylthioboronates and alkylboron dichlorides; hence they cannot be obtained in high yield even with equimolecular ratios of the reactants.

Reaction of alkylboron dibromides with ethylmercaptan in a 1 : 1 ratio leads to ethyl alkylbromothioborinate in yields of 65–75 per cent:[64] In a similar way phenylboron dibromide reacts with ethyl mercaptan.[64]

$$RBBr_2 + C_2H_5SH \longrightarrow RBBr(SC_2H_5) + HBr \tag{28}$$

The bromothioesters are much more thermally stable than their chlorine counterparts and can be distilled in vacuum without decomposition.

Esters of alkylchloroborinic acids can also be obtained by double decomposition reactions between alkylboron dichlorides and alkylthioboronates.[64] Thus, if an equimolar mixture of n-propylboron dichloride and diethyl n-propylthioboronate is allowed to react at room temperature for 20 hours, ethyl n-propylchloroborinate is isolated in 50 per cent yield based on equation (29).

$$n\text{-}C_3H_7BCl_2 + n\text{-}C_3H_7B(SC_2H_5)_2 \longrightarrow 2\,n\text{-}C_3H_7B(SC_2H_5)Cl \tag{29}$$

The difference in stability between the esters of halothioorganoboron acids and their oxygen analogs should be mentioned. The tendency to undergo disproportionation is observed only for esters of fluoro-organo-boron acids,[4] while esters of chloro- and bromo-organoboron acids are distilled without change.[76, 77]

Another method for the preparation of alkylhaloborinates consists in treating thioborinates with boron halides at room temperature.[64] Thus, n-butyl isoamylthioboronate and boron tribromide gave n-butyl isoamyl-bromothioboronite (30).

$$3 \text{ i-}C_5H_{11}B(S\text{-n-}C_4H_9)_2 + BBr_3 \xrightarrow[80\%]{} 3 \text{ i-}C_5H_9B(S\text{-n-}C_4H_9)Br + B(S\text{-n-}C_4H_9)_3 \quad (30)$$

Dialkyl-[6, 69] and diarylboron halides[61] are converted to the corresponding dialkylthio- and diarylthioborinates by heating with mercaptans. For this purpose the mercaptans also can be replaced by the mercaptides of various metals. For instance, reaction of phenylboron dichloride with sodium n-amylmercaptide gave diisoamyl phenylthioboronate.[103]

Phenylboron dibromide[125] and alkylboron dibromides[118] on heating with hydrogen sulfide in an inert solvent (benzene, carbon disulfide) are converted into the cyclic trimeric "anhydrides" of phenyl- or alkylthioboronic acids.

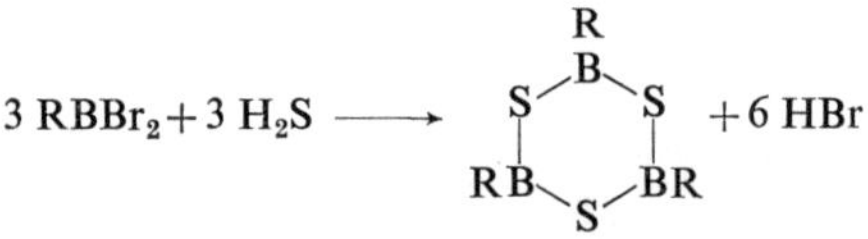

$$3 \text{ RBBr}_2 + 3 \text{ H}_2S \longrightarrow \underset{\substack{R \\ | \\ B \\ S \diagup \diagdown S \\ RB \diagdown \diagup BR \\ S}}{} + 6 \text{ HBr} \quad (31)$$

Reactions of organoboron dihalides with thio derivatives of benzene lead to cyclic esters of thioorganoboron acids. Thus, 2-phenylbenzo-1,3,2-thiazaborolane (32) and 2-phenylbenzo-1,3,2-dithiaborolane (33) were prepared from phenylboron dichloride and 2-aminothiophenol or dithio-catechol (12).

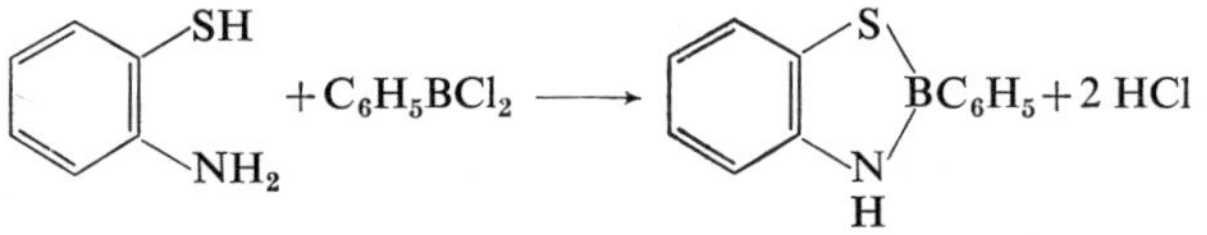

$$(32)$$

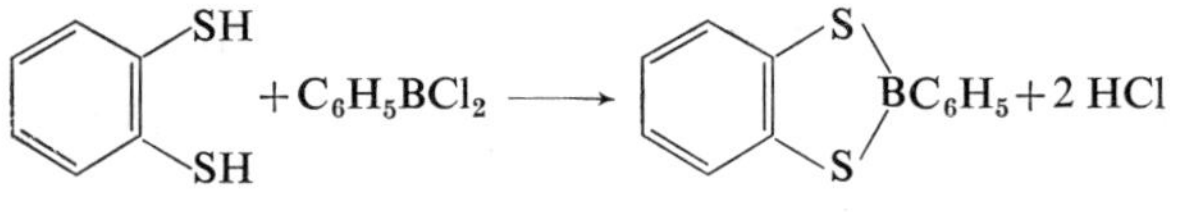

$$(33)$$

2. Synthesis of Thioboronates and Thioborinates From Trialkylboranes

(a) *Reaction of trialkylboranes with mercaptans.* A simple method for the preparation of thioborinic esters is the reaction of trialkylboranes and mercaptans.[24, 37, 100] Reaction begins at room temperature and proceeds with the evolution of heat; towards the end, heating for a short time at 110–160° is required to bring it to completion.

$$BR_3 + RSH \xrightarrow[72-87\%]{} R_2BSR + RH \tag{34}$$

A detailed study showed that this reaction is initiated by peroxides (H_2O_2, ROOH, ROOR, R_2BOOR, etc.) and by ultraviolet light.[42] Thus, only freshly distilled trialkylboranes which practically always contain traces of the peroxides $>$BOOR, begin to react at room temperature. In contrast, trialkyboranes freed of peroxides by heating at 100° for several hours or being allowed to stand at room temperature for several months do not react with mercaptans below 150°. If a small amount (2–10 ml) of air is introduced into such a mixture of "deactivated" trialkylborane and mercaptan or if porcelain chips or crushed diatomaceous brick are added, reaction begins with the formation of the thio ester R_2BSR and alkane. The rate of reaction is still greater if a drop of hydrogen peroxide or alkyl hydroperoxide is added.

The specific behavior of mercaptans towards trialkylboranes, in sharp contrast to that of other nucleophilic reagents such as water, alcohols and amines, with which the trialkylboranes react at 150–200°,[100] is apparently due to the ability of mercaptans to participate in radical reactions induced by organoboron peroxides.

Contrary to mercaptans, thiophenol reacts with trialkylboranes on heating[37, 100] to give olefins and hydrogen besides saturated hydrocarbons (35) as is common with other nucleophilic reagents[37]

$$BR_3 \xrightarrow[150-170°]{C_6H_5SH} R_2BSC_6H_5 + RH + \text{olefin} + H_2 \tag{35}$$

Apparently the formation of saturated hydrocarbons proceeds via a four-center transition state, whereas olefin and hydrogen are produced by a six-center mechanism (36).

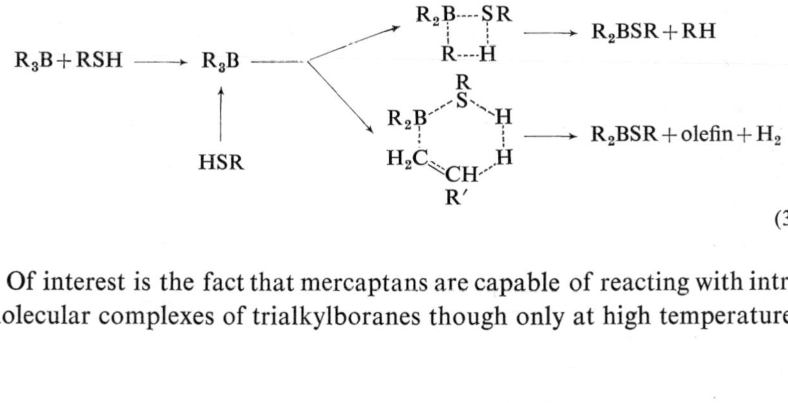

$$\tag{36}$$

Of interest is the fact that mercaptans are capable of reacting with intramolecular complexes of trialkylboranes though only at high temperatures.

Thus, butylmercaptan reacts with (γ-diethylaminopropyl)-di-n-butylborane to give n-butyl (γ-diethylaminopropyl)-n-butylthioborinate.[101]

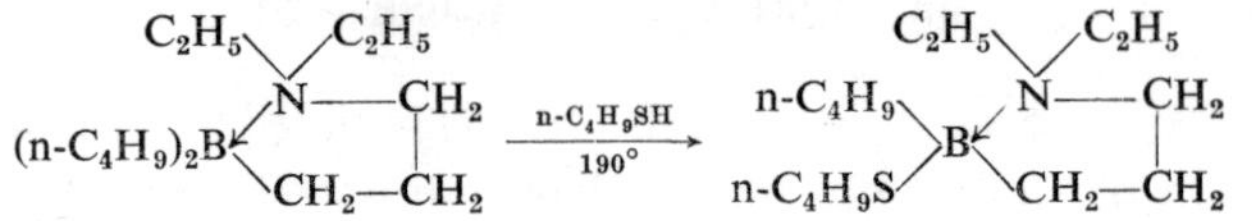

$$(37)$$

The intramolecular γ-aminopropyl-di-n-butylborane complex also reacts with mercaptans at high temperature; however, the resultant ester of γ-aminopropyl n-butylthioborinic acid immediately eliminates mercaptan, transforming into 2-n-butyl-1,2-azaborolidine.[101]

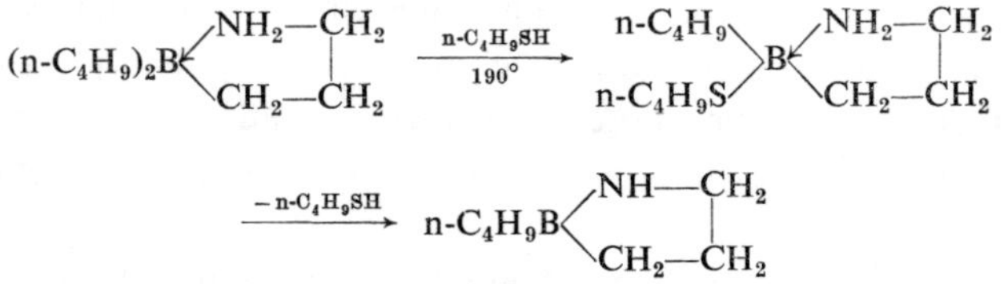

$$(38)$$

An enhanced tendency to exchange the radical for an alkylthio group is manifested by triallylborane on treatment with mercaptan.[91, 92]

Triallylborane reacts with ethyl mercaptan at -10 to $-15°$; with equimolar reactant ratio, ethyl diallylthioborinate and propylene are formed.

If one mole of triallylborane is treated with two moles of ethylmercaptan, the main reaction product becomes diethyl allylthioboronate, although some ethyl diallylthioborinate and 3-ethylthiopropylthioborinate are also formed (39).

$$B(CH_2CH{=}CH_2)_3 + 2\,C_2H_5SH \longrightarrow (CH_2{=}CHCH_2)B(SC_2H_5)_2 + 2\,C_3H_6$$

$$\downarrow {\scriptstyle C_2H_5SH}$$

$$C_2H_5SCH_2CH_2CH_2B(SC_2H_5)_2$$

$$(39)$$

Diethyl 3-ethylthiopropylthioboronate can be obtained directly from ethylmercaptan and diethyl allylthioboronate.

The high reactivity of triallylborane towards mercaptans and various other nucleophilic reactants[41, 88–90] is apparently associated with a continuous allylic rearrangement.[30] It was shown by a p.m.r. study[33] that the allylic rearrangement in triallylborane proceeds at a rate of 4.7 c/s at $-25°$ and 3.3 kc/s at $88°$ (on a n.m.r. time scale), the lifetime of each state thus varying in the above temperature range from 2×10^{-1} to 3×10^{-4} sec.

$$(C_3H_5)_2B\overset{\cdot}{C}H_2CH{=}CH_2 \xrightleftharpoons{} \overset{\text{transition}}{\underset{\text{state}}{}} \xrightleftharpoons{} (C_3H_5)_2BCH_2CH{=}\overset{\cdot}{C}H_2 \qquad (40)$$

(b) *Reaction of trialkylboranes with sulfur.* A characteristic of the trialkylboranes is their ability to react with oxygen with the formation, depending on conditions, of dialkylborinates or alkylboronates. It was found that on heating trialkylboranes also can react with sulfur to form dialkylthioborinates.[35, 36]

$$BR_3 + S \xrightarrow[40\%]{140°} R_2BSR \qquad (41)$$

By analogy with oxidation one may assume that evidently the trialkylborane first forms a molecular compound with sulfur which then reacts with a second molecule of borane.

(c) *Reaction of trialkylboranes with thioborates.* At elevated temperatures trialkylboranes participate in double decomposition reactions with thioborates. For example, prolonged heating of trimethylborane and methyl thioborate in the presence of traces of diborane yielded methyl dimethylthioborinate:[6]

$$2 B(CH_3)_3 + B(SCH_3)_3 \xrightleftharpoons{} 3 (CH_3)_2BSCH_3 \qquad (42)$$

Trialkylboranes with ethyl thioborate in the absence of the diborane catalyst gave diethyl alkylthioboronates in 65–75 per cent yields:[81]

$$BR_3 + 2 B(SC_3H_5)_3 \xrightarrow[65-75\%]{250°} 3 RB(SC_2H_5)_2 \qquad (43)$$

Under similar conditions reactions between trialkylboranes and n-butyl thioborate proceed less smoothly. Reaction products include 10–15 per cent of dialkylthioborinic esters and some starting material which is ascribed to reversibility of the reactions (44) and (45).

$$BR_3 + B(SR')_3 \xrightleftharpoons{} R_2BSR' + RB(SR')_2 \qquad (44)$$

$$R_2BSR' + B(SR')_3 \xrightleftharpoons{} 2 RB(SR')_2 \qquad (45)$$

The equilibrium lies more to the right in the case of triethylthio borate than tributyl thioborate.

The use of a tetraalkyldiborane as a catalyst greatly lowered the reaction time and temperature and resulted in a 73 per cent yield of di-n-butylthioboronate.[94]

As in the case of double decomposition reactions between trialkylboranes and orthoborates,[93, 94, 96, 97] the catalytic action of the tetraalkyldiborane and other compounds with B—H bonds is ascribed to their ability to form dimeric intermediates with three-center B—H—B bonds, for example:

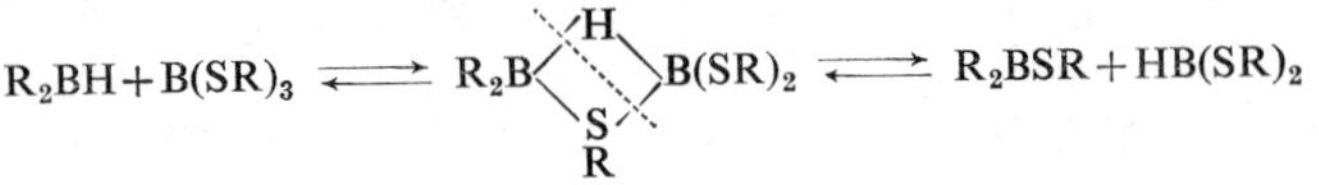

$$R_2BH + B(SR)_3 \rightleftharpoons R_2B\underset{S,R}{\overset{H}{<}}B(SR)_2 \rightleftharpoons R_2BSR + HB(SR)_2 \qquad (46)$$

$$BR_3 + HB(SR)_2 \rightleftharpoons R_2B\underset{R}{\overset{H}{<}}B(SR)_2 \rightleftharpoons R_2BH + RB(SR)_2 \qquad (47)$$

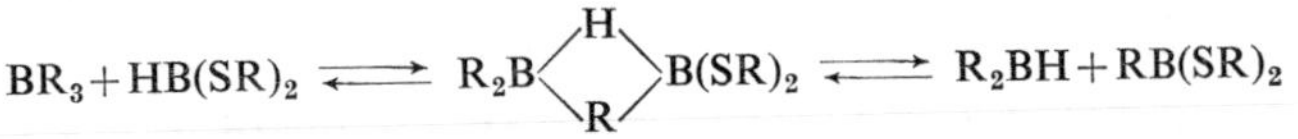

In the presence of B—H compounds 1,4-di-(1-boracyclopentyl)-butane can be smoothly transformed into 1-ethylthioboracyclopentane (48) and the latter into an ester of butane-1,4-dithioboronic acid (49).[65]

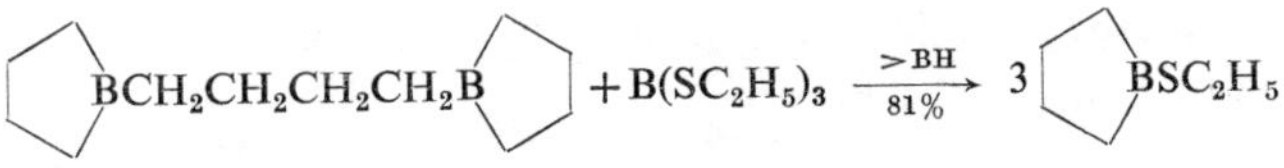

$$\overset{\frown}{BCH_2CH_2CH_2B} + B(SC_2H_5)_3 \xrightarrow[81\%]{>BH} 3 \overset{\frown}{BSC_2H_5} \qquad (48)$$

$$\overset{\frown}{BSC_2H_5} + B(SC_2H_5)_3 \xrightarrow[64\%]{>BH} (C_2H_5S)B(CH_2)_4B(SC_2H_5)_2 \qquad (49)$$

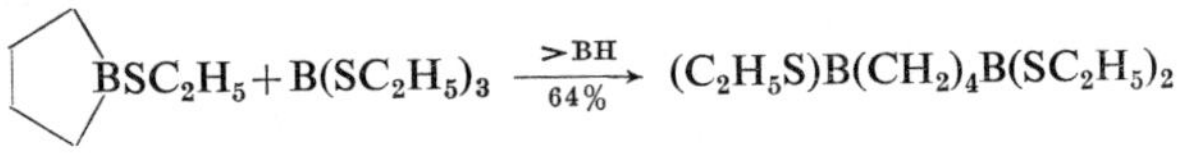

3. Synthesis of Thioboronates and Thioborinates from Esters of Organoboron Acids

A simple and highly universal preparative route to thioboronates and thioborinates is the double decomposition of the esters of organoboron acids and thioesters (50–54), which takes place smoothly even in the absence of B—H catalysts.[95-99]

$$3 RB(OR')_2 + 2 B(SR'')_3 \rightleftharpoons 3 RB(SR'')_2 + 2 B(OR')_3 \qquad (50)$$
$$3 R_2BOR' + B(SR'')_3 \rightleftharpoons 3 R_2BSR'' + B(OR')_3 \qquad (51)$$
$$RB(OR')_2 + R''B(SR''')_2 \rightleftharpoons RB(SR''')_2 + R''B(OR')_2 \qquad (52)$$
$$RB(OR')_2 + 2 R_2''BSR''' \rightleftharpoons RB(SR''')_2 + 2 R_2''BOR' \qquad (53)$$
$$R_2BOR' + R_2''BSR''' \rightleftharpoons R_2BSR''' + R_2''BOR' \qquad (54)$$

The reactions are reversible and the equilibrium can be shifted by removal of the most volatile component under vacuum at a temperature of 80–150°. After 2–3 hours under such conditions the equilibrium is practically completely shifted towards formation of the lowest boiling species and the yields of the desired compounds range 90 per cent or higher. This is an invaluable procedure for preparing esters of low molecular weight from esters of higher molecular weight. For instance, the corresponding ethyl esters were obtained in 87–97 per cent yields from the reactions of butyl n-propylthioboronate or di-n-butylthioborinate and ethyl thioborate.

The underlying mechanism of the reaction involves intermediate formation of dimeric compounds followed by their decomposition in an energetically advantageous direction (55).

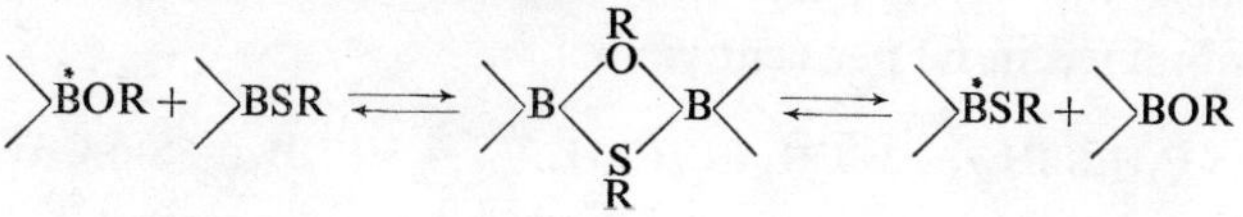

$$\text{>}\overset{\centerdot}{B}OR + \text{>}BSR \rightleftharpoons \text{>}B\underset{\underset{R}{S}}{\overset{\overset{R}{O}}{<}}B< \rightleftharpoons \text{>}\overset{\centerdot}{B}SR + \text{>}BOR \qquad (55)$$

The method has been successfully applied to alkanediboronic and triboronic compounds. Thus, on heating the tetramethyl ester of propane-1,3-diboronic acid with methyl thioborate, the former is converted into the tetramethyl ester of propane-1,3-dithioboronic acid.[70]

$$3\ (CH_3O)_2B(CH_2)_3B(OCH_3)_2 + 4\ B(SCH_3)_3 \xrightarrow[76\%]{} 3\ (CH_3S)_2B(CH_2)_3B(SCH_3)_2 +$$
$$+ 4\ B(OCH_3)_3 \qquad (56)$$

Tetraethyl butane-1,4-dithioboronate has been obtained from tetramethyl butane-1,4-diboronate and triethyl thioborate.[98]

$$2\ (CH_3O)_2B(CH_2)_4B(OCH_3)_2 + 4\ B(SC_2H_5)_3 \longrightarrow$$
$$\longrightarrow (C_2H_5S)_2B(CH_2)_4B(SC_2H_5)_2 + 4\ B(OCH_3)_3 \qquad (57)$$

The corresponding 1,1,5,9,9-pentaalkylthio-1,5,9-triboranonanes were obtained from 1,1,5,9,9-pentamethoxy-1,5,9-triboranonane and trimethyl thioborate or triethyl thioborate.[73]

$$3\ (CH_3O)_2B(CH_2)_3\underset{\underset{OCH_3}{|}}{B}(CH_2)_3B(OCH_3)_2 + 5\ B(SR)_3 \xrightarrow[72-77\%]{}$$
$$\xrightarrow[72-77\%]{} 3\ (RS)_2B(CH_2)_3\underset{\underset{SR}{|}}{B}(CH_2)_3B(SR)_2 + 5\ B(OCH_3)_3 \qquad (58)$$
$$R = CH_3,\ C_2H_5$$

The polymer formed in the hydroboration of dimethyl allylthioboronate affords tetramethyl propane-1,3-dithioboronate on heating with methyl thioborate in the presence of tetrapropyldiborane.[74]

4. Synthesis of Thioboronates and Thioborinates from Alkylthioboranes and Unsaturated Hydrocarbons

Di-n-butylthioborane reacts with 1-hexene or 1-octene to form esters of the corresponding thioboronic acids.[80]

$$(n\text{-}C_4H_9S)_2BH + CH_2{=}CHR \xrightarrow[83-89\%]{} RCH_2CH_2B(S\text{-}n\text{-}C_4H_9)_2 \qquad (59)$$

The reactions with propylene or isobutylene are complicated, giving mixtures of esters of thioboronic and thioborinic acids and butyl thioborates. The reaction of polymeric n-butylthioborane with ethylene or propylene is also multidirectional; whereas, with 1-hexene, n-hexyl thioborinate is obtained in 60 per cent yield.[85]

$$(n\text{-}C_4H_9SBH_2)_x + 2x\ CH_2\!\!=\!\!CHC_4H_9 \longrightarrow x\,(C_6H_{13})_2BS\text{-}n\text{-}C_4H_9 \qquad (60)$$

The methyl esters of diethylborinic, di-n-hexylborinic and dicyclohexylborinic acids were obtained in high yields from the reactiom of polymeric methylthioborane and ethylene, 1-hexene and cyclohexene respectively.[85]

1-Alkylthioboracyclanes can be synthesized from polymers of alkylthioboranes and dienes. Thus, 1-methylthioboracyclopentane,[85] 1-n-butylthioboracyclopentane,[82, 114]   3-methyl-1-methylthioboracyclopentane[85] and 3-methyl-1-n-butylthioboracyclopentane[114] were obtained by reaction of polymeric methylthioborane or n-butylthioborane and dienes (1,3-butadiene or isoprene) in ether solution.

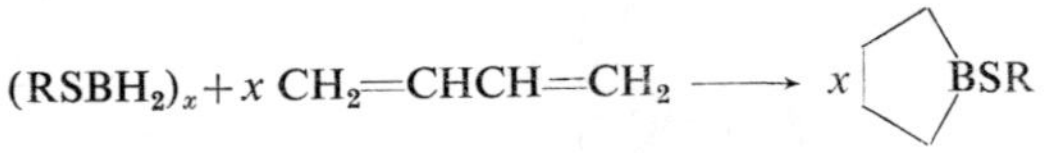

$$(RSBH_2)_x + x\ CH_2\!\!=\!\!CHCH\!\!=\!\!CH_2 \longrightarrow x \begin{array}{c}\\ BSR\end{array} \qquad (61)$$

$$(RSBH_2)_x + x\ CH_2\!\!=\!\!\overset{\displaystyle CH_3}{\underset{\displaystyle |}{C}}CH\!\!=\!\!CH_2 \longrightarrow x \begin{array}{c}CH_3\\ BSR\end{array} \qquad (62)$$

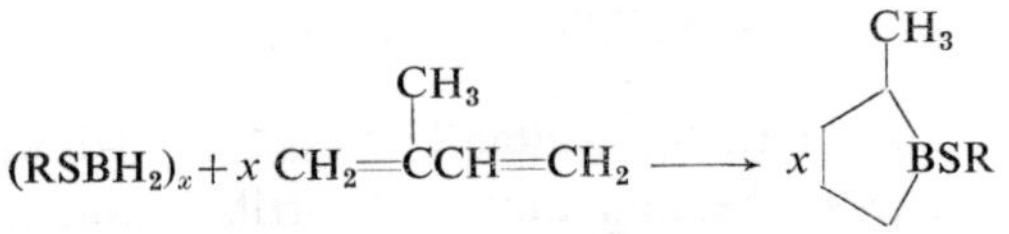

Polymers of n-butylthioborane react with 1,4-pentadiene or 1,6-hexadiene to give a mixture of alkylthioboracyclanes:[114]

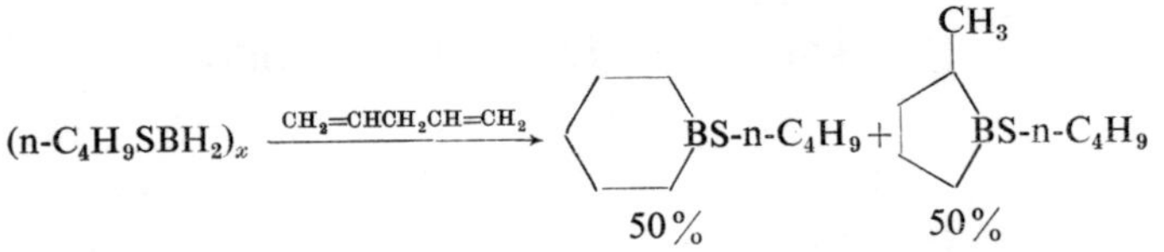

$$(n\text{-}C_4H_9SBH_2)_x \xrightarrow{\ CH_2=CHCH_2CH=CH_2\ } \begin{array}{c} BS\text{-}n\text{-}C_4H_9\end{array} + \begin{array}{c}CH_3\\ BS\text{-}n\text{-}C_4H_9\end{array} \qquad (63)$$

50%   50%

$$(n\text{-}C_4H_9SBH_2)_x \xrightarrow{\ CH_2=CHCH_2CH_2CH=CH_2\ } \begin{array}{c} BS\text{-}n\text{-}C_4H_9\end{array} + \begin{array}{c}CH_3\\ BS\text{-}n\text{-}C_4H_9\end{array} \qquad (64)$$

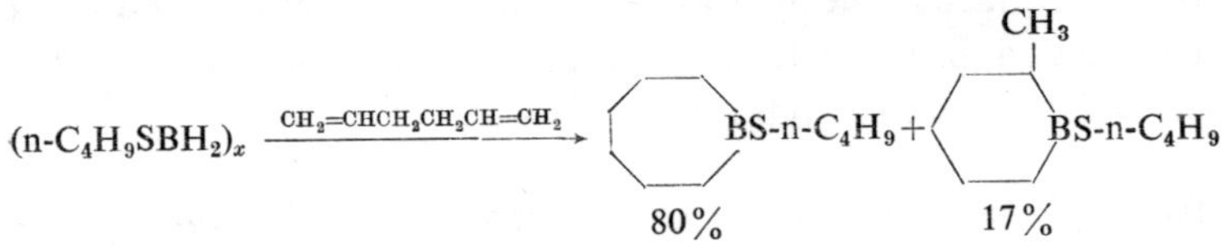

80%   17%

5. Synthesis of Thioboronates and Thioborinates from Alkyldiboranes or Aminoalkyl(aryl)boranes

Dialkylthioborinic esters are formed by the action of mercaptans on tetraalkyldiboranes.[7, 28] The reactions are complicated by the formation of small amounts of by-products as a result of disproportionation of the initial products.

$$R_4B_2H_2 + 2\ RSH \xrightarrow[75-83\%]{} 2\ R_2BSR + 2\ H_2 \qquad (65)$$

Intracoordinated (γ-diethylaminopropyl)-n-butylborane with n-butyl mercaptan at 190° results in the ester of (γ-diethylaminopropyl)-n-butyl-borinic acid:[101]

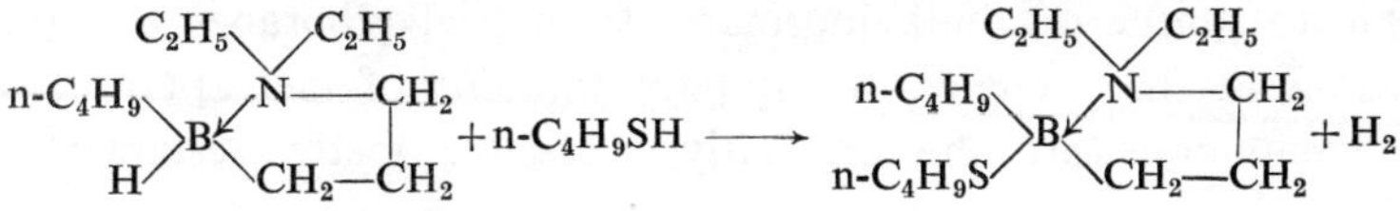

$$\qquad (66)$$

Alkylthioboronic esters have been prepared by heating trimethylamine-alkylboranes with mercaptans.[19, 20]

$$RBH_2 \cdot N(CH_3)_3 + 2\ R'SH \xrightarrow[29-72\%]{60-100°} RB(SR')_2 + 2\ H_2 + N(CH_3)_3 \qquad (67)$$

Treatment of dialkylamine-arylboranes with mercaptans leads to esters of dialkylaminoarylboronic acids.[48]

C. Properties of Thioboronates and Thioborinates

1. Thermal Transformations

Both thioboronates and thioborinates are quite heat resistant. Thus, n-propyl dipropylthioborinate undergoes no change after 4 hours heating at 160°.[36] However, on prolonged heating disproportionation of the thioesters takes place. Methyl methylthioboronate disproportionates to the extent of 10 per cent to methyl thioboronate and methyl dimethyl-borinate when kept for 34 days at 84°;[6] dibutyl isobutylthioboro-nate transforms into butyl thioborate and n-butyl diisobutylthioborinate to the extent of 30 per cent after 10 hours at 240–250°.[81]

$$2\ i\text{-}C_4H_9B(S\text{-}n\text{-}C_4H_9)_2 \underset{}{\overset{30\%}{\rightleftharpoons}} (i\text{-}C_4H_9)_2BS\text{-}n\text{-}C_4H_9 + B(S\text{-}n\text{-}C_4H_9)_3 \qquad (68)$$

Esters of butane-1,4-dithioboronic acids on heating are converted to thioborates and 1-alkylthioboracyclopentanes.[28]

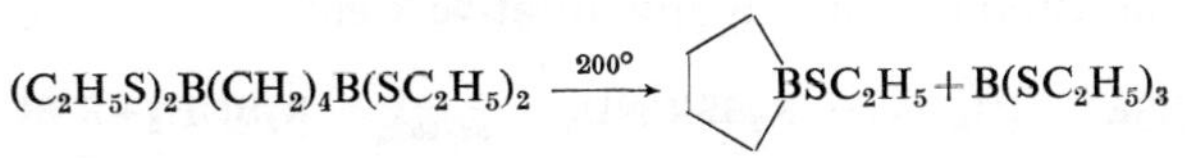

$$\qquad (69)$$

2. Hydrolysis and Alcoholysis

Esters of thioborinic acids are readily hydrolyzed to borinic acids and mercaptans.[36]

$$R_2BSR' + H_2O \longrightarrow R_2BOH + R'SH \qquad (70)$$

They also readily afford borinates on treatment with alcohols.[36] 1-Alkylthioboracyclanes react with alcohols to give 1-alkoxyboracyclanes.[82]

The rapid and smooth reaction of thioesters with alcohols makes it possible to synthesize dialkylborinates from trialkylboranes under mild conditions in the presence of catalytic amounts of mercaptan; among such compounds are the difficultly accessible methyl esters.[38] The

$$BR_3 + CH_3OH \xrightarrow[65-70\%]{RSH} R_2BOCH_3 + RH \qquad (71)$$

mercaptan forms an alkyl thioborinate with the trialkylborane; the product then reacts with the methanol to give the alkyl borinate; in turn the liberated mercaptan reacts once again with the trialkylborane. In this manner, in the presence of a small amount of mercaptan, tris-(3-dimethoxyborylpropyl)borane can be converted into 1,1,5,9,9-pentamethoxy-1,5,9-triboranonane and methyl propaneboronate.[72]

$$B[CH_2CH_2CH_2B(OCH_3)_2]_3 + CH_3OH$$

$$\xrightarrow[85\%]{C_2H_5SH, \ 100°} CH_3OB[CH_2CH_2CH_2B(OCH_3)_2]_2 + n\text{-}C_3H_7B(OCH_3)_2 \qquad (72)$$

Esters of alkylthioboronic acids react with alcohols to give dialkyl boronates. With the diol from 2,2-diphenylpropane, alkyl or aryl thioboronates form heterocyclic compounds.[66]

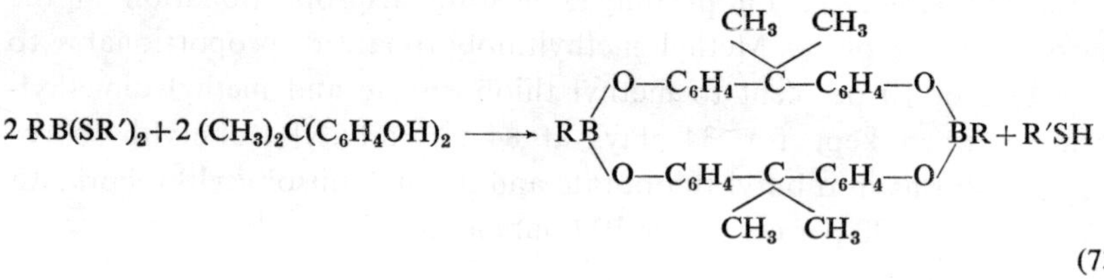

$$2\ RB(SR')_2 + 2\ (CH_3)_2C(C_6H_4OH)_2 \longrightarrow RB\underset{O-C_6H_4-\underset{CH_3}{\overset{CH_3}{C}}-C_6H_4-O}{\overset{O-C_6H_4-\underset{CH_3}{\overset{CH_3}{C}}-C_6H_4-O}{}}BR + R'SH$$

$$(73)$$

3. Reactions with Ammonia and Amines

Esters of borinic acids react with ammonia to give aminodialkylboranes.[35, 38, 39] An adduct forms first on introduction of the ammonia to the cooled ester; at 20° the adduct decomposes into an aminodialkylborane and mercaptan in nearly quantitative yield:

$$R_2BSR' + NH_3 \longrightarrow R_2BSR'NH_3 \xrightarrow[89-98\%]{} R_2BNH_2 + R'SH \qquad (74)$$

Aminodialkylboranes can be prepared directly from trialkylboranes and ammonia in the presence of small amounts of mercaptans which act as catalysts.[38, 39] The thioester, formed in the first stage of the reaction between the trialkylborane and mercaptan, on reaction with ammonia, regenerates the mercaptan, which once again reacts with the trialkylborane. The process repeats until complete disappearance of the trialkylborane (75). Esters of diarylborinic acids also readily react with ammonia to form aminodiarylboranes.[61]

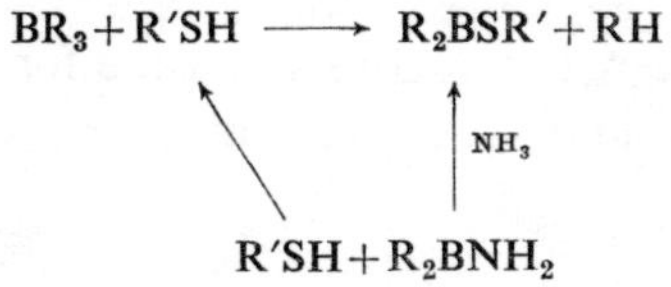

$$BR_3 + R'SH \longrightarrow R_2BSR' + RH$$

$$\nwarrow \qquad \uparrow NH_3 \qquad\qquad (75)$$

$$R'SH + R_2BNH_2$$

With amines, thioborinates form alkylaminodialkylboranes:[35, 36, 39]

$$R_2BSR' + NH_2R'' \xrightarrow[70-95\%]{} R_2BNHR'' + R'SH \qquad (76)$$

As in the synthesis of aminodialkylboranes the alkylamino derivatives can be obtained directly from trialkylboranes and amines in the presence of small amounts of mercaptans.[38, 39]

$$B(n\text{-}C_3H_7)_3 + NH(C_2H_5)_2 \xrightarrow[92\%]{n\text{-}C_3H_7SH, \, 90°} (n\text{-}C_3H_7)_2BN(C_2H_5) + C_3H_8 \qquad (77)$$

Esters of diarylthioborinic acids react similarly with aliphatic[61] or aromatic[54] amines. The same reaction is also undergone by 1-alkyl-thioboracyclopentane:[67, 82]

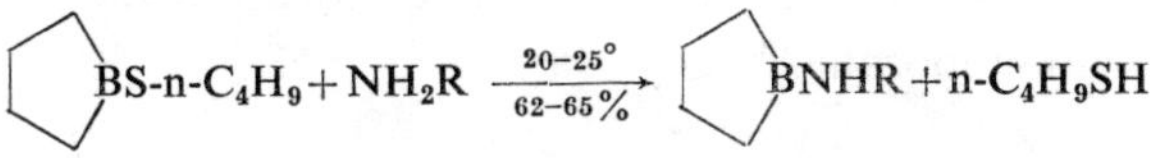

$$\text{\Large$\bigcirc$}BS\text{-}n\text{-}C_4H_9 + NH_2R \xrightarrow[62-65\%]{20-25°} \text{\Large$\bigcirc$}BNHR + n\text{-}C_4H_9SH \qquad (78)$$

With hexamethylenediamine, thioborinates form N,N'-bis(dialkyl-boryl)-1,6-diaminohexanes.[35] Hydrazine leads to

$$2\,R_2BSR' + NH_2(CH_2)_6NH_2 \xrightarrow[88\%]{160-180°} R_2BNH(CH_2)_6NHBR_2 + 2\,R'SH \qquad (79)$$

1,2-bis(dialkylboryl)hydrazines and phenylhydrazines are converted to 1-phenyl-2-(dialkylboryl)hydrazines:[35]

$$2\,R_2SR' + NH_2NH_2 \xrightarrow[80\%]{150-180°} R_2BNH\text{—}NHBR_2 + 2\,R'SH \qquad (80)$$

$$R_2BSR' + H_2N\text{—}NHC_6H_5 \xrightarrow[70\%]{160-200°} R_2BNH\text{—}NHC_6H_5 \qquad (81)$$

22*

The reactions of alkyl(aryl)thioboronates with ammonia or amines have made available a variety of nitrogen-containing compounds of boron. Trialkylborazoles are readily obtained by the reaction of ammonia with dialkyl thioborinates.[61, 62]

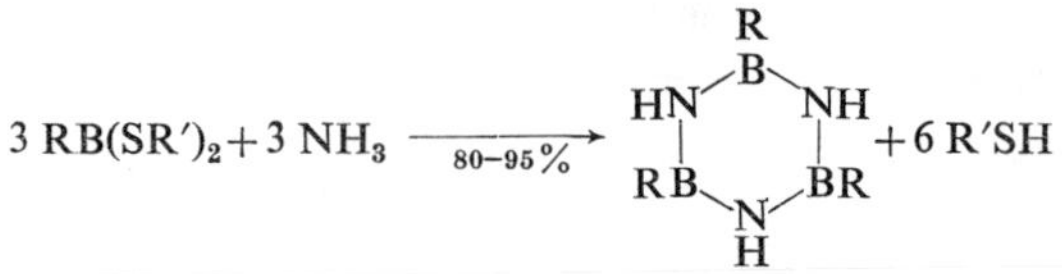

$$3\ RB(SR')_2 + 3\ NH_3 \xrightarrow[80-95\%]{} \begin{array}{c} R \\ \diagdown B \diagup \\ HN \diagdown \qquad \diagup NH \\ | \qquad\qquad | \\ RB \diagdown \qquad \diagup BR \\ N \\ | \\ H \end{array} + 6\ R'SH \qquad (82)$$

The first stage of the reaction consists of the formation of amino ester which then apparently undergoes condensation according to the following scheme:

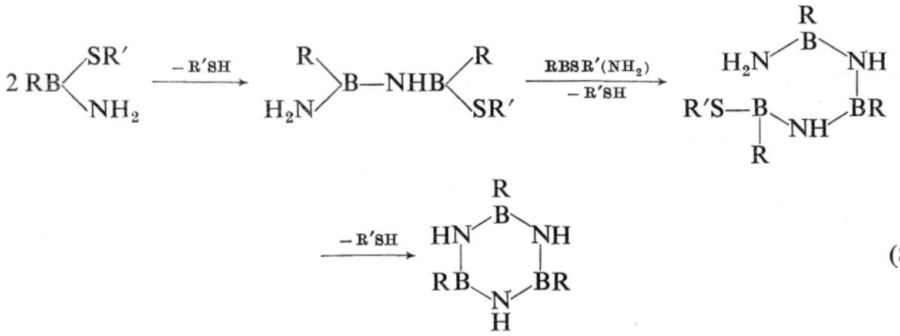

$$2\ RB \diagup_{NH_2}^{SR'} \xrightarrow{-R'SH} \diagup_{H_2N}^{R} B-NHB \diagup_{SR'}^{R} \xrightarrow[-R'SH]{RBSR'(NH_2)} \begin{array}{c} R \\ \diagdown B \diagup \\ H_2N \diagdown \qquad \diagup NH \\ | \qquad\qquad | \\ R'S-B \diagdown \qquad \diagup BR \\ NH \\ | \\ R \end{array}$$

$$\xrightarrow{-R'SH} \begin{array}{c} R \\ \diagdown B \diagup \\ HN \diagdown \qquad \diagup NH \\ | \qquad\qquad | \\ RB \diagdown \qquad \diagup BR \\ N \\ | \\ H \end{array} \qquad (83)$$

The reaction also an efficient route to B-triallylborazole.[92]

In contrast to the reaction with ammonia, on treatment with two moles of primary aliphatic amines, alkylthioboronates substitute both of their alkylthio groups for alkylamino groups to give bis(alkylamino)-alkylboranes.[63] Diethyl allylthioboronate forms with n-butylamine allyldi-n-butylaminoborane.[92]

$$RB(\text{n-}C_4H_9)_2 + 2\ NH_2R \longrightarrow RB(NHR)_2 + 2\ \text{n-}C_4H_9SH \qquad (84)$$

With equimolar quantities of the reactants, a mixture of bis(alkylamino)-alkylborane, aminothioester and initial thioester results.

Alkylthioboronates behave differently with secondary aliphatic amines (diethylamine). In this case only one alkylmercaptan residue undergoes substitution by an alkylamine group to give completely unsymmetrical products (I).[63]

$$RB(\text{n-}C_4H_9)_2 + NH(C_2H_5)_2 \longrightarrow RB \diagup_{N(C_2H_5)_2}^{S-\text{n-}C_4H_9} + \text{n-}C_4H_9SH \qquad (85)$$
$$\text{(I)}$$

Both the thio and amino residues of (I) are capable of reactingwith ethylamine when the latter is in excess resulting in the formation of bis-(alkylamino)boranes.[63]

$$RB\begin{smallmatrix}S\text{-n-}C_4H_9\\\\N(C_2H_5)_2\end{smallmatrix}+2\,NH_2C_2H_5 \xrightarrow{85-96\%} RB(NHC_2H_5)_2+\text{n-}C_4H_9SH+NH(C_2H_5)_2$$

(86)

The different behavior of alkyl(diethylamino)butylthioboranes (I) towards ethylamine and diethylamine is apparently due to the lower tendency of diethylamine to form coordination compounds.

The behavior of aryl(dialkylamino)thioboranes towards primary amines is similar to their behavior towards analogous compounds of the aliphatic series, namely, exchange of both thio and amino groups for alkylamino groups to form diamino(aryl)boranes.

The aromatic derivatives (II) are less prone to react with secondary amines. However, if a secondary amine of a relatively high boiling point is used and the liberated mercaptan is removed from the reaction mixture, the bis-secondary amino derivative is formed (87).

$$C_6H_5B\begin{smallmatrix}N(C_2H_5)_2\\\\S\text{-n-}C_3H_7\end{smallmatrix}+2\,HNC_5H_{10} \longrightarrow C_6H_5B(NC_5H_{10})_2+NH(C_2H_5)_2+\text{n-}C_3H_7SH$$

(II)

(87)

Diethyl allylthioboronate reacts with diethylamine in a specific manner, giving rise to almost equal amounts of allyl(diethylamino)ethylthio-borane and 3-ethylthiopropyl(diethylamino)ethylthioborane. The first stage of the reaction consists in the substitution of an ethylthio group by a diethylamino group (88); this is followed by addition of the liberated mercaptan to the initial reactant (89) and the final substitution of a diethyl-amino group for an ethylthio group (90).

$$CH_2=CHCH_2B(SC_2H_5)_2+NH(C_2H_5)_2 \longrightarrow CH_2=CHCH_2B\begin{smallmatrix}N(C_2H_5)_2\\\\SC_2H_5\end{smallmatrix}+C_2H_5SH$$

(88)

$$CH_2=CHCH_2B(SC_2H_5)_2+C_2H_5SH \longrightarrow C_2H_5SCH_2CH_2CH_2B(SC_2H_5)_2$$ (89)

$$C_2H_5SCH_2CH_2CH_2B(SC_2H_5)_2+NH(C_2H_5)_2$$

$$\longrightarrow C_2H_5SCH_2CH_2CH_2B\begin{smallmatrix}N(C_2H_5)_2\\\\SC_2H_5\end{smallmatrix}+C_2H_5SH$$

(90)

As with the alkylthioboronates[63] the products of reactions (88) and (90) are not further substituted by diethylamino groups.

Di-n-butyl phenylthioboronate forms with diethylamine n-butylthio-(diethylamino)phenylborane.[48]

The course of the reaction of an alkyl thioboronate with a diamine is dependent on the nature of the diamine. With ethylenediamine 2-alkyl-1,3,2-diazaborolidines are produced:[61]

$$RB(SR')_2 + H_2NCH_2CH_2NH_2 \xrightarrow[85\%]{150°} \begin{array}{c} CH_2-NH \\ | \\ CH_2-NH \end{array}\!> BR + 2\ R'SH \qquad (91)$$

Slight heating of di-n-butyl isoamylthioboronate with two moles of hexamethylenediamine followed by removal of n-butylmercaptan and unreacted diamine under vacuum gave a linear benzene-soluble polymer (III). The polymer had a molecular weight of about 900 and consisted of four $R\text{---}BNH(CH_2)_6NH$ units and an $NH_2(CH_2)_6NH$ terminus:[66]

$$H_2N(CH_2)_6NH\text{---}\!\!\left[\!\!\begin{array}{c} \text{---}BNH(CH_2)_6NH\text{---} \\ | \\ i\text{-}C_5H_{11} \end{array}\!\!\right]_4\!\!\text{---}H$$

(III)

With equimolecular amounts of reactants the cyclic polymer (IV) was obtained.

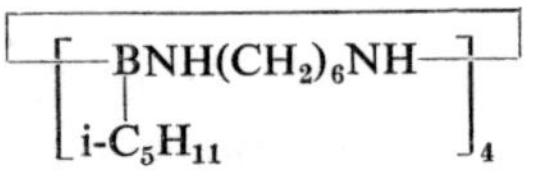

$$\left[\!\!\begin{array}{c} \text{---}BNH(CH_2)_6NH\text{---} \\ | \\ i\text{-}C_5H_{11} \end{array}\!\!\right]_4$$

(IV)

A similar polymeric compound is formed by reaction of hexamethylenediamine and dibutyl n-butylthioboronate.

With hydrazine, alkylthioboronates afford polymers possibly with a netlike cyclic structure (V), in which each unit contains three RB groups for every two $>$N–NH–groups. The number of units was shown to be ten by molecular weight determination.

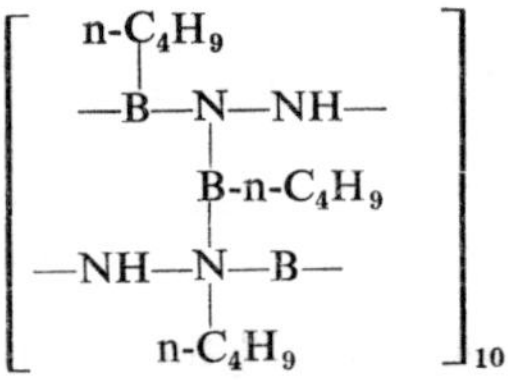

$$\left[\begin{array}{c} n\text{-}C_4H_9 \\ | \\ \text{---}B\text{---}N\text{---}NH\text{---} \\ | \\ B\text{-}n\text{-}C_4H_9 \\ | \\ \text{---}NH\text{---}N\text{---}B\text{---} \\ | \\ n\text{-}C_4H_9 \end{array}\right]_{10}$$

(V)

A different course is taken in the reaction between hydrazine and butyl phenylthioboronate to give 1,4-diphenyl-2,3,5,6-tetraaza-1,4-diborinane.[66]

$$2 \, C_6H_5B(SC_4H_9)_2 + 2 \, H_2NNH_2 \longrightarrow C_6H_5B \underset{NH-NH}{\overset{NH-NH}{\diagup\hspace{-0.3em}\diagdown}} BC_6H_5 + 4 \, C_4H_9SH \quad (92)$$

A study of the reaction of propylene-1,3-bis(diethylthioborane) with ammonia and primary and secondary amines showed that in the case of ammonia it yielded 1,3-diamino-2-aza-1,3-diboracyclohexane (VI);

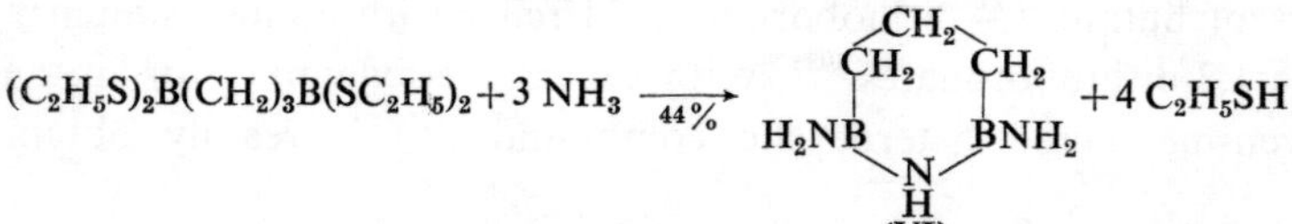

$$(C_2H_5S)_2B(CH_2)_3B(SC_2H_5)_2 + 3 \, NH_3 \xrightarrow{44\%} \quad + 4 \, C_2H_5SH \quad (93)$$

whereas, with primary amines (methylamine or ethylamine) the cyclic compounds (VII) were formed (94).

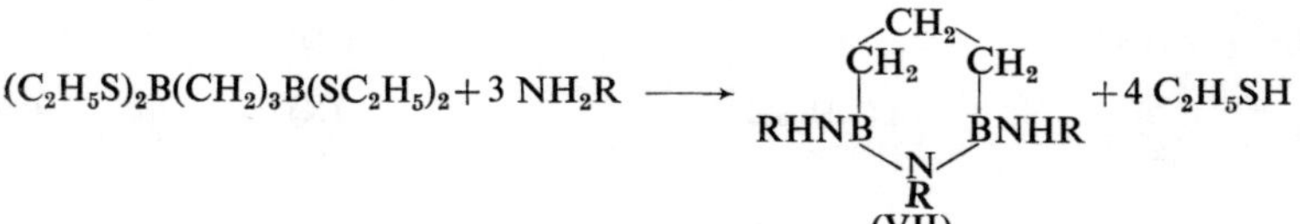

$$(C_2H_5S)_2B(CH_2)_3B(SC_2H_5)_2 + 3 \, NH_2R \longrightarrow \quad + 4 \, C_2H_5SH \quad (94)$$

In order to elucidate the mechanism of formation of the cyclic compounds (VII) the thioester was treated with an equimolar amount of methylamine. As a result it was found that in the first stage of the reaction the thioester exchanges one alkylthio group for an alkylamino group to give the linear compound (VIII) (95) which then undergoes ring closure with elimination of mercaptan to form 2-methyl-1,3-diethylthio-2-aza-1,3-diboracyclohexane (IX). Compound (IX) then exchanges the remaining alkylthio groups for alkylamino groups to afford compounds of type (VII).

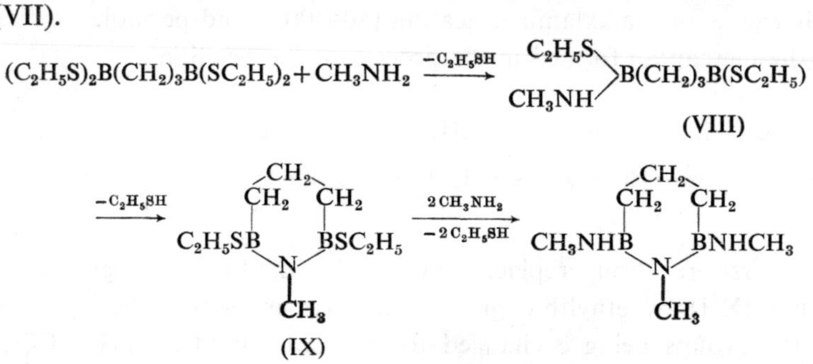

$$(C_2H_5S)_2B(CH_2)_3B(SC_2H_5)_2 + CH_3NH_2 \xrightarrow{-C_2H_5SH} \underset{CH_3NH}{\overset{C_2H_5S}{\diagup\hspace{-0.3em}\diagdown}} B(CH_2)_3B(SC_2H_5)$$

(VIII)

$$\xrightarrow{-C_2H_5SH} \quad \xrightarrow[-2C_2H_5SH]{2CH_3NH_2} \quad (95)$$

(IX)

Propylene-1,3-bis(diethylthioborane) with dimethylamine gives the 1,3-bis(dimethylaminoborane) derivative.[71] With diethylamine, like with the esters of alkylthioboronic acids, mixed alkylamino-alkylthio products are formed.

$$(C_2H_5S)_2B(CH_2)_3B(SC_2H_5)_2 + 2\ NH(C_2H_5)_2 \longrightarrow \underset{(C_2H_5)_2N}{\overset{C_2H_5S}{>}}B(CH_2)_3B\underset{N(C_2H_5)_2}{\overset{SC_2H_5}{<}} +$$

$$+ 2\ C_2H_5SH \tag{96}$$

Esters of butane-1,4-dithioboronic acid react with amines similarly to propane-1,3-dithioboronates.[98] With excess methylamine, for instance, the seven-membered heterocyclic compound (X) is readily obtained.

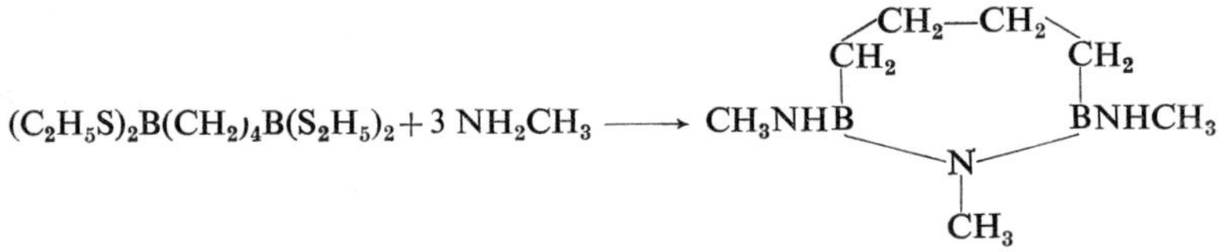

$$(C_2H_5S)_2B(CH_2)_4B(S_2H_5)_2 + 3\ NH_2CH_3 \longrightarrow$$

$$+ 4\ C_2H_5SH \qquad\qquad (X) \tag{97}$$

With secondary amines two alkylthio groups are very easily replaced to give the mixed derivative (XII).[98]

$$(C_2H_5S)_2B(CH_2)_4B(SC_2H_5)_2 + 2\ NHR_2$$

$$\longrightarrow \underset{R_2N}{\overset{C_2H_5S}{>}}B(CH_2)_4B\underset{NR_2}{\overset{SC_2H_5}{<}} + 2\ C_2H_5SH \tag{98}$$

$$(XI)$$

Further exchange of alkylthio groups in (XI) is achieved by using considerable excess of dialkylamine, heating (80–100°), and periodic distillation of the mercaptan formed in the process of the reaction.

$$(C_2H_5S)_2B(CH_2)_4B(SC_2H_5)_2 + 4\ NH(CH_3)_2 \xrightarrow[87\%]{}$$

$$\longrightarrow [(CH_3)_2N]_2B(CH_2)_4B[N(CH_3)_2]_2 + 4\ C_2H_5SH \tag{99}$$

$$(XII)$$

The reverse reaction, replacement of the dialkylamino groups in compound (XII) by ethylthio groups, also occurs with difficulty, only two of the groups being exchanged to give compound (XI) (R=CH₃).

1,1,5,9,9-Pentaalkylthio-1,5,9-triboranonanes (XIII) react energetically with dimethylamine, all the alkylthio groups being replaced by dimethylamino groups.[73]

$$(RS)_2B(CH_2)_3\underset{\underset{\displaystyle SR}{|}}{B}(CH_2)_3B(SR)_2 + 5 \ NH(CH_3)_2$$

$$\longrightarrow [(CH_3)_2N]_2B(CH_2)_3\underset{\underset{\displaystyle N(CH_3)_2}{|}}{B}(CH_2)_3B[N(CH_3)_2]_2 + 5 \ RSH \qquad (100)$$

$$(XIII)$$

The reaction of with methylamine leads to a compound having either structure (XIV) or (XV).

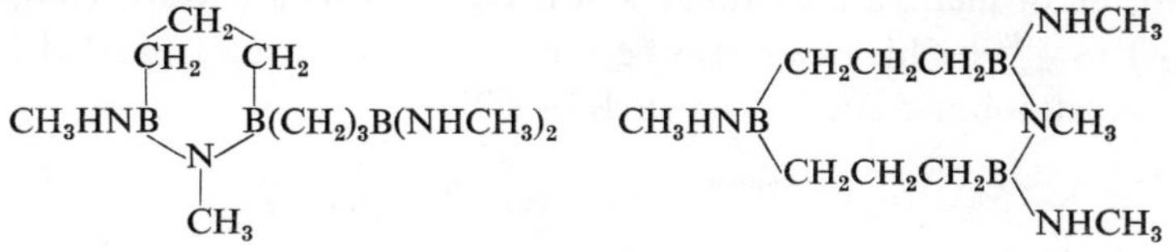

(XIV)   (XV)

Alkyl(halo)thioalkylboranes exchange their halogen with a diethylamino group to give products which are resistant to further reaction with diethylamine.[64] Hence in compounds of type RB(SR)X chlorine and bromine atoms are more labile than the alkylthio group.

$$RB(SR')X + 2 \ NH(C_2H_5)_2 \longrightarrow RB \Big\langle \begin{matrix} SR' \\ N(C_2H_5)_2 \end{matrix} + (C_2H_5)_2NH_2X \qquad (101)$$

$$X = Cl, \ Br$$

4. Reactions with Mercaptans, Hydrogen Sulfide and Hydrocyanic Acid

Heating with higher mercaptans causes alkyl thioborinates to undergo realkylation.[37] The equilibrium can be completely displaced to the right by removing the low-boiling mercaptan from the mixture reaction.

$$(C_2H_5)BSC_2H_5 + n\text{-}C_4H_9SH \rightleftharpoons (C_2H_5)_2BS)\text{-}n\text{-}C_4H_9 + C_2H_5SH \qquad (102)$$

Diethyl allylthioboronate adds ethyl mercaptan to form diethyl 3-ethylthio-n-propylthioboronate.[91, 92]

When hydrogen sulfide is passed into an alkyl thioborinate heated to 140–180°, thioborinic acids and mercaptans result:[34]

$$R_2BS\text{-}n\text{-}C_4H_9 + H_2S \xrightarrow[50-85\%]{} R_2BSH + n\text{-}C_4H_9SH \qquad (103)$$

Dialkylthioborinic acids withstand distillation in vacuum. Prolonged heating at 180–250° decomposes dibutylthioborinic acid to hydrogen sulfide, hydrogen, butylene, butane and organic boron–sulfur compounds of an unknown nature.

Reactions of alkylthioboronates with hydrogen sulfide, which could be expected to lead to anhydrides of alkylthioboronic acids, have not been investigated. One such compound, namely the trimer of methylthioboronic anhydride has been synthesized by reaction between trimethylborane and metathioboric acid.[124]

$$2 \text{ HSBS} + \text{B(CH}_3)_3 \longrightarrow (\text{CH}_3\text{BS})_3 + \text{H}_2\text{S} \tag{104}$$

The trimer is stable only in the solid state, the anhydride displaying a higher degree of association in solution. In solution the polymeric anhydride of methylthioboronic acid is oxidized by air to the compound $(\text{CH}_3\text{OBS})_3$,[124] which can also be obtained by treating trimethyl borate with metathioboric acid[120] or $(\text{BrBS})_3$.[121]

$$(\text{HSBS})_3 \xrightarrow[-\text{B(SH)}_3]{\text{B(OCH}_3)_3} (\text{CH}_3\text{OBS})_3 \xleftarrow[-\text{BBr}_3]{\text{B(OCH}_3)_3} (\text{BrBS})_3 \tag{105}$$

n-Butyl dialkylthioborinates react with hydrocyanic acid in the cold or on slight heating to form the more or less associated cyanodialkylboranes.[34]

$$\text{R}_2\text{BSC}_4\text{H}_9 + \text{HCN} \longrightarrow [\text{R}_2\text{BSC}_4\text{H}_9\cdot\text{HCN}] \longrightarrow \text{R}_2\text{BCN} + \text{C}_4\text{H}_9\text{SH} \tag{106}$$

5. Reactions with Vinyl Alkyl Ethers, Cyclohexylidene Aniline and Nitriles

Esters of dialkylthioborinic acids are capable of reacting with various types of unsaturated compounds. Similarly to the thioborates, with vinyl alkyl ethers they give vinyl alkyl sulfides and dialkylborinates.[13] It is to be assumed that as with the thioborates the reaction proceeds with addition of the thioester to the carbon–carbon double bond followed by β-elimination.

$$\text{R}_2\text{BSC}_4\text{H}_9 + \text{CH}_2\!=\!\text{CHOR} \longrightarrow \left[\text{R}_2\text{BCH}_2\text{CH}\!\!\begin{array}{l}\diagup\text{OR}\\\diagdown\text{SC}_4\text{H}_9\end{array} \right]$$
$$\longrightarrow \text{CH}_2\!=\!\text{CHSC}_4\text{H}_9 + \text{R}_2\text{BOR} \tag{107}$$

On heating butyl di-n-propylthioborinate with N-cyclohexylidene aniline an aminoborane and mercaptan are formed.[13] Apparently the aniline reacts in the tautomeric form of an enamine.

$$(\text{C}_3\text{H}_7)_2\text{BSC}_4\text{H}_9 + \text{C}_6\text{H}_5\text{N}\!=\!\!\!\bigcirc \xrightarrow{70\%} (\text{C}_3\text{H}_7)_2\text{BN}\!\!\begin{array}{l}\diagup\text{C}_6\text{H}_5\\\diagdown\end{array} + \text{C}_4\text{H}_9\text{SH} \tag{108}$$

Esters of dialkyl- and diarylborinic acids can add to[52] acetonitrile at room temperature to give the ring compounds (XVI).

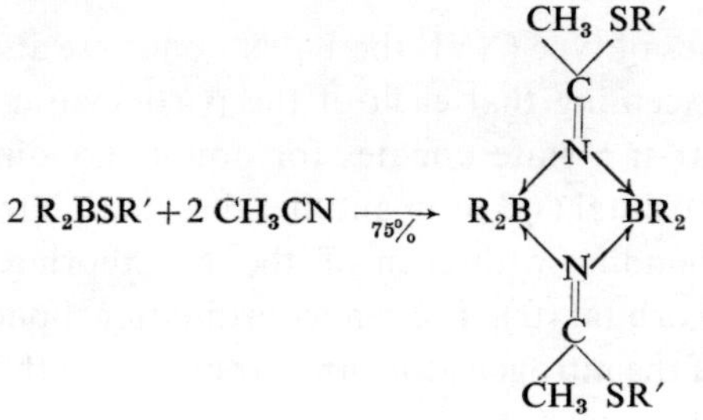

$$2\ R_2BSR' + 2\ CH_3CN \xrightarrow{75\%}$$

(109)

(XVI)

It was shown by infrared spectroscopy that the first stage of the reaction consists of formation of the adduct $R_2BN{=}CCH_3(SR')$, which then dimerizes to compound (XVI). The dimers are stable in air at room temperature and are not decomposed by alcohols, water or hydrochloric acid.

The nature of the bonds in the dimeric derivatives (XVI) is worthy of note. In these compounds, as in the dimers of aminoboranes (cyclodiborazanes), trimers of aminoboranes (cyclotriborazanes) and other such compounds (trimeric phosphinoboranes and alkylthioboranes) all bonds of the boron atom with the heteroatom (N, P, S) are equivalent. The equivalency of the bonds in such compounds as 1,1,3,3-tetramethylcyclodiborazane

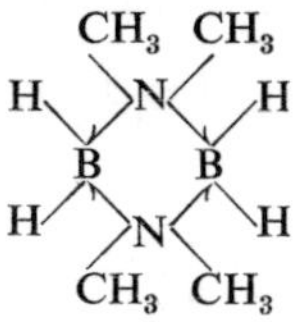

(XVII)

(XII) is due to formation of a peculiar type of coordinate link between the boron and heteroatoms. In such coordination each boron atom donates one tetrahedral (sp^3) orbital and one electron for covalent bonding with the hydrogen atoms and one sp^3 orbital and $\frac{1}{2}$ electron for bonding with each nitrogen atom. In turn each nitrogen atom donates an sp^3 orbital and one electron for bonding with each carbon atom and an sp^3 orbital and 1.5 electrons for bonding with each neighboring boron atom. Such bonding can be conveniently termed semicoordinative and depicted by means of a half-arrow. The semicoordination bond should be weaker than a covalent bond but stronger than an ordinary coordination bond, since the energy losses associated with the value of the ionization potential of the donor atom are playing a very important part in the general energy balance of formation of the coordination bond, and in this case refers to two, rather than one bond.

In cycloborazanes of type (XVI) the B—N bonds are also of a semicoordinative nature, excepting that each of the participating nitrogen atoms, existing in the tr^2 tr tr π-state donates for double bonding with a carbon atom one trigonal orbital (sp^2 or in another designation tr) and one π-orbital; whereas for bonding with each of the neighboring boron atoms it donates a trigonal orbital (tr). The semicoordination bonds formed by the trigonal orbitals of the nitrogen atom are stronger than those formed by the tetrahedral orbitals.

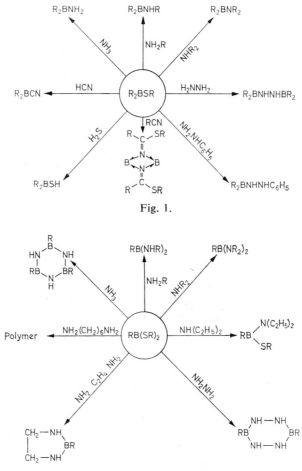

Fig. 1.

Fig. 2.

From the above discussion it can be concluded that thioboronates and thioborinates can be used in the synthesis of a wide variety of boron compounds, as outlined in Figs. 1 and 2.

IV. B-ALKYLTHIOBORAZINE

A. *Preparation of B-Alkylthioborazines*

B-Trialkylthioborazines (XVIII) were synthesized by refluxing *B*-trichloroborazine with lead mercaptide in benzene solution.[57]

$$
\begin{array}{c}
\text{Cl} \\
\underset{\text{ClB}\diagdown_{\underset{\text{R}}{\text{N}}}\diagup\text{BCl}}{2\,\text{RN}\diagup\overset{\text{B}}{\diagdown}\text{NR}}
\end{array}
+ 3\,\text{Pb(SC}_4\text{H}_9)_2 \xrightarrow[\text{70–90\%}]{}
\begin{array}{c}
\text{SC}_4\text{H}_9 \\
\underset{\text{C}_4\text{H}_9\text{SB}\diagdown_{\underset{\text{R}}{\text{N}}}\diagup\text{BSC}_4\text{H}_9}{2\,\text{RN}\diagup\overset{\text{B}}{\diagdown}\text{NR}}
\end{array}
+ 3\,\text{PbCl}_2 \quad (110)
$$

(XVIII)

B-Trialkylthioborazines served as starting materials for preparing unsymmetrical *B*-dialkylthioborazines, which formed on heating the thioborazines with tetrapropyldiborane.[60]

$$
\begin{array}{c}
\text{SR} \\
\underset{\text{RSB}\diagdown_{\underset{\text{R}}{\text{N}}}\diagup\text{BSR}}{\text{RN}\diagup\overset{\text{B}}{\diagdown}\text{NR}}
\end{array}
+ \tfrac{1}{2}\,\text{R}_4'\text{B}_2\text{H}_2 \longrightarrow
\begin{array}{c}
\text{H} \\
\underset{\text{RSB}\diagdown_{\underset{\text{R}}{\text{N}}}\diagup\text{BSR}}{\text{RN}\diagup\overset{\text{B}}{\diagdown}\text{NR}}
\end{array}
+ \text{R}_2'\text{BSR} \quad (111)
$$

B-Dialkylthioborazines also can be produced by reacting a tetraalkyldiborane with a *B*-trichloroborazine followed by treatment of the product with lead mercaptide.

The reaction between B-trialkylthioborazines and Grignard reagents gives *B*-alkyl-*B*-dialkylthioborazines in moderate yields (33 per cent, based on the borazine).[58]

However, with organolithium reagents stepwise substitution of the alkylthiogroups by alkylgroups proceeds more smoothly and *B*-alkyl-*B*-dialkylthio-*N*-trialkylborazine can be obtained in 65 per cent yields.

$$
\begin{array}{c}
\text{SC}_4\text{H}_9 \\
\underset{\text{C}_4\text{H}_9\text{SB}\diagdown_{\underset{\text{R}}{\text{N}}}\diagup\text{BSC}_4\text{H}_9}{\text{RN}\diagup\overset{\text{B}}{\diagdown}\text{NR}}
\end{array}
+ \text{R}'\text{Li} \xrightarrow[\text{65\%}]{}
\begin{array}{c}
\text{R}' \\
\underset{\text{C}_4\text{H}_9\text{SB}\diagdown_{\underset{\text{R}}{\text{N}}}\diagup\text{BSC}_4\text{H}_9}{\text{RN}\diagup\overset{\text{B}}{\diagdown}\text{NR}}
\end{array}
+ \text{C}_4\text{H}_9\text{SLi} \quad (112)
$$

B. *Properties of B-Alkylthioborazines*

B-Trialkylthioborazine, like the thioboronates and thioborinates, are very highly reactive compounds. They react with alcohols in the cold to give *B*-trialkoxy derivatives.[57]

Primary and secondary amines also undergo exchange with *B*-tri-n-butyl-thioborazines without heating, transforming the latter into the correspond-ing *B*-alkylamine derivatives.[57] For instance, *B*-tri(dimethylamino)boraz-ine is obtained 97 per cent yield from *B*-tri-n-butylthioborazine and dime-thylamine and aniline with *B*-tri-n-butylthioborazine or *B*-tri-n-butylthio-*N*-triethylborazine gives *B*-triphenylaminoborazine and *B*-triphenylamino-*N*-triethylborazine, respectively.

$$
\begin{array}{c}
\text{SR}' \\
\overset{|}{\underset{R'S\text{B}}{\overset{R\text{N}}{\diagup}}}\overset{\text{B}}{\diagdown}\underset{\text{N}}{\underset{|}{\text{BSR}'}}\text{NR}
\end{array}
+ 3\,\text{NH}_2\text{C}_6\text{H}_5 \longrightarrow
\begin{array}{c}
\text{NHC}_6\text{H}_5 \\
\overset{|}{\underset{C_6H_5NHB}{\overset{RN}{\diagup}}}\overset{\text{B}}{\diagdown}\underset{\text{N}}{\underset{|}{\text{BNHC}_6\text{H}_5}}\text{NR}
\end{array}
+ 3\,\text{R}'\text{SH} \quad (113)
$$

B-Triamino derivatives of borazine are prepared by introducing ammo-nia into a benzene solution of *B*-trialkylthioborazine.[57]

The synthesis of various unsymmetrical derivatives of borazine has been achieved with *B*-alkyl-*B*-dialkylthio-*N*-trialkylborazine as starting mat-erial.[58] By the action of methanol on *B*-n-butyl-*B*-di-n-butylthio-*N*-tri-ethylborazine, *B*-n-butyl-*B*-dimethoxy-*N*-triethylborazine is obtained.

With primary and secondary amines the products are obtained in high yields, for instance:

$$
\begin{array}{c}
\text{R} \\
\overset{|}{\underset{R'S\text{B}}{\overset{C_2H_5N}{\diagup}}}\overset{\text{B}}{\diagdown}\underset{\text{N}}{\underset{|}{\text{BSR}'}}\text{NC}_2\text{H}_5 \\
\text{C}_2\text{H}_5
\end{array}
+ 2\,\text{NH}_2\text{CH}_3 \xrightarrow[75\%]{}
\begin{array}{c}
\text{R} \\
\overset{|}{\underset{CH_3NHB}{\overset{C_2H_5N}{\diagup}}}\overset{\text{B}}{\diagdown}\underset{\text{N}}{\underset{|}{\text{BNHCH}_3}}\text{NC}_2\text{H}_5 \\
\text{C}_2\text{H}_5
\end{array}
+ 2\,\text{R}'\text{SH} \quad (114)
$$

$$\text{R} = \text{C}_2\text{H}_5,\ \text{n-C}_4\text{H}_9;\ \text{R}' = \text{n-C}_4\text{H}_9$$

B-Alkylthio derivatives of borazine can be used to make three-dimen-sional and linear polymers.[59] Polycondensation with *B*-alkylthioborazine derivatives proceeds under mild conditions so that the structure of the borazine ring is retained together with the regular alternation of the mono-meric units. Condensation of *B*-tri-n-butylthio-*N*-trialkyl-borazines with hexamethylenediamine in *o*-xylene at 125° afforded thermostable powderlike polymers (XIX), insoluble in the ordinary organic solvents and hydrolyzing on exposure to air.

$$
n \;
\begin{array}{c}
\mathrm{SR'} \\
\mathrm{RN{-}B{-}NR} \\
\mathrm{R'SB{-}N{-}BSR'} \\
\mathrm{R}
\end{array}
\; + 2\,n\mathrm{H_2N(CH_2)_6NH_2} \longrightarrow
\left[
\begin{array}{c}
\mathrm{NH(CH_2)_6NH{-}} \\
\mathrm{RN{-}B{-}NR} \\
\mathrm{B{-}N{-}B{-}NH(CH_2)_6NH{-}} \\
\mathrm{R}
\end{array}
\right]_n
\tag{115}
$$

$$R = CH_3,\ C_2H_5 \quad (XIX)$$

A thermostable polymer (XX) which is soluble in tetrahydrofuran and dioxane and does not hydrolyze in air is obtained from the reaction of B-tri-n-butylthio-N-trimethylborazole with diphenyldihydroxysilane in ether solution.

$$
2\,n \;
\begin{array}{c}
\mathrm{SR'} \\
\mathrm{RN{-}B{-}NR} \\
\mathrm{R'SB{-}N{-}BSR'} \\
\mathrm{R}
\end{array}
\; + 3\,n(C_6H_5)_2Si(OH)_2
$$

$$
\longrightarrow
\left[
\begin{array}{c}
\mathrm{O{-}} \qquad\qquad \mathrm{O{-}} \\
\mathrm{C_6H_5{-}Si{-}C_6H_5 \qquad C_6H_5{-}Si{-}C_6H_5} \\
\mathrm{O} \qquad\qquad\qquad \mathrm{O} \\
\mathrm{RN{-}B{-}NR \quad C_6H_5 \; RN{-}B{-}NR} \\
\mathrm{{-}B{-}N{-}B{-}O{-}Si{-}O{-}B{-}N{-}B{-}} \\
\mathrm{R \qquad\qquad C_6H_5 \qquad R}
\end{array}
\right]_n
\tag{116}
$$

$$(XX)\ R = CH_3$$

Linear polymers can be obtained from B-dialkylthioborazine derivatives.[59] The polymers (XXI) are formed by 1 : 1 condensation of B-n-bu-tyl-B-di-n-butylthio-N-trialkylborazines with hexamethylenediamine in benzene. They are brittle compounds less prone to hydrolyze in air than the three-dimensional polymers and more heat resistant; their decomposition temperatures are above 400°.

$$
n \;
\begin{array}{c}
\mathrm{C_4H_9} \\
\mathrm{RN{-}B{-}NR} \\
\mathrm{R'SB{-}N{-}BSR'} \\
\mathrm{R}
\end{array}
\; + n\mathrm{NH_2(CH_2)_6NH_2} \longrightarrow
\left[
\begin{array}{c}
\mathrm{C_4H_9} \\
\mathrm{RN{-}B{-}NR} \\
\mathrm{B{-}N{-}B{-}NH(CH_2)_6NH{-}} \\
\mathrm{R}
\end{array}
\right]_n
\tag{117}
$$

$$R = CH_3,\ C_2H_5 \qquad (XXI)$$

The linear polymers (XXIII) from reaction of B-n-butyl-B-di-n-butyl-thio-N-trialkylborazoles and the diol (XXII) are of relatively low melting points, but are more resistant to hydrolysis than polymers from hexamethylenediamine.

$$n \quad \begin{array}{c} \text{C}_4\text{H}_9 \\ \text{RN} \overset{\text{B}}{\diagup} \text{N} \\ \text{B}'\text{SB} \diagdown_{\text{N}} \diagup \text{BSR}' \\ \text{R} \end{array} + n\text{HO} - \underset{}{\bigcirc} - \overset{\text{CH}_3}{\underset{\text{CH}_3}{\text{C}}} - \underset{}{\bigcirc} - \text{OH}$$

(XXII)

$$\longrightarrow \left[\begin{array}{c} \text{C}_4\text{H}_9 \\ \text{RN} \overset{\text{B}}{\diagup} \text{NR} \\ \diagup \text{B} \diagdown_{\text{N}} \diagup \text{B} - \text{O} - \underset{}{\bigcirc} - \overset{\text{CH}_3}{\underset{\text{CH}_3}{\text{C}}} - \underset{}{\bigcirc} - \text{O} - \\ \text{R} \end{array} \right]_n \qquad (118)$$

(XXIII)

B-n-butyl-B-di-n-butylthio-N-tri-methylborazine gives with diphenyldihydroxysilane in boiling ether solution a viscous liquid of molecular weight ~ 1200, which corresponds to the cyclic trimer (XXIV, $n = 3$). On heating in vacuum at 200° the trimer is transformed into a brittle polymer.[59]

$$\left[\begin{array}{c} \text{C}_4\text{H}_9 \\ \text{RN} \overset{\text{B}}{\diagup} \text{NR} \quad \text{C}_6\text{H}_5 \\ -\text{B} \diagdown_{\text{N}} \diagup \text{B} - \text{O} - \overset{|}{\underset{|}{\text{Si}}} - \text{O} - \\ \text{R} \quad \text{C}_6\text{H}_5 \end{array} \right]_n \qquad \left[\begin{array}{c} \text{C}_4\text{H}_9 \\ \text{RN} \overset{\text{B}}{\diagup} \text{NR} \quad \left[\begin{array}{c} \text{CH}_3 \\ \end{array} \right. \\ \diagup \text{B} \diagdown_{\text{N}} \diagup \text{B} - \text{O} - \overset{|}{\underset{|}{\text{Si}}} - \text{O} - \\ \text{R} \quad \left. \begin{array}{c} \text{CH}_3 \end{array} \right]_5 \end{array} \right]_n$$

(XXIV)   (XXV)

The same borazine condenses with decamethyl-1,9-dihydroxypentasiloxane[59] in ether solution to form a liquid with molecular weight ~ 1100 of the probable structure (XXV, $n = 2$). On heating in vacuum the dimer is converted into an elastic polymer of m.p. 147–149°, which is not susceptible to atmospheric hydrolysis.

V. 1,2-THIABOROLANES

A. Preparation of 1,2-Thiaborolanes

1,2-Thiaborolanes are formed when 1,2-dialkyl- or 1,2-diaryldiboranes add to allylmercaptan.[53, 68]

$$\tfrac{1}{2} \text{R}_2\text{B}_2\text{H}_4 + \text{CH}_2{=}\text{CHCH}_2\text{SH} \xrightarrow[50\%]{} \text{RB} \begin{array}{c} \diagup \text{S} - \text{CH}_2 \\ \diagdown \text{CH}_2 - \text{CH}_2 \end{array} \qquad (119)$$

2-Alkyl-1,2-thiaborolanes also form on heating a mixture of trialkyl-borane, trialkylamine-borane and allylmercaptan:[53]

$$BR_3 + 2\,H_3B \cdot N(C_2H_5)_3 + 3\,CH_2{=}CHCH_2SH \xrightarrow{110°} 3\,RB{\Big\langle}\begin{smallmatrix}S{-\!-\!-}CH_2\\ |\\ CH_2{-}CH_2\end{smallmatrix} \qquad (120)$$

B. *Properties of 1,2-Thiaborolanes*

In contrast to 2-alkyl-1,2-azaborolidines, with their tendency to dimerize, 1,2-thiaborolanes are monomeric. The peculiarity of the cyclic structure of 1,2-thiaborolanes is manifested in their proneness to form complexes with amines sharply distinguishing them from the closely related esters of thioborinic acids, R_2BSR.

The enhanced tendency of 1,2-thiaborolanes (and other similar compounds such as 1,2-azaborolidines, 1,2-oxaborolanes, etc.) to form complexes is explained by the lower energy of molecular reorganization than the linear compounds on transition of the boron atom from the trigonal into the tetrahedral valency state. The energy losses associated with the reorganization are the sum of the energy differences of promotion of the boron atom from the ground state to the tetrahedral and trigonal states, which is equal to 2.1 eV[21] and the decrease in the boron-element (C, S, N, O) bond energy with decrease in s-character of the boron orbital.

In cyclic compounds of boron, in which the C—B—X (X = C, S, N, O) bond angle in the ring approaches the value for a tetrahedral angle, only one (acyclic) boron-element bond is weakened instead of three as in compounds of the type BX_3, resulting in a decrease in reorganization energy. Strictly speaking the energy of the external B—X bond in the ring compound is decreased to a greater extent on complexation than in BX_3 compounds because the s-character of the boron orbital in this bond is greater than 0.33, but the energy lapses which occur are evidently less than in the transition of two B_{tr}—X into two B_{te}—X bonds. The gain in reorganization energy on complexation of ring compounds of boron as compared with the linear analogs is adequate to the energy of strain in the ring.

2-n-Butyl-1,2-thioborolane forms a stable liquid complex with pyridine which can be distilled in vacuum without decomposition. Similar to the other coordination compounds of boron, this complex has a high dipole moment (5.2 D) whereas the dipole moment of the initial thioborolane is only 1.5 D.

Alkylthioborolanes also form stable complexes with primary and secondary amines and ammonia (XXVI) which can be distilled in vacuum without decomposition.

The infrared spectra of compounds (XXVI) (L=NHR$_2$) show N—H absorption bands at 3050–3300 cm^{-1} but no S—H bands. The infrared spectra of compounds (XXVI), (L=NH$_2$R) show a strong NH$_2$ deforma-

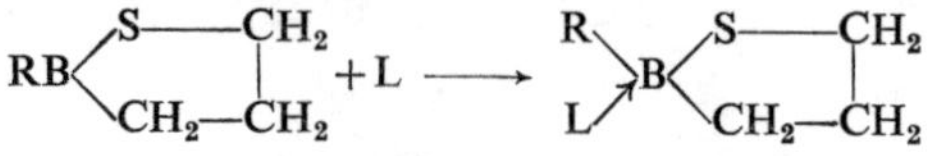

$$RB\underset{CH_2-CH_2}{\overset{S-CH_2}{<}} + L \longrightarrow \underset{L}{\overset{R}{>}}B\underset{CH_2-CH_2}{\overset{S-CH_2}{<}} \qquad (121)$$

(XXVI)

R = n-C$_4$H$_9$, i-C$_4$H$_9$;   L = NH$_3$, C$_2$H$_5$NH$_2$, (C$_2$H$_5$)$_2$NH, n-C$_4$H$_9$NH$_2$, (n-C$_4$H$_9$)$_2$NH, C$_5$H$_{10}$NH

tion band at 1582 cm^{-1}. The complexes of alkylthiaborolanes with sterically hindered amines are less stable. For instance, the complex of 2-n-butyl-1,2-thiaborolane with t-butylamine decomposes into the initial reactants on heating.

Unlike primary and secondary amines, alcohols cause rupture of the B—S bond in 2-alkyl-1,2-thiaborolanes with the formation of γ-thiopropylborinates (XXVII).

$$RB\underset{CH_2-CH_2}{\overset{S-CH_2}{<}} + R'OH \longrightarrow \underset{R'O}{\overset{R}{>}}BCH_2CH_2CH_2SH \qquad (122)$$

(XXVII)

Analogously 2-n-butylthiaborolane reacts with ethanolamine to give the crystalline ethanolamine ester of γ-thiopropyl-n-butylborinic acid.

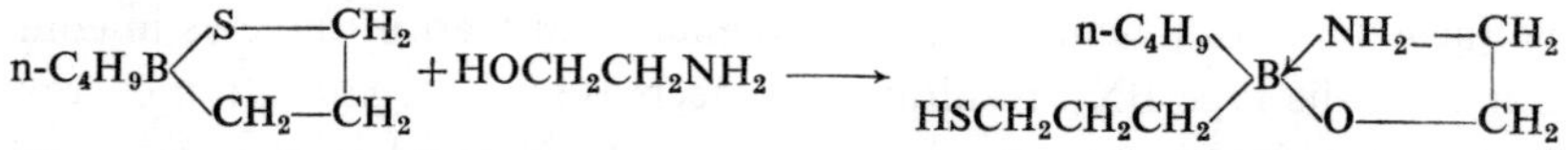

$$n-C_4H_9B\underset{CH_2-CH_2}{\overset{S-CH_2}{<}} + HOCH_2CH_2NH_2 \longrightarrow \underset{HSCH_2CH_2CH_2}{\overset{n-C_4H_9}{>}}B\underset{O-CH_2}{\overset{NH_2--CH_2}{<}}$$

(XXVII)   (123)

The infrared spectra of compounds (XXVII) display an SH absorption band at 2580 cm^{-1} and an intense band in the B—O region but no OH absorption band. The fact that in this case the reaction does not stop at the stage of complexation and that nucleophilic substitution occurs on the boron atom can be ascribed to the lower complex-forming tendency of alcohols as compared with amines and to the higher B—O than B—N bond energy.

Butyl alcohol also breaks the B—S bond in the 2-n-butyl-1,2-thioborolane pyridine complex producing pyridine and butyl (γ-thiopropyl)butyl borinate.

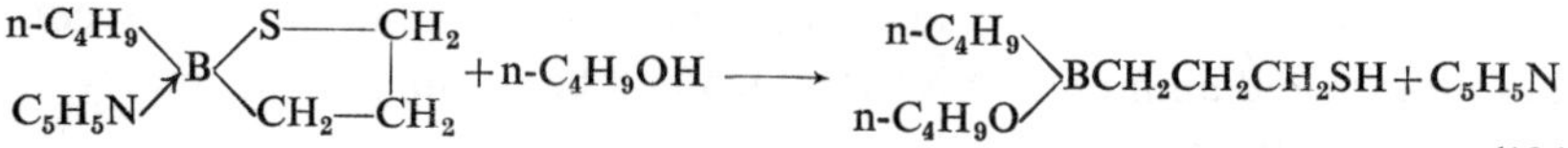

$$\underset{C_5H_5N}{\overset{n-C_4H_9}{>}}B\underset{CH_2-CH_2}{\overset{S-CH_2}{<}} + n-C_4H_9OH \longrightarrow \underset{n-C_4H_9O}{\overset{n-C_4H_9}{>}}BCH_2CH_2CH_2SH + C_5H_5N$$

(124)

The product under the influence of a strong complexating agent such as piperidene eliminates the alkoxy moiety to form once again the coordination compound of alkylthiaborolane (125).

$$\begin{array}{c} \text{n-C}_4\text{H}_9 \\ \diagdown \\ \diagup \\ \text{n-C}_4\text{H}_9\text{O} \end{array} \text{B(CH}_2\text{)}_3\text{SH} + \text{C}_5\text{H}_{11}\text{N} \longrightarrow \begin{array}{c} \text{n-C}_4\text{H}_9 \\ \diagdown \\ \diagup \\ \text{C}_5\text{H}_{11}\text{N} \end{array} \text{B} \begin{array}{c} \text{S} \!\!-\!\! \text{CH}_2 \\ | \\ \text{CH}_2 \!\!-\!\! \text{CH}_2 \end{array} + \text{n-C}_4\text{H}_9\text{OH} \quad (125)$$

VI. ALKYLTHIOBORANES

A. Introduction

A variety of types of compounds with alkylthio and hydrogen substituents on boron are known. Differing greatly in structure and properties they can still be generalized under the common name of "alkylthioboranes" because they all contain the groupings $RSBH_2$ and $(RS)_2BH$, variously coordinated with each other. Monoalkylthioboranes, $RSBH_2$, are not known in simple monomeric form; they exist as polymers (XXVIII), trimers (XXIX) or complexes with secondary and tertiary aminoacids (XXX).

$$(RSBH_2)_x \qquad (RSBH_2)_3 \qquad RSBH_2 \cdot L \qquad \begin{array}{c} \text{H} \quad\;\; \text{R} \quad\;\; \text{H} \\ \diagdown \;\, \text{S} \;\, \diagup \\ \text{B} \quad\quad \text{B} \\ \diagup \;\;\, \diagdown \;\;\, \diagdown \\ \text{H} \quad\; \text{H} \quad\; \text{H} \end{array} \qquad (RS)_2BH$$

$$(XXVIII) \qquad (XXIX) \qquad (XXX) \qquad\quad (XXXI) \qquad\qquad (XXXII)$$

Only one representative of the fourth type of compound with a $RSBH_2$ grouping formed by coordination of methythioborane with borane (XXXI) has been described.

Of the compounds of type $(RS)_2BH$ (XXXII) only di-t-butylthioborane exists in the monomeric form. The other known dialkylthioboranes are associated to a varying degree and display a high tendency to undergo disproportionation.

B. Preparation of Alkylthioboranes

The reaction of diborane with methyl mercaptan was first investigated by Burg and Wagner[7] using high-vacuum apparatus. On mixing the reagents at $-78°$ and keeping the reaction mixture at this temperature for several hours a solid polymer of methylthioborane was obtained:

$$\frac{x}{2} B_2H_6 + xCH_3SH \longrightarrow (CH_3SBH_2)_x + xH_2 \qquad (126)$$

The treatment of a partly depolymerized product of the reaction with trimethylamine gave trimethylamine-methylthioborane, $CH_3BH_2 \cdot N(CH_3)_3$. By the action of diborane on the polymer (127) or on trimethylamine-methylthioborane (128), thermally unstable methylthiodiborane was obtained, which rapidly decomposed at room temperature to diborane and the polymer $(CH_3SBH_2)_x$.

$$(CH_3SBH_2)_x + \frac{x}{2} B_2H_6 \longrightarrow xCH_3SB_2H_5 \qquad (127)$$

$$CH_3SBH_2 \cdot N(CH_3)_3 + B_2H_6 \longrightarrow CH_3SB_2H_5 + H_3B \cdot N(CH_3)_3 \qquad (128)$$

Later the reactions between diborane and mercaptans were performed in ether solution, using ordinary laboratory apparatus. Since they could thereby be carried out on a large scale sufficient quantities of the products became available to enable an extensive investigation of their properties. On adding diborane to an ether solution of two moles of mercaptan, a steady evolution of hydrogen was observed and on completion of the reaction and removal of the solvent a solid polymer (in the case of methyl or t-butylmercaptan) or a viscous product (in the case of ethyl, n-propyl or n-butylmercaptan) was formed.[83, 84]

$$\frac{x}{2} B_2H_6 + xRSH \longrightarrow \frac{x}{3} (RSBH_2)_3 + xH_2 \qquad (129)$$

Polymers of ethyl-, n-propyl- and n-butylthioborane on standing transform into alkylthioborane trimers; polymers of methyl- and t-butylthioborane transform into trimers in tetrahydrofuran solution.[83, 84]

$$(RSBH_2)_x \longrightarrow \frac{x}{3} (RSBH_2)_3 \qquad (130)$$

With the exception of t-butylthioborane trimer (m.p. 107°), the trimers of alkylthioboranes are liquids which can be distilled in vacuum without decomposition.

When n-amyl or benzyl mercaptan were reacted with diborane (2 : 1) in tetrahydrofuran or diglyme solution, two moles of hydrogen were rapidly liberated; the resultant RS-substituted diboranes were not isolated from the solution and were identified by their infrared and n.m.r. spectra.[105-6] From diborane and 1,2-dithioethylene glycol polymeric 1,2-di(borylthio)ethane is formed.[14]

$$B_2H_6 + HSCH_2CH_2SH \longrightarrow H_2BSCH_2CH_2SBH_2 + 2 H_2 \qquad (131)$$

Di-n-propylthioborane and di-n-butylthioborane were obtained by adding diborane to an ether solution of mercaptan (1 : 4) followed by distillation of the products.[78, 79]

$$B_2H_6 + 4\ RSH \longrightarrow 2\ (RS)_2BH + 4\ H_2 \tag{132}$$

A general method for preparing dialkylthioboranes consists of hydrogenation of alkyl thioborates by lithium aluminium hydride in ether solution.[112]

$$4\ B(SR)_3 + LiAlH_4 \longrightarrow 4\ (RS)_2BH + LiSR + Al(SR)_3 \tag{133}$$

Dialkylthioboranes can also be obtained from alkylthioborane trimers and thioborates when a 1 : 3 mixture of these substances is subjected to distillation.

$$(RSBH_2)_3 + 3\ (RS)_3B \longrightarrow 6\ (RS)_2BH \tag{134}$$

The synthesis of t-dodecylthioborane from aminoborane and t-dodecyl mercaptan has been described.[25]

Associated 1,3,2-dithioborolane was obtained by the action of diborane on dithioethylene glycol.[14]

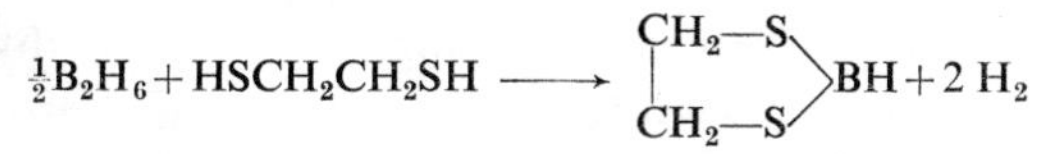

$$\tfrac{1}{2}B_2H_6 + HSCH_2CH_2SH \longrightarrow \begin{matrix} CH_2-S \\ | \quad\quad\ \ \diagdown \\ CH_2-S \diagup \end{matrix} BH + 2\ H_2 \tag{135}$$

Diborane with a 1 : 4 ratio of triphenyl thioborate in tetrahydrofuran gives diphenylthioborane,[106] which was not isolated and was identified only by n.m.r. spectroscopy (doublet at -24 ppm, $J = 140$ c/s).

C. Physical Properties of Alkylthioboranes

Polymers of alkylthioboranes are viscous liquids or solids that in most cases swell in ether and dissolve in tetrahydrofuran but not in benzene. Association of the alkylthioborane molecules occurs through formation of dative bonds between boron and sulfur atoms (XXXIII).

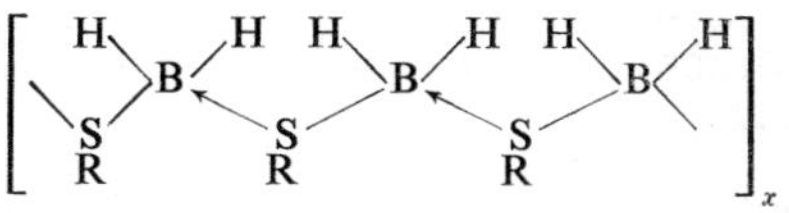

(XXXIII)

Alkylthioborane trimers are liquids of cyclic structure (XXXIV) with semicoordination bonds between the boron and sulfur atoms similar to the B—N semicoordination bonds (see p. 339).

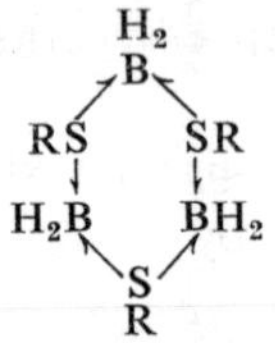

$$
\begin{array}{c}
H_2 \\
B \\
RS \quad\quad SR \\
H_2B \quad\quad BH_2 \\
S \\
R
\end{array}
$$

(XXXIV)

The dipole moments of 1,3,5-triethyl- and 1,3,5-tributylcyclotriborthianes are 2.90 and 3.16 D, respectively.

The i.r. spectra of 1,3,5-trialkylcyclotriborthianes display two intensive bands in the 2412–2420 cm^{-1} and 2465–2475 cm^{-1} regions ascribed to B—H stretching frequencies.

The B^{11} n.m.r. spectrum of 1,3,5-triethylcyclotriborthiane at 25° is a broad peak ($\delta = +31$ ppm with respect to B(OCH$_3$)$_3$ as reference). At 90° the broad peak transforms into a triplet with a $1:2:1$ intensity ratio.[102]

The p.m.r. spectrum of 1,3,5-trimethylcyclotriborthiane (T.M.S. internal reference, $\delta = 0$) displays a singlet due to methyl protons at $\delta = 2.14 \pm 0.03$ ppm and a quadruplet due to the boron hydrogens, centered at $\delta = 2.35 \pm 0.02$ ppm with $J_{BH} = 115 \pm 5$ c/s.

The properties of the dialkylthioboranes are determined by the nature of the alkyl radicals. In the i.r. spectrum of monomeric di-t-butylthioborane there is a BH absorption band at 2545 cm^{-1}. The B^{11} n.m.r. spectrum of this compound at room temperature consists of a weakly resolved doublet (chemical shift -40.7 ppm with reference to B(OCH$_3$)$_3$).

Other dialkylthioboranes which tend to disproportionate spectroscopically show, even in freshly distilled samples, the presence of about 20 per cent of thioborate and alkylthioboranes of various degrees of association. With time the process of disproportionation in the methyl and ethyl derivatives reaches equilibrium with the mixture containing about 20 per cent of dialkylthioborane.

$$2\,(RS)_2BH \rightleftharpoons (RS)_3B + RSBH_2 \tag{136}$$

Diisopropylthioborane and di-n-butylthioborane completely disproportionate on standing.

1,3,2-dithiaborolane is a monomer in the gas phase.[14] The B—H frequency falls from 2595 cm^{-1} for the vapor to 2435 cm^{-1} for the solid.

D. *Chemical Properties of Alkylthioboranes*

1. *Thermal Transformations*

The majority of the known polymers of alkylthioboranes slowly undergo spontaneous conversion at room temperature to the trimers.[83, 84] Exceptions are the solid polymers of methylthioborane and t-butythioborane, which are converted to trimers on dissolution in tetrahydrofuran.

Trimers of alkylthioboranes are thermally stable and can be distilled in vacuum with the exception of n-butylthioborane trimer, which partly decomposes on distillation. Other trimers decompose above 100°. Thus, 1,3,5-triethylcyclotriborthiane can be distilled in a small fractionation column without decomposition, but if the distillation is carried out in a highly efficient column it decomposes into volatile substances and a highly viscous distillate of the elementary composition $[(C_2H_5S)_3B_2H_3]_x$. The suggestion[102] that the polymeric product was formed by reaction of diborane with ethyl mercaptan at a ratio 1 : 3 is erroneous. Actually at the reactant ratios two hydrogens in BH_3 are replaced by ethylthio groups only to a slight extent and on distillation of the reaction products ethylthioborane trimer was obtained in high yield (up to 80 per cent).

2. *Oxidation, Protonolysis*

Trimers of alkylthioboranes are stable towards oxidants. They are practically unchanged in air and are only partially oxidized even on refluxing with hydrogen peroxide in alkaline solution.[84]

Polymers and trimers of alkylthioboranes have different hydrolytic stabilities. Polymers are easily hydrolyzed at room temperature; whereas trimers are only slowly hydrolyzed by water on heating.

Polymers react readily with alcohols in the cold; alcoholysis of trimers proceeds very slowly, but they are decomposed completely to hydrogen and alkyl borates by alcohols on boiling.[84]

$$(RSBH_2)_3 + 9\,R'OH \longrightarrow 3\,B(OR')_3 + 6\,H_2 + 3\,RSH \tag{137}$$

The trimers react with mercaptans at 110–120° to form alkyl thioborates.[84]

$$(RSBH_2)_3 + 6\,RSH \longrightarrow 3\,B(SR)_3 + 6\,H_2 \tag{138}$$

Dialkylthioboranes undergo hydrolysis to form boric acid, and alcoholysis to trialkyl borates. Substitution of hydrogen by an alkylthio group in dialkylthioboranes begins at 50–60° and proceeds energetically around 100°.[78, 79]

3. Reactions with Amines

Trimers of alkylthioboranes react with primary and secondary amines at room temperature. The reaction proceeds with rupture of the ring leading to amine complexes of the monomeric alkylthioboranes.[86, 87]

$$(RSBH_2)_3 + 3 R'R''NH \longrightarrow 3 RSBH_2 \cdot HNR_2'R'' \qquad (139)$$
$$B'' = H, \text{ alkyl}$$

Amine-alkylthioboranes are also formed in the reaction of polymers of alkylthioboranes with amines. They also are obtained in the reaction of diborane with sulfenylamines.[104]

$$\tfrac{1}{2}B_2H_6 + CH_3SNR_2 \longrightarrow CH_3SNR_2 \cdot BH_3 \longrightarrow CH_3SBH_2 \cdot NHR_2 \qquad (140)$$

Complexes of alkylthioboranes with tertiary amines are unassociated liquids, stable at room temperature, but partly decomposed on heating.[87]

Complexes of alkylthioboranes with secondary amines undergo transformations to both alkylthio substitution products, namely dialkylaminoboranes, and hydrogen substitution products, dialkylamino(alkylthio)-boranes.[87, 104]

$$RSBH_2 \cdot NHR_2' \underset{\diagdown}{\overset{\diagup}{\underset{R_2'N}{\overset{R_2'NBH_2 + RSH}{\underset{RS}{\diagdown}}}}}\!{>}BH + H_2 \qquad (141)$$

The reactions of amino-alkylthioboranes with hydrogen chloride and halogenated hydrocarbons were investigated.[56, 86] These reagents convert alkylthioborane complexes with dimethylamine to boronium salts—bis(dimethylamine)boronium chloride (142) and bromide (143).

$$2 RSBH_2 \cdot NH(CH_3)_2 + HCl \longrightarrow \{H_2B[NH(CH_3)_2]_2\}Cl + \tfrac{1}{3}(RSBH_2)_3 + RSH \quad (142)$$

$$2 RSBH_2 \cdot NH(CH_3)_2 + R'X \longrightarrow \{H_2B[NH(CH_3)_2]_2\}X + \tfrac{1}{3}(RSBH_2)_3 + RSR' \quad (143)$$

Trialkylamine-alkylthioboranes are decomposed by hydrogen chloride to the initial trimer. They enter into exchange reactions with halogenated hydrocarbons to give amine complexes of haloboranes and dialkylsulfides.

$$RSBH_2 \cdot NR_3' + R''X \longrightarrow XBH_2 \cdot NR_3' + RSR'' \qquad (144)$$

Alkylthioborane complexes with pyridine undergo a peculiar reaction with halogenated hydrocarbons. Carbon tetrachloride causes convertion bis(pyridine)boronium chloride (145). In contrast,

$$RSBH_2 \cdot NC_5H_5 + CCl_4 \longrightarrow [H_2B(NC_5H_5)_2]Cl \qquad (145)$$

alkyl halides effect on exchange of alkylthio groups and halogen atoms (146).

$$RSBH_2 \cdot NC_5H_5 + R'X \longrightarrow XBH_2 \cdot NC_5H_5 + RSR' \tag{146}$$

Of considerable interest is the reaction of dimethylamine-ethylthioborane with dimethylammonium chloride which leads to the formation of bis(dimethylamine)boronium chloride (147). In this case the complex

$$RSBH_2 \cdot NH(CH_3)_2 + (CH_3)_2NH_2Cl \longrightarrow \{H_2B[NH(CH_3)_2]_2\}Cl + RSH \tag{147}$$

salt is formed by the interaction of the ammonium cation and a molecule of the boron compound, the anion of the ammonium salt serving as anion of the complex.

Alkylthioborane trimers also readily react with primary amines; the course of the reaction differs from that with secondary or tertiary amines.[115] By the action of methyl or ethylamine on alkylthioborane trimers bis(methylamine)boronium and bis(ethylamine)boronium mercaptides are formed independently of the reactant ratio (148).

$$(RSBH_2)_3 + 6\ R'NH_2 \longrightarrow 3\ [H_2B(NH_2R')_2]SR \tag{148}$$

Bis(alkylamine)boronium mercaptides on treatment with an ethereal solution of hydrogen chloride exchange the mercaptide ion for a chloride ion (149). A similar exchange reaction takes place with benzyl chloride.

$$[H_2B(NH_2R')_2]SR + HCl \longrightarrow [H_2B(NH_2R')_2]Cl + RSH \tag{149}$$

Bis(alkylamine)boronium chlorides can be obtained directly from the alkylthioborane trimers by the action of primary amines in the presence of benzyl chloride (150).

$$(RSBH_2)_3 + 6\ NH_2R' + 3\ C_6H_5CH_2Cl \longrightarrow 3\ [H_2B(NH_2R')_2]Cl + 3\ C_6H_5CH_2SR \tag{150}$$

Dialkylthioboranes react with amines at room temperature. Primary amines lead to N-trialkylborazoles.[79]

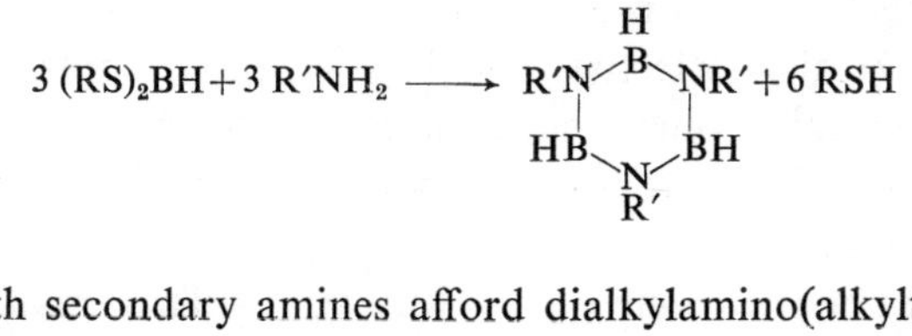

$$(151)$$

Reactions with secondary amines afford dialkylamino(alkylthio)boranes, $RS(R_2N)BH$, (152) or bis(dialkylamino)boranes (153) depending upon the reactant ratio.[50]

$$(RS)_2BH + R_2'NH \longrightarrow \underset{RS}{\overset{R_2'N}{>}}BH + RSH \qquad (152)$$

$$(RS)_2BH + 2\ R_2'NH \longrightarrow (R_2'N)_2BH + 2\ RSH \qquad (153)$$

With tertiary amines dialkylthioboranes lead to the complexes $(RS)_2BH$-amine. 1,3,2-Dithioborolane is reported to afford complexes with trimethylamine or trimethylphosphine.[14]

4. *Reactions with Ethers, Epoxides and Carbonyl Compounds*

Phenylthio-, n-amylthio- and benzylthio borane are capable of splitting ethers at room temperature.[105-6]

$$RSBH_2 + R'OR'' \longrightarrow RSR' + H_2BOR'' \qquad (154)$$

Thus phenylthioborane in diglyme at 25° for 20 hours gave thioanisole in 79 per cent yield, and in tetrahydrofuran for 3 hours gave 4-phenylthio-1-butanol in 45 per cent yield.

Epoxides react with phenylthioborane or alkylthioboranes (n-amylthio and benzylthio derivatives) with rupture of the oxide ring.[106-7] The stereochemical selectivity of these reactions depends upon the nature of the oxide.[107] With *cis* or *trans* 2-butene oxides, *trans* opening of the epoxide ring takes place, so that only threo (155) or erythro-3-phenylthio-2-butanol (156) are produced.

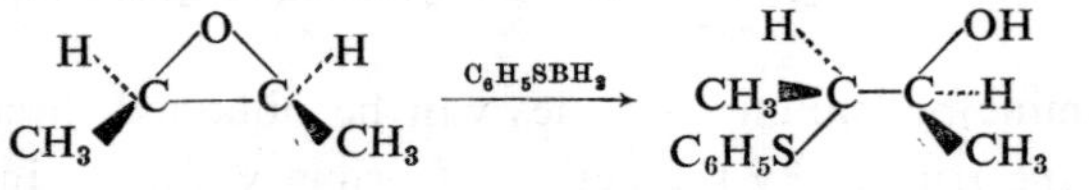

$$\text{threo} \qquad (155)$$

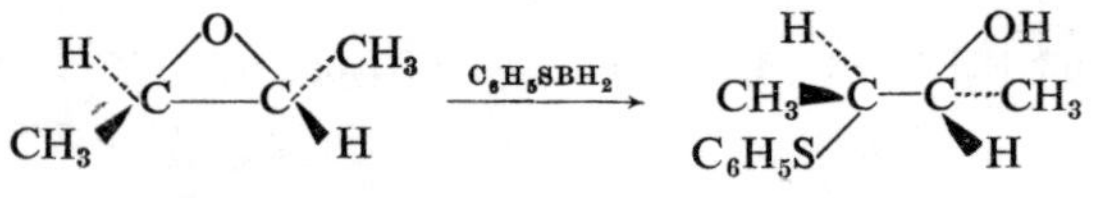

$$\text{erythro} \qquad (156)$$

The reaction of phenylthioborane with cyclohexene oxide to give trans-2-phenylthiocyclohexanol is also stereoselective.

The stereochemical selectivity of these reactions contradicts the four-center mechanism[105] and is not in accord with the postulate of carbonium ion participation. It is as yet unclear as to just how ring fissure occurs in these reactions.

The reaction of phenyl mercaptan with *trans* stilbene epoxide in diglyme proceeds with complications, only 6 per cent of the threo and erythro

1,2-diphenyl-2-phenylthioethanol derivative being formed, whereas the principal product (45 per cent) is 1,2-diphenylethanol.

Phenylthioborane, prepared from thiophenol and diborane in tetrahydrofuran, reacts with optically active L-(–)styrene epoxide to yield only L-(–)-2-phenyl-2-phenylthioethanol.

Phenylthioborane reduces carbonyl compounds and acids (benzaldehyde, acetophenone, benzoic acid and its ethyl ester, etc.) in alcohol, but does not react with benzyl chloride.

5. *Reactions with Olefins*

Alkylthioborane polymers react with olefins in ethereal solution, the course of the reaction depending upon the nature of the reactants.[85] Methylthioborane polymer, a gelatinous mass containing only a small amount of solid polymer, adds to ethylene, 1-hexene or cyclohexene on slight heating (50°, 1.5 hours) with the formation of the corresponding esters of dialkylthioborinic acids (157).

$$(CH_3SBH_2)_x + \text{olefin} \xrightarrow[75-80\%]{} R_2BSCH_3 \tag{157}$$

Hydrogenation of ethylene or propylene by n-butylthioborane polymer is accompanied by disproportionation, resulting in the formation of a considerable amount of trialkylborane and di-n-butylthioborane, besides n-butyl dialkylthioborinates.

Phenylthioborane reacts with 1-hexene in tetrahydrofuran (25°, 1–1.5 hours) to give, on methanolysis of the reaction products, methyl dihexylborinate (45 per cent) and dimethyl hexylboronate (21 per cent). A similar reaction occurs between phenylthioborane and styrene. Oxidation of the primary products yields 2-phenylethanol and 1-phenylethanol in 1 : 9 ratio.[106]

The reaction of alkylthioborane polymers with dienes leads to boron-containing heterocyclic compounds (see Section III, B, 4).

A comparison of the results of hydroboration of dienes by diborane,[1, 29, 31, 75] chloroborane[113] and alkylthioboranes shows that the orientation of the addition is determined mainly by the nature of the diene. Substitution of a hydrogen atom in borane by an electronegative halogen atom or alkylmercapto group has no essential influence on the course of the reaction.

Alkylthioborane trimers do not react with unsaturated compounds at room temperature. If, however, the reaction is carried out in boiling ether, a mixture of products is formed. Thus, in the reaction of ethylthioborane or ethylthioborane trimer (mistakingly taken for triethylthioborane) and

1-octene there are formed triethyl thioborate (10 per cent), diethyl n-octylthioboronate (20 per cent) and ethyl di-n-octylthioborinate (15 per cent).[110]

Dialkylthioboranes are less active hydroborating agents than alkylthioborane polymers. The reaction of di-n-butylthioborane with 1-hexene or cyclohexene takes place at room temperature and proceeds more rapidly on heating, the n-butyl esters of the corresponding thioacids being formed in high yields.[80]

On oxidation of the products of the reaction of diphenylthioborane with styzene, 2-phenylethanol and 1-phenylethanol are obtained in 77 per cent yield in a 95 : 5 ratio.[106]

VII. DIALKYLAMINO(ALKYLTHIO)BORANES

A. *Introduction*

A peculiar class of totally unsymmetrial organic boron–sulfur compounds are the dialkylamino(alkylthio)boranes (XXXV).

$$HB{\overset{\displaystyle NR_2}{\underset{\displaystyle SR}{<}}}$$

(XXXV)

B. *Preparation of Dialkylamino(alkylthio)boranes*

1. *Dialkylamino(alkylthio)boranes from Dialkylaminoborane Dimers*

Dialkylamino(alkylthio)boranes are obtained by heating the dimers of dialkylaminoboranes with mercaptans.[43, 44]

$$(H_2BNR_2)_2 + 2n\text{-}C_4HqSH \xrightarrow[85\%]{100°} 2HB{\overset{\displaystyle NR_2}{\underset{\displaystyle S\text{-}n\text{-}S_4H_9}{<}}} + 2H_2 \qquad (158)$$

2. *Dialkylamino(alkylthio)boranes from Dialkylamine-Boranes*

From a preparative aspect it is more convenient to synthesize dialkylamino(alkylthio)boranes by heating dialkylamine boranes with mercaptans.[43, 46] Reaction begins at 70–80° with evolution of hydrogen, is very rapid at 100°, and goes to completion on heating for a short time at 120–130°.

$$H_3B\cdot NHR_2 + R'SH \longrightarrow HB{\overset{\displaystyle NR_2}{\underset{\displaystyle SR'}{<}}} + 2H_2 \qquad (159)$$

In the first stage of the reaction, during evolution of hydrogen, the dialkyl-amine–alkylthioborane complex (XXXVI) is formed, which then rearranges into a complex of dialkylaminoborane and mercaptan (XXXVII). The latter eliminates hydrogen, forming the final product (XXXVIII).

$$H_3B \cdot NHR_2 + R'SH \xrightarrow{-H_2} \underset{\underset{SR'}{|}}{H_2B \cdot NHR_2} \longrightarrow \underset{\underset{HSR'}{\uparrow}}{H_2BNHR} \xrightarrow{-H_2} HB\begin{matrix} \diagup NHR \\ \diagdown SR' \end{matrix}$$

$$\text{(XXXVI)} \qquad \text{(XXXVII)} \qquad \text{(XXXVIII)}$$

$$(160)$$

It should be noted that complexes of borane with primary amines are converted to N-trialkylborazoles in high yield by heating with mercaptans.[45]

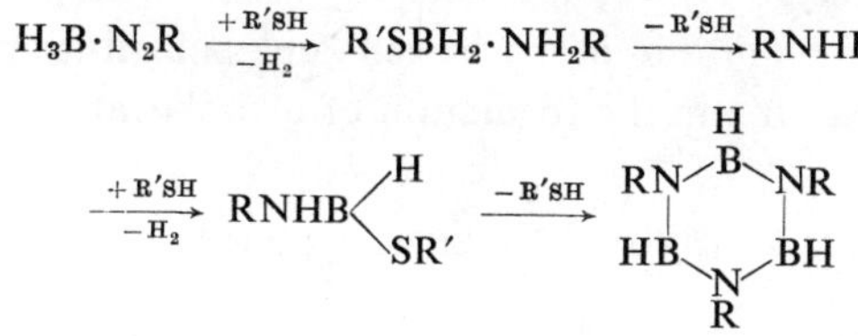

$$H_3B \cdot N_2R \xrightarrow[-H_2]{+R'SH} R'SBH_2 \cdot NH_2R \xrightarrow{-R'SH} RNHBH_2$$

$$\xrightarrow[-H_2]{+R'SH} RNHB\begin{matrix} \diagup H \\ \diagdown SR' \end{matrix} \xrightarrow{-R'SH} \begin{matrix} & H & \\ & \overset{|}{B} & \\ RN\diagup & & \diagdown NR \\ | & & | \\ HB\diagdown & & \diagup BH \\ & \underset{|}{N} & \\ & R & \end{matrix} \qquad (161)$$

Since mercaptan is regenerated in the separate stages of the reaction only a small amount of this substance is required for the syntheses of N-tri-alkylborazoles to go to completion.

3. Dialkylamino(alkylthio)boranes from Dialkylamine-alkylthioboranes

Complexes of alkylthioboranes with secondary amines on heating are converted to dialkylamino(alkylthio)boranes in 60–88 per cent yields.[87]

$$H_2BSR \cdot NHR_2 \xrightarrow[60-88\%]{100-140°} HB\begin{matrix} \diagup NR_2 \\ \diagdown SR \end{matrix} + H_2 \qquad (162)$$

The reaction is complicated by side reactions that give mercaptan and bis(dialkylamino)boranes.

4. Dialkylamino(alkylthio)boranes from Dialkylthioboranes and Amines

Dialkylamino(alkylthio)boranes are obtained on the reaction of dialkyl-thioboranes with secondary amines.[50]

$$HB(SR)_2 + NHR_2 \xrightarrow{53-85\%} HB\begin{matrix} \diagup NR_2 \\ \diagdown SR \end{matrix} + RSH \qquad (163)$$

The reaction starts at room temperature and moderate heating for a short time is required to achieve completion.

C. Properties of Dialkylamino(alkylthio)boranes

1. General

Dialkylamino(alkylthio)boranes are colorless, mobile liquids with a very unpleasant odor. They undergo hydrolysis quite rapidly on standing in air, but do not decompose or disproportionate up to 200°.

Dialkylamino(alkylthio)boranes are completely miscible with organic solvents. Their infrared spectra display B–H stretching bands in the region of 2485–2500 cm^{-1}.

2. Reactions with Alcohols and Mercaptans

Dialkylamino(alkylthio)boranes react vigorously with alcohols at room temperature with evolution of hydrogen and substitution of the R_2N and RS groups resulting in the formation of orthoborates.[47]

$$HB\diagdown_{SR'}^{NR_2} + 3\ R''OH \longrightarrow B(OR'')_3 + NHR_2 + R'SH + H_2 \qquad (164)$$

The hydrogen atoms in dialkylamino(alkylthio)boranes are not substituted by alkylthio groups on heating with mercaptans even to 200°, but when the added mercaptan has a higher boiling point than that of alkylthio residue the former displaces the latter.[37, 47]

$$HB\diagdown_{SC_2H_5}^{N(C_2H_5)_2} + n\text{-}C_4H_9SH \longrightarrow HB\diagdown_{S\text{-}n\text{-}C_4H_9}^{N(C_2H_5)_2} + C_2H_5SH \qquad (165)$$

3. Reactions with Amines

Dialkylamino(alkylthio)boranes under mild conditions exchange the alkylthio group for an arylamino group of an aromatic amine.[47]

$$HB\diagdown_{S\text{-}n\text{-}C_4H_9}^{N(C_2H_5)_2} + NH_2Ar \longrightarrow HB\diagdown_{NHAr}^{N(C_2H_5)_2} + n\text{-}C_4H_9SH \qquad (166)$$

The resultant arylamino(dialkylamino)borane in turn reacts with the aromatic amine at 60–80°, the dialkylamino grouping rather than the hydrogen atom being displaced by the arylamino group, so that a bis(arylamino)borane is formed (167).

$$HB\diagdown_{NHAr}^{N(C_2H_5)_2} + NH_2Ar \longrightarrow HB(NHAr)_2 + NH(C_2H_5)_2 \qquad (167)$$

Similarly bis(alkylamino)boranes are formed by the action of primary aliphatic amines.[44]

$$HB{\overset{\displaystyle N(CH_3)_2}{\underset{\displaystyle SR'}{}}}+2\,NH_2R \longrightarrow HB(NHR)_2+NH(CH_3)_2+R'SH \qquad (168)$$

A singular reaction is that of dialkylamino(alkylthio)boranes with secondary amines.[47, 49] When diethylamino(n-butylthio)borane is heated with diethylamine the following equilibrium is established.[47]

$$HB{\overset{\displaystyle N(C_2H_5)_2}{\underset{\displaystyle S\text{-}n\text{-}C_4H_9}{}}}+NH(C_2H_5)_2 \rightleftarrows HB[N(C_2H_5)_2]+n\text{-}C_4H_9SH \qquad (169)$$

The equilibrium can also be arrived at by heating bis(dimethylamino)-borane with n-butylmercaptan. In this case the dialkylamino group is substituted by an alkylthio group; whereas, the reverse is usually true, i.e. RS groups are replaced by R_2N groups, as, for example, in the action of amines on esters of thioorganoboron acids.[35, 63] The above-mentioned equilibrium can be shifted completely to the right if the amine employed has a higher boiling point than the liberated mercaptan which can then be distilled from the reaction mixture (170).

$$HB{\overset{\displaystyle N(i\text{-}C_5H_{11})_2}{\underset{\displaystyle S\text{-}n\text{-}C_3H_7}{}}}+NH(i\text{-}C_5H_{11})_2 \longrightarrow HBN(i\text{-}C_5H_{11})_2+n\text{-}C_3H_7SH \qquad (170)$$

Methylaniline replaces both the RS- and R_2N groups of diethylamino-(n-butylthio)borane even with equimolar reactant ratios. The reaction does not go to completion, half the initial organoboron compound remaining unconsumed. When the reactant ratio is 1 : 2, bis-(N-methyl-N-phenyl-amino)borane is obtained in 83 per cent yield.[49]

The inertness of the hydrogen atom in dialkylamino(alkylthio)boranes is also manifested in the drastic conditions required for the diethylamino-borane dimer to hydroborate 1-butene and 1-octene.[43]

It thus follows that the ease of nucleophilic substitution of the various groups on the boron atom in dialkylamino(alkylthio)boranes decreases in the order:

$$RS > NR_2 > H$$

VIII. CONCLUSIONS

From all the above it is seen that organic boron–sulfur compounds with their tendency to form complexes and to exchange the RS group for other groupings are more reactive than their boron–oxygen counterparts. This is explained on the one hand by the lower ionization potential[22] of the

TABLE 1. PHYSICAL PROPERTIES OF BORON–SULFUR COMPOUNDS

Compound	B.p. (°C)/mm, M.p. (°C)	n^{20}	d_4^{20}	Refs.
$\underline{B(SR)_3}$				
$B(SCH_3)_3$	78/1.5 m.p. 4.9	1.5778	1.098	7, 18, 84
$B(SC_2H_5)_3$	73/1	1.5473	1.019	40, 84, 110, 118
$B(S\text{-}n\text{-}C_3H_7)_3$	102/1	1.5324	0.9952	40, 79
$B(S\text{-}i\text{-}C_3H_7)_3$	80.5/1.5	1.5265	0.9791	40
$B(S\text{-}n\text{-}C_4H_9)_3$	144/2	1.5229	0.9684	19, 20, 40, 78, 79, 99
$B(S\text{-}t\text{-}C_4H_9)_3$	143/9 m.p. 120			40
$B(S\text{-}n\text{-}C_5H_{11})_3$	173/0.2		1.5145^{24}	19, 20, 108
$B(S\text{-}n\text{-}C_6H_{13})_3$	liquid			5
$B(SCH_2C_6H_5)_3$	solid			5
$B(S\text{-}n\text{-}C_{12}H_{25})_3$	decomposes			25
$B(SC_6H_5)_3$	194/0.02 m.p. 129–143			15, 23, 128
$\underline{(RSBS)_{2,\,3}}$				
$(C_2H_5SBS)_3$	heavy oil			126
$(C_2H_5SBS)_2$	m.p. 4			126
$\underline{(ROBS)_3}$				
$(CH_3OBS)_3$	m.p. 27.5			120, 121
$\underline{(RS)_2BNH(R)R}$				
$(C_2H_5S)_2BNHC_2H_5$	54/0.07	1.5122	0.9744	111
$(C_2H_5S)_2BN(CH_3)_2$	72/1	1.5147	0.9826	111
$(C_2H_5S)_2BNHC_6H_5$	disproport.	1.5882	1.0773	111
$\underline{RSB[NH(R)R]_2}$				
$C_2H_5SB(NHC_2H_5)_2$	76/3	1.4805	0.9219	111
$C_2H_5SB[N(CH_3)_2]_2$	55/1	1.4822	0.9037	111
$\underline{(RSBX_2)_2}$				
$(CH_3SBCl_2)_2$	53/25 m.p. 72.7			28
$(CH_3SBBr_2)_2$	73/8 m.p. 112.3			18, 127
$\underline{RB(SR)_2}$				
$CH_3B(SCH_3)_2$	154.9			6
$n\text{-}C_3H_7B(SC_2H_5)_2$	69/3	1.4987	0.9299	81, 99

TABLE 1. *(cont.)*

Compound	B.p. (°C)/mm, M.p. (°C)	n^{20}	d_4^{20}	Refs.
$i\text{-}C_3H_7B(SC_2H_5)_2$	64/2	1.4963	0.9303	81
$n\text{-}C_3H_7B(S\text{-}n\text{-}C_4H_9)_2$	150/13	1.4956	0.9106	20, 61, 80, 81
$n\text{-}C_3H_7B(S\text{-}n\text{-}C_4H_9)_2$	129/7	1.4940	0.9079	62, 81
$i\text{-}C_3H_7B(S\text{-}n\text{-}C_5H_{11})_2$	120/0.5			20
$CH_2{=}CHCH_2B(SC_2H_5)_2$	64/2	1.5182	0.9563	92
$C_2H_5S(CH_2)_3B(SC_2H_5)_2$	121/1.5	1.5312	1.0076	92
$n\text{-}C_4H_9B(SC_2H_5)_2$	82/3.5			20, 99
$n\text{-}C_4H_9B(S\text{-}n\text{-}C_3H_7)_2$	131/9			20
$n\text{-}C_4H_9B(S\text{-}n\text{-}C_4H_9)_2$	150/7	1.4936	0.9045	20, 61, 81, 99
$n\text{-}C_4H_9B(SC_5H_{11})_2$	132/0.5			19, 20
$n\text{-}C_4H_9B{\Big\langle}\begin{smallmatrix}S-CH_2\\|\\S-CH_2\end{smallmatrix}$	84/6			19, 20
$i\text{-}C_4H_9B(S\text{-}n\text{-}C_4H_9)_2$	110/1	1.4919	0.9020	20, 63, 80
$i\text{-}C_4H_9B{\Big\langle}\begin{smallmatrix}S-CH_2\\|\\S-CH_2\end{smallmatrix}$	83/6.5			19, 20
$s\text{-}C_4H_9B(S\text{-}n\text{-}C_4H_9)_2$	105/0.8			19, 20
$t\text{-}C_4H_9B(S\text{-}n\text{-}C_5H_{11})_2$	122/0.5			19, 20
$i\text{-}C_5H_{11}B(S\text{-}n\text{-}C_4H_9)_2$	154/6	1.4871	0.8998	61
$n\text{-}C_6H_{13}B(SC_2H_5)_2$	106/2.5	1.4960	0.9127	99
$n\text{-}C_6H_{13}B(S\text{-}n\text{-}C_4H_9)_2$	98/0.06	1.4840	0.8860	19, 20, 80, 99
$n\text{-}C_8H_{17}B(n\text{-}C_4H_9)_2$	136/0.05	1.4820	0.8770	80
$\text{(cyclohexenyl)}{-}B(S\text{-}n\text{-}C_4H_9)_2$	156/1	1.5737	0.9516	80
$C_6H_5B(S\text{-}n\text{-}C_4H_9)_2$	138.5/1.5	1.5464	0.9865	61
$C_6H_5B(S\text{-}n\text{-}C_5H_{11})_2$	144/0.3			103
(RBS)₃				
$(CH_3BS)_3$	m.p. 60			124
$(n\text{-}C_4H_9BS)_3$	119/0.8			118
$(i\text{-}C_4H_9BS)_3$	94/0.8			118
$(C_6H_5BS)_3$	m.p. 233			125
$RB{\Big\langle}\begin{smallmatrix}O-CH_2\\|\\S-CH_2\end{smallmatrix}$				
$n\text{-}C_3H_7B{\Big\langle}\begin{smallmatrix}O-CH_2\\|\\S-CH_2\end{smallmatrix}$	97/100	1.4680	0.9828	42
$n\text{-}C_4H_9B{\Big\langle}\begin{smallmatrix}O-CH_2\\|\\S-CH_2\end{smallmatrix}$	77/18	1.4674	0.9692	42

TABLE 1. *(cont.)*

Compound	B.p. (°C)/mm, M.p. (°C)	n^{20}	d_4^{20}	Refs.
$RB\begin{smallmatrix}NR_2'\\SR''\end{smallmatrix}$				
$n\text{-}C_3H_7BN(C_2H_5)_2(SC_2H_5)$	92/10	1.4720	0.8688	64
$n\text{-}C_3H_7BN(C_2H_5)_2(S\text{-}n\text{-}C_4H_9)$	97.5/4	1.4704	0.8621	63
$n\text{-}C_4H_9BN(C_2H_5)_2(S\text{-}n\text{-}C_4H_9)$	103/3	1.4702	0.8620	63
$i\text{-}C_4H_9BN(C_2H_5)_2(S\text{-}n\text{-}C_4H_9)$	106.5/4	1.4700	0.8613	63
$i\text{-}C_5H_{11}BN(C_2H_5)_2(S\text{-}n\text{-}C_4H_9)$	98/2	1.4688	0.8752	63
$RB\begin{smallmatrix}Cl(Br)\\SR''\end{smallmatrix}$				
$n\text{-}C_3H_7BCl(SC_2H_5)$	50/13	1.4665	0.988	64
$i\text{-}C_3H_7BCl(SC_2H_5)$	44/16			64
$n\text{-}C_4H_9BCl(SC_2H_5)$	42/4	1.4602	0.9545	64
$i\text{-}C_3H_7BBr(SC_2H_5)$	39.5/5	1.5065	1.287	64
$i\text{-}C_5H_{11}BBr(SC_2H_5)$	57/2	1.4945	1.1920	64
$i\text{-}C_5H_{11}BBr(S\text{-}n\text{-}C_4H_9)$	79/1	1.4890	1.1320	64
$C_6H_5BBr(SC_2H_5)$	90.5/1.5	1.5962	1.3605	64
R_2BSR'				
	vapor tens.			
$(CH_3)_2BSCH_3$	42.8/0			7
$(C_2H_5)_2BSCH_3$	44/27	1.4500	0.8344	85
$(C_2H_5)_2BSC_2H_5$	90/100	1.4557	0.8252	37
$(C_2H_5)_2BS\text{-}n\text{-}C_4H_9$	79.5/14	1.4603	0.8300	37
$(n\text{-}C_3H_7)_2BSC_2H_5$	73/13	1.4562	0.8214	37, 99
$(n\text{-}C_3H_7)_2BS\text{-}n\text{-}C_3H_7$	78/5	1.4546	0.8241	35, 36, 37, 100
$(n\text{-}C_3H_7)_2BS\text{-}n\text{-}C_4H_9$	98/11	1.4598	0.8286	28, 80, 99, 100
$(n\text{-}C_3H_7)_2BSC_6H_5$	124/11.5	1.5194	0.9252	37, 100
$(CH_2\!\!=\!\!CHCH_2)_2BSC_2H_5$	70/11	1.4719	0.8419	92
$(n\text{-}C_4H_9)_2BSC_2H_5$	47/15	1.4585	0.8269	99
$(n\text{-}C_4H_9)_2BS\text{-}n\text{-}C_4H_9$	117/8	1.4548	0.8367	28, 35, 36, 99, 100
$(n\text{-}C_4H_9)_2BSC_6H_5$	141/8	1.5136	0.9126	37
$(i\text{-}C_4H_9)_2BS\text{-}n\text{-}C_4H_9$	107/9.5	1.4572	0.8213	80, 100
$(i\text{-}C_5H_{11})_2BS\text{-}n\text{-}C_4H_9$	133/8.5	1.4551	0.8219	100
$(n\text{-}C_6H_{13})_2BSCH_3$	103/2	1.4690	0.8433	85
$\left(\langle\bigcirc\rangle\!\!-\right)_2BSCH_3$	110/1	1.5200	0.9552	85
$(C_6H_5)_2BS\text{-}n\text{-}C_4H_9$	180/7	1.5871	1.001	61
$(\alpha\text{-}C_{10}H_7)_2BS\text{-}n\text{-}C_4H_9$	247/2			61

TABLE 1. *(cont.)*

Compound	B.p. (°C)/mm, M.p. (°C)	n^{20}	d_4^{20}	Refs.
R$_2$BSH				
(n-C$_3$H$_7$)$_2$BSH	52/25	1.4423	0.8093	34
(n-C$_4$H$_9$)$_2$BSH	83/21	1.4463	0.8175	34
(i-C$_5$H$_{11}$)$_2$BSH	81/8	1.4506	0.8179	34
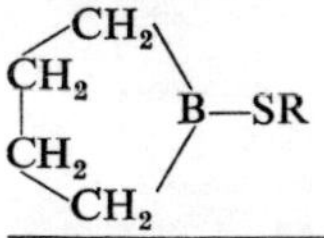				
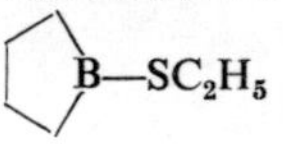 B—SC$_2$H$_5$	90/80	1.4497	0.8060	65
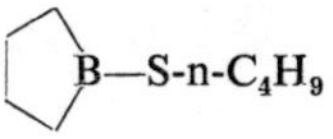 B—S-n-C$_4$H$_9$	49/2	1.5090	0.9326	82, 114
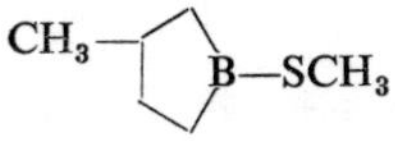 CH$_3$ B—SCH$_3$	43.5/9	1.5030	0.9277	85
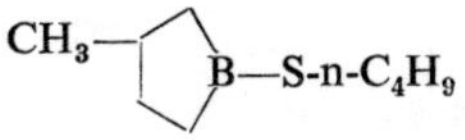 CH$_3$ B—S-n-C$_4$H$_9$	55/1	1.4822	0.8836	82
(RS)$_2$B(CH$_3$)$_n$B(SR)$_2$				
(CH$_3$S)$_2$B(CH$_2$)$_3$B(SCH$_3$)$_2$	m.p. 70			70
(C$_2$H$_5$S)$_2$B(CH$_2$)$_3$B(SC$_2$H$_5$)$_2$	144/1 m.p. 59			70
(C$_2$H$_5$S)$_2$B(CH$_2$)$_4$B(SC$_2$H$_5$)$_2$	150/0.2 m.p. 44.5			65, 98
(RS)$_2$B(CH$_2$)$_3$B(CH$_2$)$_3$B(SR)$_2$ SR				
(CH$_3$S)$_2$B(CH$_2$)$_3$B(CH$_2$)$_3$B(SCH$_3$)$_2$ SCH$_3$	203/2 m.p. 96			73
(C$_2$H$_5$S)$_2$B(CH$_2$)$_3$B(CH$_2$)$_3$B(SC$_2$H$_5$)$_2$ SC$_2$H$_5$	m.p. 63			73
1,2-Thiaborolanes				
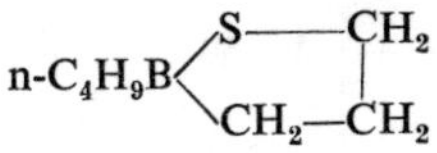 n-C$_4$H$_9$B	53/2	1.4759	0.8888	53, 68

TABLE 1. *(cont.)*

Compound	B.p. (°C)/mm, M.p. (°C)	n^{20}	d_4^{20}	Refs.
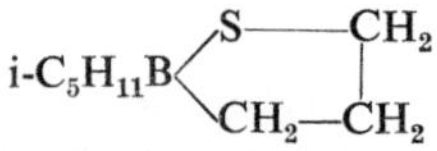 i-$C_5H_{11}B\overset{S——CH_2}{\underset{CH_2—CH_2}{\big<}}$	69/8	1.4780	0.8846	68
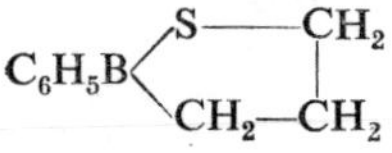 $C_6H_5B\overset{S——CH_2}{\underset{CH_2—CH_2}{\big<}}$	m.p. 41			68
(RSBH$_2$)$_3$				
(CH$_3$SBH$_2$)$_3$	73/1	1.5498	1.0210	84
(C$_2$H$_5$SBH$_2$)$_3$	96/1	1.5323	0.9772	84
(n-C$_3$H$_7$SBH$_2$)$_3$	105/0.09	1.5210	0.9508	84
(n-C$_4$H$_9$SBH$_2$)$_3$	117/0.07	1.5130	0.9376	84
$HB\overset{NR_2}{\underset{SR'}{\big<}}$				
$HB\overset{N(CH_3)_2}{\underset{S\text{-}n\text{-}C_3H_7}{\big<}}$	50/13	0.8706	1.4701	44, 46
$HB\overset{N(CH_3)_2}{\underset{S\text{-}n\text{-}C_4H_9}{\big<}}$	7/75	0.8666	1.4699	46
$HB\overset{N(C_2H_5)_2}{\underset{SC_2H_5}{\big<}}$	67/19	1.4616	0.8502	47, 87
$HB\overset{N(C_2H_5)_2}{\underset{S\text{-}n\text{-}C_3H_7}{\big<}}$	81/17	1.4628	0.848	43
$HB\overset{N(C_2H_5)_2}{\underset{S\text{-}n\text{-}C_4H_9}{\big<}}$	51/1.5	1.4636	0.849	43, 47, 50, 87
$HB\overset{N(C_2H_5)_2}{\underset{SC_6H_5}{\big<}}$	84/1.5	1.5470	0.9736	47
$HB\overset{N(i\text{-}C_5H_{11})_2}{\underset{SC_2H_5}{\big<}}$	80/1	1.4608	0.8530	87
$HB\overset{N(i\text{-}C_5H_{11})_2}{\underset{S\text{-}n\text{-}C_3H_7}{\big<}}$	94/1.5	1.4640	0.8422	46, 50
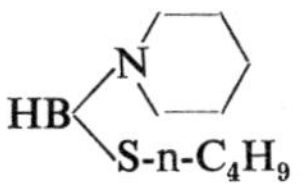 $HB\overset{N\big\rangle}{\underset{S\text{-}n\text{-}C_4H_9}{\big<}}$	74/1.5	1.4944	0.9710	46, 50

TABLE 1. *(Cont.)*

Compound	B.p. (°C)/mm, M.p. (°C)	n^{20}	d_4^{20}	Refs.
RS-derivatives of borazine				
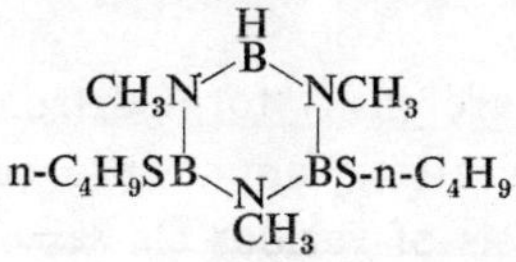	130/0.07	1.5150	0.9989	60
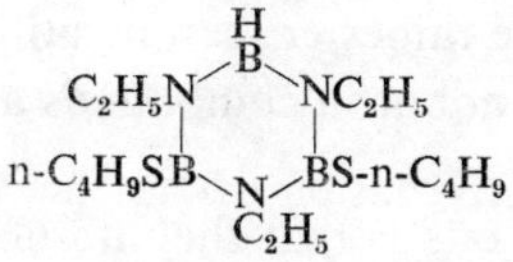	115/0.2	1.5000	0.9594	60
$(n\text{-}C_4H_9SBNH)_3$	180/0.3	1.5348	0.0356	57
$(n\text{-}C_4H_9SBNCH_3)_3$	185/0.05	1.5360	1.0290	57
$(n\text{-}C_4H_9SBNC_2H_5)_3$	135/0.1	1.5228	1.0060	57
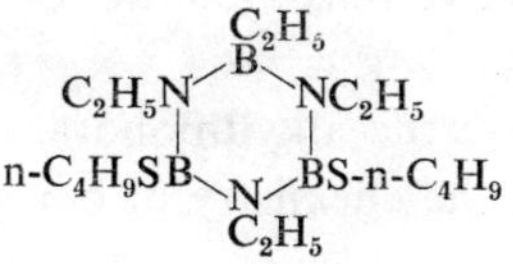	163/0.15	1.5080	0.9826	58
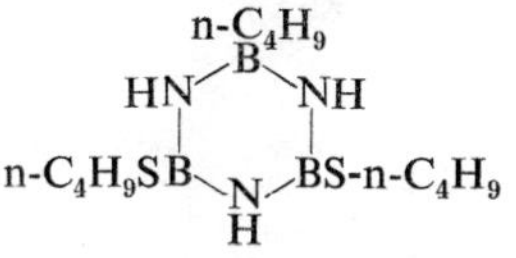	120/0.2	1.5065	0.9860	58
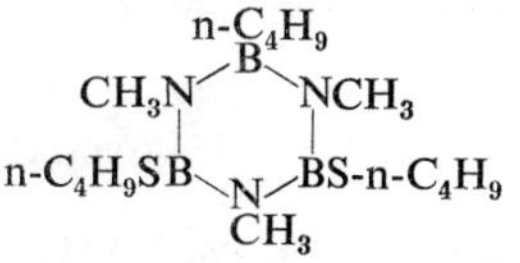	160/0.08	1.5035	0.9998	58
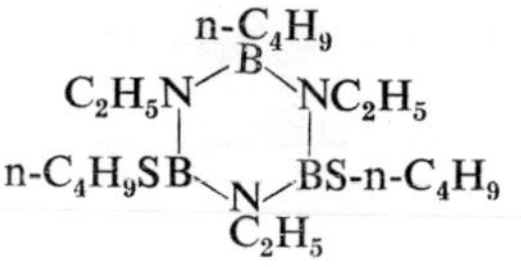	167/0.3	1.5082	0.9630	58

unshared electron pair relative to oxygen (18.70 eV), and on the other hand by the relatively low value for the B—S bonding energy (110–115 kcal/mole); whereas, both types of compounds have the same magnitudes of value for the NH (85 kcal/mole), B— (80 kcal/mole) and S—H (87 kcal/mole)[10] bonding energies. Owing to this the splitting of the B—S bonds leading to the formation of boron bonds with other elements (N, O) proves to be energetically advantageous.

The readily available alkylthio borates which are easily formed from boron trichloride and mercaptans in the presence of tertiary amines proved to be convenient starting materials for the synthesis of various classes of sulfur-containing organoboron esters. In turn these thioesters can be utilized in the synthesis of various other types of organoboron compounds as was shown in Figs. 1 and 2.

A characteristic feature of these synthetic methods is that they do not employ Grignard reagents. The starting materials are trialkylboranes produced from diboranes and olefins. The products are readily converted by a combination of new synthetic methods into reactive boron–sulfur compounds that can be employed in further syntheses.

A second source of organoboron compounds are the alkylthioboranes prepared from diborane and mercaptans which in the chemistry of boron is of an independent theoretical interest.

IX. REFERENCES

1. BEZMENOV, A. Ya., VASILIEV, L. S. and MIKHAILOV, B. M., *Izvest. Acad. Nauk SSSR, Ser. Khim.*, 2111 (1965).
2. BLAU, J. A., GERRARD, W. and LAPPERT, M. F., *J. Chem. Soc.*, 667 (1960).
3. BLAU, J. A., GERRARD, W., LAPPERT, M. F., MOUNTFIELD, B. A. and PYSZORA, H., *J. Chem. Soc.*, 380 (1960).
4. BRINDLEY, P. B., GERRARD, W. and LAPPERT, M. F., *J. Chem. Soc.*, 824 (1956).
5. BROTHERTON, R. J. and PETTERSON, L. L., U.S. Pat. 2 960 530; *C.A.*, **55**, 9345 (1961).
6. BURG, A. B. and GRABER, F. M., *J. Am. Chem. Soc.*, **78**, 1523 (1956).
7. BURG, A. B. and WAGNER, R. I., *J. Am. Chem. Soc.*, **76**, 3307 (1954).
8. COCKSEDGE, H. E., *J. Chem. Soc.*, **93**, 2177 (1908).
9. CONKLIN, G. W. and MORRIS, R. C., U.S. Pat. 2 886 575; *C.A.*, **53**, 20103 (1959).
10. COTTRELL, T. L., *The Strengths of Chemical Bonds*, 2nd ed., Butterworths Scientific Publications, London, 1958.
11. COUNCLER, C., *J. Pract. Chem.*, **18**, 371 (1878).
12. DEWAR, M. J. S., KUBBA, V. P. and PETTIT, R., *J. Chem. Soc.*, 3076 (1958).
13. DOROKHOV, A. A. and MIKHAILOV, B. M., *Izvest. Acad. Nauk SSSR, Ser. Khim.* 364 (1966).
14. EGAN, B. L., SHORE, S. G. and BONNELL, J. E., *Inorg. Chem.*, **3**, 1024 (1964).
15. FUNK, H. and KOCH, H. J., *Wiss. Z. Martin–Luther–Univ.*, **8**, 1025 (1959); *C.A.*, **55**, 11417 (1961).

16. GERRARD, W., *The Organic Chemistry of Boron*, Academic Press, London–New York, 1961.
17. GERRARD, W., LAPPERT, M. F. and MOUNTFIED, B. A., *J. Chem. Soc.*, 1529 (1959).
18. GOUBEAU, J. and WITTMEIER, H. W., *Z. Anorg. Chem.*, **270**, 16 (1952).
19. HAWTHORNE, M. F., *J. Am. Chem. Soc.*, **82**, 748 (1960).
20. HAWTHORNE, M. F., *J. Am. Chem. Soc.*, **83**, 1345 (1961).
21. HINZE, J. and JAFFÉ, H. H., *J. Am. Chem. Soc.*, **84**, 540 (1962).
22. HINZE, J. and JAFFÉ, H. H., *J. Am. Chem. Soc.*, **87**, 1501 (1963).
23. JENKNER, H., Ger. Pat. 950 640; *C.A.*, **53**, 2160 (1959).
24. LANG, K., Ger. Pat. 1 079 634; *C.A.*, **55**, 13316 (1961).
25. LANG, K., Ger. Pat. 1 092 463; *C.A.*, **55**, 24566 (1961).
26. LAPPERT, M. F. and PYSZORA, H., *Proc. Chem. Soc.*, 350 (1960).
27. MIKHAILOV, B. M., *Uspekhi Khimii*, **28**, 1450 (1959).
28. MIKHAILOV, B. M., AKHNAZARIAN, A. A. and VASIL'EV, L. S., *Doklady Acad. Nauk SSSR*, **136**, 828 (1961).
29. MIKHAILOV, B. M., BEZMENOV, A. YA., VASIL'EV, L. S. and KISELEV, V. G., *Doklady Acad. Nauk SSSR*, **155**, 141 (1964).
30. MIKHAILOV, B. M. and BEZMENOV, A. YA., *Izvest. Acad. Nauk SSSR, Ser. Khim.*, 931 (1965).
31. MIKHAILOV, B. M., BEZMENOV, A. YA. and VASIL'EV, L. S., *Doklady Acad. Nauk SSSR*, **167**, 590 (1966).
32. MIKHAILOV, B. M. and BLOKHINA, A. N., *Izvest. Acad. Nauk SSSR, Otdel. Khim. Nauk*, 1373 (1962).
33. MIKHAILOV, B. M., BOGDANOV, V. S., LOGODZINSKAYA, G. V. and POZDNEV, V. F., *Izvest. Acad. Nauk SSSR, Ser. Khim.*, 386 (1966).
34. MIKHAILOV, B. M. and BUBNOV, YU. N., *Doklady Acad. Nauk SSSR*, **127**, 571 (1959).
35. MIKHAILOV, B. M. and BUBNOV, YU. N., *Izvest. Acad. Nauk SSSR, Otdel. Khim. Nauk*, 172 (1959).
36. MIKHAILOV, B. M. and BUBNOV, YU. N., *Zhur. Obshchei Khim.*, **29**, 1648 (1959).
37. MIKHAILOV, B. M. and BUBNOV, YU. N., *Zhur. Obshchei Khim.*, **31**, 160 (1961).
38. MIKHAILOV, B. M. and BUBNOV, YU. N., *Izvest. Acad. Nauk SSSR, Otdel. Khim. Nauk*, 1872 (1960).
39. MIKHAILOV, B. M. and BUBNOV, YU. N., *Zhur. Obshchei Khim.*, **31**, 577 (1961).
40. MIKHAILOV, B. M. and BUBNOV, YU. N., *Izvest. Acad. Nauk SSSR, Otdel. Khim. Nauk*, 1378 (1962).
41. MIKHAILOV, B. M. and BUBNOV, YU. N., *Izvest Acad. Nauk SSSR, Ser. Khim.*, 1874 (1964).
42. MIKHAILOV, B. M. and BUBNOV, YU. N., *Izvest. Acad. Nauk SSSR, Ser. Khim.*, 2248 (1964).
43. MIKHAILOV, B. M. and DOROKHOV, V. A., *Doklady Acad. Nauk SSSR*, **136**, 356 (1961).
44. MIKHAILOV, B. M. and DOROKHOV, V. A., *Izvest. Acad. Nauk SSSR, Otdel. Khim. Nauk*, 1163 (1961).
45. MIKHAILOV, B. M. and DOROKHOV, V. A., *Izvest. Acad. Nauk SSSR, Otdel. Khim. Nauk*, 1346 (1961).
46. MIKHAILOV, B. M. and DOROKHOV, V. A., *Izvest. Acad. Nauk SSSR, Otdel. Khim. Nauk*, 2084 (1961).
47. MIKHAILOV, B. M. and DOROKHOV, V. A., *Zhur. Obshchei Khim.*, **31**, 3750 (1961).
48. MIKHAILOV, B. M. and DOROKHOV, V. A., *Izvest. Acad. Nauk SSSR, Otdel. Khim. Nauk*, 1213 (1962).
49. MIKHAILOV, B. M. and DOROKHOV, V. A., *Zhur. Obshchei Khim.*, **32**, 1511 (1962).

50. MIKHAILOV, B. M., DOROKHOV, V. A. and SHCHEGOLEVA, T. A., *Izvest. Acad. Nauk SSSR, Otdel. Khim. Nauk*, 498 (1963).
51. MIKHAILOV, B. M., DOROKHOV, V. A. and MOSTOVOI, N. V., *Zhur. Obshchei Khim.*, Sbornik "Problems of Organ. Synthesis", 223 (1965).
52. MIKHAILOV, B. M., DOROKHOV, V. A. and YAKOLEV, I. P., *Izvest. Acad. Nauk SSSR, Ser. Khim.*, 332 (1966).
53. MIKHAILOV, B. M., DOROKHOV, V. A. and MOSTOVOI, N. V., *Doklady Acad. Nauk SSSR*, **166,** 1114 (1966).
54. MIKHAILOV, B. M. and FEDOTOV, N. S., *Zhur. Obshchei Khim.*, **32,** 93 (1962).
55. MIKHAILOV, B. M. and FEDOTOV, N. S., *Izvest. Acad. Nauk SSSR, Otdel. Khim. Nauk*, 999 (1962).
56. MIKHAILOV, B. M., FEDOTOV, N. S., SHCHEGOLEVA, T. A. and SHELUDYKOV, V. D., *Doklady Acad. Nauk SSSR*, **145,** 340 (1962).
57. MIKHAILOV, B. M. and GALKIN, A. F., *Izvest. Acad. Nauk SSSR, Otdel. Khim. Nauk*, 371 (1961).
58. MIKHAILOV, B. M. and GALKIN, A. F., *Izvest. Acad. Nauk SSSR, Otdel. Khim. Nauk*, 619 (1962).
59. MIKHAILOV, B. M. and GALKIN, A. F., *Izvest. Acad. Nauk SSSR, Otdel. Khim. Nauk*, 641 (1963).
60. MIKHAILOV, B. M. and GALKIN, A. F., *Doklady Acad. Nauk SSSR*, **176,** 1078 (1967).
61. MIKHAILOV, B. M., KOZMINSKAYA, T. K., FEDOTOV, N. S. and DOROKHOV, V. A., *Doklady Acad. Nauk SSSR*, **127,** 1023 (1959).
62. MIKHAILOV, B. M. and KOZMINSKAYA, T. K., *Izvest. Acad. Nauk SSSR, Otdel. Khim. Nauk*, 2247 (1960).
63. MIKHAILOV, B. M. and KOZMINSKAYA, T. K., *Zhur. Obshchei Khim.*, **30,** 3619 (1960).
64. MIKHAILOV, B. M. and KOZMINSKAYA, T. K., *Izvest. Acad. Nauk SSSR, Otdel. Khim. Nauk*, 256 (1962).
65. MIKHAILOV, B. M., KOZMINSKAYA, T. K. and BEZMENOV, A. YA., *Izvest. Acad. Nauk SSSR, Ser. Khim.*, 355 (1965).
66. MIKHAILOV, B. M. and KOZMINSKAYA, T. K., *Izvest. Acad. Nauk SSSR, Ser. Khim.*, 439 (1965).
67. MIKHAILOV, B. M., KOZMINSKAYA, T. K. and TARASOVA, L. V., *Doklady Acad. Nauk SSSR*, **160,** 615 (1965).
68. MIKHAILOV, B. M., MOSTOVOI, N. V. and DOROKHOV, V. A., *Izvest. Acad. Nauk SSSR, Ser. Khim.*, 1358 (1964).
69. MIKHAILOV, B. M. and NIKOLAEVA, M. E., *Izvest. Acad. Nauk SSSR, Ser. Khim.*, 1368 (1963).
70. MIKHAILOV, B. M. and POZDNEV, V. F., *Izvest. Acad. Nauk SSSR, Otdel. Khim. Nauk*, 1861 (1962).
71. MIKHAILOV, B. M. and POZDNEV, V. F. *Doklady Acad. Nauk SSSR*, **151,** 340 (1963).
72. MIKHAILOV, B. M., POZDNEV, V. F. and KISELEV, V. G., *Doklady Acad. Nauk SSSR*, **151,** 577 (1963).
73. MIKHAILOV, B. M. and POZDNEV, V. F., *Probl. Organ. Sinteza, Acad. Nauk SSSR, Otd. Obshch. i Teklin. Khim.*, 220 (1965).
74. MIKHAILOV, B. M. and SAFONOVA, E. N., *Izvest. Acad. Nauk SSSR, Ser. Khim.*, 1487 (1965).
75. MIKHAILOV, B. M., SAFONOVA, E. N. and KISELEV, V. G., *Probl. Organ. Sinteza, Acad. Nauk SSSR, Otd. Obshch. i Teklin. Khim.*, 213 (1965).
76. MIKHAILOV, B. M. and SHCHEGOLEVA, T. A., *Izvest. Acad. Nauk SSSR, Otdel. Khim. Nauk*, 508 (1956).
77. MIKHAILOV, B. M. and SHCHEGOLEVA, T. A., *Izvest. Acad. Nauk SSSR, Otdel. Khim. Nauk*, 1080 (1957).

78. MIKHAILOV, B. M. and SHCHEGOLEVA, T. A., *Izvest. Acad. Nauk SSSR, Otdel. Khim. Nauk*, 1868 (1959).
79. MIKHAILOV, B. M. and SHCHEGOLEVA, T. A., *Doklady Acad. Nauk SSSR*, **131**, 843 (1960).
80. MIKHAILOV, B. M., SHCHEGOLEVA, T. A. and BLOKHINA, A. N., *Izvest. Acad. Nauk SSSR, Otdel. Khim. Nauk* 1307 (1960).
81. MIKHAILOV, B. M., SHCHEGOLEVA, T. A. and SHASHKOVA, E. M., *Izvest. Acad. Nauk SSSR, Otdel. Khim. Nauk*, 916 (1961).
82. MIKHAILOV, B. M. and SHCHEGOLEVA, T. A., *Izvest. Acad. Nauk SSSR, Otdel. Khim. Nauk*, 1142 (1961).
83. MIKHAILOV, B. M., SHCHEGOLEVA, T. A., SHASHKOVA, E. M. and SHELUDYKOV, V. D., *Izvest. Acad. Nauk SSSR, Otdel. Khim. Nauk*, 1163 (1961).
84. MIKHAILOV, B. M., SHCHEGOLEVA, T. A., SHASHKOVA, E. M. and SHELUDYKOV, V. D., *Izvest. Akad. Nauk SSSR, Otdel. Khim. Nauk*, 1218 (1962).
85. MIKHAILOV, B. M., SHCHEGOLEVA, T. A., SHELUDYKOV, V. D. and BLOKHINA, A. N., *Izvest. Acad. Nauk SSSR, Otdel. Khim. Nauk*, 646 (1963).
86. MIKHAILOV, B. M., SHCHEGOLEVA, T. A. and SHELUDYKOV, V. D., *Izvest. Acad. Nauk SSSR, Otdel. Khim. Nauk*, 816 (1963).
87. MIKHAILOV, B. M., SHELUDYKOV, V. D. and SHCHEGOLEVA, T. A., *Izvest. Acad. Nauk SSSR, Otdel. Khim. Nauk*, 1559 (1962).
88. MIKHAILOV, B. M. and TER-SARKISIAN, G. S., *Izvest. Acad. Nauk SSSR, Ser. Khim.*, 380 (1966).
89. MIKHAILOV, B. M. and TUTORSKAYA, F. B., *Doklady Acad. Nauk SSSR*, **123**, 479 (1958).
90. MIKHAILOV, B. M. and TUTORSKAYA, F. B., *Izvest. Acad. Nauk SSSR, Otdel. Khim. Nauk*, 1158 (1961).
91. MIKHAILOV, B. M. and TUTORSKAYA, F. B., *Izvest. Acad. Nauk SSSR, Otdel. Khim. Nauk*, 2068 (1961).
92. MIKHAILOV, B. M. and TUTORSKAYA, F. B., *Zhur. Obshchei Khim.*, **32**, 833 (1962).
93. MIKHAILOV, B. M. and VASIL'EV, L. S., *Izvest. Acad. Nauk SSSR, Otdel. Khim. Nauk*, 531 (1961).
94. MIKHAILOV, B. M. and VASIL'EV, L. S., *Doklady Acad. Nauk SSSR*, **139**, 385 (1961).
95. MIKHAILOV, B. M. and VASIL'EV, L. S., *Izvest. Acad. Nauk SSSR, Otdel. Khim. Nauk*, 2101 (1961).
96. MIKHAILOV, B. M. and VASIL'EV, L. S., *Izvest. Acad. Nauk SSSR, Otdel. Khim. Nauk*, 1756 (1962).
97. MIKHAILOV, B. M. and VASIL'EV, L. S., *Zhur. Obshchei Khim.*, **35**, 925 (1965).
98. MIKHAILOV, B. M., VASIL'EV, L. S. and BEZMENOV, A. YA., *Izvest. Acad. Nauk SSSR, Ser. Khim.*, 712 (1965).
99. MIKHAILOV, B. M. and VASIL'EV, L. S., *Zhur. Obshchei Khim.*, **35**, 1073 (1965).
100. MIKHAILOV, B. M., VAVER, V. A. and BUBNOV, YU. N., *Doklady Acad. Nauk SSSR*, **126**, 575 (1959).
101. MOSTOVOI, N. V., DOROKHOV, V. A. and MIKHAILOV, B. M., *Izvest. Acad. Nauk SSSR, Ser. Khim.*, 90 (1966).
102. MUETTERTIES, E. L., MILLER, N. E., PACKER, K. J. and MILLER, H. C., *Inorg. Chem.*, **3**, 870 (1964).
103. NIELSEN, D., McEWEN, W. and VANDERWERF, C., *Chem. and Industr.*, 1069 (1957).
104. NÖTH, H. and MIKULASCHEK, G., *Ber.*, **94**, 634 (1961).
105. PASTO, D. J., *J. Am. Chem. Soc.*, **84**, 3777 (1962).
106. PASTO, D. J., CUMBO, C. C. and BALASUBRAMANIYAN, P., *J. Am. Chem. Soc.*, **88**, 2187 (1966).
107. PASTO, D. J., CUMBO, C. C. and FRASER, J., *J. Am. Chem. Soc.*, **88**, 2194 (1966).

108. PETTERSON, L. L., BROTHERTON, R. J. and BOONE, J. L., *J. Org. Chem.*, **26**, 3030 (1961).
109. SAZIER, W. A. and SALZBERG, P. L., U.S. Pat. 2 402 591; *C.A.*, **40**, 5769 (1946).
110. SHCHEGOLEVA, T. A. and BELYAVSKAYA, E. M., *Doklady Acad. Nauk*, **136**, 638 (1961).
111. SHCHEGOLEVA, T. A., SHASHKOVA, E. M. and MIKHAILOV, B. M., *Izvest. Acad. Nauk SSSR, Otdel. Khim. Nauk*, 918 (1961).
112. SHCHEGOLEVA, T. A., SHASHKOVA, E. M. and MIKHAILOV, B. M., *Izvest. Acad. Nauk SSSR, Otdel. Khim. Nauk*, 494 (1963).
113. SHCHEGOLEVA, T. A., SHASHKOVA, E. M., KISELEV, V. G. and MIKHAILOV, B. M., *Izvest. Acad. Nauk SSSR, Ser. Khim.*, 365 (1964).
114. SHCHEGOLEVA, T. A., SHASHKOVA, E. M., KISELEV, V. G. and MIKHAILOV, B. M., *Zhur. Obshchei Khim.*, **35**, 1078 (1965).
115. SHELUDYKOV, V. D., SHCHEGOLEVA, T. A. and MIKHAILOV, B. M., *Izvest. Acad. Nauk SSSR, Ser. Khim.* 632 (1964).
116. STEINBERG, H., *Organoboron Chemistry*, Vol. 1, pp. 187–239, Interscience Publishers, New York–London–Sydney, 1964.
117. STOCK, A. and POPPENBERG, O., *Ber.*, **34**, 399 (1901).
118. WIBERG, E. and STURM, W., *Angew. Chem.*, **67**, 483 (1955).
119. WIBERG, E. and STURM, W., *Z. Naturforsch.*, **8b**, 530 (1953).
120. WIBERG, E. and STURM, W., *Z. Naturforsch.*, **8b**, 689 (1953).
121. WIBERG, E. and STURM, W., *Z. Naturforsch.* **10b**, 108 (1955).
122. WIBERG, E. and STURM, W., *Z. Naturforsch.*, **10b**, 109 (1955).
123. WIBERG, E. and STURM, W., *Z. Naturforsch.*, **10b**, 111 (1955).
124. WIBERG, E. and STURM, W., *Z. Naturforsch.*, **10b**, 112 (1955).
125. WIBERG, E. and STURM, W., *Z. Naturforsch.*, **10b**, 113 (1955).
126. WIBERG, E. and STURM, W., *Z. Naturforsch.*, **10b**, 114 (1955).
127. WIBERG, E. and SÜTTERLIN, W., *Z. Anorg. Allgem. Chem.*, **202**, 37 (1931).
128. YOUNG, D. M. and ANDERSON, C. D., *J. Org. Chem.*, **26**, 5235 (1961).

NAME INDEX

Abel, E. W.   249, 258, 259
Abeler, G.   275
Adams, R. M.   178, 180–183
Addy, L. E.   37, 43, 44
Akerfeld, S.   223
Akhnazarian, A. A.   328, 329, 360, 362
Akishin, P. A.   181
Alagy, J.   4–6, 10, 14, 21, 58
Alcock, C.   181
Alekseyenko, E. V.   37
Alentéva, Y. S.   63–65
Alpatova, V. I.   214, 219, 292
Alsobrook, A. L.   282
Amberger, E.   281
Ammar, M. M.   193
Anderson, C. D.   315, 360
Anderson, E. R.   218
Anderson, G. A.   267, 268
Andreeva, T. P.   30, 36, 38, 62
Arimoto, F. S.   18
Armitage, D. A.   249, 258, 259
Armour, A. G.   119
Armstrong, D. R.   248
Aronovich, P. M.   146
Arora, S. K.   147, 148, 168, 169, 173
Arzoumanian, H.   147–149, 160, 173
Ashby, E. C.   279
Aylett, B. J.   216, 218, 229
Azzaro, M.   233

Baccaredda, M.   45, 91
Baechle, H. T.   293, 294
Bains, M. S.   296
Balasubramaniyan, P.   167, 348, 349, 354–356
Baldwin, J. C.   294, 295
Banford, L.   242–245
Banister, A. J.   258, 265, 266
Banus, J.   253
Barker, R. S.   46, 49, 53
Bartlett, J. H.   30, 33, 38, 39, 69, 70, 75, 77–79
Bartlett, R. K.   25
Bashkirov, A. N.   2, 4, 5, 12, 26, 29, 30, 31–36, 38, 57, 59, 60, 62–66, 69, 70, 73

Bauer, H.   101, 104, 105
Baumgarten, R. J.   217
Bawn, C. E. H.   151, 213
Beachley, O. T.   236, 262
Becher, H. J.   243, 249, 293, 294
Becker, M. M.   25, 27, 46, 50, 54, 58
Bekasova, N. I.   213
Belinski, C.   217
Beller, H.   2, 12
Bellut, H.   285
Belyavskaya, E. M.   356, 360
Benedikt, G.   284, 287
Bengelsdorf, I. S.   126, 137, 150, 164, 171
Bennett, J. E.   193, 194, 197
Berkowitz, J.   182
Bertrand, R. D.   218
Bessler, E.   278, 294
Bettman, B.   118, 119
Beyer, H.   214, 216, 218, 220, 237–239, 274
Bezmenov, A. Ya.   289, 290, 326, 327, 336, 355, 363
Biallas, M. J.   230
Bianchi, G.   172
Bidinosti, D. R.   187
Bigot, J. A.   114
Billis, J. L.   189
Binger, P.   150, 165
Birnbaum, E. R.   220
Blanchard, E. P.   163
Blanchard, H. S.   9
Blangey, L.   90
Blau, J. A.   317
Blauer, J. A.   182, 187
Blokhina, A. N.   320, 327, 328, 355, 356, 361, 362, 363
Böddeker, K. W.   214, 235, 261
Bogdanov, V. S.   324
Bonnell, J. E.   318, 348–350, 354
Boone, J. L.   184, 292, 315, 316, 360
Boonstra, H. J.   15
Boulton, A.   132–134, 162
Bouman, M. G.   191
Bowie, R. A.   155

SUBJECT INDEX